Soviet/Russian Armor and Artillery Design Practices: 1945 to Present

Andrew W. Hull   David R. Markov   Steven J. Zaloga

FRONT COVER
A T-90 during a dynamic mobility demonstration in Abu Dhabi during the 1997 International Defence Exhibition.

BACK COVER
TOP: A colorful T-80UM Bars shown at Omsk in 1997.
BOTTOM: Again at Omsk was this BMD-2.

Published and distributed by
Darlington Productions
P.O. Box 5884
Darlington, Maryland 21034
USA

ISBN 1-892848-01-5

Copyright © 1999
All right reserved. No portion of this publication may be reprinted, reproduced, stored, or retransmitted in any form without the prior permission of the publisher.

The Authors assert the moral right to be identified as the authors of this work.

Contact Darlington Productions for more information on other military publications at the address above or at www.darlingtonproductions.com

INDEX

Preface	5
Introduction and Overview	6
1. Evolution of Soviet/Russian Tanks	
The Soviet Tank Force at the Beginning of WW2	7
The Revolution in Tank Design	9
Battle for the Factories	11
The Technological Imperative Revived	12
The First Post-War Medium Tank Generation: 1942-1962	21
Post-War T-34-85 Production	21
Early Post-War Medium tank Design	21
T-54 Medium Tank Design	23
T-55 Medium Tank	31
The T-62 Tank	50
Soviet Post War Heavy Tank Design	61
The IS-4 Heavy Tank	62
Heavy Tank Modernization	63
Post-War Heavy Tank Development	63
Expanding Heavy Tank Role?	68
Further Heavy Tank Development	70
Soviet Post-War Light Tanks	74
Second Generation Tank Development: Advent of Main Battle Tanks 1962 to Present	87
The New Standard: Kharkov's T-64 Tank	87
Economy Tank: The Nizhni Tagil T-72	102
The T-90 (Obiekt 188) Tank	122
Foreign T-72 Development	132
Turbine Tank: Leningrad's T-80 Tank	142
Future Trends in Russian Tank Development	164
Future Armor Developments	164
Future Russian Main Battle Tank Development: Three Scenarios	167
2. Anti-Armor Developments	
World War 2 Lessons	170
Post-War Anti-Armor Development	172
Early Soviet ATGM Development	173
Guided Tank Projectiles	196
Trends in Russian Guided Projectiles	201
3. Armored Infantry Vehicles	
Early Development Trends	207
Early Post-War Developments	208
Post War Soviet Infantry Vehicle Development	208
Wheeled Infantry Vehicles	208
The BTR-152 Armored Transporter	208
The BTR-60	213
The BTR-70 Armored Transporter	220
The BTR-80 Armored Transporter	226
The BTR-80A	227
2S23 Nona	227
BREM-K	227
BTR-80KSh	227
BTR-80 PU-12M	229
BTR-80 RKhM	229

PKNP Kushetka-B	229
GAZ-59037	229
The BTR-90 Armored Transporter	229
Tracked Armored Infantry Vehicles	232
The BMP Infantry Combat Vehicle	238
The MT-LB Armored Transporter	267
The BMP-3	269
Future Trends In Infantry Vehicle Development	284

4. Armored Airborne

Early Soviet Armored Airborne Vehicle: ASU-57	288
Early Soviet Armored Airborne Vehicle: ASU-85	291
The BMD Airborne Assault Vehicle	292
Future Russian Light AFV Development Trends	305

5. Self-Propelled Artillery

Mechanized Artillery in WW2	308
New Direction in the 1950s	309
Early Post-War Missile Development	314
Early Post-War Multiple Rocket Launcher Development	314
Mechanized Field Artillery Requirements After 1945	314
The 2S1 Gvozdika Self-Propelled Gun	315
The 2S3 Akatsiya Self-Propelled Gun	317
The 2S5 Giatsint Self-Propelled Gun	320
The 2S7 Pion Self-propelled Gun	322
The 2S19 Msta-S Self-Propelled Gun	324
The Bereg 130 mm Coastal Defense Artillery System	331
The Germes 155mm Self-Propelled Gun System	336
Post-War Self-Propelled Mortars Developments	336
The 2S4 Tyulpan Mechanized Mortar	337
The 2B9 Vasilyek 82mm Mortar Mounted on a MT-LBu	339
The 2S9 Nona-S	340
The 2S12 Truck-Mobile Mortar	342
The 2S23 Nona-SVK	343
2S31 Vena 120mm Self-Propelled Gun-Mortar	345
Future Trends in Russian Self-Propelled Artillery	350
Artillery Support Equipment	351
Multiple Rocket Launch Systems (MLRS)	358
"Artillery" Missiles	374

6. Towed Artillery and Anti-Tank Systems

Towed Artillery Systems	391
Anti-Tank Gun Systems	403

7. Final Observations

Observations About the Design Process	410
Overarching Technical Priorities	410
Continuing Applicability of These Principles	411

Appendix A	413
Appendix B	420
Endnotes	434
Index	444

Preface

In the past few years, a wealth of new material has appeared on Soviet/Russian armor and artillery developments. This has made it possible to provide a more definitive account than previous books and articles on the evolution of Soviet/Russian armor and artillery concepts since World War 2. To that end, the book offers detailed discussions of Soviet/Russian tanks, armored fighting vehicles, airborne armored fighting vehicles, anti-tank guided missiles, short-range ballistic missiles, multiple rocket launcher systems, self-propelled and towed artillery designs. It includes well over 300 photographs of prototype and series production models, many of which were never previously published in the West.

However, without the gracious assistance of numerous contributors this book would not have been possible. The authors would like to thank Mr. Andrew Marshall, Director of the Department of Defense's Office of Net Assessment for his generous support. In addition, the authors are indebted to the United States Marine Corps in general and Ms. Karin Dolan and Mr. Charles Cutshaw in specific for their support and encouragement throughout this process. The authors would also like to thank the Institute for Defense Analyses for allowing its publication.

In addition, the authors would like to acknowledge the help provided by a number of individuals. Special thanks goes to Mr. Christopher F. Foss for his review and comment on the initial manuscript. Mr. Foss is the editor of *Jane's Armour and Artillery,* considered the most current and authoritative work on military land arms and equipment published today. Special thanks also go to colleagues to numerous to name in Russia and Eastern Europe for their valuable assistance with photographs and drawings. Finally, the authors would like to thank the staff at the NIIBT Tank Museum in Kubinka for permitting access and photography of this superb collection of Soviet/Russian armored vehicles.

Lastly, the information presented in this book is the sole responsibility of the authors and does not necessarily represent the views nor official policy or position of the United States Marine Corps (USMC) or the United States Department of the Defense (DoD). Original Copyright © Institute for Defense Analyses 1997.

Introduction and Overview

The confluence of harsh economic realities within the Russian defense industrial complex, the intense competitiveness of the international arms market, and the wide-ranging openness about Russian defense matters in general offer a unique opportunity to understand how (and why) Soviet/Russian armor and artillery have evolved since World War 2.[1] For example, one can now view many past developments first-hand. Russian military museums, open to foreigners and the Russian general public for the first time, now display a wide-range of prototypes, test-beds, and model variants — many of which were never acknowledged before.

At the same time, international market pressures are forcing Russian arms makers to reveal unprecedented levels of detail about current and future tank and artillery designs. The last few years have also produced a flood of press and journal articles by Russian defense industrialists discussing past system requirements and design tradeoffs.

This study exploits such newly-available, open-source material to examine the evolution of Soviet/Russian armor and artillery designs since World War 2. The purpose of doing so is to:

1. Describe the main technical characteristics of individual model variants and prototypes which distinguish one from another;

2. Provide a chronology of developments;

3. Classify individual systems into more general "family" groups;

4. Identify (where possible) the requirements which led the Soviets/Russians to develop various types of equipment;

5. Discuss (again where possible) the underlying design tradeoffs faced by the Soviets/Russians in pursuing particular concepts;

6. Uncover basic, long-standing design principles which transcend individual systems; and

7. Assess how changing conditions will affect the continuing validity of those principles for future Russian designs.

The following discussion of these issues is organized into six system-specific chapters and one broader concluding overview. Individual chapters devoted to tanks, armored infantry fighting vehicles, airborne armored infantry fighting vehicles, anti-tank guided missiles, self-propelled, and towed artillery concentrate on addressing issues one through five above while a concluding chapter takes up the questions of identifying broad design principles and assessing their continuing validity. These chapters (when taken collectively) provide valuable insights into the "whats," "whys," and "wherefores" of past Soviet/Russian armor and artillery designs as well as offer a general guide for anticipating the broad directions of future Russian developments in these same areas.

1
EVOLUTION OF SOVIET/RUSSIAN TANKS

Soviet Tank Development: Lessons of World War 2

Introduction

To understand the development of Soviet tank technology during the Cold War, it is important to have a general understanding of the lessons of World War 2. This section provides a brief overview of the development of Soviet tanks in World War 2. It surveys the general trends in tank design during the war, as well as related issues such as the evolution of the design bureau system, and the trade-offs between production and tank effectiveness.[1] An underlying issue is the genesis of the "tank mentality" that shaped Soviet post-war tank engineering.

When the Soviet Union finally imploded in August 1991, the Soviet Army had 77,000 tanks, more than the rest of the world combined.[2] The experiences of World War 2 had convinced the Soviet Army of the need to maintain a substantial numerical superiority in key weapons over all potential adversaries, such as NATO, a phenomenon dubbed "the tank mentality" by Russian defense reformers. This "tank mentality" was a deeply rooted set of attitudes that set the basic parameters for post-war Soviet tank design. We now take it for granted that Soviet tank design is willing to trade off qualitative advantages for quantity. But this was not always the case, and it is quite possible that the contemporary Russian Army will revert to the pre-war objective of meeting or surpassing Western tank design in quality, with resulting effects on quantity.

The Soviet "tank mentality" is seldom if ever explicitly stated, especially not in an unclassified context. It was implicitly understood that in order to win on the battlefield, the Soviet Army needed significantly more tanks than its opponents. Russian qualitative inferiority to Western armies is not a new phenomenon. Indeed, it can be argued that this sense of inferiority is an age-old feature of Russian military culture. But because of the "tank mentality," to have explicitly discussed this issue would have raised the embarrassing question of why the wartime Red Army lost tanks at a rate three to seven times as great as their German opponents. Any explicit admission of Soviet tactical or technical inferiority was politically unacceptable for over four decades after the war as undermining the honor and prestige of the Soviet Army at the moment of its greatest victory. When a civilian defense reformer attempted to broach the subject in 1991 by pointing out the disproportionate losses of Soviet tanks in World War 2, he was roundly criticized in the official military press.[3] Ironically, his lack of access to official documents at the time actually led him to seriously underestimate the relative scale of Soviet tank losses.

The Soviet Tank Force at the Beginning of World War 2

The Red Army possessed the world's largest tank force in 1941. As was the case fifty years later, the Soviet tank inventory of 1941 was larger than that of the rest of the world combined, numbering some 23,106 tanks (the US tank force at this time was about 2,000 tanks).[4] The German tank force in June 1941 had 5,262, tanks of which 3,671 were committed to the invasion of the USSR on 22 June 1941.[5]

The bulk of the Soviet tank force was made up of two basic types: the T-26 light tank and the BT fast tank. In total about 8,300 BT tanks and 12,000 T-26 tanks were manufactured from 1933 to 1940.[6] This followed the British and French pattern of developing two distinctly different tanks for two roles — a slow infantry tank for

T-37 amphibious scout tank.

7

T-26 infantry tank.

close support of foot soldiers and a fast tank, variously called a cavalry tank or cruiser tank, for exploitation and other maneuver missions.

Early Soviet tank technology was derivative of Western technology. The T-26 infantry support tank was an improved copy of the British Vickers 6-ton light export tank. The BT cavalry exploitation tank was a licensed copy of the American Christie tank. They were both excellent designs for their day, and remarkably, neither of these tanks was produced in quantity in their native countries. The Soviets made important improvements in these designs. Both were fitted with new turrets which permitted them to carry a 45mm tank gun; the Vickers 6-ton had been armed with a short 37mm gun, while the Christie was armed only with machine guns. This Soviet 45mm gun was the best general purpose tank gun in common service in the mid-1930s, firing an excellent anti-tank projectile that could defeat nearly any existing tank, and also firing a useful high explosive round. Many tank guns of the period did not show similar versatility and could either fire a good anti-tank projectile or a good high explosive projectile, but not both. For example, contemporary French infantry tanks were armed with a short 37mm gun with little anti-armor capability; the contemporary German PzKpfw I was armed only with two 7.62mm machine guns.

These Soviet light tanks continued to be viable weapons well up to World War 2. During the Spanish Civil War, they dominated their German and Italian counterparts; in the fighting with Japan at Lake Khasan in 1938 and Khalkin-Gol in 1939, they completely outclassed their Japanese opponents. The main Soviet advantage in these cases was their superior firepower. This helped establish a trend to adopt tank guns of superior caliber and performance compared with their Western counterparts, a trend that the Soviets have attempted to maintain for nearly sixty years.

The remainder of the 1941 tank force included light amphibious scout tanks, including the T-37 (2,400 produced); T-38 (1,200 produced) and the new T-40 (222 produced). There were very small numbers of older medium and heavy tanks, including the archaic T-28 (600) and T-35 (40).

T-28 medium tank.

T-40 amphibious scout tank.

The Revolution in Tank Design

The most technically significant elements of the 1941 tank force were two new designs, the T-34 medium tank and the KV heavy tank. The T-34 was intended to replace the BT cavalry tank. It had benefited greatly from Soviet experiences in tank warfare in Spain in 1937 to 1938 and against the Japanese in 1938 to 1939. These tank battles, although small scale compared later with World War 2 fighting, had important technological lessons for the Soviet tank designers. It made clear that the existing levels of armor protection, little changed since World War 1, were completely inadequate when faced with contemporary tank and anti-tank guns. This prompted the Soviets to adopt "shell proof armor" on both tanks, so that when production started in 1940, they were unquestionably the best armored tanks in the world. Secondly, the fighting convinced the Soviets of the soundness of equipping their tanks with a good dual-purpose gun of superior performance to the opposition. Anticipating foreign improvements, the new tanks leap-frogged forward in armament, going from a 45mm gun to a 76mm gun. Once again, this made both tanks the best armed of all contemporary tanks. Finally, the wartime experiences convinced the Soviets of the vulnerability of gasoline engines on tanks, leading them to adopt a new diesel engine. This engine, the V-2, has been the standard powerplant of Soviet medium tanks since then, having been used (in improved form) through the 1990s up to the T-90 tank.

Although Soviet histories of tank design have portrayed the T-34 as emerging from Red Army requirements, in fact, it was more the result of innovative ideas of the design bureau that were not initially supported by either the Red Army or the military industry. Had the Red Army's tactical-technical requirements been precisely fulfilled, the design would have been considerably inferior in armor, firepower and mobility and would have resembled contemporary British cruiser tanks. Instead, it became the tank which set the pace for all later World War 2 tank development. Although there is a widespread perception in the West that Soviet tanks stem from a for-

T-34 Model 1942 medium tank.

KV-1 heavy tank.

T-50 infantry tank.

KV-2 heavy support tank.

mal process steered by Soviet army doctrine and tactics, in the case of most tank innovations over the past half century, this has not been the case.

The origins of the KV heavy tank were similar. The original requirement for the KV called for a multi-turreted monstrosity similar to the pre-war T-35. This conception was resisted by the design team. According to legend, it was Stalin who authorized the team to pursue a more modern design against the wishes of the Red Army bureaucracy. In any event, the KV represented the most thickly armored and heavy tank of its day. This tank was developed based on the lessons of the wars of the 1930s which showed that all existing tanks could be defeated by existing anti-tank guns like the ubiquitous German 37mm gun. The aim of the program was to deploy a "shell-proof" tank that could assist the more numerous cavalry tanks in securing a breakthrough against enemy infantry protected by modern anti-tank weapons like the German 37mm PaK 36. It was not intended specifically for tank fighting, so its gun was no better than that on the new T-34 cavalry tank. Since the KV was not intended for the exploitation role, mobility was not a major concern; it was powered by the same V-2 diesel as the cavalry tanks, even though it was ten tons heavier. This tank was conceptually similar to British infantry tanks, though larger, more heavily armored, and better armed.

A third tank was also in development, the T-50 infantry tank, intended to replace the T-26. This was a very sophisticated light tank roughly comparable to the German PzKpfw III. It had severe engineering shortcomings, and so was not available when war broke out. This twist of fate had enormous implications for Soviet tank programs in World War 2. With the T-50 not ready for mass production, the T-34 cavalry tank was pressed into both cavalry and infantry tank roles, much to the benefit of the Soviet armored force.

The new T-34 and KV tanks were available in substantial numbers in June 1941 with some 508 KV and 967 T-34 in service. The best German tanks were the PzKpfw III, armed with a 37mm gun, and the PzKpfw IV, armed with a short 75mm gun with poor antiarmor performance. There were 1,449 PzKpfw III and 517 PzKpfw IV available in June 1941. They were inferior to the new Soviet tanks in armor, firepower and mobility. The revolutionary combination of armor, firepower, and mobility of the T-34 tank established it as the technological pace-setter of World War 2 tank design. The focus of tank technology shifted from its traditional center in England and France, eastward to Germany and the Soviet Union as Germany responded to the challenge of matching the T-34. The technological arms race between Germany and the USSR, prompted by the revolutionary T-34, set the pace for worldwide tank development throughout World War 2.

In spite of substantial numerical superiority, and important qualitative superiority, the Soviet tank force was decisively defeated by the smaller and more modestly equipped German tank force in the summer of 1941. The roots of this defeat are connected mainly in the Red Army's lack of preparedness for war, exacerbated by the corrosive influences of the purges of the officer ranks in the late 1930s.[7] From a technological standpoint, the defeat highlighted shortcomings in Soviet tank design philosophy, some of which continue to be trademarks of Soviet tank design. The T-34 tank design stressed the "Big-3" —armor, firepower, and mobility — to the exclusion of other key tank-fighting features. Crew layout was poor; the turret only accommodated a crew of two, and so the commander could not execute his command functions and had to double as a gunner. The commander was not provided with adequate vision devices, and the hatch design made it impossible for the tank commander to ride with his head outside the tank as was the German practice. Soviet tank commanders, already hampered by inadequate training, were overwhelmed with the simple mechanics of operating the tank. They were unable to develop an awareness of the terrain or the location and status of friend and foe around them.

Soviet tank crews were thus hindered in carrying out cooperative battlefield tactics, making them vulnerable to the better coordinated German tank units. The Soviets did not understand the revolutionary implications of radio technology on the command and control of tank units and few tanks had radios. They had developed a mistrust of radio communications because of the disastrous results of poor Russian radio security in the 1905 war with Japan and the 1914 battles with Germany. These early failures discouraged proper tactical radio doctrine in the army in the 1930s, and this deficiency was further exacerbated by the backwardness of the Russian electronics industry. The radio shortcomings had a synergistic effect with the poor command and control features of the tank, leading to abysmal tank tactics. Soviet tank units were very vulnerable to the more experienced German tank and anti-tank units. Total Soviet tank losses from June to December 1941 were 20,500; German losses from 22 June 1941 through the end of February 1942 were only 3,402, a 6:1 exchange ratio. While the causes of the high Soviet losses were more clearly attributable to strategic and tactical failures, technical design flaws aggravated these problems.

The Soviet tank designers had some appreciation for these shortcomings, but as is evident in recent Russian writings on the subject, there was little contact between the designers and combat tank crews. The Soviet tank design process was to filter combat experience through the central Red Army bureaucracy (the Tank and Armored Directorate known by the acroynm GBTU) and pass the assessments to the design bureaus in the form of design requirements. The Soviet designers had some projected improvements on the drawing boards in 1941 to ameliorate these technical problems, but the exigencies of the time led to their abandonment.

It should be recognized that the following discussion of wartime tank development focuses on the technological imperatives of tank-versus-tank fighting. The Soviets considered it essential that their medium and heavy tanks be capable of defeating opposing tanks, and this requirement often dominated tank design. However, the Soviets also recognized that tank-vs-tank combat is not that common, and that a tank had to be armed with a gun firing a useful high explosive round to deal with anti-tank guns, enemy infantry, and other typical targets. This attitude stands in contrast to several armies (notably the British) which often built designs stressing one firepower capability to the exclusion of the other.

Battle for the Factories

In the summer of 1941, Soviet leaders faced the critical decision of whether to leave the tank factories in place and risk losing them to the advancing German forces, or halt tank production in spite of the heavy battlefield losses and move them to the safety of the Urals. Stalin gambled and ordered the factories moved, sacrificing short-term production. It was a bold, and ultimately successful, decision. It had dramatic effects on tank design, however, since it forced the Soviets to freeze any further qualitative improvements for more than eighteen months.

The main Soviet design bureau for the T-34 tank was located in Kharkov, Ukraine, as part of the Kharkov Locomotive Plant (KhPZ Zavod Number 183). At the time, the locomotive plant was the only manufacturer of the T-34, though efforts were already underway before the war to establish a second plant at Stalingrad (now Volgograd) at the site of the Stalingrad Tractor Plant. The T-34 design bureau, headed by Aleksandr Morozov, was ordered to evacuate Kharkov along with the staff and equipment of the locomotive plant. It was reestablished in Nizhni Tagil at the site of the Urals Railcar Plant (Uralvagonzavod Number 183). This plant has since become the largest of all Soviet (and Russian) tank plants. The first T-34 tank from the new production plant was not ready until 20 December 1941. To compensate for the temporary loss of the Kharkov plant, in July 1941 the Krasnoye Sormovo plant in Gorki was ordered to begin preparing to manufacture the T-34; the first were delivered to the Moscow Front in November 1941.

Light tank design and heavy tank design was centered in Leningrad at the Kirov Plant; tank production was co-located here as well. As was the case of Kharkov, the design bureau, headed by Zhozef Kotin, and the associated production facility were ordered to Chelyabinsk in the Urals where they formed the new Tankograd (Tank City) complex.

T-60 light tank.

Plans to build the T-50 infantry tank as a replacement for the old T-26s were canceled since it was almost as expensive to produce as the more effective T-34 cavalry tank. Instead, Plant Number 174 in Omsk, earmarked for T-50 production, was switched to the manufacture of T-34 components.

The T-60 light reconnaissance tank, intended to replace the overly sophisticated T-40 amphibious scout tank and the obsolete T-37 and T-38 tanks, was the third type of tank to remain in production. Production of the T-60 had begun in Moscow but was evacuated to the GAZ automotive plant in Gorki and the Plant Number 38 in Kirov. Further design of the light tanks was assigned to Nikolai Astrov's team. While the KV and T-34 required locomotive and railroad plants capable of handling very heavy subcomponents, the light tanks could be built at automotive plants with less resources.

With the Red Army barely surviving from the winter of 1941 to 1942, every effort was made to increase tank production. Efforts to improve the T-34 and KV were frozen. The Morozov design bureau had already developed an improved T-34, called T-34M, which circumvented many of the problems mentioned before by increasing the turret size, adding a commander's vision cupola and improving the suspension by transitioning from the Christie style of springs to a torsion bar system. The Kotin heavy tank design bureau planned a new version of the KV, the KV-3, armed with an 85mm gun. This would have given the Red Army a tank comparable in performance to the later German Tiger I, but a year earlier. Super-heavy tanks were also considered, but they did not reach prototype stage.

Instead of continuing the technology race, the tank designers were told to freeze their designs and concentrate on making the tanks easier and cheaper to manufacture. For example, the original 1941 version of the T-34 76.2mm tank gun had 861 parts; the 1942 production version had only 614. Production time of the T-34 was cut in half and the cost was driven down from 269,500 rubles in 1941 to 193,000 rubles in 1942.

While Soviet design stagnated because of production pressures, the Germans took the opposite approach and began an intensive effort to field a superior new tank. In the short term, the PzKpfw IV was rearmed with a more effective long 75mm gun making it capable of penetrating the T-34. Work on the new Tiger I heavy assault tank was accelerated, and it appeared on the Eastern Front in January 1943. The Tiger was a wild over-reaction to the tank panic that had set in after the first encounters with Soviet T-34 and KV tanks in the summer of 1941. The new Soviet tanks were impervious to most German tank and anti-tank guns.

The Germans set out to trump the Red Army by fielding a tank even heavier and better armed. However, the resulting Tiger was so expensive that it could never be manufactured in quantity. Only 1,354 were produced during the entire war, equal to less than a month of T-34 production. As a lower cost alternative to the Tiger I, the Germans developed the Panther, ostensibly a medium tank, but in fact nearly double the weight of the original T-34. It would be manufactured in larger numbers than the Tiger (5,976), but still not enough to entirely replace the outdated PzKpfw IV, which remained the most numerous German tank through the war.

The Soviet concentration on production paid off. The Soviet tank inventory rose from 7,700 tanks in January 1942 to 20,600 tanks at the beginning of 1943, inspite of massive combat losses in 1942 due to the inept tactical use of the new tank corps. German tank inventories also rose during the same period from 4,896 in January 1942 to 5,648 in January 1943. But discounting obsolete types, the combat-ready inventory actually fell slightly, from 4,084 at the end of 1941 to 3,939 at the end of 1942. The year 1942 saw the German and Soviet armored forces as equal as they had ever been. The Soviet numerical advantage was slight, and its technological edge was gradually worn away by German technical improvements. The Germans continued to display a great deal more tactical finesse in the employment of armored formations and anti-tank defense. Soviet tank losses in 1942 were 15,000, while German losses (on all fronts) were 2,648, or an exchange ratio of over 6:1 — nearly as bad as the 1941 disaster.

The Technological Imperative Revived

The 1942 defeat at Stalingrad was the high water mark of the German advance on the Eastern Front. Although often called the turning point of the war, Germany retained the strategic initiative until the battle of Kursk in

T-70 light tank.

the summer of 1943. Technically, the Soviet tank force in the summer of 1943 was not significantly different from its condition in mid-1942. By now, the Soviet factories were turning out T-34s in increasing numbers, and it had become the staple of the Red Army's tank and mechanized corps. It was still armed with the same 76mm gun as in 1942 and protected by the same level of armor.

The KV-1 had proven a dissappointment in 1942. It lacked a firepower advantage over the T-34 and its armor was no longer invulnerable to German anti-tank weapons as it had been in 1941. Its weight caused tactical mobility problems without conferring relative invulnerability, and it was plagued by lingering technical problems, especially a poor transmission. As a result, its armor thickness was actually reduced in 1942, and the KV-1 was removed from the tank corps and segregated into separate tank regiments for infantry support. A portion of the production lines at Chelyabinsk shifted to T-34 production and consideration was given to ending heavy tank production completely in favor of the T-34.

In the area of light armor, the T-60 light tank was being joined by the modestly improved T-70 light tank. Both types would have dissappeared but for the fact that their factories did not have heavy machinery capable of turning out T-34s. The main improvements in the Soviet armored force in 1942 were tactical, not technical, particularly the maturation of the new tank and mechanized corps.

In June 1942, the Red Army authorized the development of a "universal tank." The idea was to combine the better armor of the KV with the superior mobility of the T-34; the tank gun remained the same. The universal tank would replace both the T-34 and the KV-1. The heavier armor was necessary as the advent of the German long 75mm gun in May 1942 had made the T-34 vulnerable for the first time to the standard German tank at normal combat ranges. Two designs were competitively developed in Nizhni-Tagil and Chelyabinsk, an uparmored and improved T-34 called the T-43, and a reduced-weight KV called the KV-13. While the idea of standardizing on a single, universal type was desirable, the focus on armored protection proved to be a mistake.

The T-43 or KV-13 might have entered production in the summer of 1943 but for the arrival of excellent new

T-43 universal tank.

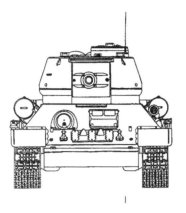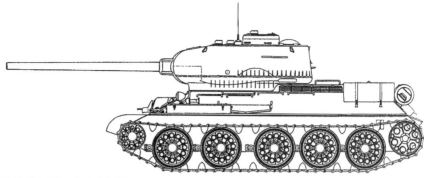

T-34-85 Model 1944

German tanks. Tiger Is were encountered in small numbers on the Leningrad front starting in January 1943 and one was promptly captured by the Soviets and examined. Although clearly superior to the KV in armor, mobility, and firepower, few were encountered in combat before the summer of 1943. But during the battle of Kursk in June 1943, the Germans introduced the first Panthers, and the numbers of Tigers dramaticaly increased. The Panther had been specifically developed to deal with the T-34 and was intended to become the standard tank of the German armored force. Soviet tank formations were decimated at long ranges in many lop-sided encounters. For the first time during the war, tank panic set in amongst the Soviet units, and the tank force demanded tanks with "longer arms" to be able to deal with the new German designs.

The priority assigned to Soviet tank production was clearly needed given the Soviet rates of loss, but it distracted the tank design bureaus from preparing for technological improvements by the Germans. The inadequately armed "universal tank" requirement indicated that the Soviets did not anticipate the German shift towards heavier, better armed tanks and were unprepared with a new tank gun capable of dealing with them.

This failure by the Red Army bureaucracy helped give more influence to the tank designers who had been pushing for larger guns, a traditional preference that had been suppressed by the production managers from 1942 to 1943. A crash program was instituted in the late summer of 1943. A new 85mm tank gun, derived from the 85mm anti-aircraft gun, was adapted into a new three-man turret for the T-34-85. This ignored the increased armor sought on the T-43 universal tank, but did recognize the command deficiencies of the earlier T-34s. This tank entered service in small numbers in February 1944 as the T-34-85, and proved an immediate success. Although its gun was not as effective as either the Panther's long 75mm gun or the Tiger I's 88mm gun, it restored a measure of balance in the technolgical arms race since it could defeat either tank under the right circumstances. Furthermore, being based on a virtually unchanged T-34 chassis, it did not upset production to the extent that the costly new Panther had upset German industry.

The KV tank was substantially redesigned after its poor showing at Kursk in 1943. Had the Soviets allowed the KV-3 to replace the KV-1 as planned in 1941, it would have remained viable and could have served as a counterweight against the Tiger and Panther at Kursk. The Red Army was still concerned about the vulnerability of the new T-34-85 to enemy tank fire, and desired to have a heavy tank that was less vulnerable to enemy anti-tank guns. There was a clear recognition that such a tank could not replace the T-34 because it placed greater demand on industrial resources, and so a mix of medium and heavy tanks was continued. Due to the disgrace of the namesake of the KV tank, Marshal Klimenti Voroshilov, in the 1941 battles, the evolutionary improvement was renamed the IS (Joseph Stalin) heavy tank. It was derived from the cancelled KV-13 universal tank, but it retained the KV's longer hull. The most pressing issue was the matter of armament.

The capture of the German Tiger tank in January 1943 led a series of ballistic trials against the captured tank. The Tiger was subjected to fire from 76mm tank guns

T-34-85 Model 1944 medium tank.

IS-2 Stalin heavy tank.

Eastern Front Tank Balance 1941-1945: Critical Indices

	June 1941	1942	1943	1944	1945	
Soviet Tank strength	22,600	7,700	20,600	21,100	25,400	
German Tank strength*	5,262	4,896	5,648	5,266	6,284	

	1941	1942	1943	1944	1945	Total
Soviet Tank production	6,274	24,639	19,959	16,975	4,384	72,231
German Tank production	3,256	4,278	5,966	9,161	1,098	23,759
Production ratio	1:2	1:5.6	1:3.3	1:1.85	1:4	1:3

	1941	1942	1943	1944	1945	Total
Soviet Tank losses	20,500	15,000	22,400	16,900	8,700	83,500
German Tank losses	2,758	2,648	6,362	6,434	7,382	25,584
Tank exchange ratio**	1:7	1:6	1:4	1:4	1:3	1:4.4

*As of January of each year except for 1941, which is for 22 June 1941. German strength is entire strength, not only Eastern Front. In July 1944, the Germans had over 1,500 tanks in Normandy, and several hundred in other theaters such as Italy and the Balkans. Likewise, the Soviets kept about 3,000 tanks in the Far East through much of the war.

**German tank losses here include all fronts; tank exchange ratio is based on estimated of German losses to Anglo-American forces versus losses to Red Army, and so reflects only Soviet-German loss ratio

using new ammunition, from 122mm howitzers, 85mm anti-aircraft guns and 122mm corps guns. Both the 85mm anti-aircraft gun and 122mm corps gun gave good performance against the Tiger, and there were hopes that new anti-tank projectiles could further improve their performance. The effort to adapt the 85mm anti-aircraft gun to tanks was handed over to General F. Petrov's artillery design bureau in Sverdlovsk. As a stopgap, the 85mm gun was installed on a modifed KV-1 tank called the KV-85, and about 130 were completed at Chelyabinsk in September, 1943. The heavy tank design bureau in Chelyabinsk was apparently unaware that the T-34 design bureau in Nizhni-Tagil was already working on adapting the same gun on to the T-34. This would lead to the same imbalance as before with both the T-34 and the new heavy tank being armed with the same weapon.

Firing trials of the new D-5T 85mm gun proved disappointing. Several captured German Tiger I tanks were shipped to Chelyabinsk, where they were subjected to 85mm fire from various angles. The 85mm gun could not reliably penetrate the Tiger I except at ranges well within the lethal envelope of the Tiger I's own 88mm gun. The solution was to mount a heavier gun in the new IS tank. Petrov's bureau favored the new D-10 100mm gun being developed by his bureau specifically for tank fighting. (This gun would later arm the SU-100 tank destroyer and the T-54/T-55 rank.) However, it was unlikely to be ready in time, and ammunition supply would also be a problem since it represented a new gun caliber for the Red Army. Ammunition for the 122mm corps gun was already in the Red Army's supply network, and there was surplus industrial capacity for the 122mm gun, so this weapon was selected. This project was approved in November 1944, just after the first batch of IS-85 tanks with the inadequate 85mm gun were coming off the production lines. A total of sixty-seven IS-85s were completed by the end of 1943 and forty more at the beginning of 1944, but nearly all were rearmed with 122mm guns before being issued to the troops. Production of the new version with the 122mm gun, called the IS-2, began immediately afterward in January 1944.

The Soviet selection of the 122mm gun for the IS is illustrative of Soviet tank design philosophy during the war. Unlike the Germans, who were willing to incur a continual string of production and logistics difficulties to acquire modest firepower, mobility, and armor advantages, the Soviet designers were forced to compromise in order to ensure ease of production, high production rates, and logistical harmony with the supply system. This experience was not unlike that faced by US tank design in World War 2. It was only in late 1943, with the war turning clearly in favor of the USSR, that the heavy tank designers were allowed to bring their costly new tank to the production stage. Even then, the design contained many common elements with the previous KV tank, including a very similar engine, a common track and many other identical components.

Soviet light tank production was abandoned after Kursk. There were several reasons for this. To begin with, light tanks, armed with 45mm guns, were inadequate in tank fighting or infantry support due to their poor firepower. This view was shared in many other armies of

Soviet Wartime Tank Production*

Light Tanks	1941	1942	1943	1944	1945	Total
T-40	41	181				222
T-50	48	15				63
T-60	1,818	4,474				6,292
T-70		4,883	3,343			8,226
T-80			120			120
Subtotal	1,907	9,553	3,463			14,923

Medium Tanks	1941	1942	1943	1944	1945	Total
T-34	3,014	12,553	15,529	2,995		34,091
T-34-85			283	11,778	7,230	23,661
T-44					200	200
Subtotal	3,014	12,553	15,812	14,773	7,430	53,582

Heavy Tanks	1941	1942	1943	1944	1945	Total
KV-1	1,121	1,753				2,874
KV-2	232					232
KV-1S		780	452			1,232
KV-85			130			130
IS-2			102	2,252	1,500	3,854
Subtotal	1,353	2,533	684	2,252	1,500	8,322
Total	**6,274**	**24,639**	**19,959**	**17,025**	**8,930**	**76,827**

*1941 figures are for last six months of the year; 1945 figures are for first six months. There are some discrepencies in published totals, probably due to the inclusion of prototypes in some figures and their omission in others.

the time. The US Army, for example, abandoned the 37mm armed M-5A1 light tank for the 75mm armed M-24 in 1944 for much the same reason. Secondly, need for light tanks for reconnaissance units were being fulfilled by Lend-Lease supplies of Canadian Valentine tanks, which became the predominent light scout tank in Soviet units in 1944. Finally, the industrial resources allotted to the T-70 were better used building a light assault gun version of the T-70, armed with the 76mm ZIS-3 divisional gun, the SU-76M. Indeed, the SU-76M became the second most common Soviet armored vehicle of the war (after the T-34). It was used in Soviet infantry units for close support, much like the pre-war T-26 light tank.

The Red Army was shipped 1,683 light tanks and 5,488 medium tanks from the United States; and 5,218 tanks from Britain and Canada under Lend-Lease. This amounted to sixteen percent of Soviet wartime tank production. The Red Army held a generally disparaging view of Allied tanks, comparing them very unfavorably to the T-34. However, British and American light tanks were no worse than the T-60/T-70, and the M-4A2 Sherman was not significantly inferior to comparable models of the T-34. These tanks were widely used in Soviet units, and in 1943, about twenty percent of Soviet tank brigades were of mixed Soviet/Lend-Lease composition while about fifteen percent were equipped entirely with Lend-Lease types. US and British tank design had little technical impact on Soviet tank design after the war except for some subcomponents, such as tank gyrostabilizers. In contrast, US truck design was enormously influential, forming the basis for most post-war Soviet military truck design, and much of Soviet post-war automotive engine design. German tank technology had far more impact, particularly in engine, transmission, and suspension design.

Seeds of Post-War Armor

With the war turning decisively in the Soviet Union's

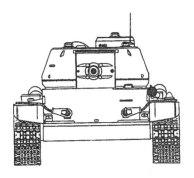 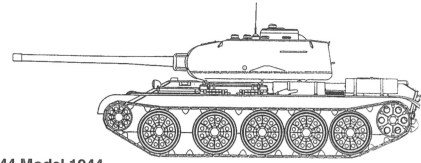

T-44 Model 1944

favor, the Morozov design bureau began to develop the new T-44 tank, originally codenamed Obiekt 136, which embodied the technical lessons of the wartime tank design effort. Free from the production constraints of the mid-war years, the design represented the first generational break from the T-34. It would form the basis for nearly twenty years of Soviet medium tank production, from the T-44 of 1944 through the T-62 of 1962. The Obiekt 136 used a turret and gun system virtually identical to that on the T-34-85. The main departure in the design came in the hull. The hull form was extremely simple, its compact size made possible by a radically different transverse engine layout. The powerplant was a derivative of the wartime V-2 diesel that powered the T-34, KV and IS, but mated to a new transmission. The suspension externally resembled the T-34, but internally, torsion bars had replaced the Christie-style spring suspension in order to provide more internal volume. The first trials series of the tank entered production in 1944. No large-scale production of the tank was undertaken during World War 2 because of severe teething problems with the design, especially its new powertrain. This tank represented the culmination of Soviet wartime design, with an impressive mixture of design simplicity and high combat effectiveness for a 30-ton tank. It is interesting to note that the Soviets did not feel obliged to match the combat capabilities of the German Panther in terms of firepower or armor, yet were able to come very close in a design that weighed only about sixty-five percent as much.

The final heavy tank of the war to appear was the IS-3 Stalin, originally codenamed Kirovets-1. General Nikolai Dukhov of the Chelyabinsk bureau led the de-

T-44 (Obiekt 136) medium tank.

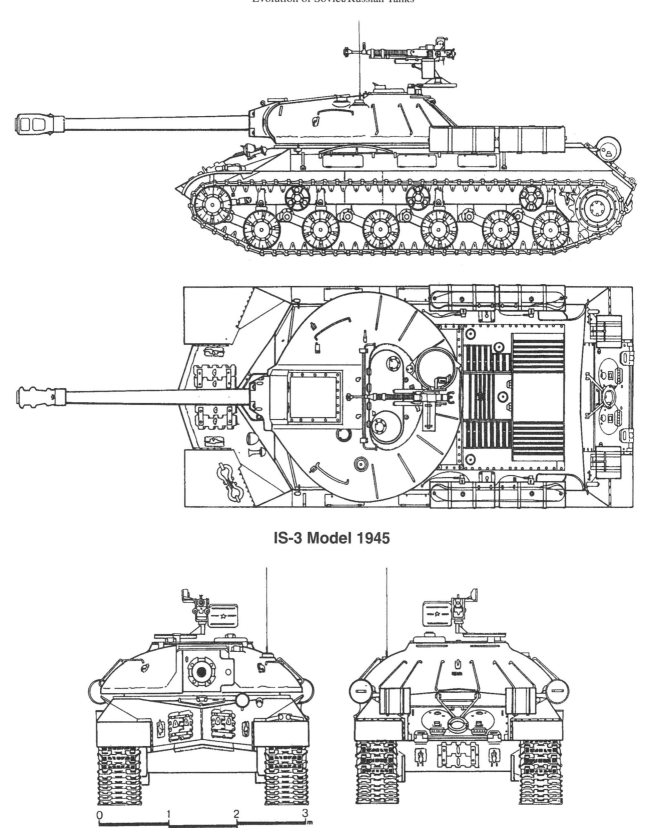

IS-3 Model 1945

sign team, with an objective to modify the IS-2 armor layout to be more resistant to the fire from the long 88mm gun on the German Tiger II. One of his engineers had conducted a study on tank vulnerability that concluded that the hits on the turret front were most often the cause of tank loss, followed by hits on the hull front. As a result, a radical new shape was designed. The turret was a simple hemisphere with a thickly armored gun mantlet faired cleanly into the shape. The hull panels were cleverly laid out to increase their effective thickness to fron-

T-44M medium tank.

tal attack by heavily angling them. To accommodate the large turret, the upper hull sides actually sloped inward, a feature hidden by attaching thin metal tool stowage bins along the upper sides. Internally, the IS-3 was essentially similar to the IS-2. It also used the same main gun.

The prototype of the IS-3 was completed in October 1944, and it was accepted for series production in parallel to the IS-2M. Production took place at Chelyabinsk from 1945 and concluded in 1951, by which time about 1,800 had been manufactured. The IS-3 design had been prematurely rushed into production, and the tank was beset with scores of mechanical problems. Large numbers of IS-3 tanks were sidelined with mechanical problems. Also, the welds on the thick armor plates on the front of the hull tended to crack open after service use. As a result, no significant number of IS-3s were ready before the end of the war in Europe. The IS-3 was first publicly displayed at a victory parade in Berlin on 7 September 1945 which involved fifty-two IS-3 tanks from the 2nd Guards Tank Army.

The hull and turret configuration of the IS-3 was enormously influential for this sleek simplicity. In the Soviet Union, the shape was adopted on later Soviet medium tanks such as the T-54A and retained as standard until today's T-72B and T-80. In the West, the shape influenced designs such as the American M-48, German Leopard 1, and French AMX-30. Once again, Soviet tank engineering was setting the pace of world tank design due to its innovative design.

Conclusion

The T-34-85 and IS-2 were the last significant tanks to enter Red Army service during the war years. In the end, neither tank was as effective as its comparable German opponent (the Panther and Tiger I) in terms of armor, firepower or mobility. But the technological disparity was not great enough to substantially affect battlefield performance in any meaningful way. To their credit, the Soviet tanks were considerably simpler and cheaper to produce, allowing the Soviet Union to continue to build up a substantial quantitive advantage over the German armored forces from 1944 to 1945. In 1943, the Germans had been able to maintain a combat equilibrium on the Eastern Front by offsetting their numerical weakness by modest technological advantages and superior crew and unit performance. In 1943, they were still destroying about four Soviet tanks for every one of their own lost, thereby dulling the impact of Soviet numerical advantages. However, in 1944, the Germans were not able to maintain the equilibrium because of a revival in Soviet tank design, substantial armor transfers to western Europe in the spring of 1944 to deal with the forthcoming Allied invasion, and a diminishing disparity in German versus Soviet tank crew tactical skills. It is worth pondering whether the German industrial policy to manufacture small quantites of high quality tanks was not one of the root causes for the German's reverses from 1943 to 1944. While it is outside the scope of this short survey to provide an answer, clearly the Soviet lesson was that modest technological disadvantages were acceptable when counterbalanced by numerical superiority. While the Soviets were always loathe to admit it, significant quantitative advantages and near parity in weapons technology remained a neccessity to defeat Western armies.

The First Post-War Medium Tank Generation: 1942-1962

At the conclusion of the Second World War, the Red Army (renamed the Soviet Army in 1947) was continuing to reequip its tank force with a new generation of vehicles. The T-34 Model 1943, armed with a 76mm gun, was still in widespread use; it was gradually withdrawn into war reserve as the T-34-85 replaced it. The legendary T-34 was reaching the end of its developmental potential, and the new T-44 medium tank had entered limited production in parallel. In 1945, production of the T-34-85 was underway at three plants: Nizhni Tagil Number 183, Omsk Number 174, and Gorkiy Number 112. The former Kharkov Locomotive Plant (KhPZ Number 183) was liberated in the summer of 1943, and it began small-scale production of the T-34-85 after a hiatus of four years.

Post-War T-34-85 Production

The T-34-85 was numerically the most important Soviet tank produced from 1945 to 1950. US Army intelligence estimated that 22,700 T-34-85s were manufactured from July 1945 to 1950 in the USSR, in addition to the 23,611 manufactured during the war. In 1952, T-34-85 production was initiated at the Stalin Plant, in Martin, Czechoslovakia. This tank was essentially identical to the standard Soviet type, though the turret casting was somewhat different in detail. The Czechoslovaks are believed to have produced about 3,000 T-34-85s, many for export to the Middle East. Likewise, Poland began license manufacture of the T-34-85 at a new plant in Labedy; total production from 1951 to 1954 was 1,380 tanks which were mostly used by the Polish Peoples Army (LWP). Yugoslavia considered manufacturing a locally modified version, but this program came to naught except for a few prototype vehicles.

Early Post-War Soviet Medium Tank Design[8]

At the end of the war, medium tank development efforts were still concentrated under the Morozov design bureau at Nizhni-Tagil in the Urals industrial region. The pace of development there slackened considerably in 1945, in part due to the end of the war, and it part due to the return of many engineers to their original hometown of Kharkov in Ukraine, where the Morozov design bureau had been located until 1941. Work was underway simultaneously on two tank programs: making the T-44 (Obiekt 136) suitable for series production, and developing a better armed version, the T-54 (Obiekt 137).

The need for a better weapon than the ZIS S-53 85mm gun on the T-34-85 became evident during the war. The most common ammunition type was the BR-365K AP-T. With an initial muzzle velocity of 792 meters per second, it could penetrate 102 millimeters of armor at 1,000 meters. This round was supplemented by the the BR-365P HVAP-T round, which had a tungsten carbide core. With an initial muzzle velocity of 1,030 meters per second, it had a penetration of 130 millimeters at 1,000 meters. The German Panther tank had turret front armor 110 millimeters thick and a glacis plate 80 millimeters thick (at 55 degrees). As a result of the armor layout, the Panther could not be easily penetrated using the standard BR-365K in a frontal engagement at 1,000 meters, and was only marginally vulnerable to the less common BR-365P ammunition.

The Soviet Union had been provided with a single example of the US Army M-26 Pershing tank through Lend-Lease. This tank had thicker armor than the Panther: a glacis plate of 100 millimeters (at forty-six degrees) and turret frontal armor of 114 millimeters. Both the Panther and M-26 were considerably more vulnerable to attack by the 85mm gun from side angles of course. The Soviet Union was probably unfamiliar with armor details of the British Centurion tank. Although the German Panther, American Pershing, and British Centurion tanks were heavy tanks by Soviet standards, it had become evident to Soviet tank designers that these forty-five-ton tanks would become the medium tank class, of the post-war generation. The 85mm gun on the T-34-85 and T-44 tanks was still a viable weapon in the short-term, but there was clearly the need for an improved weapon in the long-term to deal with improved derivatives of the M-26 and Centurion.

Such a weapon was already available in the form of the 100mm D-10 rifled gun, developed by General F. Petrov's design bureau in Sverdlovsk at Artillery Plant Number 9. This weapon had originally been developed in 1944 to arm the SU-100 tank destroyer, based on the T-34 chassis. It had been considered for arming the IS-2 but rejected because, at the time, the ammunition was only beginning to enter large-scale production for the army. With the war over, however, the logistics concerns were no longer of paramount importance. The D-10 gun offered a substantial improvement in anti-armor performance. The BR-412 AP-T round had an initial muzzle velocity of 1,000 meters per second and an armor penetration of 185 millimeters at 1,000 meters. This was more than adequate to defeat any existing tank short of the rare German King Tiger heavy tank.

The Petrov D-10 100mm gun was already in production, along with the towed 100mm BS-3 anti-tank gun designed by the rival V. Grabin design bureau. For unknown reasons, a third gun was considered for the new tank gun, called the LB-1.[9] The initial attempts were directed at mounting the D-10 and LB-1 in the T-44 tank. However, the engineers were skeptical of this effort because of the enormous size of the D-10's breech block and the relatively small space available inside the T-44 turret. A testbed of the T-34, armed with the 100mm gun as the T-34-100 had already been built. As a result, a parallel program called the Obiekt 137 was started using a new turret on the T-44 chassis to provide the added space.

In December 1944, Morozov's bureau had completed initial design work on Obiekt 137, also known as the B-40. The new turret was more thickly armored than the turret on the T-44, and indeed its frontal armor of 200 millimeters was as thick as that on the German King Tiger tank. The new turret was designed from the outset to carry the D-10 100mm gun. The new turret required hull extensions added on either side to increase the turret race diameter. The turret was significantly different than on subsequent models, with a conventional, wide mantlet, and prominent turret overhang both front and rear. It was fitted with two SG-43 machine guns mounted in boxes on either fender for defense against infantry; curiously enough, the US Army also experimented with fender-mounted machine guns around this time to replace the separate machine gun station in the hull so typically found on wartime tanks. Other internal changes of the T-54 were the improved V-54 engine, and a planetary gear box to overcome problems on the T-44 powertrain. On 20 May 1945, the final drawings of the vehicle were transferred to the Nizhni Tagil factory for prototype construction. Prototypes of this tank were completed in 1946, and it became known as the T-54 Model 1946. T-44 production was halted in 1946 in order to permit small-scale production of the T-54 Model 1946 for operational trials. In 1948, Morozov and his design bureau (including A. Kolesnikov, V. Matyukhin, P. Vasiliev, and N. Kucherenko) received a state prize for its new design.

T-54 Model 1947 (Obiekt 137).

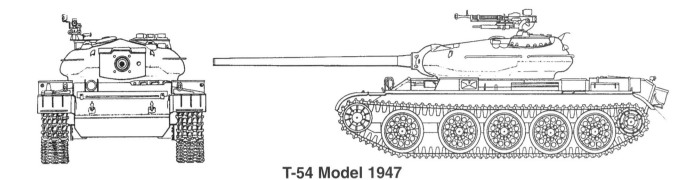

T-54 Model 1947

The new T-54s were tested by tank units in the Byelorussian Military District. The trials were a fiasco with many faults discovered during operation, especially poor reliability. The situation was so serious that complaints were sent directly to the Politbureau. The plan had been to begin production in 1947, but instead, the assigned factories halted production while awaiting the modifications to be implemented.[10]

T-54 Medium Tank Designs

T-54 Model 1949

The poor showing of the T-54 led to a heavy redesign. The turret was completely changed with the wide mantlet being replaced by a narrow gun mantlet. The new turret also avoided the shot-traps at the front corners by adopting a hemispherical shape. An improved anti-aircraft heavy machine gun ring mounting with the 12.7mm DShK was added to the turret roof. The fender-mounted machine guns were removed as being impractical, and a single SGMT 7.62mm machine gun was added near the driver; it would fire through a hole in the glacis plate. These initial production vehicles introduced a multistage centrifugal oil-bath air filter for the engine, and an oil-supply preheater for cold weather starting. A new track was introduced which was widened eighty millimeters to a total of 580 millimeters for better flotation on poor ground. This version was designated the T-54 Model 1949 and went into full-scale production in 1950 at the three main tank plants at Nizhni Tagil, Omsk, and Kharkov. This was the first version of the T-54 built in significant numbers. The initial T-54 Model 1946 tanks were later rebuilt to at least this standard.

T-54 Model 1951

The T-54 underwent continual modernization through the early 1950s. In 1951, the turret casting was changed, probably to simplify manufacturing. The overhang at the rear of the turret was completely removed, giving the

T-54 Model 1949 medium tank.

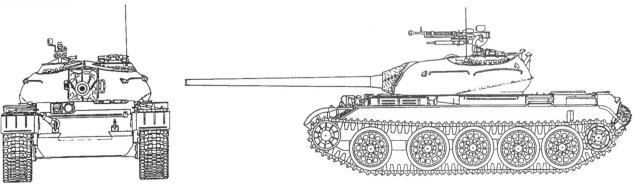

T-54 Model 1949

T-54 Model 1951 its classic hemispherical shape, resembling half an egg shell. This version also replaced the previous TSh-20 gunner's sight with the TSh-2-22. Turret bearings were improved for greater durability, and the vehicle electronics were better sealed from dust. This version also was the first to regularly introduce mounting attachments for mine-rollers (initially the PT-3, and later the PT-54).

T-54A Model 1951

The Soviet Union received some experience with gun stabilizers on the M-4A2 Sherman tanks provided under Lend-Lease. The first Soviet tank to introduce a gun stabilization system was the T-54A. The T-54 had a probability-of-hit of only three percent while the tank was moving; the objective was to increase this to thirty percent by adding a one-axis stabilizer. Two bureaus competed on the project, V.G. Grabin's design bureau, and a small design team under I.V. Pogozhev. As the T-54 was armed with the 100mm gun from his arch rival, Petrov, Grabin offered a new 100mm gun with a new stabilization system. Pogozhev adapted his system to the existing D-10T gun, which won state approval since it would prove to be a less costly solution. The modified gun with the new Gorizont (Horizon) vertical-plane gun stabilization system was designated D-10TG and was incorporated into the new T-54A tank. The D-10TG 100mm gun was modified with a new fume extractor at this point.

The fume extractor was another probable American influence, as in the autumn of 1950, the North Korean Army and Chinese Volunteer Army had captured several M-46 tanks during the retreat from Pyongyang. Other internal improvements included a new multistage air cleaner and radiator vane controls for improved engine

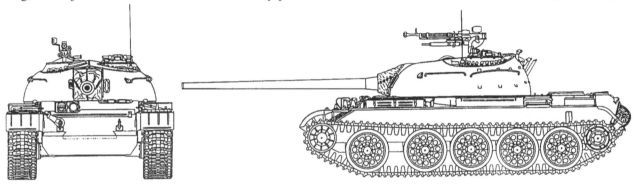

T-54 Model 1951

T-54 Model 1951 medium tank.

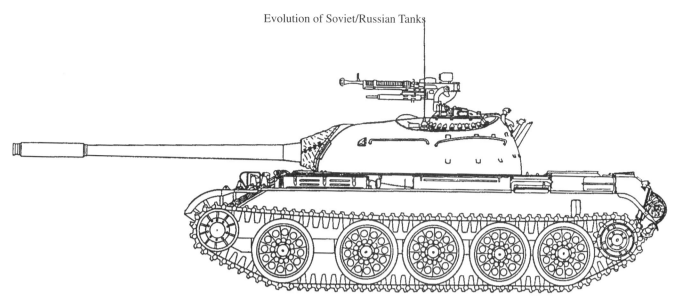

T-54A Model 1951

performance.

The Red Army had been experimenting with deep fording tanks since the late 1930s as a means to hastily cross rivers. Work continued in the early 1950s, with experiments on T-34-85s and other tanks. The T-54A was selected as the first tank to be regularly fitted with this equipment, as the hull configuration made it easier for sealing than the T-34-85. This was the first version of the T-54 series regularly fitted with the OPVT river fording equipment.[11]

The OPVT system consisted of vehicle sealing equipment and a snorkel tube system. Although part of the sealing system was permanently fitted to the vehicle in the factory, attaching the snorkel and adding the additional sealing took about 1.2 hours to complete. The OPVT system allowed a tank to ford rivers to a maximum depth of five meters, a maximum width of 700 meters and a maximum river flow speed of 1.5 meters per second. The tank can fire its main gun about thirty seconds after having left the river, but it takes ten to fifteen minutes to completely remove the sealing equipment from the tank. The tank is steered underwater using the

T-54A Model 1951 medium tank.

GPK-48 gyrocompass. During underwater fording, the crew usually wears a special escape breathing apparatus. In the event that the tank is stranded mid-stream, the tank interior is gradually flooded, and once filled, the hatches are opened, and the crew escapes. Deep wading is a very dangerous process since river bottoms are irregular, and the tank has a certain measure of buoyancy which degrades traction and steering. Deep wading operations are usually prepared by special river reconnaisance teams, and often a tank recovery vehicle is the first vehicle sent across in order to help tow out stranded tanks. Nevertheless, this feature allows Soviet tank units to cross river obstacles, unaided by engineer bridging equipment. The OPVT system was retrofitted to many older tanks, including the T-34-85, and thereafter was included as a standard feature of Soviet tanks beginning with the T-54A. For training purposes, a second tube was developed which was wide enough to permit the tank crew to evacuate the vehicle if it became bogged down on the river bed.

The T-54A was also the first Soviet tank fitted with a driver's night vision system. This consisted of an infrared headlight and a metascope periscope that could be substituted for the normal daylight periscope.

Limited production of the T-54A began in the USSR in the autumn of 1954. The stabilizer proved troublesome, and production improvements were continually added over an eighteen-month period, finally resolving the difficulties. Shortly after series production began, a new style of road-wheel was introduced with a distinctive starfish pattern. This subsequently became the standard style on the T-54A and the subsequent T-55 and T-62 as well. The original spoked wheel became uncommon, as the new wheel was added to older T-44s and T-54s during their periodic rebuilding.

T-54B Model 1952

The first Soviet experience with tank night fighting occurred in Hungary in January to February 1945 when the 1st Guards Mechanized Corps, equipped mostly with Lend-Lease M-4A2 Sherman tank, was engaged on several occasions by German Panther tanks equipped with an early active infrared night fighting system. The German system consisted of an infrared searchlight mounted on a supporting armored half-track vehicle; the Panther tanks carried a metascope for viewing the infrared illumination. The Soviets captured samples of this equipment later in 1945. It is not known if they experimentally deployed the German system, but they waited for a more practical autonomous system before making a major investment in night fighting equipment.

Some T-54As were fitted with an active infrared night driving system consisting of an FG-100 infrared headlight and a TVN-2 driver's night periscope. The first Soviet tank regularly equipped with night fighting equipment was the T-54B Model 1952, with the prototypes completed at Nizhni Tagil in 1952. This version introduced a Luna L-2 infrared searchlight on the turret front, an OU-3 IR searchlight on the commander's cupola, and a TPN-1 gunner's day/night sight in place of the simpler MK-4 periscope in front of the gunner's station which included a night channel with metascope.

The T-54B also introduced the Tsiklon (Cyclone) two-axis stabilization system as part of the improved D-10T2S

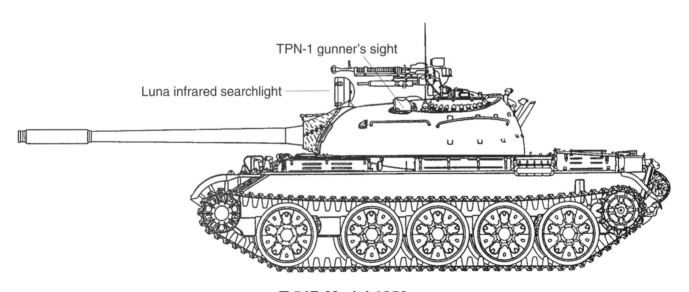

T-54B Model 1952

T-54B Model 1952 medium tank.

100mm gun. The previous Gorizont stabilization system was only in one-axis. The new stabilization system, based around the STP-2, was of a fairly elementary nature and did not offer real fire-on-the-move capability. Rather, it kept the main gun pointing with a rough accuracy so that once the tank halted for firing, only minor aiming adjustments would be needed. The T-54B was also fitted with an electric power traverse to supplement the normal manual turret traverse. It enabled the turret to be completely rotated in under thirty seconds.

T-54A (rebuilt to T-54B standards).

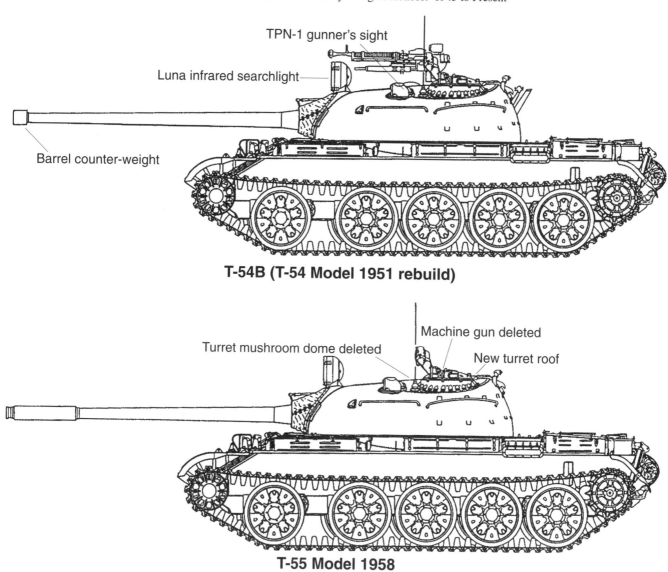

T-54B (T-54 Model 1951 rebuild)

T-55 Model 1958

The Polish equivalent of the T-54B was designated T-54AM, a designation which sometimes erroneously is used in the West for this type whether Soviet or Polish. Series production of the T-54B version began in the USSR in early 1957.

T-54M

Developed under the code name Obiekt 140 towards the end of 1953, the T-54M was a major modernization of the T-54, with a new, long barreled D-54 100mm smoothbore gun designed by A.A. Barikhin at the Petrov design bureau in Sverdlovsk. The new gun had superior armor penetration to the D-10T and the tank could carry fifty rounds of ammunition instead of the usual thirty-four. This vehicle was supposed to carry the 14.5mm KPVT machine gun in lieu of the usual DShK, but what few photos exist of it show the normal anti-aircraft machine-gun mount. The engine was an improved V-54-6. This vehicle never entered quantity production, and at a later date, the designation T-54M was applied to the modernized T-54 with the D-10T2S gun found on the T-54B. The D-54 gun served as the basis for the later 115mm U-5T on the T-62 tank.

T-54K Model 1954 Command Tank

This was the first standard command version of the T-54A, built in small numbers. The additional R-112 command radio was incorporated into the turret by reducing the ammunition storage. The only external indication of its command function was the addition of an antenna stowage tube on the hull rear. Nearly all subsequent Soviet tank types had command derivatives which had an additional radio along with necessary antenna stowage.

The T-54AD was the Polish command version equivalent of the T-54K, the "D" indicating dowodca, or "commander." This version has a modified turret with a slight extension on the turret rear to provide space for the command radios. This vehicle is used by regimental commanders and regimental chiefs-of-staff. A similar Polish T-55 derivative was also built.

OT-54 Flamethrower Tank

This tank for engineer troops substituted an ATO-1 automatic flamethrower for the usual co-axial machine gun. The original prototype, the Obiekt 483, had the flamethrower replacing the 100mm gun. This approach was not favored, however, since it was quite obvious that the tank was a flamethrower; this could lead to special attention from enemy forces. Instead the vehicle was redesigned to be armed with both the 100mm gun and flamethrower. The bow ammunition stowage was modified to permit carrying 460 liters of flammable liquid using compressed gas for propulsion. Maximum range was 160 meters, and the system could fire fifteen to twenty bursts per minute.

T-54 Rebuilt Tanks

A confusing aspect of identifying T-54 tanks by their subtype was the regular Soviet practice of periodically rebuilding older tanks to newer standards. The Soviet Union established several rebuilding plants (remzavod: remontniy zavod, rebuilding plant) specifically for this purpose, the largest being in Kiev. The criteria for rebuilding varied through time: every 7,000 kilometers, every 500 engine operating hours, every ten years, or due to a special modernization program. As older versions of the T-54 tank were gradually returned for their periodic rebuilding, some of the new features of the T-54A and T-54B were added including the new wheels, infrared night-fighting equipment, OPVT deep fording fittings, and other features. It is very difficult to identify the various versions of the T-54 on the basis of external appearance. For example, a T-54 Model 1951, rebuilt with a new D10-T2S gun, new gunner's sight, IR equipment, and new wheels, is virtually indistinguishable from a T-54B. NATO sometimes designates these vehicles as T-54(M) and T-54A(M); the Soviet Union generally designated the tank on the basis of its rebuilding standard, so that a rebuilt T-54 Model 1949 was considered a T-54B if all the new features were incorporated.

Chinese T-54 Variants

Type 59 (WZ120) — The Soviet Union provided the Peoples' Republic of China with T-54 Model 1951 tanks in the late 1950s. In the mid 1950s, the Soviets also helped to set up a production line for the T-54A tank at Baotou in the Beijing military region. Production of this vehicle began in 1957, and it was designated Type 59 in PRC service (it is also known by its industrial designator of WZ120). Production was later extended to a second plant in the industrial region of Inner Mongolia and Shangxi. The PRC gradually modernized the design. Later improvements include the addition of a laser rangefinder developed by the Yangzhou 2nd Electronics Plant over the barrel, and the addition of side skirts.

Type 59-II (WZ120B) — A small series batch of Type 59s were produced in the 1980s using a 105mm gun, believed to be based on Israeli technology and essentially similar to the standard NATO gun.

Type 62 — Although externally very similar to the Type 59, this is in fact a scaled-down light tank patterned after the Type 59. It is armed with an 85mm gun and is more thinly armored.

Type 69-I (WZ121) — This is product improved ver-

Chinese Type 59 medium tank (T-54 license copy).

sion of the Type 59 tank, entered production in the late 1970s. It was first seen in September 1982 at Zhangiakou. The Type 69-I version is armed with a new 100mm smoothbore gun which is distinguishable by its bore evacuator, located further back on the tube than the usual 100mm rifled gun.

Type 69-II (BW121A) — This more common version of the Type 69 family was produced since the late 1970s. Unlike the -I version, it is equipped with the normal 100mm rifled gun. The Type-69-II most closely resembles the Soviet T-54B as it incorporates the TSFCS two-axis gun stabilization, a new Type 70 gunner's sight, full active infrared night fighting equipment analogous to the Soviet Luna system, and an engine smoke discharger. It differs from its Soviet equivalent in many details, including the provision of an NBC protective system, a Yangzhou laser rangefinder, and many small details. The Type 69-II can be distinguished from the earlier Type 59 family by the rear engine plate, which has a small elliptical bulge at the bottom for the cooling fan, somewhat reminiscent of the Soviet T-62. This version also has the driver's headlights mounted in two pairs on the fenders, not on the hull glacis plate as on the Type 59. The basic Type 69-II has a standard turret arrangement, but later production batches have "Boom Shield" standoff armor baskets on the turret. Some vehicles have only a rear mounted set, while on others the Boom Shiled extends fully around the turret. There is a wide range of detail differences on this vehicle type since China offers a large variety of options to its export clients.

Type 79 (WZ121D) — This version was first called Type 69-III and is a product improved Type 69 with a 105mm gun. In its export variant it is usually called Type 79.

Type 80 — The Type 80 appeared in the mid-1980s and is an upgraded Type 79 with a new suspension. The suspension uses wheels reminiscent of American MBTs. An improved version with automotive modifications, the Type 80-II, has also been developed.

Assessing the T-54A

The T-54A was first seen close-up by Western intelligence in 1956 when a Soviet tank was driven into the grounds of the British embassy in Budapest during the Hungarian Uprising. The thickness of the armor on the T-54A was a shock to NATO and prompted the adoption of the new British L7 105mm gun on later US, UK, and German tanks. Compared with NATO tanks of the period, the T-54A was impressive in many respects. It was much smaller than NATO tanks of the period like the US M-48 Patton or the British Centurion, but was comparably armored. Its 100mm gun was similar in performance to the 90mm gun on the US M-48 (though actual anti-armor penetration was heavily dependent on the type of ammunition fired not just on the gun design). Smaller size meant lighter weight, and lower overall cost; it also meant that more T-54s were produced than all other NATO tanks of the period combined. The T-54A had mobility comparable to that of the M-48 Patton tank, although the ride was a bit harder. Soviet tanks of the period were more apt to have infrared night fighting equipment than NATO tanks, and more thought had been given to river crossing requirements with the provision of OPVT equipment. The Soviet choice of a mature diesel engine also gave the T-54A excellent range without refueling, and the Soviets were not plagued to the range restraints that constrained gasoline-powered NATO tanks like the M-48 Patton and Centurion at the time.

On the debit side, the T-54A had mediocre fire controls by NATO standards. It was a good tank in close terrain where engagements would take place at ranges of under 1,000 meters. At greater ranges, it was outclassed by NATO tanks which could take advantage of better fire controls and optics, better gunlaying, and gun stabilization equipment. The Soviets had opted for a tank fire control system optimized for shorter range engagements. Soviet studies of central European terrain as well as actual combat experience from 1944 to 1945 had shown that much of the terrain was fine grained, and that long-range engagements were likely to be the exception, rather than the rule. For example, a later West German study of the inter-German border region found that only six percent of the terrain offered sighting ranges over 2.5 kilometers, only ten percent over two kilometers, seventeen percent over 1.5 kilometers and forty-five percent over 0.5 kilometers. In fact, over fifty-five percent of the terrain surveyed had sighting ranges under 500 meters. For this reason, the Soviets favored simpler stadia rangefinders instead of the coincidence rangefinders used on US tanks of the period.

Although stadia rangefinders were less accurate at longer ranges, they offered comparable accuracy at 1,000 meters when used with APDS ammunition, compared with US tanks using coincidence rangefinders, but firing HEAT ammunition. The US Army and the German Bundeswehr preferred to use HEAT ammunition at the time because of its very high anti-armor penetration. HEAT rounds have a pronounced ballistic arc which requires good rangefinding for proper elevation correction, while APDS rounds have a very flat ballistic arc at ranges of 1,000 meters and are therefore not as dependent on accurate rangefinding. The Soviets did use HEAT ammunition but favored APDS for tank fighting, even if its armor penetration characteristics were inferior to HEAT during this period.

In close-range tank melees, the T-54A suffered from a very cramped interior which meant less ammunition carried, poorer layout of ready ammunition, and slower and more difficult reloading. The T-54A was not as maintainable as NATO tanks. For example, to replace the engine or transmission, the turret had to be removed first. Soviet tank regiments of the time did not have equipment to do this in the field. NATO tanks of the period were designed to facilitate repairs at as low a level as practical.

The T-54 tank was in production in the Soviet Union from 1946 to 1958. Total production of the T-54 in the USSR is believed to have been about 35,000 tanks. In the period from 1957 through 1958, T-54 production was initiated at one plant in Czechoslovakia and Poland, and at two plants in China. Total Czechoslovak production is believed to have been about 2,500 tanks, and total Polish production about 3,000 tanks. The Chinese manufactured the T-54A as the Type 59, and modified it later as the Type 69. Production is still continuing and is believed to have reached 16,000 tanks by 1985.

The T-54 tank and its later derivatives, the T-55 and the T-62, are the most significant tanks of the post-World War 2 period in terms of numbers manufactured and extent of combat service. They are the most widely manufactured tanks of all times. First entering production in 1946, this family remains in production today in foreign derivatives such as the Romanian TR-85 and the Chinese Type 69. In total, production of the T-54, T-55 and foreign derivates exceeds 100,000 tanks, with an additional 20,000 of the closely related T-62 tank. Not only have they been produced in enormous numbers, but they have also seen widespread combat use. The largest tank battles of the post-war years—the 1967, 1973, and 1982 Middle East Wars, the Indian-Pakistan Wars, the Vietnam Wars—the Gulf Wars, all saw the extensive use of tanks of these types.

The T-55 Medium Tank

With development of the T-54A tank completed in 1951, the Kremlin allowed Aleksandr Morozov to move the main medium tank design bureau back to Kharkov, where it had been based prior to World War 2. However, a small design bureau, under Leonid Kartsev, remained in Nizhni Tagil at the "Vagonka" (Uralmashvagon Zavod) plant. Morozov's bureau was assigned to the development of a new generation tank to replace the T-54, while Kartsev's bureau was responsible for the less challenging task of updating the T-54 as necessary. In 1953, Kartsev was shown the technical-tactical requirements (TTZ) for the new Obiekt 430 being developed by Morozov's bureau. Kartsev was convinced that an upgraded T-54B could satisfy the requirement; so began the competition between the two components of the wartime medium tank design bureau. The original plan, called Obiekt 140 or T-54M, was a heavily reconfigured design with a new long-barreled D-54 100mm rifled gun. The improved V-54-6 engine was developed at the Barnaul engine plant after Trashutin's main diesel engine design bureau in Chelyabinsk refused to assist Kartsev in this venture. Many other improvements were envisioned for the type, including a new transmission, a modified supension and enhanced night vision equipment. Although a prototype was built in 1954, the changes were too extensive to win government approval. Kartsev turned his attention to less elaborate improvements.

In the 1950s, a state commission for certifying the technical documentation of tanks was held each fall, attended by representatives of GVTU (Main Military Technical Administration), the Mintransmash, the main tank plants and the main subcomponent manufacturers. The meeting was chaired by the production administration of the GBTU Main Armored Directorate. This commission usually approved any plans to upgrade tanks. During the

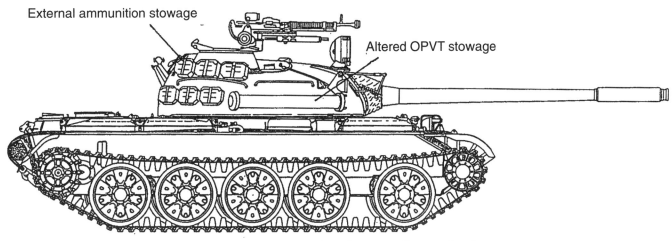

T-55 (final production configuration)

October 1955 commission session, Kartsev proposed a simultaneous set of upgrades for the T-54B tank to the commission head, General Aleksandr M. Sych. Sych approved the proposal, but the plant head, I. V. Okunev, suggested that the upgraded version be designated as a new tank. The project was designated Obiekt 155 and became the T-55 when approved for service use.

One of the most significant improvements in the T-55 were the first stages of a nuclear protective system. With the death of Joseph Stalin in 1953, the Soviet Army began examining the implications of tactical nuclear weapons on the conventional battlefield. By the late 1950s, Soviet tactical doctrine was in a state of flux, the so-called "revolution in military affairs," with growing emphasis being placed on the need to prepare for fighting on the nuclear battlefield. The Soviets began to study the technical effects of nuclear weapons on tanks from their earliest nuclear bomb trials. On 14 September 1954, they tested a nuclear bomb in the Totskoye region of the Urals during a special wargame to examine the psychological impact on troops of a nearby explosion. Soviet concern over the consequences of nuclear weapons on the battlefield was given greater urgency by the US Army's gradual adoption of tactical nuclear weapons as a counterweight to Soviet numerical superiority in central Europe. By the mid-1950s, Soviet doctrine was beginning to recognize that nuclear weapons would be present on the contemporary battlefield, and that modern weapon systems would have to be able to survive the after-effects of nuclear weapons.

The first manifestation of this doctrinal evolution in tank design came with the new T-55 tank which was adopted in 1958. A major innovation in the new design was a PAZ nuclear contamination detection and filtration system and an improved system to seal the tank from contaminated outside air.

The PAZ system was based around an RPZ-1M gamma ray detector system. This system, once triggered by a nuclear blast or nuclear contamination, turned off the engine to alert the crew, and turned on an overpressure system to keep out airborne radioactive contaminants. The mild overpressure of 0.0015 kilobar per cubic centimeter obviated the need for complicated sealing equipment. The PAZ system became a standard feature on all subsequent Soviet tanks, though the later systems were often more elaborate. The T-62A PAZ system included an explosive squib system to automatically close louvres and vent covers. Later PAZ systems also incorporated chemical protection features.

The T-55 closely resembled the T-54B externally. The most noticeable external changes were the deletion of the large vent dome cover in front of the loader's hatch on the turret, and the new flush loader's hatch. The uprated V-55 engine had been improved by Trashutin's diesel engine design bureau in Chelyabinsk, and the engine deck was modified. The fuel tanks were reconfigured so that ammunition could be stored in cavities in the forward tank; this saved internal volume and permitted more ammunition to be carried. The practice of fitting BDSh smoke canisters externally on the rear of the hull was omitted in favor of a TDA smoke system which created smoke by injecting fuel into the engine manifold. This freed up space on the rear plate to permit mounting of fifty-five gallon external fuel drums. An automatic fire extinguishing system for the engine compartment (UAPPO) was incorporated for the first time. An AK-150TS air compressor was substituted for compressed air bottles for engine starting. Other changes had been adopted including an improved transmission, a revolving turret floor, improved oil filtration, and a bilge pump to support the OPVT deep wading system. Three prototypes were built in 1956 and on the successful completion of trials, the new tank was accepted as the T-55 in the middle of 1957 with production scheduled to begin on 1 January 1958.

The T-55 continued to evolve with modest improvements and was produced for many years in parallel to both the T-62 and the T-64. Indeed, the T-55 outlasted the T-62 on the production lines up to 1979 but was produced in diminishing numbers for Soviet export clients. Total Soviet production of the T-55 probably totaled about 27,500 tanks. Czechoslovak and Polish production is believed to have been about 10,000 tanks combined. This brought total T-54 and T-55 production to about 100,000 vehicles, making this series the most widely produced tank type of all times. Besides the basic tank versions, the T-54 and T-55 formed the basis for air defense gun vehicles like the ZSU-57-2, armored recovery vehicles like the BTS-1, BTS-2, BTS-3 and BTS-4, and assault guns like the SU-122.

T-55 Model 1958

Instead of manufacturing the uparmed T-54M (Obiekt 140), the Soviets selected the T-55 (Obiekt 155) to replace the T-54B. The T-55, although closely resembling the T-54, was in fact a substantially redesigned tank with many internal and external differences. Externally, the T-55 used a new turret, similar in shape to the T-54A turret, but without the prominent roof-top "mushroom" ventilator dome, and as a result, two enlarged "D" roof panels. Internal improvements include the new V-55 engine, increase of ammunition to forty-three rounds, and other features. Unlike the T-54B, the T-55 dropped the loader's 12.7mm DShK anti-aircraft machine gun. This version had the initial stages of the PAZ chemical/radiation pro-

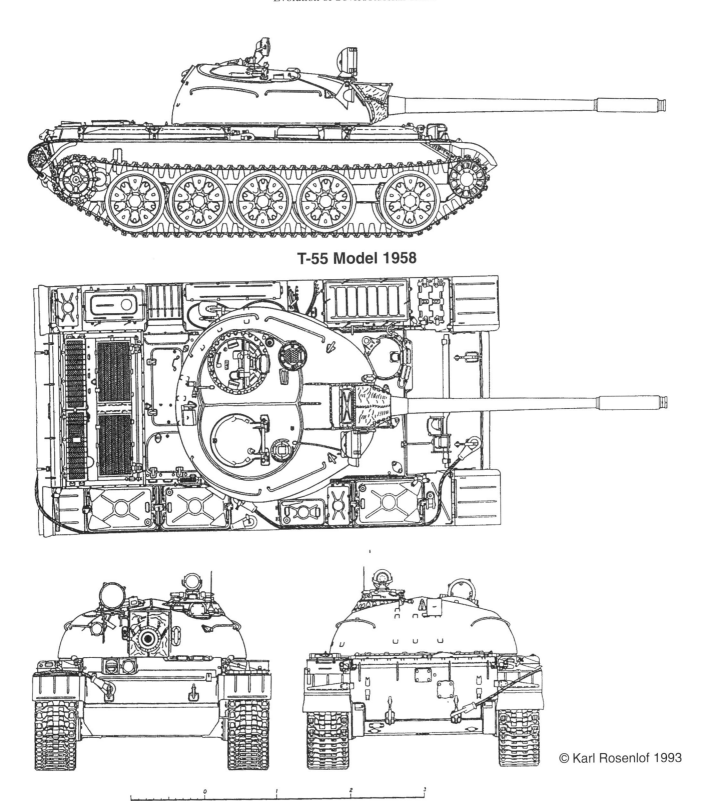

T-55 Model 1958

© Karl Rosenlof 1993

tective system. The T-55 was accepted for service in the summer of 1957 and placed into production in the USSR on 1 January 1958.

T-55A

The T-55A was primarily developed to incorporate a new antiradiation lining and the full PAZ/FVU chemical filtration system. The T-55A finally dispensed with the fixed, hull-mounted machine gun. One of the major internal additions was the use of a plasticized lead sheeting for antiradiation protection. This was evident externally due to the use of an enlarged driver's hatch and enlarged combings over the commander's and loader's hatch to accommodate the new material. The Polish version is sometimes called the T-55A(P) by other armies which

T-54 Model 1958 medium tank.

T-55A medium tank.

T-55A

use it, such as the former East German NVA. Originally, the T-55 and T-55A did not carry the 12.7mm DShK antiaircraft machine gun on the turret, as had been typical of the T-54.

T-55A Model 1970

Beginning in 1970, T-55A tanks began to receive a new turret fitting for the 12.7mm DShK antiaircraft machine gun. The revival of the turret antiaircraft machine gun resulted from the growing importance of NATO anti-tank helicopters, as evidenced by the appearance of the AH-1G Cobra with TOW anti-tank missile shortly after the end of the Vietnam war. This upgrade was later adopted on the T-62M in 1972. Besides its use on new-production T-55A tanks, this feature was retrofitted on earlier T-55 and T-55A tanks as well. This version is sometimes called T-55M or T-55AM in NATO. Two different machine gun fittings were used. On vehicles with the larger antiradiation hatch collars, a collared mounting was used. In the late 1970s, a low-cost export type was produced, with a simpler collar and a hatch arrangement than was used on the T-62M.

T-55, T-55A Model 1974

Beginning in 1974, T-55s began to be retrofitted with

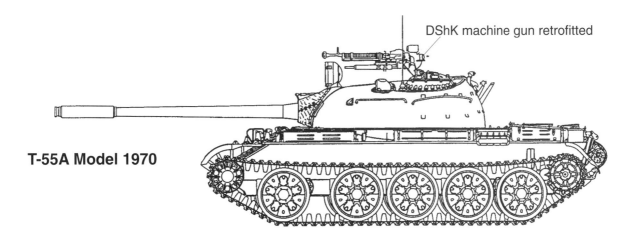

T-55A Model 1970

T-55A Model 1970 medium tank.

the KTD-2 (kvantoviy tankoviy dalnomer) laser range finder as part of a fire control upgrade. This was mounted immediately above the main gun in an armored box. It is not clear whether these initial rangefinder modifications included the more elaborate Volna fire control upgrade mentioned below.

T-55M, T-55AM Model 1983

In 1983, the T-55 upgrade program entered a new phase, featuring a comprehensive rebuilding of the T-55. The upgraded T-55s were designated T-55M and upgraded T-55As were designated as T-55AM. This upgrade included passive applique armor, the Volna fire control upgrade, optional fitting of the new Bastion guided projectile, automotive upgrades, and many small improvements.

The applique armor package was developed by the Scientific Research Institute of Steel (NII Stali) in Moscow, the main Russian research institute for metalurgical

T-55 Model 1974 medium tank.

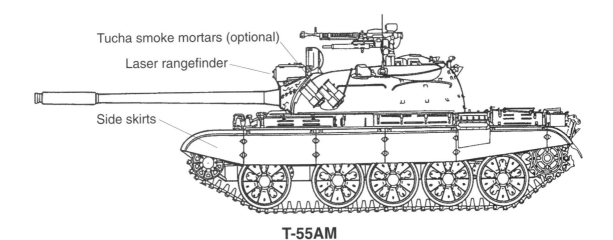

T-55AM

and armor research. The applique did not receive an official name, although some documents refer to it by the acronym "BDD." NII Stali unofficially called this applique armor "brow armor" because of the two rounded panels added on either side of the gun tube. The glacis applique armor consists of a welded steel box formed from thirty millimeter of steel sheet. Inside are six layers of five millimeter steel plate spaced thirty millimeters apart with the cavity between filled with penapolyurethane. The turret applique is a similar hollow construction, but the cast shell is thicker, about sixty millimeters on the outer side. The new turret armor increases the tank's protection from its basic 210 millimeters of steel to the equivalent of 380 millimeters against APFSDS and 450 millimeters against HEAT according to Russian sources. The hull applique raises the effective protection from 200 millimeters equivalent to 410 millimeters against APFSDS and 380 millimeters against HEAT. The total package adds about 2.2 metric tons to the weight of the tank. (During the 1991 Gulf War, a small number of Iraqi tanks appeared with an improvised armor applique which was a crude duplicate of this type of armor, which uti-

T-55M with KTD-2 laser rangefinder.

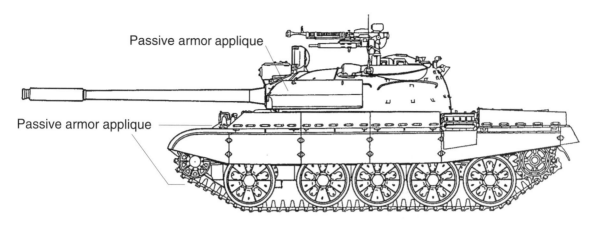

T-55AM (with Brow armor)

lized alternating layers of rubber/steel and may have been based on Russian advice.)

Side skirts were added to provide additional protection to the suspension and hull sides against shaped-charge warheads. This consists of a ten millimeters-thick, steel-reinforced rubber material. There is an optional anti-mine package consisting of a twenty millimeter armored plate mounted on frames under the hull to protect the driver against anti-tank mines. For enhanced anti-radiation protection, improved antiradiation lining was added to these tanks, and the crews were issued with IPZh-1 anti-radiation protective vests.

The Volna fire control system was based around a KTD-2 laser rangefinder, mounted externally on the gun mantlet in an armored box. The laser rangefinder has an effective envelope from 500 to 4,000 meters. The Volna system includes the BV-55 analog ballistic computer, the first time a ballistic computer was used on the T-54/T-55 series. The gunner's normal TSh-2M-32P sight was replaced by the improved TShSM-32PB, which has a continuously stabilized field of view in the vertical axis. The previous Tsiklon two-axis stabilization system was replaced by the upgraded Tsiklon M1 gun stabilization system.

The T-55M/T-55AM were fitted with the new 9K116 Bastion 100mm guided projectile system developed by the Shipunov design bureau in Tula. This system consisted of a 3UBK10-1 100mm ammunition round which encases the 9M117 missile. This round resembled a conventional 100mm round and was handled and loaded in the same fashion. To guide this laser-beam riding projectile, the gunner's normal sight was replaced by the 1K13 sight in place of the normal TPN-1 sight. Other necessary additions were the 9S831 transformer, a control panel and a new electronic panel. The missile was fired like a conventional round, with a rocket engine igniting 1.5 seconds after the round fired. The sustainer rocket engine in the missile would burn for six seconds. The missile has a flight time of twenty-six to forty-one seconds, at which point it self-detonated if it had missed its target. The missile had an effective penetration of 1,000 millimeters against homogenous steel armor and an effective range of four kilometers against both helicopters and tanks. Generally, the tank would carry four to six of the missiles in addition to its usual combat load.

The T-55M upgrade also included automotive improvements. The V-55U engine was substituted for the previously standard V-55V. The V-55U offered an increase of forty horsepower to 620 total horsepower with inertial supercharging and partitioned intake manifolds.

Roadwheel dynamic travel was increased from 135 to 149 millimeters to 162 to 182 millimeters through the use of new torsion bars. These were produced using an electroslag process which also doubled their effective life. The T-55M and T-55AM were gradually refitted with the RShM track (developed for the T-72) which necessitated a different drive sprocket; this new track added a half ton to overall vehicle weight.

Some T-55M and T-55AM also had other improvements, including a new side skirt made of metal reinforced rubber sheeting for protection against shaped-charge projectiles. The Type 902B Tucha smoke discharger system began to be introduced at this time, which

9K116 Bastion guided projectile.

T-55AM-1 medium tank with "brow" armor appliqué.

fired the 3D6 81mm smoke grenade at ranges of 200 to 350 meters. The grenades supplemented the TDA smoke discharge system and their cloud lasts 60 to 130 seconds. Four 3D6 grenades create a smokescreen 100 to 120 meters wide and eight meters high.

Other improvements on the T-55M/T-55AM include the substitution of a R-173 radio for the older R-123, and incorporation of features to defend the tank against napalm attack.

T-55M-1, T-55AM-1

While the T-55M modernization program was continuing, it was decided to incorporate the V-46 engine from the T-72 into the T-55 and T-62. The version used was designed V-46-5. Modified T-55s with this engine were designated with a "1" suffix, such as T-55M-1, T-55AM-1, T-55AD-1, and T-55MV-1, etc.

T-55AM-2

Both the Polish and Czechoslovak tank factories began a T-55 upgrade program paralleling the Soviet T-55M/T-55AM upgrade called the T-55AM2. However, not all of these upgrades included the 9K116 Bastion missile system. The basic T-55AM2 does not fire the Bastion missile. The version fitted with the 1K13 fire control system capable of guiding the 9K116 Bastion missile is designated T-55AM2B (Czechoslovak built) and T-55AM2P (Polish built).

The Czechoslovak equivalent of the Soviet Volna fire control upgrade uses the locally designed Kladivo laser rangefinder, mounted over the gun tube and a Czechoslovak wind sensor mounted centrally on the turret rear. The Polish equivalent uses the Polish Merida fire control system with the laser rangefinder integrated into the

Cross section of laminate appliqué armor on T-55AM2.

T-55AM-2 medium tank (Czechoslovak).

1K13 gunner's sight for T-55AM-2.

gunner's sight, and a hammer-head wind sensor forward on the turret.

T-55MV, T-55AMV

In 1983, the NII Stali in conjunction with the NIIBT Research Institute of the Main Armor Directorate at Kubinka completed the development of the Kontakt explosive reactive armor for tanks. Reactive armor is generally called "dynamic protection" (dinamicheskaya zashchita) in Russian. Although the Soviets had examples of Israeli Blazer armor from examples captured by the Syrians in the 1982 Lebanon war, the Soviet type was of a different type according to Russian sources. This reactive armor has undergone several generational changes; this is not immediately apparent as the same external boxes have been used.

In Russian service, the ERA bricks are called kostek (dice). T-55M retrofitted with Kontakt explosive reactive armor are designated T-55MV, the "V" signifying vzryvnoi, or "explosive." The T-55AM was designated T-55AMV when fitted with reactive armor. The use of dynamic protection appears to have been less common on the T-55 series than the passive "brow" armor. The Soviet Naval Infantry is known to use dynamic armor on their T-55 tanks. In 1993, the Czech Republic showed its own local version of reactive armor on a T-55 tank at the IDET trade fair.

In 1994, NII Stali displayed a photograph of a T-55 tank fitted with its third generation of applique armor, Kontakt-5, originally developed for the T-80U. This is a combined passive/reactive armor package intended to provide enhanced protection against both APFSDS and HEAT. At the Niznhi-Novgorod arms show, a NII Stali placard compared the armor protection offered by the

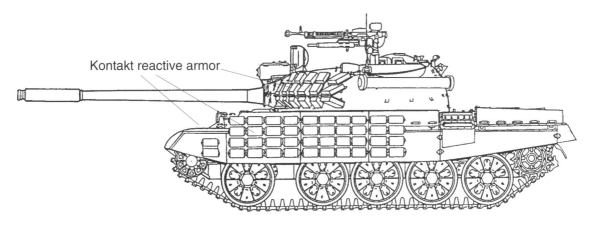

T-55AMV

T-55AMV medium tank.

three generations of applique on the T-55 tank as shown in the chart on page 51.

T-55AD Drozd

As an alternative to passive or reactive armor defense systems against antitank rockets and weapons, the Tula Machine Design Bureau under A. Shipunov developed the world's first active tank defense system, called the KAZT Drozd (Kompleks aktivnoi zashchity tanka: Tank active defense system-Thrush). Drozd entered development in 1977 on the basis of an army requirement and the first prototypes were ready by 1978. The Army lost interest in the concept, but the program continued due to support from the Soviet Naval Infantry.

The Soviet Naval Infantry was having difficulty replacing its T-55 tanks as later types such as the T-62 and T-72 posed weight and size problems in standard navy

41

T-55AM2 with "brow" armor appliqué.

T-55 with Kontakt-5 appliqué armor.

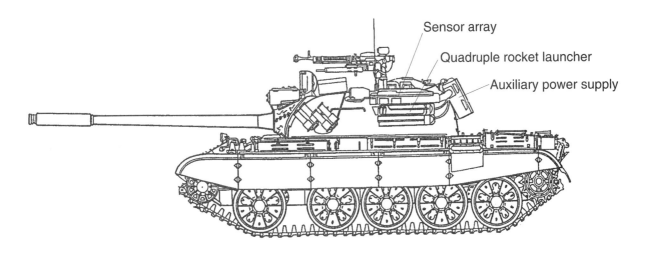

T-55AMD Drozd

amphibious ships and landing craft. The development costs for Drozd, are believed to have been around $170 million. This represented a much smaller investment than the cost of developing a new tank specifically tailored for the needs of Soviet Naval Infantry.

Production of Drozd began on a limited scale from 1981 to 1982. This system was first fitted to T-55M or T-55AM tanks, which were then redesignated T-55AD. When fitted to the upgraded T-55M1 or T-55AM1 tanks with the improved V-46 diesel engine, the designation became T-55AD1.

The Drozd tanks are not fitted with the appliqué armor, but do carry the 9K116 Bastion guided missile system. The total Drozd production run was small at less than 300 tank systems. It does not appear to have been commonly deployed by the Soviet Navy Infantry, probably due to the extensive security measures taken to protect the system's existence. By the late 1980s, when Kontakt-1 reactive armor became available, the Soviet Naval Infantry switched to T-55 tanks with reactive armor as a more practical alternative to the cumbersome and relatively expensive Drozd system. The Drozd system has been exported in small numbers to several European countries, China, and to one undisclosed Middle East client.

The Drozd system consists of three main elements, two launcher arrays on either side of the turret and an auxiliary power unit on the rear of the turret. Each launcher array consists of four launch tubes with a Doppler radar sensor array mounted above the tubes. The radar sensor arrays actively emit a radio frequency beam forward of the tank. The auxiliary power unit can provide power to the system and can surge up to a maximum of 800 watts for short-periods of time.

T-55AD tank with Drozd active defense system.

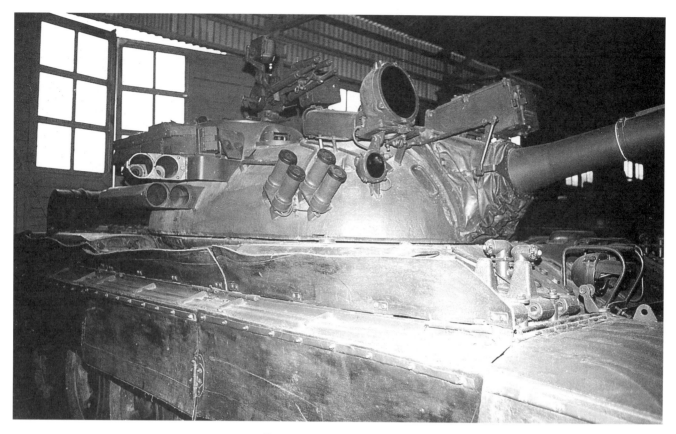

Close-up of Drozd launcher.

T-55 AD Drozd: Active Protection System

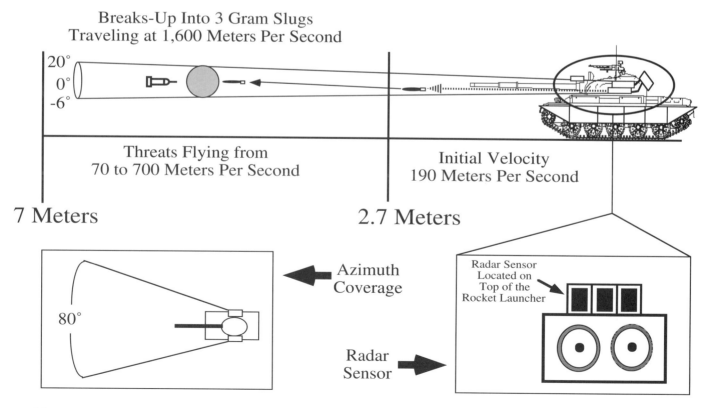

The radar is gated to acquire targets moving at speeds of between 70 to 700 meters per second; this gating process avoids engaging the system against small arms and other high-speed projectiles (e.g., long-rod armor penetrators). On acquiring the incoming slow-moving projectile, the Drozd's analog computer determines which of the eight KAZ projectiles to launch. The radar system determines the range of the incoming missile or rocket, and the computer calculates when to fire the KAZ projectile, and does so automatically. The KAZ is launched to intercept the enemy missile at seven meters from the tank.

The KAZ projectile is 107 millimeters in diameter, weighs nine kilogram and is rocket boosted out of the tube with an initial velocity of 190 meters per second. The fuze detonates the warhead at seven meters in front of the tank, and there is no guidance aboard the KAZ projectile after launch. The high explosive warhead has a prefragmented steel casing which on detonation breaks up into three-gram slugs traveling at 1,600 meters per second. From a Russian promotional video on Drozd, it would appear that the KAZ projectile explodes over the incoming missile, with a directed downward blast. The launcher arrays are configured to cover eighty degrees in azimuth, and -6 to +20 degrees in elevation. The Drozd system includes a rearward pointing light system for warning nearby friendly units when it will fire. The Drozd launchers take around ten minutes to reload from an on-board supply of rockets.

This system has been fitted on prototype testbeds of the T-62, T-72, and most recently the T-80U.

T-55K Command Tank

The T-55K is fitted with two additional radios (originally an R-112 and R-113; later an R-123 and an R-124), an additional AB-1P/30 generator for the radios, and a special ten-meter aerial carried in a tube at the hull rear. To accommodate the added equipment, the T-55K had the hull SGMT machine gun deleted, and it carried six fewer rounds of ammunition. Eventually, at least three versions of the T-55K were developed. The K1 and K2 versions both carry two R-123 or R-123M and an R-124; the T-55K3 carries one R-130M, an R-123M, an R-124, and a 10-meter antenna. Variants based on later chassis are designated accordingly: T-55AK, T-55MK, etc.

The Poles build their own version of the T-55 command vehicle, called the T-55AD. This vehicle differs from the Soviet type in that it is fitted with a slight rear turret bulge to accomodate the added communication equipment.

OT-55 Flamethrower Tank

This is an engineer flamethrower tank based on the basic T-55, called OT-55 or TO-55. This version can be

OT-55 flamethrower.

Polish T-55A combat engineer tank with KMT mine rake and PW-LWD minefield breaching system.

distinguished by the large pig's head mantlet over the ATO-200 barrel. The ATO-200 flamethrower uses a unique revolver system fitted with twelve charges to ignite the flammable liquid. A total of 460 liters of flamethrower fuel is carried, with thirty-five liters used in each burst. The ATO-200 can fire at a rate of seven bursts a minute and has a maximum range of 200 meters. The maximum effective range of the system is 200 meters, with the stream having an intial muzzle velocity of 100 meters per second. The TO-55 is fitted with the same D-10 gun as a normal tank but carries considerably fewer rounds of ammunition. The distribution of the OT-55 is not known, but it probably equips special combat engineer tank regiments, rather than being used in normal tank regiments.

T-55 Engineer Tanks

In 1959, a portion of the new production T-55s were modified to accept the PT-55 mine-roller. These versions are externally identifiable by the attachment fittings on the hull front. The PT-55 mine roller weighed six to seven metric tons and was fitted to the tank only when in mine clearing operations. Shortly afterwards, additional fittings were developed for th BTU and BTU-55 bulldozer blades. The BTU-55 was an improved and lightened type (1.4 versus 2.3 tons) and was eventually the more common of the two.

Romanian T-55 Variants

TR-580 — This Romanian derivative of the T-55 uses a lengthened hull and a new suspension system with spoked wheels. It has also been called the TR-77, which may be an export designation.

TR-85 — This drastically modified Romanian version of the T-55, with a new German diesel engine and a completely redesigned suspension. The turret, although resembling the T-55 was in fact, completely new. This vehicle has been called the TR-80 and TR-800, which may be export designations.

Romanian TR-85 (local T-55 variant).

T-54/T-55 Armor Protection

Location	T-54A (millimeters)	T-55 (millimeters)	Angle (degrees)
Upper glacis	110	100	60
Lower bow	100	100	55
Side	80	80	0
Rear upper	60	50	17
Rear lower	30	20	70
Hull top	20 and 30	16 and 33	90
Hull belly	20	20	90
Turret front	210	205	various
Turret mantlet	210	210	various
Turret sides	95 to 110	120 to 130	various
Turret rear	60	60	various
Turret top	30	30	various

T-54/T-55 Comparative Technical Data

	T-54	T-54A	T-55	T-55A	T-55AM
Length overall (millimeters)	9,000	9,000	9,000	9,000	9,000
Width (millimeters)	3,270	3,270	3,270	3,270	3,520
Hull length (millimeters)	6,040	6,040	6,200	6,200	6,200
Height overall (millimeters)	2,750	2,750	2,350	2,700	3,010
Height w/o MG (millimeters)	2,400	2,400	2,350	2,350	2,350
Ground clearance (millimeters)	425	425	425	425	350
Combat weight (metric tons)	36.0	36.0	36.0	36.0	40.5
Gun type	D10-T	D10-TG	D10-T2S	D10-T2S	D10-T2S
Caliber	100mm	100mm	100mm	100mm	100mm
Elevation (degrees)	-4 +17	-4 +17	-5 +18	-5 +18	-5 +18
Ammo stowage*	34	34	43	43	38+5 TUR
Stabilization	-	Gorizont	Tsiklon	Tsiklon	Tsiklon M1
Anti-aircraft machine gun	DShKM	DShKM	-	DShKM	DShKM
Ammunition stowage	200	200	-	200	200
Co-axial machine gun	SGMT	SGMT	SGMT	PKT	PKT
Glacis machine gun	SGMT	SGMT	SGMT	-	-
7.62mm ammo stowage	3,500	3,500	3,500	3,500	1,250
TC's day sight	TPK-1	TPK-2	TPK-2A	TPKU-2B	PNK
TC's night sight	-	-	TKN-1	TKN-1S	TKN-1S
TC's Infrared searchlight	-	-	OU-3G	OU-3G	OU-3G
Gunner's telescope	TSh2-22	TSh2B-22	TSh2B-22B	TSh2B-32P	TShSM-32PV
Gunner's day sight	TNP-165	TNP-165	TNP-165	TNP-165A	1K13 BOM
Gunner's night sight	-	-	TPN-1M-22	TPN-1M-22	1K13 BOM
Main infrared searchlight	-	-	L-2 Luna	L-2G	L-4
Loader's sight	MK-4	MK-4U	MK-4U	MK-4M	MK-4M
Driver's day Sight	F Blok	F Blok	F Blok	F Blok	F Blok
Driver's night sight	-	-	TVN-2B	TVNO-2	TVNO-2
Radio	R-113	R-126	R-123	R-123M	R-173
Intercom	R-120	R-120	R-124	R-124	R-124
Engine type	V-54	V-54G	V-55	V-55A	V-55U
Horsepower	520	520	580	580	620
Power-weight (horsepower/ton)	14.4	14.4	16.1	16.1	14.75
Maximum speed (kilometers/hour)	50	50	50	50	50
Integral fuel (liters)	812	817	960	965	965
Basic range (kilometers)	510	510	460	460	385
External fuel (liters)	-	400	400	400	400
Range+2x 200 liter drums	-	720	650	650	545
Ground pressure	0.81	0.81	0.81	0.81	0.91
Chemical/Biological/Radiological	no	no	PAZ	PAZ/FVU	PAZ/FVU
Smoke	BDSh	BDSh	TDU	TDU	TDU
Fording (meters)	1.4	1.4	1.4	1.4	1.4
Deep wading (meters)	5.0	5.0	5.0	5.0	5.0

*Usual ammunition stowage for T-54 (T-55) was: 11 (15) AP-T; 6 (7) HEAT; 17 (21) HE-Frag

T-55 Armor Modernization

	Basic Tank (no applique)	Variant 1 (Brow armor)	Variant 2 (Kontakt 1)	Variant 3 (Kontakt 5)
Added weight (kilograms)	0	2,190	1,320	2,800 to 3,000
Equivalent steel protection (millimeters)				
(versus APFSDS)	200	330	200	450 to 480
(versus HEAT)	200	400 to 450	700 to 900	700 to 900

T-54/T-55 (100mm) Tank Ammunition

Round (kilograms)	Projectile (kilograms)	Type (millimeters)	Round wt.	Projectile wt.	Armor penetration
53-UBR-412B	53-BR-412B	AP-T	30.1	15.9	185
3UBR3	53-BR-412B	AP-T	30.6	15.9	185
53-UBR-412D	53-BR-412D	AP-T	30.6	16.0	185
	3BM8	AP-T		4.13	
	3BM25	APFSDS			
3UBK4	3BK5	HEAT	22.6	12.37	380 (190)
3UBK4	3BK5M	HEAT	22.6	4.4	380 (190)
	3BK17	HEAT			380 (190)
3UBK10-1	9M117	Missile	24.5	17.6	550* (255)
53-UOF-412	53-OF-412	HE-Frag	30.3	15.6	
53-UOF-412U	53-OF-412	HE-Frag	26.7	15.6	
53-UOF-412ZhU	53-OF-412Zh	HE-Frag	26.7	15.5	
53-UO-415	53-O-415	HE-Frag	30.1	15.4	
3UOF10	3OF32	HE-Frag	30.4	15.4	
3UOF11	3OF32	HE-Frag	26.8	15.4	

*Some Russian promotional material claims the penetration to be 660 millimeters.
** At 1,000 meters, 0 degrees, and 60 degrees of obliquity.

The T-62 Tank

Introduction

Following the development of the T-55 tank in 1957, the Vagonka design bureau under Leonid Kartsev proposed to continue the evolution of this design by increasing its combat performance. Kartsev noted that the T-55 did not improve the medium tank in either firepower or armor. As in the case of the aborted Obiekt 140 (T-54M) effort, the new design would focus on an improved gun. The gun selected was the D-54TS, an improved variant of the D-54 100mm gun tested on the T-54M. Two approaches were considered, a tank with only the turret enlarged to accommodate the new gun and a tank with a lengthened hull and enlarged turret to ensure that the ammunition load did not decrease compared with the T-55. The latter approach was approved by the defense ministry, and it was designated Obiekt 165. Three prototypes were completed in November 1958. A 0.43 meters (seventeen inches) section was added in the center of the T-55 hull to accommodate a larger diameter (1.825 to 2.245 meters) turret.

While work was progressing on this design, the Main Missile and Artillery Directorate (GRAU) was completing development of a new smoothbore 100mm antitank gun. This was displayed in 1958 to Nikita Khrushchev, who insisted that it be adapted to tanks. Kartsev complained that the projectile, which was 1.2 meters (3.9 feet) long, was too long to fit into existing tanks. After long debate, the GRAU changed its specifications, reducing the round's length to 1.1 meters, and increasing its diameter to 115mm. This gun, which eventually became the U-5T (2A20), was selected as the next armament for the evolved T-55. In 1959, the Military Industrial Commission (VPK) of the Council of Ministers approved this plan and authorized development of the new tank, designated Obiekt 166. The first prototypes were completed in 1959. The new tank closely resembled the Obiekt 165 and used the new Meteor gun stabilizer first developed for the D-54TS. The first set of tests at Kubinka in the autumn of 1960 were successful, but some faults were uncovered. The Meteor stabilization system needed a more powerful generator, and the new gun's larger ammunition created problems inside the fighting compartment during test firing because of their large size. The later problem was solved by designing a system to expel the spent casings out of the turret through a small hatch in the roof after each firing.

Although the Obiekt 166 was deemed successful, it became mired in a political struggle within the GBTU Main Armor Directorate. The chairman of the NTK Scientific-Technical Commission, General A. V. Radus-Zenkovich, delayed authorization of production. The problem stemmed from the fact that the GBTU was sponsoring the Kharkov bureau's Obiekt 430 as the next Soviet main battle tank, and it was still armed with the same D-54TS 100mm gun as the abandoned Vagonka Obiekt 165. It was considered imprudent to authorize production of the Obiekt 166 with its superior 115mm gun since this would upset the GBTU's modernization plans. Instead, Kartsev was told to develop an improved Obiekt 166, the Obiekt 177, with an uprated engine with supercharger being developed by L. A. Vaysburd team at the Chelyabinsk engine plant. Work also began on a modified Obiekt 166 with a new suspension, the Obiekt 167.

In December 1960, the new American M-60 tank was issued to US troops in Europe for the first time. The initial evaluation of the design was made available to senior Soviet officers in January 1961. The commander of the Soviet Ground Forces, Marshal V.I. Chuikov, was infuriated to learn that the tank was armed with a 105mm gun, when the standard Soviet medium tank of the period, the T-55, was armed with a 100mm gun. Furthermore, the thicker frontal armor of the M-60A1 turret could not be penetrated by the standard Soviet 100mm tank gun. This created a crisis atmosphere in the tank industry. Chuikov called the head of the GBTU Main Armor Directorate, Marshal Pavel P. Poluboyarov, into his office in Moscow for a severe dressing down. Poluboyarov mentioned that Vagonka had developed a tank with a 115mm gun but that there had been problems with the stabilizer. Chuikov, the commander of Soviet forces at Stalingrad in World War 2, was notoriously short tempered and foul mouthed. He screamed at Poluboyarov, "Why are you jerking me around over this stabilizer? I don't care if it's mounted on a pig! Just come up with this gun!"

The director of the Nizhni Tagil tank plant, Ivan V. Okunev, was still upset with the 1960 decision against starting production of the Obiekt 166. To give them a hard time, he refused to begin production of the Obiekt 166, insisting instead that they wait until the improved Obiekt 167 with its new suspension complete its trials. This incident highlights a feature of Soviet military-industrial relations seldom appreciated in the West; the plant directors reported to their industrial ministry, not directly to the military and so had some power in the outcome of new weapons programs. Okunev was forced to back down in a July 1961 VPK meeting in Moscow, chaired by D. F. Ustinov, since the VPK could direct both the military and the industry. Shortly after the Obiekt 166 was declared as standard by the Ground Forces and designated T-62. In the later half of 1961, twenty-five pre-series T-62s were manufactured for trials purposes with an operational unit at Novgorod Volynskiy. After adjustments, series production of the T-62 began at Nizhni Tagil in July 1962.

In many respects the T-62 program closely resembled the contemporary American M-60 program, where an existing design (the US M-48 or Soviet T-55) had a new weapon system added while retaining many features of the preceding model. Likewise, the T-62 was accepted for production over the more revolutionary Obiekt 430, just as the M-60 was accepted instead of the more revolutionary T-95. There were differences, however, while the T-95 program was terminated, the Obiekt 430 program continued, eventually leading to the new T-64 design that was produced in parallel to the T-62.

The T-62 closely resembled the T-55, except for the extension of its hull. This permitted it to carry forty rounds of the much bulkier 115mm ammunition compared with forty-three on the T-55. The extremely cramped conditions of Soviet tank interiors led to some unique features in the T-62's design. The interior was too small to permit spent shell casings from ejecting inside, so a unique system was incorporated to automatically eject the spent casings out a hatch in the turret rear. This system slightly slows the gun firing cycle, since the gun must be depressed after each firing, eject the casing, and then reorient back to the original elevation. There have also been some problems with the spent casings failing to clear the rear port, and careening into the turret, injuring the crew. Nevertheless, this system highlights the Soviet predilection to keep overall tank size very compact.

The T-62 remained in production from 1962 to about 1975 in the USSR and from 1973 to about 1978 in Czechoslovakia. Total production in the USSR was probably about 20,000 tanks, and Czechoslovak production was probably 1,500 tanks. One of the enigmas of the T-62 is that no significant number were deployed with Warsaw Pact forces outside of the Soviet Union itself, even though Czechoslovakia eventually manufactured it. In addition Bulgaria purchased a small number. The main

T-62 Model 1962 medium tank.

drawback to its export was probably cost. The price of the T-62 was reportedly 250,000 rubles, which was about 50 percent greater than the cost of the T-55 at the time. Having just retooled to produce the T-55 from 1960 to 1964, Poland and Czechoslovakia probably were uninterested in another expensive reconversion, and additional licensing fees to the USSR. The only significant advantage offered by the T-62 was superior anti-armor penetration with the new gun. Czechoslovakia's entire T-62 production run was exported to the USSR and to foreign clients in the Middle East.

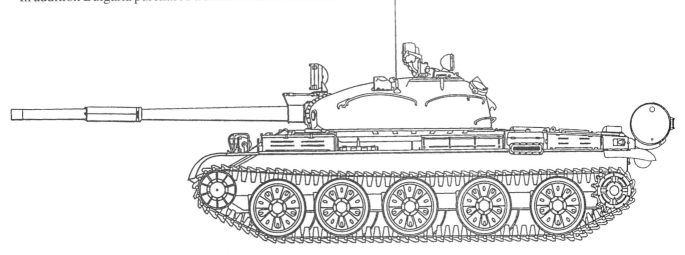

T-62 Model 1962

T-62 Model 1967 medium tank.

T-62 Model 1962

This was the initial production model of the T-62. A total of twenty-five trial tanks were built in 1961 and full-scale series production began on 1 July 1962.

T-62 Model 1967

In 1967, the T-62 was modernized. The basic T-55 pattern engine deck configuration was changed by altering the access hatches to the engines.

T-62 Model 1972

In 1972, as part of the general upgrade affecting the T-55 fleet as well, the turret of the T-62 was modernized to permit a 12.7mm DShKM anti-aircraft machine gun to be fitted over the loader's hatch. This changed the shape of the right turret side. This version is commonly called T-62A in NATO. Other changes on this version included a new stowage arrangement for the OPVT (on the turret side), a larger vehicle identification light, and improvements in the fording attachment. Unlike the T-55, the practice of adding a new DShK mounting to older T-62s was not done.

T-62 Model 1975

The final upgrade of the T-62 prior to its withdrawal from production involved the incorporation of a KTD-1

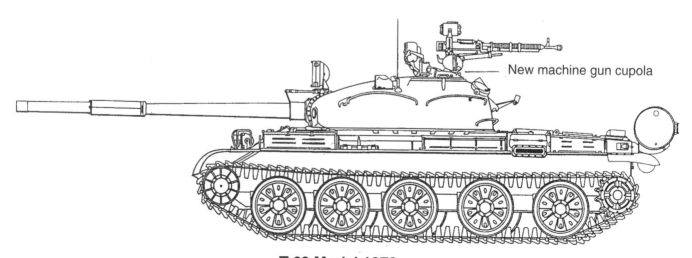

T-62 Model 1972

T-62 Model 1972 medium tank.

laser rangefinder over the main gun mantlet to increase accuracy in long range engagements.

T-62K

This is the command tank variant of the T-62. Like earlier T-55K types, it has one additional R-122 radio, a TNA-2 navigation system, a four-meter antenna, but less ammunition. The corresponding command tank for the T-62M is the T-62MK.

T-62M

The T-62 series was subjected to the same type of upgrade program as the T-55 series starting in 1983. This involved the addition of BDD passive applique "brow" armor, engine upgrades, gun fire control upgrades, and a new guided projectile. The Volna fire control upgrade

Sheksna 115 mm round with 9K116-1 guided projectile.

Initial T-62M with KTD laser rangefinder.

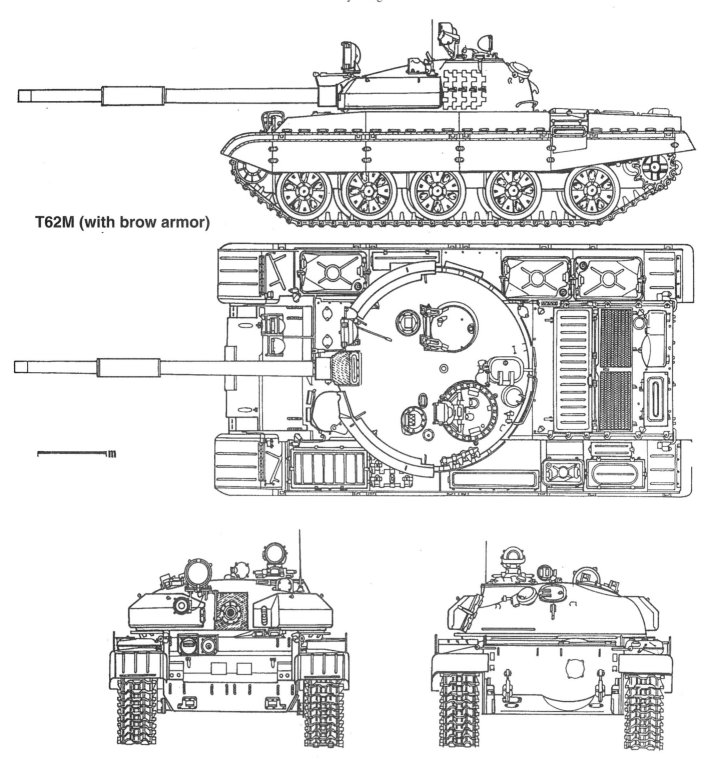

T62M (with brow armor)

on the T-62M included the KTD-2 laser rangefinder, the BV-62 ballistic computer, the TShSM-41U gunner's sight and the Meteor M1 gun stabilization system. A thermal sleeve was also added to the gun barrel to reduce barrel warpage. The T-62M was also fitted with a 115mm relative of the 9K116 Bastion, designated as the 9K116-1 Sheksna. This fired the same 9M117 guided projectile as the 9K116, but the projectile was modified with suitable bands to fit the larger diameter gun tube. As in the case of the T-55M, the 1K13 gunner's sight was used on vehicles carrying the Sheksna system. Some of the upgraded T-62M tanks were not fitted with the Sheksna system; these were designated T-62M1.

The "brow" armor package fitted to the T-62M was essentially the same as on the T-55M upgrade, with two curved panels on either side of the turret, a glacis panel and an optional underbelly anti-mine panel. In the late 1980s, however, some T-62Ms were fitted with an addi-

T-62M tank with "brow armor" appliqué.

tional bar-armor applique. This replaced the reinforced rubber skirts over the suspension, and included an added panel on the rear corners of the turret. This package was first deployed in Afghanistan, but was also fitted on tanks in the USSR.

The T-62M incorporated the same automotive improvements as the T-55M, namely the V-55U engine, improved torsion bars, and new RShM track. The total package adds 3.9 tons to the weight of the vehicle. The T-62M also had two extra hydraulic shock absorbers mounted on the second roadwheel station to accommodate the added weight of the upgrades. Other improvements included the substitution of the new R-173 radio for the older R-123M. As in the case of the T-55M and T-55M1 improvements, later T-62 upgrades had the V-46-5M engine used instead of the V-55U. These were tanks were designated T-62M-1.

To further confuse matters, a variety of packages were fitted to the T-62M1 tanks without the Sheksna missile. The T-62M1-1 was a T-62M1 with the V-46-5M engine and was essentially the same as the T-62M-1, but without the Sheksna. T-62M1-2 was a version of the T-62M1, but lacking the glacis applique armor. T-62M1-2-1 was a T-62M1-2, but fitted with the improved V-46-5M engine.

T-62MV

Instead of the passive applique armor package, some T-62Ms were fitted with the Kontakt reactive armor package developed by NII Stali. These had all of the other modifications of the T-62M, including the Volna fire controls, Sheksna system and V-55U engine upgrade. Likewise, later T-62MV upgrades had the improved V-46-5M engine susbstituted for the V-55U; these are designated T-62MV-1.

T-62M medium tank with Kontakt reactive armor.

T-62D

As in the case of the T-55, the Drozd system was offered as an alternative to passive and reactive armor arrays for tank defense against antitank missiles. The Drozd was fitted to tanks with the other T-62M upgrades including the Volna fire control system, automotive improvements and the Sheksna missile system. T-62 equipped with Drozd were designated T-62D; those with the later V-46-5M engine upgrade were designated T-62D-1. This particular conversion does not appear to have been undertaken in any significant numbers.

Missile Armed T-62 Evolution

During the late 1950s, Nikita Khrushchev became convinced that the days of the tank were numbered due to the advent of the antitank guided missile (ATGMs). As a result, as early as 1956, Soviet tank designers were ordered to begin to develop armored vehicles armed with ATGMs. This did not prove practical until the early 1960s. Further details of the missile programs are contained in the anti-armor section of this study.

The Kartsev design bureau at Nizhni Tagil began work on a missile-armed derivative of the T-62, designated the Obiekt 150. The Obiekt 150 was based on a standard T-62 hull with a modified turret. The turret contained a missile launcher which popped up from a roof hatch. A magazine of twelve to fifteen missiles was contained within the turret, along with the associated launcher guidance system. The 3M7 Drakon (Dragon) missile was developed by the A.E.Nudelman OKB-16 design bureau in Moscow, and the missile guidance system by A. A. Raspletin's KB-1 design bureau. A small test series of Obiekt 150 were ordered as the IT-1 (Istrebitel tankov= Tank destroyer). In 1964, Khrushchev had been removed from power and so the pressure to adopt missile tanks dissappeared. In 1965, the IT-1s were used to form two tank destroyer battalions, one manned by tankers, the other by artillery troops to test the concept. One battalion served with a motor rifle division in the Carpathian Military District, the other in the Byelorussian MD. Small-scale production of the vehicle took place from 1968 to 1970. However, the IT-1 was found to be impractical in service and the vehicles were converted to recovery vehicles. A turbine-powered version, the Obiekt 150T, was also developed for further trials.

Final Evolution of the T-62

In order to continue development of the T-62, Kartsev adopted the simple expedient of adding a simple external launcher for 9M14 Malyutka (AT-3 Sagger) missiles on further T-62 upgrade proposals to placate Khrushchev. The most significant of these was the Obiekt 167. In 1961 the Vagonka design team began developing of an improved version of the T-62 using a new suspension to improve road speed. The new suspension used return rollers and smaller diameter main road wheels. In addition, the vehicle was fitted with a new 700 horsepower

IT-1 (Obiekt 150) tank destroyer.

Obiekt 167 (upgraded T-62) medium tank.

supercharged engine. Three launchers for the 9M14 missile were fitted on the right rear of the turret. In 1962, an improved version, the Obiekt 167GTD, was developed equipped with an 800 horsepower Isotov GTD-800 gas turbine engine. These vehicles were demonstrated to the Soviet government in competition with the Morozov design bureau T-64. Although some officials of the Soviet Ministry of Defense supported the Obiekt 167 (as the T-67 tank), the GBTU tank administration favored the more expensive and radical T-64 design. Following the cancellation of the Obiekt 167 program, the Vagonka design bureau experimented with the new 2A26 125mm gun in a T-62, and also developed an autoloader. In 1968, work began on Obiekt 172, using the suspension from the Obiekt 167, the 125mm gun from the T-64, the autoloader from the experimental T-62 variants, and an improved engine and transmission from the T-62. This resulted in the T-72.

Interior view of T-62 gunner's station.

Obiekt 167 9M14 (AT-3 Sagger) external launcher.

T-62 115mm ammunition.

T-62 Comparative Technical Data

	T-62	T-62A	T-62M
Crew	4	4	4
Weight (metric tons)	36.3	37.5	41.5
Length overall (meters)	9.335	9.335	9.335
Hull length (meters)	6.63	6.63	6.63
Width (meters)	3.3	3.3	3.52
Clearance (millimeters)	430	430	350
Armament	U5-TS (2A20) 115mm smoothbore gun		
Gun elevation (degrees)	-5 to +18	-5 to +18	-5 to +18
Rate of fire (rounds per minute)	3-5	3-5	3-5
Stabilization	Meteor	Meteor	Meteor M1
Effective range (50% P_h) (meters)	1,600	1,600	5,000 (Sheksna)
Ammunition stowage*	40	40	40
Guided missile system	no	no	9K116-1 Sheksna
Gunner's telescope	TSh2B-41U	TSh2B-41U	TShSM-41U
Gunner's night sight	TPN-1-41-11	TPN-1-41-11	1K13-1
Rangefinding	stadiametric	stadiametric	KTD-2 laser
Ballistic calculations computer	manual	manual	BV-62 analog
Coaxial machine gun	7.62mm PKT	7.62mm PKT	7.62mm PKT
Rate of fire (rounds per minute)	200 to 250	200 to 250	200 to 250
Ammunition stowage	2500	2500	2500
AA machine gun	no	12.7mm DShK	no
Ammunition stowage	n/a	300	n/a
Commander's sight	TKN-3	TKN-3	TKN-3
Driver's night sight	TVN-2	TVN-2	TVN-2
Infrared searchlight	Luna L-2G	Luna L-2G	Luna L-4
Night vision range (meter)	800	800	800
Engine	V-55	V-55V	V-55U
Power (horsepower)	580	580	620
Transmission	synchronized, constant mesh; planetary final drive; 5 forward, 1 reverse		
Maximum speed (kilometers/hour)	50	50	45
Paved road range (kilometers)**	450/650	450/650	450/650
Dirt road range (kilometers)**	320/450	320/450	320/450
Maximum grade (degrees)	30	30	30
Maximum bank angle (degrees)	30	30	30
Ditch crossing (meters)	2.85	2,85	2.85
Wall (meters)	0.8	0.8	0.8
Unprepared fording depth (meters)	1.4	1.4	1.4
Prepared fording depth (meters)	5.5	5.5	5.5
Internal fuel (liters)	675	675	675
External fuel (liters)	285 + 400	285 + 400	285 + 400
Radio	R-113	R-123	R-173
CBR protection	PAZ	PAZ	PAZ

*standard load is 14 APFSDS, 20 HE-Frag, 6 HEAT; T-62M substitutes 4-5 missile rounds
**with basic fuel load/with two external 200 liter fuel tanks)

T-62 Armor Protection

Location	Thickness (millimeters)	Angle (degrees)
Upper glacis	100	60
Lower bow	100	55
Side	80	0
Rear upper	45	0
Rear lower	20	30
Hull top	16 and 30	90
Hull belly	20	90
Turret front	230	various
Turret mantlet	230	various
Turret sides	120	various
Turret rear	60	various
Turret top	30 to 34	various

T-62 Armor Modernization

	Basic Tank (no applique)	Variant 1 (Brow armor)	Variant 2 (Kontakt-1)	Variant 3 (Kontakt-5)
Added weight (kilograms)	0	2,190	1,320	2,800 to 3,000
Equivalent steel protection (millimeters)				
(versus APFSDS)	200	330	200	450 to 480
(versus HEAT)	200	400 to 450	700 to 900	700 to 900

T-62 (115mm) Tank Ammunition

Round	Projectile	Type	Round Weight (kilograms)	Projectile Weight (kilograms)	Armor penetration* (millimeters)
3UBM3	3BM3	APFSDS	22.3	4.0	
3UBM4	3BM4	APFSDS	22.5	4.2	
3UBM5	3BM6	APFSDS	22.5	4.0	228/199
3UBM9	3BM21	APFSDS	23.5	3.9	
3UBM13	3BM28	APFSDS	24.0	5.4	
3UBK3	3BK4	HEAT	26.6	13.1	495/248
3UBK3	3BK4M	HEAT	26.6	13.1	
	3BK15	HEAT			
3UBK10-1	9M117	Missile	28.0	17.6	550/275
3UOF1	3OF11	HE-Frag	28.1	14.7	
3UOF6	3OF18	HE-Frag	30.75	17.7	
3UOF37	3OF27	HE-Frag	30.75	17.7	

*At 1,000 meters, 0 degrees, and 60 degrees of obliquity.

Soviet Post-War Heavy Tank Design

In the immediate post-war years, the evolution of Soviet heavy tanks proceeded in a very different fashion from heavy tank development in NATO. In the US Army, production of the M4A3 medium tank was halted, and the M26 Pershing tank became the forerunner of future main battle tanks. Likewise in Britain, the Centurion tank continued in production as the main battle tank, and tanks like the Churchill infantry tank and Cromwell cruiser tank simply disappeared. In both cases, this was a reaction to the German Panther tank, which set the benchmark for late World War 2 tank development.

The Soviet Union maintained the medium tank/heavy tank distinction much later than any other major army, and this has had significant impact of the nature of its main battle tank design.[12] Current Russian main tanks such as the T-64, T-72, and T-80 have all evolved out of medium tank designs and have tended to be much lighter and smaller than their American and British counterparts, which evolved out of World War 2's definition of heavy tanks.

At the conclusion of World War 2, the Soviet Union had two heavy tanks in production, the IS-2M and the IS-3. Both shared a common powertrain and armament system; the principal difference was in the hull and turret castings with the IS-3 embodying the latest trends in Soviet design. As a result, the IS-3 replaced the IS-2M at the sole production facility at the Chelyabinsk Tractor Plant Number 100 (Tankograd) by late 1945.

No significant number of IS-3s were ready before the end of the war in Europe, and the IS-3 was first publicly displayed at a victory parade in Berlin on 7 September 1945.

The IS-3 design had been pushed into production much too quickly. Large numbers of IS-3 tanks were sidelined with mechanical problems. The thick armor plates on the front of the hull proved worrisome as the welds tended to crack open after service use, probably from the vibration of rough cross-country travel and the shock of gun firing. From 1948 to 1952, a redesign effort was undertaken to correct the problems. This included strengthening the hull, improving the final drive and reinforcing the engine mounts. Production of the IS-3 at Chelyabinsk lasted until 1951 by which time about 1,800 had been manufactured.

IS-4 Stalin heavy tank.

The IS-4 Heavy Tank

While the IS-3 program had been underway in 1944, the design team under Troyanov began work to further evolve of the IS-2 design under the code name Obiekt 701. Several alternatives were proposed on paper, and three designs were presented to the Red Army's GBTU tank directorate: the Obiekt 701-2, armed with the S-34 100mm gun, the Obiekt 701-5 with a different armor configuration, and the Obiekt 701-6 armed with the standard D-25T 122mm gun. The latter was accepted for further development. There were three significant changes in the Obiekt 701-6 design: thicker armor, a lengthened hull, and an uprated engine. The armor basis for the hull was

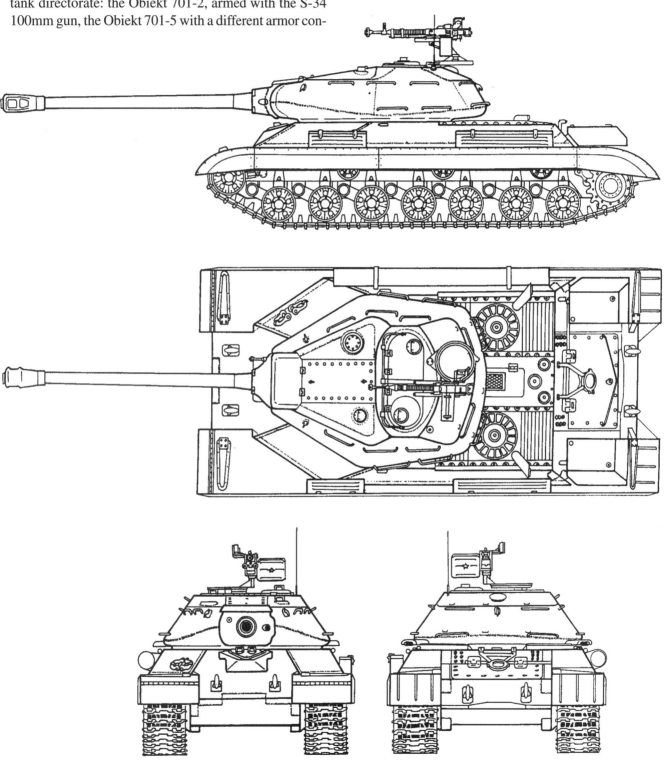

IS-4

increased to 160 millimeter and the turret to 250 millimeter. The 750 horsepower V-12 engine used a revised cooling system influenced by the layout of German Panther tanks, with the radiators under a pair of circular fans. The design was accepted for quantity production as the IS-4 tank in 1947.

After a short production run of only 200 tanks, IS-4 production was halted. The main criticism was that the speed and the mobility of the vehicle were inadequate. In November 1950 after the outbreak of the Korean War, nearly all of the IS-4 regiments were shipped to the Far East. They were deployed to form the shock force for a tank army that Stalin was organizing to intervene in the Korean conflict. In spite of intense pressure from the Chinese, Stalin decided against intervening in Korea for fear it would result in the outbreak of a general war with the nuclear-armed American armed forces. The IS-4s remained in the region and in the late 1950s, they were modernized along the same lines as the IS-3M and kept in service into the 1960s.

Heavy Tank Modernization

Due to the significant existing inventories of heavy tanks, both the IS-2M and IS-3 underwent modernization programs in the post-war years. In 1954, the IS-2s began to be rebuilt as the IS-2M. This program included an increase in the amount of main gun ammunition from twenty-eight to thirty-five rounds, an improved driver's periscope, the improved V-54K-IS engine, a modernized engine cooling and oil flow system, new radios, and intercoms. Externally, the IS-2M had stowage increased by adding tool bins on the front hull side; dust skirts were also added.

The parallel program for the IS-3M began in 1960 and included additional hull reinforcement, replacement of the DShK 12.7mm machine gun with the DShKM, addition of a TVN-2 night vision device for the driver, substitution of the V-54K-IS engine, incorporation of the Multitsiklon air filter system and many other small changes. New wheels were added from the T-10 heavy tank which, had improved ball-bearings. In addition, external stowage was improved and dust skirts were added over the suspension.

In the postwar Soviet Army, heavy tanks were deployed much like the World War 2 pattern. They were deployed in independent tank regiments (twenty-one tanks) or independent heavy tanks brigades (sixty-five tanks) under army control and could be allotted to provide long-range fire support for tank divisions. During the 1947 reorganization, a special heavy armored regiment was added to the tank and mechanized divisions, containing forty-four to forty-six heavy tanks and twenty-one of the related ISU-122 or ISU-152 assault guns. This formation was added due to Soviet recognition that the standard Soviet tank regiments, still primarily equipped with the T-34-85, could not deal with the newer British or American tanks like the M-26 Pershing or the Centurion. This deployment pattern remained in effect until the major reorganizations of 1958 through 1959.

Post-War Heavy Tank Development

Soviet post-war heavy tank development paralleled the situation with the medium tank design bureau; the bulk of the bureau returned from its wartime exile in Chelybinsk and reestablished at its prewar location at the Kirov Plant in Leningrad. At the same time, small elements of the bureau remained in Chelyabinsk, where they would eventually form a competitive team. No fewer than three separate design efforts were pursued.

The main design bureau under General Zhozef Kotin in Leningrad was working on a project called Obiekt 703. The design used components from the IS-4, but the main focus of the design was to study the advantages of electric transmissions. Electrical transmissions had significant theoretical advantages in transferring more usable power from the engine to the tracks, and also promised

IS-3M Stalin heavy tank.

ISU-122 heavy assault gun.

ISU-152 heavy assault gun.

IS-6 Stalin heavy tank.

to offer improved steering and mobility. The concept had first been tried on the French St. Chamond tank in World War 1, and in World War 2, the US Army tested the concept on the T-23 medium tank and the German's Porsche competitor to the Henschel Tiger I and on the Elefant tank destroyer. The Kotin team was undoubtedly most familiar with the German approach, as several Elefants had been captured at Kursk, and the NIIBT armored research center at Kubinka had done an extensive study of the transmission system. During the war years, a testbed called the IS-1E was built on a spare IS-1 chassis. The new transmission for the Obiekt 703 was based around a DK-305A 385 kilowatt generator. The Obiekt 703 was redesignated the IS-6, but the design never proved reliable enough for production. The day of its first trials, the prototype burst into flames hardly thirty yards outside of the assembly hanger. It was soon found that the electrical transmission required too much cooling. The addition of sufficient fans added unacceptable weight to the design and well as drawing off power. An attempt was made to retreive the design by substituting the mechanical transmission from an IS-4. This vehicle, built as the izdeliye 252, offered no advantages over existing tanks and was canceled.

In the meantime, Nikolai Shashmurin's team had begun work on a completely new tank design, the IS-7, which bore no immediate connection to any previous Soviet heavy tank. The design was envisioned as a counterpart to the German Tiger II heavy tank in terms of armor and firepower. Shashmurin took advantage of the transfer back to Leningrad, examining a number of components developed by Soviet naval research institutes in the city. This included a 1,050 horsepower marine diesel which would be needed to power such a heavy tank, and a 130mm gun derived from the naval 56-SM gun. The gun fired a 36.5 kilogram projectile at 945 meters per second, which made it the most powerful weapon ever mounted in a Soviet tank up to that point. The co-axial weapon was the powerful 14.5mm KPVT heavy machine gun, and no fewer than six other 7.62mm machine guns were provided, two co-axially in the mantlet, two on the right side of the hull, and two more on either side of the turret in small armored barbettes. Another 14.5mm KPVT was fitted in a remote control mount on the roof for air defense. Given its thick armor, the IS-7 was the heaviest tank ever built in the Soviet Union, weighing sixty-eight metric tons. Inspite of its weight, its powerful engine gave it a higher road speed than previous Soviet heavy tanks.

The first prototype of the IS-7 was ready for testing in 1948. The crews were not entirely happy with the internal layout, which was extremely cramped even by

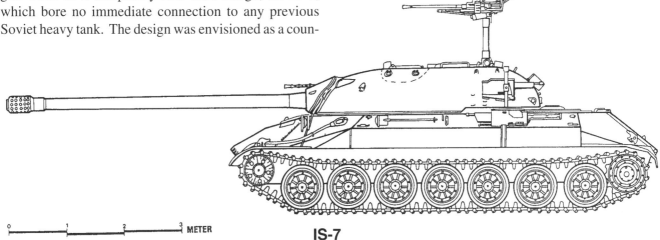

IS-7

IS-7 Stalin heavy tank.

Soviet standards. The ammunition was very heavy and difficult to load under these circumstances. The machine gun armament was viewed as being excessive, and the location of the ammunition trays made it impossible to reload these weapons in combat.

The suspension used internal shock absorbers in the wheels patterned after those on the German Tiger II. Unfortunately, these were subject to rapid wear and led to breakdowns when the tank was operated near its top speed. The Soviet Army's GBTU armored vehicle directorate was very unhappy about the vehicle's weight for two reasons. On the one hand, the extreme weight would tax the limited Soviet road and rail-network. The tank would be very difficult to employ tactically since there were few bridges that would withstand its weight. Secondly, heavy weight implied high cost, both in the purchase price and in regular operations and maintenance. In the end, only a small series of test vehicles were manufactured. The IS-7 remains the heaviest tank ever built in the USSR.

In some respects, it was ahead of its time. The level of firepower and armor on the IS-7 was very similar to that found on NATO tanks in the 1960s, such as the US Army M-60A1 or the British Chieftain. On the other hand, its main gun and fire controls were very much constrained by the limits of 1940's technology and lacked the accuracy of later tank gun systems.

The T-10 Heavy Tank

The experimental heavy tanks of the 1940s helped the Soviet Army define more clearly what it sought in a heavy tank design. In the end, weight and cost constraints led to the conclusion that an updated IS-3 design would satisfy these needs. As a result, development of the IS-8 began in 1948 in Leningrad.

The IS-8 used components from many of the experimental tanks. For example, the electrical turret traverse and elevation were derived from the IS-7, as was the short torsion bar suspension. The V-12-5 engine was a derivative of the type used in the IS-4 and IS-6, and the track also came from the IS-4. The D-25TA gun was a slightly improved version of the same gun used on the IS-2 and IS-3. The new BR-472 ballistic-capped 122mm projectile offered better penetration than the wartime BR-417B projectile. The turret resembled the IS-3 turret, but the armor basis was raised to 200 millimeter. Engine cooling was improved by boosting the airflow through the radiators using exhaust gases. The added weight of the IS-8 and the improved engine cooling system led to a lengthened hull which had an additional set of roadwheels similar to the IS-4 configuration.

Production of the IS-8 began in late 1950 or early 1951 at Chelyabinsk. There are also some reports that indicate that production may have taken place at the tank

T-10 (IS-8) Stalin heavy tank.

T-10A heavy tank.

plant in Omsk as well. Following Stalin's death in 1953, the IS-8 was redesignated the T-10 as part of the de-Stalinization program.

The T-10A and T-10B Heavy Tanks

As the pace of T-10 production picked up in the early

T-10B heavy tank.

1950s, further improvements were added. The T-10A version incorporated the new D-25TS gun, which had a stabilization system added in the vertical axis, as well as a bore evacuator. One of the main problems of the Soviet heavy tanks had been the heavy weight of the projectiles. As a result, the T-10A added a simple rammer. The gunner loaded the projectile and casing on a special tray, and the rammer pushed them into the breech. On the T-10A, the TSh-2-27 telescopic sight was replaced by a new TPS-1 periscopic sight with stadiametric ranging and a TUP telescopic sight. Other improvements on the T-10A included a TVN-1 night vision device for the gunner and a GPK-48 gyrocompass for navigation.

In the mid-1950s, this was followed by the T-10B. This version added a two-axis stabilization system for the gun and new T2S-29 fire control sights, but externally it was very similar to the T-10A.

The T-10M Heavy Tank

The final variant of the series, the T-10M, was introduced in 1957. The most important change was the addition of the longer M-62-TS gun, which had better antiarmor performance than the earlier D-25, about 185 versus 160 millimeters of armor penetration at 1,000 meters using the normal armor piercing round. In addition, the gun could fire the BP-460A HEAT projectile, which offered penetration of about 300 millimeters. The M-62-TS gun was fitted with the Liven two-axis stabilization system and could be easily distinguished by its distinctive multi-slotted muzzle brake. The T-10M version also substituted KPVT 14.5mm heavy machine guns for both the 12.7mm DShK co-axial and antiaircraft machine guns. The KPVT was a closer ballistic match for the new M-62 gun, and so could be used for rough ranging. It used the uprated V-12-6 engine with an output of 750 horsepower. By the time that production ended in 1962, about 8,000 T-10s of all versions had been manufactured, making it numerically the most significant heavy tank of the Stalin series. In the early 1960s, the T-10s underwent a factory rebuild to improve their transmission. A new six speed transmission and main clutch were installed in all the vehicles to replace this perennially troublesome item.

Expanded Heavy Tank Role?

In the post-war years, Soviet heavy tank units underwent several reorganizations. During the 1947 reorganization of the mechanized forces, a composite heavy tank/assault gun regiment was added to each tank or mechanized division. These regiments contained forty-four to forty-six heavy tanks and twenty-one heavy assault guns of either the ISU-122 or ISU-152 types. The composite regiment was intended to provide added firepower to the division, particularly in breakthrough operations. With the production of the T-10 in full swing, the Soviet Army began experimenting with heavy tank divisions.

These divisions were organized like conventional tank divisions but had two heavy and one medium tank regiment, instead of the three medium tank regiments in a normal division. Two of these divisions, the 13th and 25th Guards Heavy Tank Divisions, served in the Group of Soviet Forces-Germany in the 1950s and 1960s. It is believed that an additional two or three of these divisions existed, at least one in the Far East. These divisions were

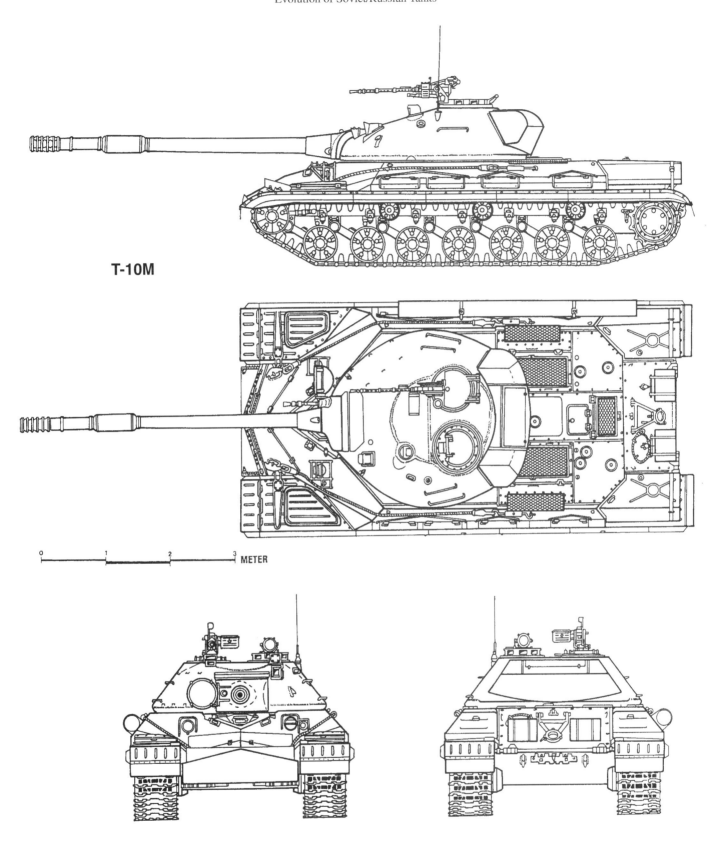

T-10M

intended to act as the shock forces of the tank armies during offensive operations. During the 1958 to 1959 reorganization, the composite heavy tank/assault gun regiment was replaced by a homogenous heavy tank regiment. These were a formidible formation, equipped with one hundred IS-3M or T-10 tanks.

The main advantage of the heavy tanks were their armor, not their firepower. Indeed, the 100mm D10-T gun on the T-55 was superior in anti-tank performance to the 122mm D-25 and comparable in performance to the later M-62. This was not a major cause for concern since the role of the heavy tanks since the later part of World War 2 had been to assist in breakthrough operations by providing an assault tank more capable of engaging enemy antitank defenses.

The thick armor of the IS-3 and T-10 tanks caused considerable consternation in both the US Army and the British Army. This led both armies to develop antidotes—the Conqueror which entered service in 1956, and the M-103, which entered service in 1958. Both were armed with very long 120mm guns with substantial armor penetration. The M-103 could penetrate 221 millimeters of armor at thirty degrees at 1,000 yards, 196 millimeters of armor at 2,000 yards using armor piercing ammunition, and 330 millimeters of armor at 2,000 yards using HEAT (High-explosive anti-tank) projectiles. Both vehicles were heavier than their Soviet counterparts, the Conqueror at sixty-five tons, and the M-103 at sixty-two tons.

Had there been a conflict between NATO and the Warsaw Pact in the late 1950s, the tactical balance would not have been all that dissimilar from the German-Soviet fighting of 1944. NATO tanks of the late 1950s, such as the US Army M-48, would have had a tough time dealing with the T-10 frontally at ranges of 1,000 meters, much as the Panther had difficulty with IS-2s in 1944. The new HEAT ammunition used with the M-48's 90mm gun could not penetrate the thickest sections of the T-10's armor. The M-48 could have gotten penetrations on the side of the vehicle only. The T-10's 122mm gun could have penetrated nearly any NATO medium tank of the period from 1,000 meters. In a duel between the T-10 and NATO heavy tanks such as the Conqueror and M-103, the NATO tanks would have enjoyed a modest technical advantage. The M-103 and Conqueror had sophisticated coincidence rangefinders that offered excellent accuracy at long ranges in the hands of a well trained crew. The T-10 relied on a stadiameric rangefinder which was not particularly efficient at ranges over 1,000 meters. The HEAT projectile fired by the M-103 could penetrate the T-10 frontally, and its shaped charge warhead had the same penetration regardless of range. The Soviets never deployed a heavy tank with the type of elaborate fire controls found on the NATO vehicles, though work in this area was underway in the late 1950s.

The only significant post-war combat use of Soviet heavy tanks took place during the 1967 Middle East war. The Egyptian Army acquired about one hundred IS-3M tanks from the USSR as part of its modernization program. During the 1967 war, one battlion with twenty-one IS-3Ms was stationed with the 11th Infantry Brigade (7th Infantry Division) at Rafah; the 125th Tank Brigade of the 6th Mechanized Division at Kuntilla also was equipped with about sixty IS-3M.

The IS-3M was the most feared tank in Egyptian service due to its thick armor. Israeli infantry units and paratrooper units had considerable difficulty when the Stalins were encountered, as the existing bazookas and other antitank weapons could not penetrate its frontal armor. The Israeli tanks also had problems, especially the various models of Shermans. Even the more modern designs, such as the M-48A2 Patton with its 90mm gun, could not easily penetrate its armor at normal battle ranges.

There were a number of engagements between M-48A2 Pattons of the Israeli 7th Armored Brigade and IS-3M regiment supporting the Egyptian positions near Rafah, with at least three Pattons being knocked out in the fighting. The Israeli tankers were usually able to overcome the armor problem by better training and tactics. The Egyptian tanks were most dangerous when firing from ambush positions. But when confronted in tank-versus-tank fighting, the slow rate of fire of the Stalin and its rudimentary fire controls proved to be a problem. In total, the Egyptian Army lost seventy-three IS-3Ms in the 1967 war. At least one regiment of IS-3M tanks was still in Egyptian Army service during the 1973 war apparently did not see much fighting. The T-10 heavy tank was never exported.

By the 1960s, the T-10s armor advantage dissappeared. New NATO main battle tanks such as the M-60A1 and Chieftain had similar armor protection to the Soviet heavy tanks and their 105mm and 120mm guns could penetrate the frontal armor of the T-10 at standard combat ranges.

Further Heavy Tank Development

In 1955, a competitive program was begun to examine alternative heavy tanks for the future, the Obiekt 277, Obiekt 279, and Obiekt 770. The main feature shared in common was an advanced version of the 130mm gun developed for the ill-fated IS-7 heavy tank, mated to optical rangefinders for greater long-range accuracy. The new turret included an ammunition assist system for the loader, and was fitted with infrared night fighting equipment as well. Although these tanks shared a very similar armament system, they had significantly different chassis.

The Obiekt 277 used a modernized hull evolved out of the T-10 chassis, lengthened with an additional set of roadwheels. The most significant innovation in the design was the incorporation of a gas-turbine engine. Kotin's design bureau at the Kirov Plant in Leningrad was the first Soviet tank bureau to experiment with gas turbine engines for tanks. It has continued to use them through the present, most notably on their T-80 tank design. From 1948 to 1949, the Turbine Group of the Kotin design bureau was commissioned by the Armored Force

Obiekt 277 heavy tank.

Directorate of the Soviet Army to develop a gas turbine for tanks. The program did not meet its requirements and was eventually rejected by the army. The Obiekt 277 program merged the turbine engine requirement with the new heavy tank effort.

The team headed by L. S. Troyanov in Leningrad came up with a far more radical approach, however. Troyanov's Obiekt 279 was designed to fight on the nuclear battlefield. Beginning in 1953, the Soviet Ground Forces had participated in a number of nuclear tests, placing several tank designs in the blast area. In September 1954, a special atomic bomb test was conducted in the Totskoye region with selected army units taking part. It became evident at the tests that tanks near the nuclear blast area were often knocked over by the shock waves that ensued. Although overturned, the tanks were intact and still functional. This led to the requirement for a tank that could survive a nearby explosion of a tactical nuclear weapon. Troyanov's approach to the requirement was to increase the surface area of the track and reduce the center of gravity. In addition, the hull was aerodynamically shaped to withstand the high speed wind that accompanied the blast's shock wave as well as to reduce radiation effects to the crew inside. The vehicle was fitted with hydropneumatic suspension so that the entire chassis could be lowered closer to the ground to further resist overturning.

The prototype of the Obiekt 279 was completed in 1957. It had a remarkable appearance. The suspension was modular with four sets of running gear. Each pair of running gear attached to a central core which also served as the main fuel supply. The hull was oval in shape when viewed from above, with very steeply angled sides. The hull configuration permitted a great deal of dead space that could be used to contain antiradiation material. It was powered by a 1,000 horsepower, twelve-cylinder diesel engine which gave it excellent mobility for a sixty-ton tank, and was capable of speeds of up to thirty-five miles per hour. The turret and armament of the Obiekt 279 were essentially similar to those on the Obiekt 277, with a 130mm gun mounted in a large, conventional three-man turret, with a coincidence rangefinder. The tests on the tank were successful in view of the novelty of the design. But the hull construction made the tank extremely expensive to produce, and it was never accepted for service.

The third heavy tank developed in the late 1950s was the product of a new design team, headed by young designer at Chelyabinsk, Pavel Isakov. The experimental tank, labeled the Obiekt 770, attempted to incorporate the heavy armor and heavy firepower of the 1950s tanks in a more compact and lighter design. The Obiekt 770 weighed fifty-five tons and was powered by a 1,000 horsepower engine. The prototype was completed in 1957 and

Obiekt 279 heavy tank.

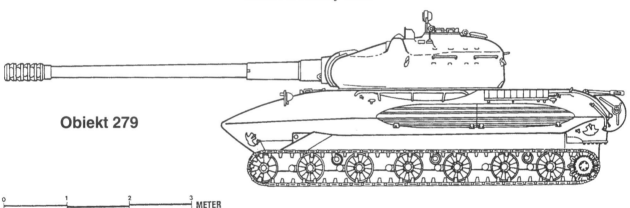

sent to state trials. Although it was widely regarded to be a technical success, its future was doomed from the start because of political developments in Moscow.

By the mid-1950s, Nikita Khrushchev had emerged as Stalin's heir. Khrushchev was faced by many of the same problems as Gorbachev in the 1980s, namely a bloated military and a stagnant economy. Khrushchev was determined to solve both problems in a related fashion. Major cuts for the armed forces were planned, both in manpower and equipment. The Soviet Union would change its strategic concepts of defense, and place more emphasis on strategic nuclear-armed missiles, with less emphasis on the conventional forces. As is discussed in other sections of this study, Khrushchev became infatu-

ated with missiles as a technological panacea for the Soviet Union's defense problems. He regarded the larger forms of common weapon systems — battleships in the case of the navy, heavy bombers in the case of the air force, heavy tanks in the case of the tank force, and superheavy artillery in the case of the artillery — as evidence of reactionary resistance to his efforts and contrary to the spirit of the "revolution in military affairs" brought about by nuclear weapons. As a result, he ordered all of these categories of weapons eliminated. In the case of heavy tanks, he saw anti-tank missiles as a superior solution because he misconstrued the primary role of the heavy tanks as long range anti-tank fighting.

On 22 July 1960, Khrushchev attended a presenta-

Obiekt 770 heavy tank.

tion of the latest tank technology at Kubinka, including the new Obiekt 150 missile armed tank. Present with Khrushchev were several missile designers. General Zhozef Kotin tried to win approval for the production of his new heavy tank design, but the proposal only aggravated the situation. Instead, Khrushchev told the assembled defense officials that heavy tank development would not be allowed to continue, and that T-10 heavy tank production would be brought to a close.[13] Heavy tank production did not finally end until 1962 because the existing five-year industrial plan.

Heavy tanks did not immediately disappear from the Soviet Army. The heavy tank division remained in service until 1969, when one of the two heavy tank regiments in the division were replaced with medium tanks. Finally, in 1970, the heavy tank divisions disappeared and the separate heavy tank regiments were gradually disbanded. But as late as 1978, about 2,300 heavy tanks were still in service, mainly in the Far East. Most of the heavy tanks were not actually melted down, but merely withdrawn into inactive reserves.

Some heavy tanks remain in service to this day, though not in a mechanized role. Due to tensions with China in the early 1960s, border defenses were substantially beefed up. One manner of doing this was to add pillboxes along the frontier. One of the most cost-effective approaches was to bury outdated tanks as improvised pillboxes, with only their turrets showing. This was first done with IS-2 and IS-3 tanks, but in the 1970s, the process extended to the T-10. Many, if not most, of these tank pillboxes are still in place today.

Soviet Post-War Light Tanks

The nature of the light tank in the Soviet army changed completely during World War 2. Before the war, the T-26 light tank was the most numerous tank type and was the primary type used to provide infantry support. But because of the escalation in tank and anti-tank armament during the war, the light tank gradually became less viable. It's gun was inadequate to deal with either longer-ranged enemy antitank guns or more thickly armored enemy tanks, and its protection was woefully inadequate when faced with contemporary antiarmor weapons. As its combat capabilities withered, its role narrowed to that performed by pre-war light reconnaissance tanks like the T-37 and T-38.

During World War 2, reconnaissance vehicles took over many of the roles formerly undertaken by the cavalry. Although the primary role of reconnaissance vehicles was nominally scouting for tank and mechanized units, such vehicles performed a much wider range of tasks than their name would suggest. They were often used to provide flank security for mechanized units, or to mop up pockets of resistance behind the frontlines. The Red Army used two main classes of vehicles in the reconnaissance role, light tanks, and armored cars. Light tanks were used to perform more demanding missions, like scouting in front of a mechanized unit, or providing flank security. The armored cars and armored transporters were used for less demanding roles, such as liaison between commanders, mop-up operations or scouting in less contested areas. The light tanks included both domestic designs like the T-60 and T-70, as well as Lend-Lease designs like the British and Canadian Valentine. Armored cars and transporters included the BA-64B armored car as well as a number of Lend-Lease types, especially the American M-3A1 scout car and the British Universal Carrier.

T-80 Model 1944 light tank.

The wartime generation of scout vehicles was not highly regarded. Many of the designs had been selected as much for their low cost and ease of production as for their combat utility. The BA-64B was very poorly armed, and it lacked mobility. The T-70 suffered from a very poor engine configuration and was not well armed. Any serious efforts to replace these vehicles was delayed until after the war.

The PT-76 Scout Tank

One of the main drawbacks of the T-70 light tank and its Lend-Lease counterparts, like the Valentine, was that they lacked amphibious capability. Rivers inhibited their ability to conduct their main mission of scouting. Soviet tank designers had realized the importance of this feature in their pre-war scout tank designs like the T-37, T-38, and T-40, but wartime pressures caused the amphibious capability to be dropped because of its the added cost and design complexity.

In 1949, the Ministry of Transport Machine Building began a more formal effort to develop an amphibious light tank. Four design teams were given the assignment: N. Astrov's bureau in Mytishchi (which had been responsible for wartime light tanks); Colonel A. F. Kravtsev at the VRZ Number 2 tank rebuilding plant near Moscow in conjunction with the Moscow Engineering Facility; and two design teams from the heavy tank design bureau at the Kirov Plant in Leningrad, headed by headed by L. Troyanov, N.Shashmurin and the Balzhiy design team at Chelyabinsk. From the outset, it was anticipated that new chassis would be used to develop a variety of other light armored vehicles, including armored transporters and armored utility vehicles. The designs were all fairly similar, with three of the designs using a conventional propeller for water propulsion. Shashmurin's entry used a hydro-jet system instead, developed by N. Zhukovskiy and N.Konovalov.

Kravtsev's K-90 tank was the first completed in 1950. Kravtsev proposed a family of light tracked vehicles using components from the T-70 light tank and M-2 light artillery tractor. These included the K-73 amphibious self-propelled gun, the K-75 armored transporter, the K-78 amphibious armored transporter and the K-90 light amphibious tank. The K-78 amphibious armored transporter and K-90 were related designs, using boat-shaped hulls manufactured of ten to fifteen millimeter armor plate. Propulsion was by means of a pair of propellers, powered off the normal 140 horsepower gasoline engine. It was armed with a LB-70T 76mm gun mounted in a welded turret.[14] Prototypes of both vehicles were assembled beginning in 1949 and were completed in 1950. With this armor, the vehicles were relatively light; the K-90 am-

K-90

phibious tank weighed only ten metric tons.

Shashmurin's design used a conical turret similar to that of the K-90, but it was armed with the D-56 76mm gun developed by Petrov's design bureau in Sverdlovsk. It's hull was considerably longer and lower than the K-90's. It employed an entirely new suspension. Details of the Astrov and Troyanov light tank competitors are lacking, and it is possible that neither reached prototype stage. Kravtsev's design showed the designer's lack of experience in tank engineering. The hull was insufficiently buoyant, the crew compartment was poorly laid out, and the prototype suffered from significant technical failings. As a result, the state commission terminated any further work on the K-78 and K-90 and approved the Shashmurin design instead. The new tank was designated PT-76 (plavayushchiy tank: swimming tank). The design was left on the shelf for over a year, apparently due to Stalin's inattention. Authorization for production apparently came in 1953 after Stalin's death, based on Nikita Khrushchev's decision.

The PT-76 was a fairly unconventional tank in many respects. No NATO army had an amphibious scout tank. At the time, the US Army was adopting the M-41 Walker Bulldog light tank, and the French Army the AMX-13 light tank, neither of which was amphibious. The PT-76 was unusually large for a light tank of this period, mainly because of the buoyancy requirements for its swimming capability. The hydrojet system allowed the PT-76 to travel across rivers with stream speeds no greater than about eight km/h. It had a water speed of about ten km/h. The engine selected for the PT-76 was based on half a T-54's V-54 engine, with six-cylinders instead of twelve. The PT-76 chassis served as the basis for a variety of other armored vehicles including the BTR-50 armored transporter, the Luna (FROG) nuclear missile launch vehicle, the Penguin arctic research vehicle, and the ASU-85 airborne assault gun. It also indirectly spawned a later generation of light vehicles, including the ZSU-23-4 Shilka air defense gun vehicle.

The initial production model of the PT-76 was armed with the D-56T gun with its distinctive multi-baffle muzzle brake. The standard production model of the PT-76 was armed with the D-56TM gun, which had a more conventional muzzle brake and a fume extractor. In the late 1950s, the PT-76 underwent significant internal redesign in keeping with the general trend in the Ground Forces to reconfigure equipment for suitability on a nuclear battlefield.

This led to the PT-76B. The PT-76B had an NBC filtration system installed and was fitted with a new sta-

PT-76 scout tank.

bilized gun system, the D-56TS. The added weight of this version forced the designers to heighten the hull by 130 millimeters (5.1 inches) to increase the hull buoyancy. Other improvements on this version included a provision to employ additional external fuel tanks. The PT-76B entered service in 1962.[15]

The last significant variant of the PT-76 was the PT-76M, developed specifically for the Soviet Naval Infantry. This version used a modified hull with rounded extensions on either side to provide more buoyancy in the water. Although two prototypes were built, naval trials found them to be unacceptable. Instead, the Navy adopted the PT-76B with several modifications, including alterations in the rear deck air intakes and a snorkel-like device to prevent the tank from becoming flooded.

The PT-76 was used in both tank and motor rifle divisions. Each tank and motor rifle regiment was autho-

PT-76 Technical Data

Crew	3
Overall length (meters)	7.62
Hull length (meters)	6.91
Width (meters)	3.14
Height (meters)	2.19
Weight (metric tons)	14
Main gun	D-56T 85mm rifled gun
Elevation (degrees)	-4 to +30
Effective range (meters)	650
Main gun ammunition	40
Gunner's sight	TShK-66 telescope
Co-axial machine gun	7.62mm SGMT
Engine	V-6 240 hp diesel
Internal fuel (liter)	250
Maximum road speed (km/hour)	44
Water speed (kilometers per hour)	10
Maximum road range (kilometers)	260
Maximum gradient (degrees)	38
Armor thickness (millimeters)	13 to 15

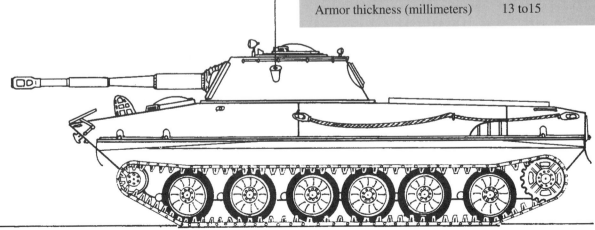

PT-76B scout tank.

PT-76M scout tank.

rized a reconnaissance company with five PT-76 and three armored transporters (usually the related BTR-50). Besides these regimental scout units, each division had a separate reconnaissance company with an additional five PT-76. In the 1960s, the companies were reduced in strength to only three PT-76 per company. The PT-76 was also widely used in the Naval Infantry. During the 1961 to 1964 reorganization, each fleet was allotted at least a regiment of naval infantry. The new regiments included a tank battalion which consisted of three light tank companies (ten PT-76 each) and one medium tank company (ten T-55). In the 1980s, the enlarged Naval Infantry Regiments had a larger complement of T-55 tanks (forty per brigade), while the PT-76 component shrank to twenty-six per brigade. Some brigades (such as the 810th Naval Infantry Brigade at Sevastopol) have no PT-76s.

Production of the PT-76 continued until 1967 (or 1969 according to some estimates). By 1967, the PT-76 was obsolete. It was a very large, lightly armored, and conspicuous target, and it was not well armed. Its archaic communications and sensor equipment did not meet the reconnaissance demands of the modern battlefield. Nevertheless, US forces encountered the PT-76 in combat in Vietnam in 1968, and the PT-76 has seen extensive use in combat around the world including the 1965 and 1971 Indo-Pakistan wars, the 1973 Middle East war, and the 1980 to 1988 Iran-Iraq war. It is still in service in dwindling numbers, and is being supplanted by a derivative of the BMP, the BRM.[3] A total of about 7,000 PT-76s were built, of which about 2,000 were eventually exported to at least twenty-eight countries. The PT-76 was a staple export item in the 1960s and 1970s, being cheap and easy to maintain. The People's Republic of China began to manufacture a PT-76 derivative, the Type 60 light tank, in 1966, followed shortly afterwards by the heavily reconfigured Type 63 light tank.

PT-85

Although not widely known in the West at the time, evolution of the PT-76 continued after the PT-76B. In 1963, further development of the PT-76 was entrusted to the I. Gavalov design bureau at the Volgograd Tractor Plant, where the PT-76 was being manufactured. The design transfer came in the wake of Khrushchev's decision to stop development of heavy tanks, a decision which led to a temporary suspension of design work by the heavy

PT-85 (Obiekt 906) light amphibious tank.

Obiekt 940 armored command vehicle.

tank design bureau in Leningrad which had been responsible for the original PT-76 design.

The revised PT-76 was designated Obiekt 906, unofficially called the PT-85. The overall appearance of the vehicle remained much the same, but the hull was modified to accommodate an improved suspension with return rollers. The most dramatic change was the new armament, consisting of an 85mm D-58 gun with a fifteen-round autoloader in the turret bustle. The tank was fitted was the standard array of active night fighting equipment, most notably a Luna infrared searchlight. The Obiekt 906 was powered by a UTD-20 engine, the same type being adopted at the time by the BMP-1 armored infantry vehicle. Ultimately, the Obiekt 906 was not accepted for production, but it is unclear whether the reasons were technical or were related to lack of a requirement for a better armed alternative to the PT-76B.

Light Tanks in the 1970s

There was a lull in Soviet light tank design in the late 1960s and early 1970s. At this time, the PT-76 was probably viewed as an acceptable vehicle for the reconnaissance role, even though its main armament was completely inadequate to defend itself against contemporary main battle tanks. In the mid-1970s, an attempt to develop a replacement began. It is unclear what requirements prompted this new program, as the Russians have not written anything on this program. Indeed, this program was completely secret until the early 1990s when visitors were finally allowed to inspect prototypes of these light tanks now housed at the Kubinka armor museum.

The effort appears to have been a competitive program, as both 1975 designs share many features in common. The Obiekt 934 was another attempt by the Volgograd design bureau (headed by A. Shabalin after Gavalov's retirement), which had worked on the earlier Obiekt 906. Obiekt 934 appears to have been an effort to develop a new family of light armored vehicles. Besides the Obiekt 934, a concurrent armored command vehicle was also designed in 1976 (Obiekt 940), and other variants may not have been revealed yet. The competitive Obiekt 685 was developed by the Kurgan Machine Plant design bureau, best known for the BMP series of armored infantry vehicles. Both designs made extensive use of existing components from the BMP.

The Obiekt 934 and Obiekt 685 were officially designated as light airborne amphibious tanks (legkiy plavayushchiy aviadesantiruyemiy tank). It seems that this was not an attempt to replace the BMD series, but an attempt to replace both the PT-76 in tank and motor rifle divisions and naval infantry brigades, and the ASU-85 in airborne divisions. Both light tank designs shared a common 2A48-1 100mm rifled gun. Beyond the armament similarity, the designs were very close in overall characteristics; that is, weight, size and basic configuration. Both designs employed autoloaders for the main gun, though the configuration of both systems was different given the turret configuration differences between both designs. Both designs also had mounting plates at the top of the

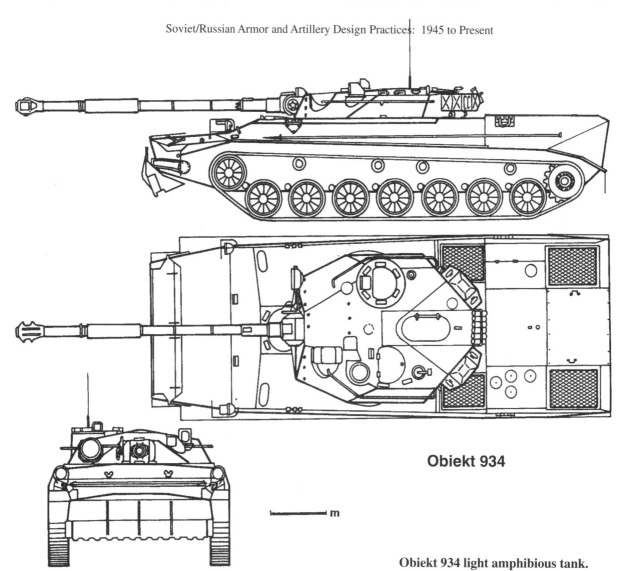

Obiekt 934

Obiekt 934 light amphibious tank.

Obiekt 685 light amphibious tank.

gun mantlet which may have been intended for a laser rangefinder/designator for use of tube-fired ATGMs like the Bastion.

The Obiekt 934 employed a relatively large turret diameter (about 2.65 meters) and a significantly larger turret than the Obiekt 685. The turret design is reminiscent of the style adopted on the French AMX-10RC: very large, multi-faceted, low, flat, and capable of withstanding up to a 23mm cannon round.[16] The Obiekt 934 chassis was subsequently used as the basis for the 2S31 Vena mortar vehicle prototype.

The gunner sat on the left side of the turret and had the usual active infrared night vision device immediately in front of his position, and the gunner's primary sight was in a housing further forward on the left turret side roof. The commander sat in the right half of the turret and was provided with a very typical Soviet pattern cupola. There was a large port at the rear of the turret roof, to eject spent casings. The hull was quite spacious, probably due to the buoyancy needs for amphibious capability. The suspension used typical Soviet-style stamped wheels, with return rollers. One unusual aspect of the design was the relatively large number of road wheels for so light a vehicle. The suspension could be collapsed like the BMD for loading on airborne PRSM pallets. Typical pallet loading lugs were located on the hull side.

The Obiekt 685 employed a relatively small turret of unusual shape, narrowing dramatically at the front. The turret diameter was considerably smaller than the Obiekt

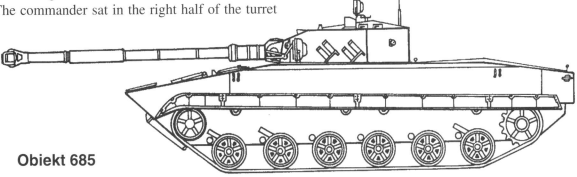

Obiekt 685

934, about 2.35 meters. The turret layout was essentially the same as the Obiekt 934, with gunner to the right, commander to the left and a shell casing ejection port on the rear turret roof. Unlike the Obiekt 934, which had the smoke mortars on the turret rear (like the BMP-1P), the Obiekt 685 had them mounted in two pairs on the turret side. The hull design is reminiscent of the later Obiekt 688 (BMP-3), and the Obiekt 685 helped shape the basic conception of the BMP-3 hull. The tank was powered by a 2V-06 400 horsepower multifuel diesel engine.

Neither the Obiekt 934 nor Obiekt 685 was accepted for production. One of the lingering questions about this program was the relation of these two light tanks to the concurrent Kurgan Obiekt 676 program that resulted in the BRM and BRM-1 reconnaissance vehicles. It is not clear whether the development was actually a three-way competition, or whether the light tanks were intended for the specialized role in support of the VDV airborne and naval infantry requirement and the Ground Forces reconnaissance vehicle requirement. Compared with the light amphibious tanks, the focus in the BRM design was in sensors and communications rather than firepower. The Obiekt 676/BRM was armed with the same 73mm 2A28 low pressure gun as the BMP but was fitted with a sophisticated array of day and night sights, navigation aids, and a battlefield surveillance radar in the case of the BRM-1 version.

The BRM Reconnaissance Vehicles

The BRM (Boevaya razvedyvatnaya mashina) is a scout version of the BMP that first appeared in 1976 (hence its NATO code name, BMP M1976/1). The basic BRM has a large, two-man turret with 2A28 Grom 73mm gun and the associated 1PN22M2 gunner's sight. Unlike the later BRM-1K, it has no radar. The commander, who is seated in the turret, is provided with a day/night sight and a DKRM-1 (1D8) ruby-laser rangefinder. The navigator, who sits behind the driver in the hull, is provided with a TNPK-240A observation device.

The basic communications package includes R-123M, R-130 and dismountable R-148 transceivers, as well as a R-014D teletype. The BRM series uses the TNA-1 Kvadrat 1 navigation device, 1G11N gyro compass and 1T25 land navigation device. Two scouts are seated in the back of the vehicle for dismount operations and to protect the vehicle from rear and side attack during mobile operations. BRMs are issued on a scale of one per each motor rifle or tank regiment, and three per divisional reconnaissance battalion.

The BRM-1K is an upgraded version of the BRM and was called BMP M1976/2 by NATO. The most significant change on this vehicle was the addition of a PSNR-5K (1RL133-1) battlefield surveillance radar (NATO: Tall Mike) which operates in the 16.0 to 16.3 Ghz frequency band. It is fitted with a telescoping fifty-meter antenna stowed over the hull rear above the exit doors.

Whither the Russian Light Tank?

The Obiekt 934 has had recent offshoots, notably as the testbed for the 2S31 Vena self-propelled gun, developed by the TsNII Tochmash in Klimovsk. The prototype vehicle was based on the same hull as the Obiekt 934, and it is actually mounted on one of the surplus prototype Obiekt 934 hulls. The new 120mm weapon is the 2A80 and is a rifled gun that fires the same ammunition as to the 2A60 on the 2S9 and 2S23.[17] The new weapon has a longer tube and new fume extractor, probably to extend the effective range of the weapon beyond the current 8.7 kilometer maximum range of the 2A60 when firing HE rounds. The new turret bears some design similarities to that on the Obiekt 934, but it is noticeably higher. The choice of the Obiekt 934 hull for the 2S31 was not a deliberate attempt to revive the Obiekt 934, but rather an expedient until a new universal tracked chassis based on the BMP-3 was available. According to representatives from the Tochmash design bureau which developed the 120mm weapon, a second prototype of this vehicle which was under construction uses the definitive BMP-3 production hull.

The appearance of the 2S31 Vena raises interesting questions about the future of Russian light AFV development. The role of a light armored gun vehicle is different in a post-Cold War world where antiarmor capability is no longer a primary requirement. As the Soviets discovered in Afghanistan, general purpose guns which fire a large HE projectile are far more useful than tank guns when confronting irregular opponents who do not have a significant number of armored vehicles. In this case, a conventional tank gun such as the 2A48 on the 1970 light tanks fires a 13.3 kilogram projectile with 1.7 kilogram high explosive fill; the 120mm mortar bomb is 16 kilogram with double the high explosive, 3.2 kilogram. These weapons also fire a wide range of other useful projectiles including smoke, illumination, and white phosphorus. The Russians are also examining alternative precision strike munitions such as the Germes (Hermes) 155mm low-pressure guided missile gun, mounted on a modified BMD-3 hull.

At the moment, the Russians have no fewer than three suitable light armored vehicle hulls for a future light tank: BMP-2, BMP-3, and BMD-3. However, the critical is-

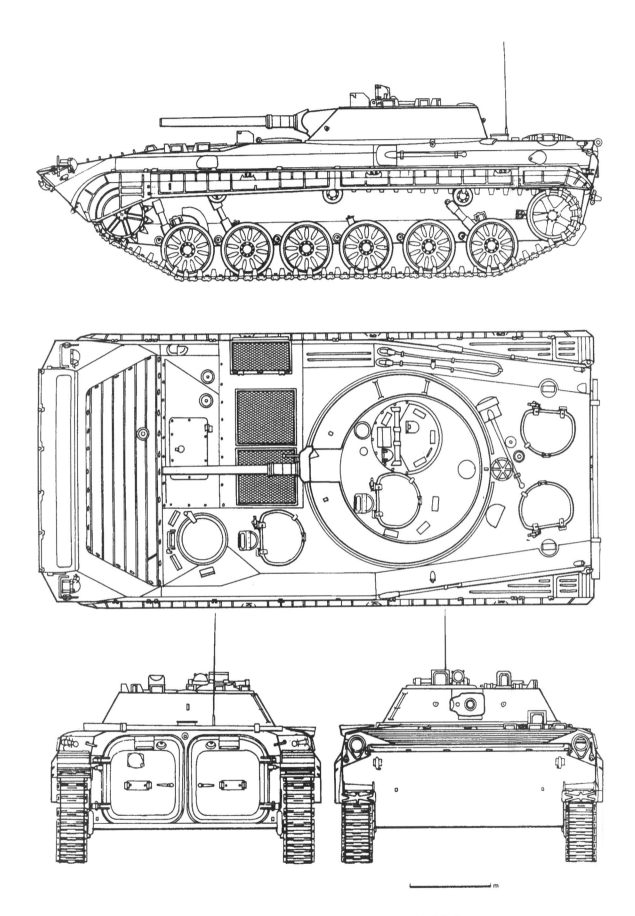

BRM Armored Reconnaissance Vehicle

BRM-1K turret with radar erected.

sue will not be the automotive requirements of the chassis, but the armament requirement. As of the time this is being written, the direction that the Russian Army will take is not at all clear. The French AMX-10RC wheeled reconaissance vehicle also uses a long 105mm gun. But with the Soviet experience with the 2S9 Nona 120mm armored gun system in Afghanistan in mind, the Russians could take an alternate direction.

The second issue confronting Russian light tank designers will be: "Who is the customer?" This is not as obvious as it sounds. Until 1991, Soviet tank designs were based entirely on domestic requirements. Today, with relatively few Russian army contracts compared with their Soviet heydays, Russian designers are beginning to turn their attention to the export market. If the design emphasis is on the overseas market, the armament balance between explosives versus penetration could swing back in favor of conventional tank guns. Many third world countries employ light tanks as surrogate main battle tanks and seek good antitank performance. An example of this trend is Thailand's selection of the Cadillac Gage Stingray in 1992. The second "client" issue will be which service acquires the vehicle. With the increasing emphasis on rapid deployment forces, it seems likely that the VDV Airborne Assault Force will be the most likely candidate for a future light tank. This being the case, the

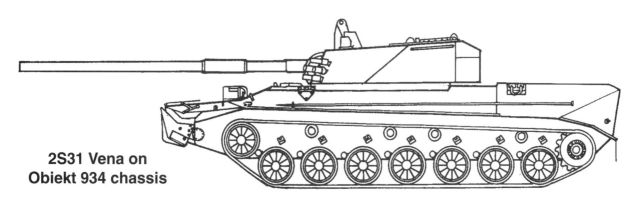

2S31 Vena on Obiekt 934 chassis

2S31 Vena (on Obiekt 934 chassis).

VDV is likely to insist on the adoption of a vehicle based on the BMD-3 chassis.

Finally, the Russians may take the French mix-and-match approach: offer one or more common light chassis (BMD-3 or BMP-3) alongside a variety of turret options such as a tank gun, gun/mortar, and air defense system. In this case, the Russians could very well opt for gun-howitzer combination for their own VDV use, while at the same time offering an optional high-performance anti-tank gun for their export clients.

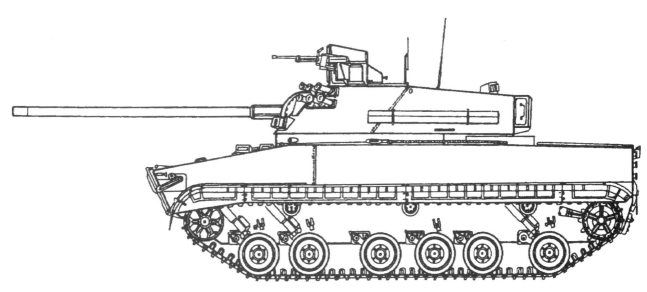

2S31 Vena on BMP-3 chassis

2S9 Nona-S 120mm self-propelled weapon.

Light Tank Prototype Comparative Technical Data

	Obiekt 906	Obiekt 685	Obiekt 934	2S31
Design bureau	Volgograd	Kurgan	Volgograd	Tochmash
Crew	3	3	3	3
Hull length (meters)	7.73	9.4	9.4	9.4
Overall length (meters)	10.1	10.4	10.6	10.6
Width (meters)	2.95	3.15	3.15	3.15
Height (meters)	2.25	2.6	2.4	2.7
Weight (metric tons)	15.0	16.5	17.5	21.0
Main gun	D-58T	2A48-I	2A48-I	2A80
Main gun caliber	85mm	100mm	100mm	120mm
Autoloader?	yes	yes	yes	yes
Ammunition stowage	40	40	40	70 + 10
Engine (horsepower)	300	400	400	400
Maximum road speed (kilometers per hour)	75	70	70	70
Maximum road range (kilometers)	500	600	600	600
Amphibious?	yes	yes	yes	yes

Second Generation Tank Development: Advent of Main Battle Tanks 1962 to Present

Khrushchev's cancellation of the entire Soviet heavy tank program in 1960 continues to have a profound impact on Russian tank development to this day. The demise of the heavy tank meant an end of a central and long-standing principle of Soviet tank design philosophy — the building of special classes of tanks for specific missions. The old formula of using medium tanks for most tank roles, supplemented with heavy tanks for missions where long range firepower or heavy armor was needed, was now dead. Henceforth, a new main battle tank concept (in Russian — osnovnoi tank or standard tank) would gradually emerge.

As the following discussion indicates, this proved far harder to achieve in practice than in theory. This post-1960s search for an optimal main battle tank was complicated by two conflicting schools of thought: (1) the long standing desire within the medium tank design centers and much of the armored force bureaucracy to develop a "universal" tank, and (2) Khrushchev's radical, and ultimately impractical idea of replacing conventional gun tanks with revolutionary missile-armed tanks.

Furthermore, the technical aspects of the universal tank were shaped by Khrushchev's heavy tank cancellation. Any new tank design approaching Soviet heavy tanks in size or weight was liable to be branded a "heavy tank" by Khrushchev, and hence cancelled. This forced Soviet tank designers to develop a universal tank within a very tight envelope that mandated radical solutions in armor, crew size, engine design, and other features to cut weight and size. The resulting design, the T-64, was plagued with technical problems caused by these design constraints.

The perverse long term consequence of Khrushchev's heavy tank cancellation was that in combination with the general Soviet desire to minimize tank size and weight, the follow-on designs to the T-64, namely the T-72 and T-80, were forced into the same straight-jacket of size and weight constraints.

The New Standard: Kharkov's T-64 Tank

The T-64 tank was the most radical advance in world tank technology in the 1960s, as much a breakthrough as the T-34 tank had been in 1940. The T-64 was the first series manufactured main battle tank to employ combined ceramic/steel armor, the first to use an autoloader, and the first to exceed thirty-five miles per hour on the road. In spite of its modest weight of thirty-eight metric tons, it was armed with a very potent 125mm gun, at a time when fifty-five metric ton NATO tanks were still armed with 105mm guns. Indeed, all Soviet main battle tank design until the early 1990s was only an evolutionary outgrowth of the T-64 design. The T-64 stemmed from several sources. The head of the Kharkov design bureau, Aleksandr Morozov, had dreamed of developing a universal tank ever since the aborted T-43/KV-13 competition of 1943. Morozov's dream would come true in the early 1960s with the Soviet abandonment of heavy tanks. But the progress towards this goal was contorted and difficult due to the politics of weapons acquisition in the Khrushchev years.

Even considering its revolutionary nature, the development of the T-64 was unusually prolonged by Soviet standards. Morozov was transferred back to Kharkov in December 1951, and the Kharkov tank design bureau did not begin testing prototypes of the new tank until 1960. The considerable delay was probably attributable to a variety of factors. The posting of Vyacheslav Malyshev to head the nuclear weapons program after World War 2 led to some significant diversions of engineering talent out of the tank programs and into the atomic bomb program, diversions which have remained secret until recently.[18] For example, Nikolai Dukhov, one of the two main heavy tank designers in World War 2, was appointed to head the atomic bomb development institute at Arzamas-16 by Malyshev. Also, many young tank engineers were diverted into the strategic weapons program. For example, Nikolai Shomin, who would eventually become the program head for the T-64 effort at Kharkov, was assigned to a team developing high speed fuel pumps for strategic missile silos in the late 1950s. Besides the diversion of engineering talent into the strategic weapons program, the Kharkov facilities were significant in the USSR's post-war rebuilding efforts, notably in the production of locomotives and turbine generators.[19] Engineering development of the new tank did not begin in earnest until 1958.

Obiekt 430

Little has been written in Russian of the technical-tactical requirements that guided the development of the T-64 design. The approach appears to have focused on improving the firepower, armor, and mobility of the T-54, all within a vehicle not significantly larger or heavier than the T-54. The initial prototype of the new tank was designated Obiekt 430. The weapon selected for the Obiekt 430 was the D54-TS 100mm gun, a rifled gun developed by the Petrov design bureau in Sverdlovsk with a longer tube than the standard D10-T used on the T-54 and offering improved anti-armor penetration. The gun

Obiekt 430 (T-64 prototype).

was stabilized in both axes by the new Metel stabilization system. The internal ammunition stowage was nearly fifty percent greater than on the T-54A, with fifty rounds carried. The fighting compartment was extremely cramped, as the Obiekt 430 retained the traditional three-man turret crew. One of the main objectives of the program was to develop an advanced fire control system which gave the tank a high first hit probability.[20] This was accomplished by incorporating an optical rangefinder instead of the simple stadiametric rangefinders previously used on Soviet tanks.

A second technical requirement was resistance to 100mm projectiles at ranges of 1,000 meters or more. As NATO tanks of the early 1960s, especially the US Army, tended to rely on HEAT projectiles with shaped charge warheads, a special focus of research was on the development of armor resistant to such weapons. The Soviets had begun to work on explosive reactive armor as early as 1949, with an article in a restricted journal by engineers from the NII Prometei research institute suggesting the use of explosives to break up shaped charge jets. By the late 1950s, actual experiments had been conducted, and X-ray photography showed the process of disrupting the shaped-charge's jet. These experiments led to an initial understanding of the effect of an added metal layer to enhance the disruption. In the early 1960s,

a design team under A. I. Platov from the NII Stali (Scientific Research Institute for Steel) was assigned to develop an explosive reactive armor system for the T-64. During trials, there was a serious accident with an ERA equipped tank, and the T-64's chief designer, Aleksandr Morozov, rejected any further use of reactive armor on his tank.[21]

The method finally selected was the use of combination armor using conventional steel with additional layers of ceramic armor inside, also developed at NII Stali.[22] A central focus of the design team was to minimize the internal hull volume in order to permit maximum armor and minimum overall weight. Internal hull volume rose only one percent, from 11.3 to 11.5 meters cubed, even though internal ammunition stowage rose fifty percent. This was accomplished primarily by decreasing the volume of the engine and powertrain. Compared with the T-54A, the fighting compartment volume rose by four percent to 8.37 meters cubed, but the engine compartment volume actually decreased by four percent to 3.13 meters cubed.

The reduction in the volume of the engine compartment was brought about by the adoption of the first major new Soviet tank engine in over two decades, the 5TD. Developed by L.L. Golents at the Kharkov Diesel Engine Plant, the 5TD was a five-cylinder, flat opposing piston design, with common combustion chambers for the opposing pistons utilizing dual crankshafts. This design was probably influenced by American Fairbank-Morse locomotive engines sent to the USSR through Lend-Lease in World War 2. Curiously enough, Britain selected a similar approach, the six-cylinder Leyland Motors L60, to power its contemporary Chieftain tank.

The Missile Tank Controversy

As mentioned in the preceding section on the T-55, in 1956, Khrushchev directed the four major tank design bureaus (Morozov in Kharkov, Kartsev in Nizhni Tagil, Dukhov in Leningrad and Isakov in Chelyabinsk), to begin to develop a missile-armed tank. Of the four bureaus, only Morozov managed to avoid the assignment. This was probably due to the mutual antipathy of the Main Tank Administration and Morozov himself towards the Khrushchev proposal. This was one of a number of Khrushchev plans at the time that were resisted, including an effort to develop a tank mounted on an air-cushion chassis.

The bureau which took Khrushchev's proposal most seriously was the Pavel Isakov bureau in Chelyabinsk. Isakov's bureau was a counterpart of the Nizhni Tagil/Kharkov spin-off in the medium tank field. In the case of heavy tanks, the main bureau under General Zhozef Kotin returned to Leningrad's Kirov Plant in the early 1950s, leaving behind a small design bureau eventually

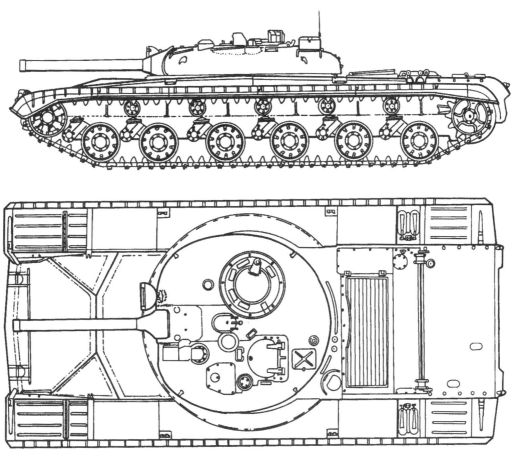

Obiekt 775 missile tank

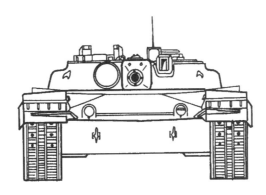

89

headed by Isakov. Isakov's missile tank was designated the Obiekt 775. The chassis used the wheels and tracks from Morozov's Obiekt 430 but incorporated a more compact external hydropneumatic suspension. The design was even more radical than the Obiekt 430, reducing the crew to only two: the commander/gunner and driver, both of whom were located in a low profile turret. The Obiekt 775 also benefited from the powerplant design of the Obiekt 430, using the compact 5TD diesel engine. The armament of the Obiekt 775 was equally radical, consisting of a short-tubed 125mm gun/missile launcher with an autoloader. This particular approach was favored by the head of the GBTU at the time, General Lebedev, who was not keen on arming tanks with conventional low-speed, rail-launched anti-tank missiles. The gun-fired rocket boosted projectiles included the Rubin radio command guided projectile and the unguided Bur high explosive rocket projectile. A total of twenty-four guided and forty-eight unguided projectiles were carried in the hull. In many respects, this design paralleled the contemporary American/German MBT-70, except that the MBT-70 was substantially larger and heavier.

The Kotin bureau at the Kirov Plant in Leningrad took another unconventional approach to the requirement. Instead of a conventional turret, Kotin Obiekt 287 design had a combined armament turret. In the center of the turret was a pop-up Taifun (Typhoon) missile launcher similar to the Drakon system selected for the Obiekt 150/IT-1 of the Kartsev bureau in Nizhni Tagil. Only fifteen radio-command guided missiles were stowed internally because of their large size. To provide firepower against unarmored targets, the turret had two unmanned sub-turrets armed with the new 73mm Grom 2A28 low-pressure guns, developed by A. Shipunov's design bureau in Tula for the BMP-1 infantry fighting vehicle. These smoothbore 73mm guns fired a finned, rocket-propelled anti-tank grenade. The turrets housed thirty-two rocket rounds in an automated magazine along with 3,000 rounds of 7.62mm ammunition for the co-axial machine guns. The crew of two sat in the hull front. This was a significant problem since it required a complicated fire control system for the commander/gunner to aim the various weapons mounted in the traversing turret. Like the Obiekt 775, the Obiekt 287 made extensive use of components developed for the Kharkov bureau's Obiekt 430.

As mentioned in the earlier section on the T-55, the Nizhni-Tagil bureau developed the least exotic of the missile tanks, a simple T-62 tank variant which substituted a pop-up missile launcher for the new 3M7 missile in place of the conventional gun. As it would transpire, this was the only one of the three designs to actually reach production. The Obiekt 775 prototype was completed in 1962 but was plagued with problems. Radio command guidance of the projectiles proved to be more difficult than anticipated. In addition, the need to link the driver's controls through a complicated universal joint mounting (due to the location of the driver in the turret) created severe mechanical problems in the powertrain.

On 14 September 1964, Khrushchev witnessed a firing demonstration of the Nizhni Tagil Obiekt 150 missile tank at Kubinka, where three moving tank targets were destroyed in quick sequence with the new 3M7 missile. The next day at a party conference, he said: "Yesterday, I saw how effectively tanks are destroyed even on the move. When such anti-tank missiles are ready, tanks will become unnecessary." Khrushchev had already ordered a halt to heavy tank development, and there was real fear within the Ground Forces that he might order a cancellation of medium tanks as well. As it was, the technical problems with the radical missile tanks were not ironed out before October 1964, when Khrushchev was ousted from power by Leonid Brezhnev. The Brezhnev administration leaned much more favorably toward the traditionalist views of the military industry and the armed forces, and put an end to many of the more radical weapons ideas fostered under Khrushchev. The missile tank idea was one of these, and both the Obiekt 775 and Obiekt 287 programs never proceeded beyond the testbed stage.

Rearming the Obiekt 430

The technical problems of the missile-tank competitors to the Obiekt 430 could not hide its own serious armament problems. It is unclear why Morozov had selected a 100mm gun for the new tank when his bureau had traditionally sought more powerful guns in each succeeding generation. As mentioned earlier, in February 1961, a crisis developed in the Soviet Ground Forces when the first details of the US Army's M-60A1 became evident to the top leadership. A crash program was undertaken to adopt the new U5-T 115mm gun on the T-55, resulting in the T-62.

A similar effort was undertaken on the Obiekt 430 in late 1961. However, this was a more complicated procedure than on the T-62 because the fighting compartment was extremely limited in volume. The rearmed Obiekt 430 was designated Obiekt 432. The new gun was the 115mm D-68 (2A21) developed by the Petrov bureau in Sverdlovsk. To solve the crew compartment problem, the Morozov design bureau decided to adopt an autoloader; this had the obvious advantage of eliminating a single crewman. The configuration of the autoloader was ingenious. Instead of the magazine style used in the French AMX-13 (or the later Swedish S-Tank), the autoloader was incorporated into the revolving turret basket. The ammunition was stowed in two parts (projectile

Obiekt 432 (T-64 prototype).

and propellant casing) in a configuration nicknamed the korzhina (wicker basket); the autoloader held thirty of the tank's forty-five rounds. The disadvantage of this approach is that it required a new family of 115mm ammunition from the type carried on the T-62. The ammunition family used the same types of projectiles as the T-62's U5-T 115mm gun, but a new case had to be designed for the split ammunition. After the gun fired, the spent casing was returned into the autoloader tray.

The new Obiekt 432 was completed in 1963. It was otherwise very similar to the Obiekt 430 with a few other modest improvements. For example, it used the uprated 5TDF engine, which provided 700 horsepower compared with 600 horsepower on the Obiekt 430, thereby increasing the road speed to sixty-five kilometers per hour. The running gear was modernized, and the number of road wheels rose from five to six per side. Yet the suspension remained very lightweight, only about fifteen percent of total vehicle weight.

The Obiekt 432 did not entirely end the armament question. In 1962, a meeting of the tank technical council of the VPK was held, attended by the main tank designers.[23] Leonid Kartsev, who was finalizing the uparmed T-62 of the time, was critical of the decision to arm the Obiekt 432 with only the 115mm gun, arguing that by the time it entered service, it would already be approaching obsolesence. The technical council agreed that work should begin on a new gun (the 125mm 2A26), but approval was received to proceed with the 115mm D68-T gun on the Obiekt 432. The Obiekt 432 enjoyed two major firepower advantages over the T-62 in spite of similar gun ballistics. The Obiekt 432 used an autoloader, giving it a theoretical rate of fire of ten rounds per minute, versus four rounds per minute for the T-62. Secondly, the Obiekt 432 was fitted with a TPD-43B coincidence rangefinder, giving it superior accuracy at longer ranges. Compared with the T-62, the effective range while moving was 1.1 kilometers verses 0.9 kilometers with APFSDS kinetic energy projectiles, while for HEAT shaped charge projectiles it was 900 versus 600 meters.

The rivalry between Kharkov and Nizhni Tagil continued. In October 1962, at a demonstration of new tank technology for Khrushchev at Kubinka, Kartsev attempted to convince him that the modernized T-62 with new suspension, the Obiekt 167, would be a more prudent choice for the new tank than the "raw" Obiekt 432 (see discussion Final Evolution of the T-62). Kartsev indicated that the Obiekt 167 could be built at Nizhni Tagil and other plants without any major new capital investments. Kartsev's position was supported by the commander of the Soviet Ground Forces, Marshal Chuikov, but it was vigorously resisted by the GBTU, which sought a more sophisticated new tank. Trials of the Obiekt 432 in 1963 made it very clear that the design was far from mature,

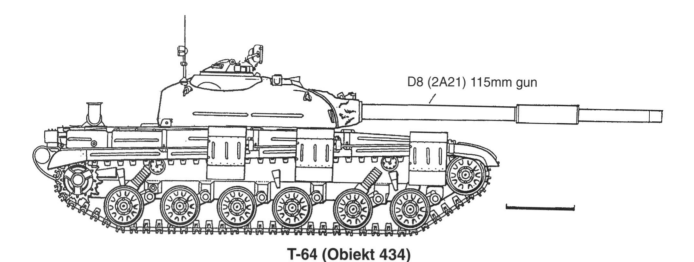

T-64 (Obiekt 434)

but the program continued to have strong support in the GBTU. One of the most significant advantages of the Obiekt 432 was its armor. The "Combination K" mixed steel-ceramic armor offered superior protection against HEAT shaped charge projectiles, which at the time were the preferred type of ammunition used by NATO tanks. The T-62's angled steel armor offered the equivalent of 200 millimeters of steel against APFSDS and HEAT; the T-64 armor was equivalent to 410mm of steel against APFSDS and 450 millimeters of steel against HEAT warheads. Another unusual innovation in the design was the incorporation of folding armor panels over the suspension, which were sprung outward during combat to reduce the effectiveness of HEAT projectiles when fired against the suspension area.

In spite of its advantages over the T-62, the trials had revealed a wide range of shortcomings that needed immediate attention. The Obiekt 432's track, reminiscent of the narrow 500 millimeter track from the original T-54, was inadequate given the high road speeds possible with the design. A new double pin track was developed, based on the track being designed for Isakov's BMP design. Two suspension testbeds were developed (Obiekt 433?) to refine the running gear. In order to carry the maximum amount of armor, a lightweight suspension system was used, employing internal shock absorbers instead of the traditional rubber rim, a type of wheel first used on the KV-1 tank of 1939. The suspension used short co-axial torsion bars, with the first, second and sixth roadwheels having an additional telescopic hydraulic shock absorber. This suspension saved several tons of weight compared with more conventional suspensions.

The Obiekt 432 hull was completely redesigned to simplify production and to incorporate the new antiradiation subsystems already adopted on the T-55A and T-62. The design collective under Nikolai Shomin

T-64 (Obiekt 434).

in Kharkov incorporated the improvements into the Obiekt 434 prototype which was completed in 1964. Trials began in 1964, but before they were completed, Khrushchev authorized production.

T-64 Teething Problems

Prior to series production of the T-64, a pre-series batch of at least ten tanks was built at Kharkov for state operational trials. The trials were conducted in October 1967 in Byelorussia with officer cadets from the 100th Guards Tank Training Regiment. A state commission headed by A.I. Kritsin, the deputy minister of the defense industry, was assigned to examine the results. The trials were widely regarded as a failure due to serious engine problems. The problem stemmed from the engine's air filter system, which clogged easily and led to unexpected engine shut-downs. The transmission suffered from oversteering problems that led to severe power losses when turning. The autoloader, though extremely fast (ten

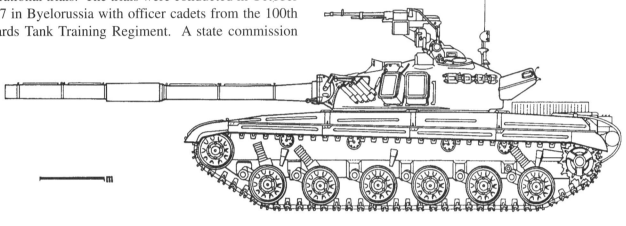

T-64A

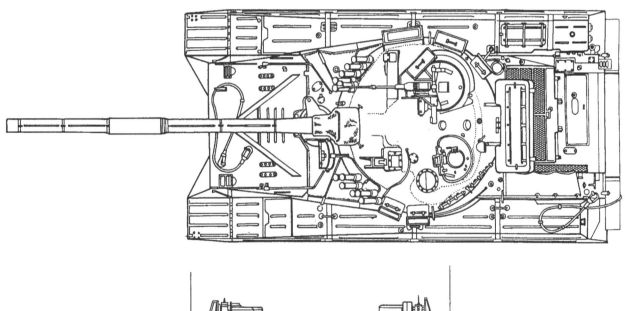

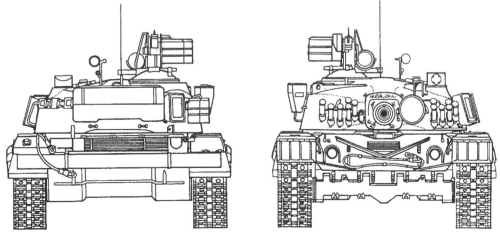

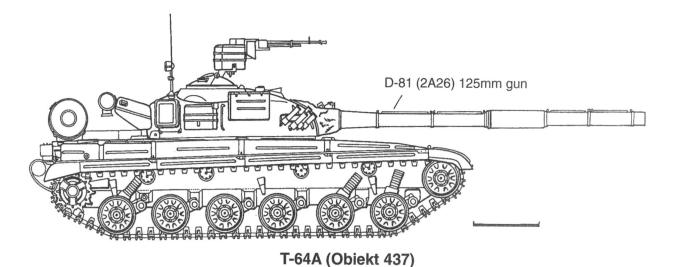

T-64A (Obiekt 437)

rounds per minute) was not adequately shielded from the crew, leading to injuries. Low rate production of the T-64 was authorized while improvements were undertaken, a process which lasted until 1973. At least 600 of the initial version of the T-64 were produced, and it was first deployed with the 41st Guards Tank Division.

T-64A Tank

At the 1962 VPK meeting, the tank technical council agreed to consider rearming the T-64 at the earliest practical opportunity. Development of a new 125mm gun was undertaken by Petrov's bureau in Sverdlovsk, the D-81 (2A26).[24] A prototype of a T-64 armed with this new weapon, the Obiekt 437, was completed at Kharkov in 1969. The Obiekt 437 had a number of improvements over the Obiekt 434/T-64 besides the main gun. To accommodate the new weapon, a new fire control system and gun stabilization system were required. As was occurring in much of the Soviet tank fleet, an anti-aircraft machine gun mounting was developed for the tank. This was more elaborate than on any other tank, consisting of a remote control PZU-5 mounting with the NSVT 12.7mm machine gun. Turret armor protection was also increased. Other improvements included an improved engine fire extinguishing system, a self-entrenching blade on the

T-64A (modernized).

T-64AK (Obiekt 437K).

T-64B (Obiekt 447).

lower bow, attachment points for the new KMT-6 mine clearing system, the Brod tank sealing system to improve OPVT deep-fording, high altitude accessories for the engine and turn signals.[25] A small pre-series batch of tanks was produced in 1972 for trials, and production began in 1973.

The initial production batches had a ZIP tool stowage bin on the right front side and no rear turret stowage bin. On the standard production batches, the ZIP tool stowage box on the right front fender was replaced by an additional fuel cell, and a thermal sleeve was added to the gun barrel. A substitute ZIP tool stowage box was added behind the turret.

In the 1980s, many T-64s were rebuilt with later features. This includes substitution of a reinforced rubber side skirt for the fold-out armor, addition of glacis plate hull armor applique, stowage improvements, the Type 902 Tucha smoke mortars and other small changes.

T-64AK (Obiekt 437K) Tank

Like all Soviet main battle tanks, the T-64A had a corresponding command tank version, the T-64AK. This included an additional HF radio, a dismountable radio antenna, additional navigation equipment, and an auxiliary battery charging unit. The T-64AK was not fitted with the PZU-5 12.7mm NSVT machine gun.

T-64B (Obiekt 447) Tank

With production of the T-64A underway in 1973, the Morozov bureau turned its attention to other potential improvements, under a program designated Obiekt 447. Since the early 1960s, several armies had been working on guided tank projectiles, including France with its ACRA and the United States with its MGM-51A Shillelagh. As mentioned earlier, the Soviet Army had actually led the world in exploring this technology with the 125mm radio-command guided projectile on the revolutionary Obiekt 775 tank. Work had apparently continued on this technology in spite of its lack of favor by the GBTU. The Soviet decision to revive this technology, in spite of the lingering distaste for such such systems dating back to the Khrushchev feuds in the early 1960s, was probably due to the US decision to arm the M551 Sheridan and M60A2 tank with 152mm guided missile projectiles.

The system selected for the Obiekt 447 was the 9K112-1 Kobra. KB Tochmash design bureau in Moscow designed the Kobra system. The challenge posed by this system compared to the American Shillelagh was that it had to be small enough to fit into the existing autoloader on the T-64A. (The Shillelagh was manually loaded and was significantly larger than conventional tank ammunition.) The resulting 9M112 Kobra guided missile (NATO codenamed AT-8 Songster) was a two-piece missile with the shaped charge warhead and sustainer rocket in the forward portion, and the dry cell battery power source, flight control system and rearward pointing guidance link in the rear component. Both sections were automatically joined during the autoloading process.

Guidance for the 9M112 was semi-automatic radio command. The GTN-12 (9S461-1) command antenna was mounted in an armored box placed immediately in front of the commander's station on the right-hand side of the tank, while the guidance electronics were placed behind the gunner's station. There are three firing modes for the weapon. In the primary mode, the gun is elevated before firing to minimize the amount of dust created on firing. The radio command system then captures the missile and brings it back down to the gunner's line of sight. The gunner keeps the cross-hairs on the target, while the fire control system, through the gunner's 1G42 rangefinder sight, optically tracks the missile by its modulated rearward pointing light source and sends course corrections via the command antenna. There is an auxiliary launch mode in situations where dusty or smoky conditions for-

T-64B (Obiekt 447)

ward of the tank may interfere with the command link. In this mode, the gunner must insert target range data into the fire control system. After launch, the missile proceeds to the immediate vicinity of the target in an elevated ballistic path, and is steered into the gunner's line of sight in the final moments of the flight. A third, emergency, mode is provided for the missile. In the event that the missile is already loaded and a target appears at a range under the standard 1,000-meter minimum, the gunner can activate an emergency mode. The gun barrel is only slightly elevated, and the missile is gathered into the line of sight after eighty to one hundred meters after launch.

The standard combat load for this missile system is six rounds. It has an effective range of 4,000 meters and a probability of hit of eighty percent at full range. The Kobra missile was not intended to be the primary means of tank fighting for the new tank. Rather, it could counter NATO's growing number of long-range anti-tank missile platforms, both ground and helicopter-based.

During the development of the Kobra system, Colonel General Yu. Potapov, one of the most influential Soviet tank commanders and later the commander of the Soviet Armored Force in 1977, proposed to incorporate the Kobra tanks in an expanded tank battalion, enlarged from thirty-one tanks to thirty-five tanks. Of these, four were to be Kobra tanks in a separate platoon. This platoon could provide long-range fire support for the rest of the battalion. The Potapov plan was rejected by the then-head of the armored force, Marshal A. Kh. Babadzhanyan, due to the usual Soviet disfavor with specialized units of this sort.

The 9K112-1 missile system was part of an extensive fire control upgrade on the Obiekt 447. The new fire control system was designated 1A33. Other innovations in the system included the substitution of a 1G42 laser rangefinder sight for the optical rangefinder found in previous models of the T-64. The new version also incorporated a ballistic rangefinder. The newer 2A46-2 gun was employed, which had a quick-change barrel for easier replacement.

The Obiekt 447 also introduced a new generation of combination armor which offered the same level of protection as on the T-64A, but was less bulky. A new metal-reinforced rubber side skirt was added for additional protection against infantry anti-tank rockets. A new system was introduced to reduce the vulnerability of the tank to napalm strikes, based on experiences in the Middle East wars. The torsion bar suspension system was also improved, increasing the movement dynamics of the roadwheels. Prototypes of the Obiekt 447 were completed in 1976 and the tank was accepted for service use as the T-64B, and it was nicknamed the Beshka in service. Series production was probably begun in 1979. The initial production T-64B lacked the Type 902B Tucha smoke mortars and was called SMT M1980/2 by NATO; when later fitted with smoke mortars, the T-64B was dubbed SMT M1981/1.

The T-64B1 was a downgraded version of the T-64B without the 9K112-1 Kobra missile system; it does have all the other upgrades characteristic of the T-64B. The proportion of T-64B to T-64B1 indicates that the T-64B1 is the less common type; of the 1,600 T-64B variants in service west of the Urals in 1990, 400 were T-64B1 compared with over 1,200 T-64B. As in the case of nearly all other Soviet main battle tanks, a command version, T-64B1K, was built without the Kobra system.

Final Evolution of the T-64

The T-64 had a troubled history. The technical problems with the initial T-64 (Obiekt 434) and its high unit

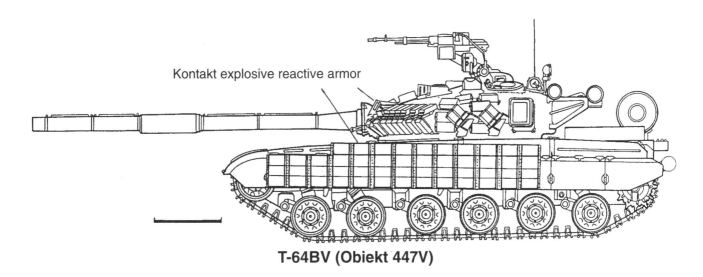

T-64BV (Obiekt 447V)

T-64BV1 (without Kobra antenna).

cost led the Soviet Army to authorize production of a low-cost alternative, Nizhni Tagil's T-72, in 1971. To add insult to injury, in 1976 the Soviet Army authorized production of Leningrad's T-80 in parallel to the T-64A.

The T-80 was not a low-cost alternative to the T-64, but rather a more mobile alternative, sharing the armament system of the contemporary T-64A, but powered by a 1,000 horsepower turbine engine. The T-80's development paralleled the T-64's, and in 1980, the T-80B was authorized for production in parallel to the T-64B, fitted with the advanced fire controls and missile system of the T-64B.

In an attempt to compete with the T-80, the Morozov design bureau (now headed by Nikolai Shomin), improved the mobility of the T-64B by introducing the 6TD engine. This six-cylinder version of the 5TDF on the T-64A/T-64B, offered 1,000 horsepower. This version was produced in modest numbers, but in 1987, all T-64 production ceased with the Soviet Army standardizing on the T-72 and T-80. An unclassified US DIA estimate placed T-64 strength in 1984 at 7,900 tanks.

Modification of the earlier T-64 versions continued. In the mid-1960s, design work continued on reactive armor at the NII Stali and NIIBT. With the advent of plasticized explosives, the problems with detonation of the earlier cast explosive reactive armor that had been rejected for use on the T-64 earlier in the decade were finally overcome. A practical form of explosive reactive armor was available by the 1970s but was not adopted on Soviet tanks. The effective use of Blazer reactive armor by the Israelis in the 1982 Lebanon war led to the Ground Forces' decision to adopt the NII Stali Kontakt design, despite the objections of the tank design teams who felt that the added 1.5 metric tons was too much given that the new reactive armor protected only against HEAT warheads. A state commission ordered the fitting of the Kontakt ERA on 15 January 1983, and the first T-64 tanks received Kontakt ERA packages beginning in September 1983. The T-64B with reactive armor were designated T-64BV, the T-64B1 as T-64BV1 and the command version as T-64BV1K, the "V" indicating vyzryvnoi, or "explosive."

T-64R Rebuild Tanks

In the 1980s, the original T-64 (Obiekt 434) underwent a systematic modernization program, with the D-68 115mm gun and its associated autoloader and fire control system replaced with the 2A46-2 125mm gun, 1A33 fire control system, 9K112-1 Kobra missile system and other T-64B upgrades. This version is designated T-64R (R= remontirovanniy- rebuilt).

T-64BV with Kontakt ERA.

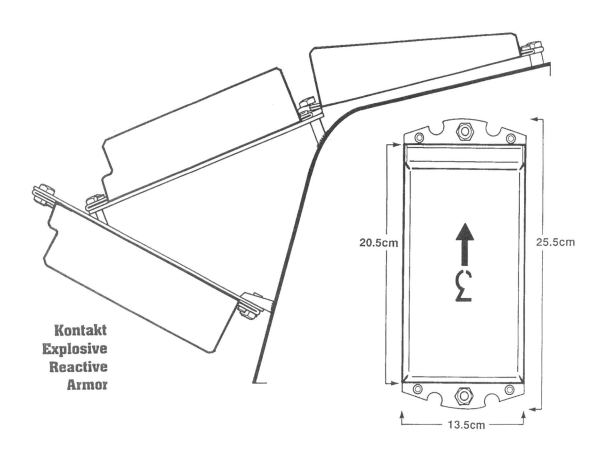
Kontakt Explosive Reactive Armor

Future T-64 Upgrades

The future of the T-64 in the Russian Army is clouded by the relative age of many of the vehicles as well as by its relatively small numbers. As of 1990, the USSR reported that there were 3,982 T-64s west of the Urals. Of these, 2,091 were in Ukraine. As a result, the total number of T-64s in European Russia is modest, and so there may not be as much incentive to modernize this tank as opposed to the more numerous T-72 and T-80. The total CFE inventory of T-64s in European Russia on 1 January 1994 was 625, on 1 January 1995 was 449, on 1 January 1996 was 161, and on 1 January 1997 was 186.

T-64 Comparative Technical Data

	T-64	T-64A	T-64B
Crew	3	3	3
Combat weight (metric tons)	36.7	38.6	40.3
Horsepower per ton ratio	19.1	18.4	17.7
Ground pressure (kilograms/cm^2)	0.815	0.84	0.86
Overall length (meters)	8.95	9.23	9.23
Hull length (meters)	6.43	6.45	6.45
Width (meters)	3.41	3.41	3.41
Height (meters)	2.15	2.17	2.17
Clearance (millimeters)	456	450	450
Main gun	D-68T (2A21)	D-81T (2A26M2)	D-81T (2A46-2)
Main gun caliber (millimeters)	115mm	125mm	125mm
Rate of fire (rounds per minute)	10	8	8
Elevation (degrees)	-6 + 14	-6 + 14	-6 + 14
Main gun ammunition/Autoloader capacity	40 / 30	37 / 28	36 / 28
Gun stabilization	2E18 2-axis	2E36	2E36M
Maximum effective range (meters)	1,100	1,100	1,900/4,000
Effective range (night, meters)	800	800	800
Guided missile	no	no	9K112-1 Kobra
Guided missile storage	no	no	6
Co-axial machine gun	PKT 7.62mm	PKT 7.62mm	PKT 7.62mm
7.62mm ammunition	2,000	2,000	1,250
Anti-aircraft machine gun/mounting	no	NSVT 12.7mm/PZU-5	NSVT 12.7mm
12.7mm ammunition	n/a	300	300
Gunner's rangefinder sight	TPD-43B	TPD-2-49	1G42 (laser)
Gunner's night sight	TPN1-432	TPN1-49	TPN1-49-23
Ballistic computer	no	no	1V517
Infrared searchlight	Luna L-2AG	Luna L-2AG	Luna L-4A
Commander's day/night sight	TKN-3	TKN-3	TKN-3V
Driver's day sight	TNPO-168	TNPO-168	TNPO-168
Driver's night sight	TVN-2BM	TVN-2BM	TVN-2BM
Engine	5TDF	5TDF	5TDF
Power (horsepower)	700	700	700
Internal fuel/External fuel(liters)	815 / 330	815 / 330	815 / 330
Maximum speed (kilometers per hour)	60	60	60
Maximum road range (kilometers)	650	550	500
Maximum unpaved road range(kilometers)	370	310	285
Grade/sideslope (degrees)	30 / 30	30 / 30	30 / 30
Trench crossing (meters)	2.85	2.85	2.85
Wall (meters)	0.8	0.8	0.8
Unprepared fording (meters)	1.8	1.8	1.8
Radio/Intercom	R-123 / R-124	R-123 / R-124	R-123M / R-124

T-64R (rebuilt T-64).

T-64 Tank Ammunition

115mm Tank Ammunition (D-68 gun)

Round	Projectile	Type	Round Weight (kilograms)	Projectile Weight * (kilograms)
3VBM1	3BM5	APFSDS	18.0	5.3
3VBM5	3BM5	APFSDS	18.0	5.3
3VBK4	3BK8M	HEAT	22.5	13.0
3VBK8	3BK8	HEAT	22.5	13.0
3VBK8	3BK8M	HEAT	22.5	13.0
3VOF23	3OF17	HE-Frag	27.4	17.8
3VOF24	3OF17	HE-Frag	27.0	17.8
3VOF38	3OF28	HE-Frag	27.0	17.8

125mm Tank Ammunition (D-81 gun)

Round	Projectile	Type	Round Weight (kilograms)	Projectile Weight (kilograms)	Armor Penetration* (millimeters)
3VBM3	3BM9	APFSDS	19.5	3.6	
3VBM6	3BM12	APFSDS	19.7	3.8	
3VBM7	3BM15	APFSDS	19.7	3.8	150
3VBM8	3BM17	APFSDS	19.7	3.8	
3VBM9	3BM22	APFSDS	20.2	6.9	
3VBM13	3BM32	APFSDS	20.4	7.1	250
3VBM17	3BM42	APFSDS	20.4	7.1	
3VP6	3P6	Training	18.5	5.2	
3VBK7	3BK12	HEAT	28.5	19.0	
3VBK7	3BK12M	HEAT	28.5	19.0	
3VBK10	3BK14M	HEAT	28.5	19.0	220
3VBK16	3BK18M	HEAT	28.5	19.0	260
3VBK17	3BK21B	HEAT	29.0	19.5	
	3BK29	HEAT	28.4	18.9	ERA + 300
3VP5	3P11	Training	28.5	19.0	
	9M112	Missile			
3VOF22	3OF19	HE-Frag	33.0	23.0	
3VOF36	3OF26	HE-Frag	33.2	23.2	

* At 1,000 meters, 0 degrees, and 60 degrees of obliquity.

Economy Tank: The Nizhni Tagil T-72 Ural

The Russian T-72 Ural tank is the most widely deployed main battle tank of the current generation. Not only is it used by the armies of the former Warsaw Pact and Soviet Union, but it has been exported in large numbers to many of the confrontation states in the Middle East, including Syria, Libya, Iraq, and Iran. Indeed, it is the only one of the modern Russian main battle tanks — T-64, T-72, and T-80 — to have been exported in any significant numbers. The T-72 has seen combat in many recent conflicts, including the 1982 Lebanon war, the 1980-1988 Iran-Iraq War, the 1991 Gulf War and the Yugoslav civil war. It is currently being manufactured by at least five countries, more than any other current tank. Curiously enough, the T-72 was produced in parallel to the T-64A tank, and later, the T-80 tank. It has long been an enigma why the USSR manufactured several main battle tanks simultaneously. As has become apparent in recent Russian writing, the answer to this question lies more in the bureaucratic and political inefficiencies of the Brezhnev years than in any dedicated defense policy.

Domestic T-72 Development

In 1967, the Soviet army adopted the T-64 as its next new main battle tank. At the time, two other tanks were in production, the T-55 and T-62. The production of the T-55 was intended mainly for export as, until that point, the T-62 had never been exported. The year 1967 was an important one for Soviet military industrial policy. The war between Israel and its Arab neighbors and the ensuing Arab defeat led to an arms race in the region that has not entirely abated to this day. The Soviet Union played the willing role of arms merchant to the Arab side, and a significant fraction of total Soviet tank production in the ensuing twenty-five years would be earmarked for the Middle East. For example, from 1972 to 1983, the Soviet Union exported the equivalent of forty-four percent of its tank production to the developing world, the vast majority to the Middle East. Aside from the geostrategic implications of these sales, they also had important effects on Soviet industrial policy as weapons sales became the second largest hard currency export item (after raw materials). The Soviet tank industry began to broaden its production to accommodate these export requirements beyond its traditional role of supplying the Soviet army. The decision to produce the T-64 did not sit well with the tank design bureau at the Vagonka (Uralvagonzavod) in Nizhni Tagil. From 1951 to 1967, Nizhni Tagil had been the primary center for the further development of Soviet medium tanks; with the demise of the heavy tanks in 1960 it had become the most significant design bureau in terms of actual product. The rival Morozov design bureau in Kharkov still held enormous prestige, having originated the T-34, T-44, and T-54 designs. Aleksandr Morozov was still held in great esteem by the Main Armored Directorate (GBTU), and it is not altogether surprising that his bureau was assigned the task of the developing the second post-war generation of Soviet medium tanks.

Nevertheless, the selection of the Kharkov T-64 design in 1967 was not greeted with universal acclaim in the Soviet army. The army had become used to the evolutionary developments of the T-54/T-55/T-62 series. The conscript base of the Soviet army made the leadership inherently conservative in the adoption of any radical new technology.

The greater the change, the greater the burden in assimiliating the new technology by the conscript force of young short-term tankers. While there was also a long tradition of pride in Soviet superiority in tank design amongst the tank force, the T-64 was a mixed blessing. To the average tank officer, its 115mm gun offered no more firepower than the T-62, as its higher rate of fire and greater accuracy were not particularly evident due to the stingy peacetime allotments of training ammunition. Its armor was still so secret that the average tank officer had no appreciation for its advance over the T-62. Its mobility, while potentially superior to the T-62's, was in fact significantly inferior due to the teething problems with the new powerplant and the higher maintenance demands of the engine, tracks and suspension. It is not altogether surprising that the commander of Soviet Ground Forces, Marshal Chuikov, had favored the evolutionary Obiekt 167 over the revolutionary T-64.

On the industrial side, the selection of the T-64 left the experienced Vagonka design bureau out of work. While there was still the need for the design team to continue evolutionary improvements on the T-55 and T-62 while they remained in production, this would not absorb the full attention of a bureau of this size.

It is often supposed that there was little room for initiative in the Soviet system, but this often was not the case. In 1965, after news of the Council of Ministers' resolution approving T-64 production, the Vagonka design bureau in Nizhni Tagil began to work on adapting an automatic loader to the T-62. The design bureau chief, Leonid Kartsev, was not happy with the "korzhina" autoloader, feeling that it was a mistake to cut off the loader from the rest of the crew.[26] He instructed the Kovalev and Bystritskiy design bureau to develop an alternative autoloader.

In the meantime, the Council of Ministers decided to shift production at Nizhni Tagil from their own T-62 tank

Obiekt 167T.

to Kharkov's new T-64 tank. Production was slated to begin after the Morozov bureau in Kharkov had adapted the T-64 to accept the new 125mm gun. Kartsev was convinced that this would take some time, so he directed his bureau to begin to adapt the 125mm gun with the new Kovalev/Bystritskiy autoloader into a T-62 chassis. To provide some cover for this unauthorized activity, the design was tied to the upcoming 50th anniversary of the October Revolution in October 1967.

In November 1967, Sergei A. Zverev, the Minister for Transportation Machinery, visited the Nizhni Tagil plant as part of the anniversary festivities. While being given a tour of the experimental workshop, he was shown the upgunned T-62 tank. Zverev exploded and accused Kartsev of "intriguing against Kharkov again." Kartsev placated the minister by pointing out that the United States and Germany had an active program for modernizing their series production tanks. He slyly asked the minister why they were forbidden from doing so. This calmed down the argument, and Zverev was given a demonstration of the new autoloader. After having heard of the difficulties with the T-64 autoloader, Zverev was impressed with the smooth performance and speed of the Vagonka system, as well as the fact that it was already adapted to the new 125mm gun. Zverev suggested that the new system be incorporated into the Kharkov T-64. Kartsev agreed but suggested that the modification include the addition of a new diesel engine from Trashutin's team at Chelyabinsk, yet another evolutionary development of the long-serving V-2 diesel of 1939. At a meeting with the plant manager, I. V. Okunev, on 6 November, Zverev agreed to permit the Nizhni Tagil design bureau to build six prototypes of a T-64 variant with the new autoloader and new diesel. This agreement was reached without the participation of the army, or even of the GBTU tank administration.

Development work on the new tank, designated the Obiekt 172 Ural, began in December 1967. The name "Ural" was chosen by the Vagonka bureau to highlight the origins of the design and to distinguish it from the rival Kharkov bureau's design.[27] There was clearly an intent from the outset to substantially redesign the tank along the lines of other Nizhni Tagil designs. Kartsev intended to replace the T-64's complicated suspension with the suspension developed for the Obiekt 167 (improved T-62). The Vagonka design bureau was not happy with the transmission since it had been designed for another engine. As a result, the bureau called in transmission designers from other design bureaus and ordered a new transmission design.

Obiekt 172.

Zverev found out about the proposed changes and ordered Kartsev to Moscow on 15 January 1968 for a thorough dressing down. Kartsev agreed to limit the proposed changes to the T-64 but continued to insist that the Kharkov running gear was inadequate "and would have to be replaced eventually." Zverev apparently agreed to allow the Nizhni Tagil team to develop a prototype with the minimal changes and another prototype with a full set of changes. The basic type was a largely unaltered T-64A but with a Trashutin V-45 diesel with an ejector cooling system developed at Kharkov. The other protoypes were based on the T-64 but were fitted with the Obiekt 167 suspension, using the new RMSh track that was being developed to upgrade the T-55 and T-62. These vehicles also had the Vagonka autoloader, and a V-45 with a Vagonka fan cooling system similar to those employed in the T-62. Ironically, at the end of 1967, Kartsev was offered the position of head of the Kharkov tank design bureau to replace the retiring Aleksandr Morozov. After the competition of the preceding decade, he decided against it, and instead recommended his fellow academy student Nikolai Shomin. Instead, he was assigned as deputy chairman of the tank research committee of the GBTU in Moscow. His position as head of

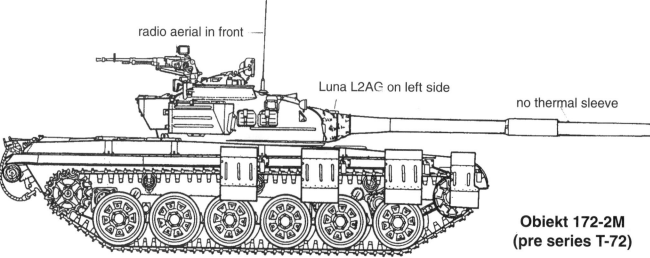

Obiekt 172-2M (pre series T-72)

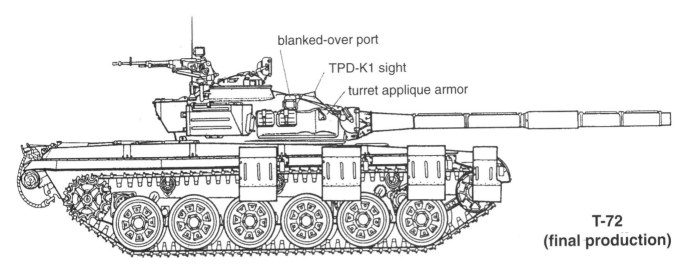

T-72 (final production)

the Vagonka design bureau was taken by V.N. Venediktov. Venediktov is usually given credit as the chief designer of the T-72, even though it was Kartsev who saw it through most of its gestation.

Competitive trials of the Obiekt 172 prototypes began in 1968 including tests at the Kubinka proving ground. Desert trials in Central Asia were conducted in the late summer of 1969 on two Obiekt 172 prototypes. The vehicles with the more elaborate improvements demonstrated the best results in the trials. However, the new features added four tons of weight to the tank, leading to the decision to employ a more powerful engine. Once again, the chosen solution was a V-2 variant, the V-46 780 horsepower engine, evolved from the V-45. The trials resulted in the decision to replace the drive train which had proven unreliable. Prototypes of this vehicle were first completed in November 1969 and were designated Obiekt 172-2M. In February 1971, the Obiekt 172-2M tanks were subjected to cold weather trials in the Zabaikal region of Siberia. This was the final set of trials for the

T-72 (Obiekt 172-2M).

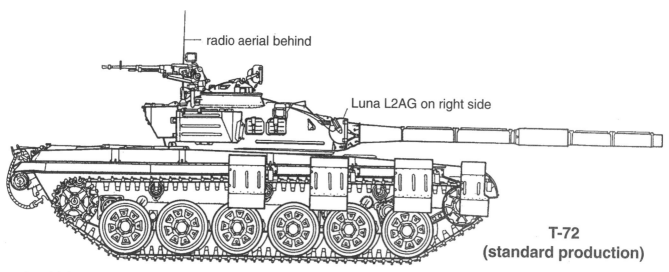

T-72 (standard production)

tank, which was accepted for production in 1971 as the T-72.

The decision to produce the T-72 in parallel to the T-64A was a compromise between the GBTU, which favored the most sophisticated tank possible, and the Ground Forces, which continued to favor designs that permitted numerical superiority. The decision was an odd one since it meant that the armed forces would be equipped with two tanks fitted with different engines, tracks and suspensions. Many spare parts were completely different. The T-72 adopted a simple pintle-mounted 12.7mm anti-aircraft machine gun instead of the elaborate remote control machine gun on the T-64A. The autoloaders were both completely different, and the fuel tanks were significantly different as well. In terms of supplies, the only common items were the ammunition.

The rationale for the T-72 decision was offered by Nikolai Shomin, the T-64's chief designer in a recent interview:

. . . The T-64 embodied all of the progressive ideas that allowed it to be modernized and upgraded through its life. But the first series of production vehicles had a lot of problems. Think about what the world situation was at the time. The Vietnam war was continuing, Israel was carrying out its aggression against its Arab neighbors and nearly all the governments of the Mid East, we were estranged from China. And the result was that a direct military confrontation was a possibility. Combined with that situation, the Ground Forces realized that with our industrial capacity, it would take five to ten years before we would have a sufficient number of such (T-64) tanks. In the eyes of the Ground Forces, we would cause them to forfeit numerical superiority in armored vehicle equipment. It was only for that reason that they arrived at the decision to immediately begin design of a new tank that would combine the basic qualities of the T-64 with the traditional tank engine: a modification of the same diesel that had been used on the T-34.[28]

T-72 standard tank.

T-72A Ural tank.

The T-72 Ural Tank

Initial production of the Obiekt 172-2M tank began in 1972 at Nizhni Tagil. These vehicles were intended for operational trials with regular Soviet Army tank units prior to official service acceptance. They differed from subsequent production vehicles in a number of details, such as the positioning of the main L2AG Luna infrared searchlight. The early service trials of the Obiekt 172-2M regiments led to an immediate improvement program to rectify flaws in the design. This improved vehicle was designated Obiekt 172M. The Obiekt 172M had the Luna searchlight switched to the right side of the main gun, the radio antenna was moved from in front of the commander's station to a position behind it, and additional external stowage was provided on the turret. The Obiekt 172M was accepted for Soviet Army service in 1973 and entered series production in 1974; it was called the T-72. Production of the T-72 was initially limited to Soviet Army use. In general, the T-64A was deployed

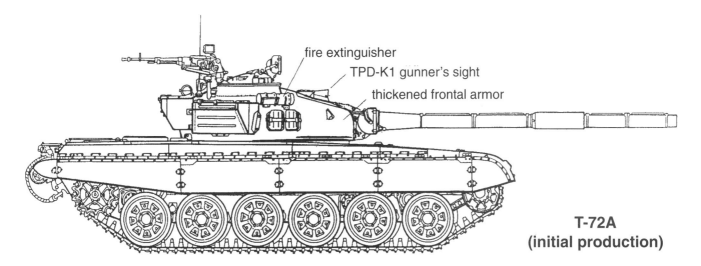

T-72A (initial production)

T-72 ammunition layout.

BM15 steel penetrator rod from 3VBV7 125mm ammunition.

with the front-line units with Group of Soviet Forces-Germany, while the T-72 was deployed in the second echelon with the Groups of Forces in the Warsaw Pact countries, and in the western Soviet military districts such as Byelorussia and western Ukraine.

The T-72 Model 1975

One of the secondary objectives of the T-72 program was to provide a tank that could be exported to prime clients in the Middle East. In 1975, a special export model was developed that had a revised turret frontal armor (probably homogenous steel), a downgraded PAZ (anti-radiation filter), and a different ammunition configuration. This may also have been the type intially provided to the Warsaw Pact countries.[29]

The T-72A Ural Tank

A major block improvement program for the T-72 continued in the 1970s by the Uralvagon KB as the Obiekt 172M-1. The T-72's TPD-2 coincidence rangefinder was expensive to produce, complicated to employ, and not particularly accurate in low light conditions. The new TPD-K1 laser rangefinder was developed to replace it, leading to the deletion of the optical port on the right side of the turret. A new gunner's sight, the TPN-3-49, was added at the same time. A new version of laminate steel/

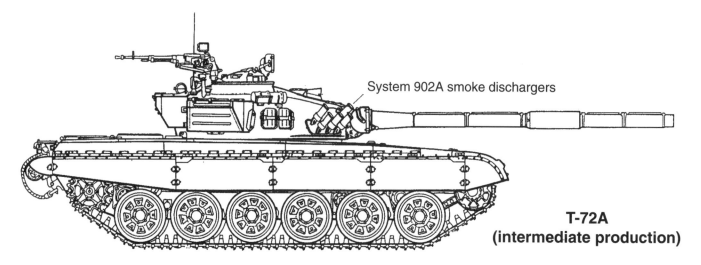

T-72A
(intermediate production)

ceramic turret armor was introduced as well, boosting the effective protection of the turret from APFSDS attack from about 410 millimeter of steel equivalent to 500 millimeters equivalent.[30] Protection against HEAT shaped charge was less dramatic, from about 500 millimeters equivalent to 560 millimeters equivalent. The engine was improved, as well as the communications suite, night vision equipment and ammunition stowage.

A number of other improvements were also incorporated into the design, including the L-4 infrared searchlight, the 2A46 version of the D-81T tank gun, napalm protection features, turn-signals, greater dynamic travel for the suspension arms, and the uprated V-46-6 engine.

Prior to series production of the new design, a final batch of T-72 was produced, incorporating some of these features; an additional layer of armor was added to the turret front, the right optical port of the coincidence rangefinder was blocked off and the TPD-K1 was substituted. These may have been pre-series production testbeds. On entering production in 1979, the new version was designated as the T-72A.

The T-72A was externally distinguishable from the basic T-72 by the lack of a coincidence rangefinder port in front of the commander's station on the right side of the turret. The T-72 used the T-64-style "gill armor" flip out panels on the hull side to prematurely detonate enemy shaped charge projectiles, rockets and missiles; on the T-72A, conventional skirts made of a metal-reinforced rubber fabric were used instead. On the original T-72, the frontal quadrants of the turret sides were inclined at an angle of about fifty-five to sixty degrees, while on the T-72A, the frontal turret armor was almost vertical at most points. The thickened appearance of the turret frontal armor of the T-72A led to the unofficial US Army nickname "Dolly Parton" for this variant, after the buxom American country singer. For some time, this version was mistakenly called the T-74 based on a mistaken association of it with the Obiekt 174 prototype.

The T-72A continued to evolve. By 1980, the turret had been modified by the addition of twelve 902B Tucha smoke grenade dischargers on the turret front and two tubular containers for OU-2 fire extinguishers. This version was initially misidentified by the US Defense Intelligence Agency (DIA) as the T-80 tank in its 1983 edition of Soviet Military Power, and later was referred to as Soviet Medium Tank (SMT) 1980/2. The final 1981 production style of the T-72A had a modified turret with a layer of antiradiation cladding on the roof and an additional stowage box in the 7 o'clock position. This version was called SMT 1981/3 by NATO.

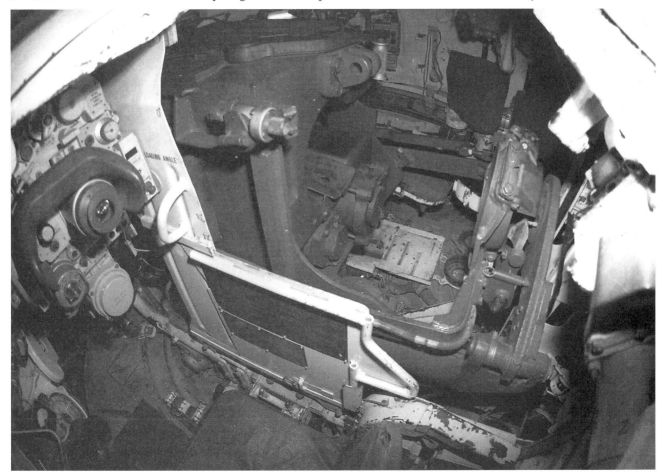

D-81 gun breech in T-72M1 tank.

T-72M1 commander's station.

T-72M1 gunner's station

Production of the T-72 was first undertaken by the Ural Railcar Plant (Uralvagonzavod) in Nizhni Tagil. Production was later extended to the the Chelyabinsk Machinery Plant once it had ceased producing the T-55 and T-62 for export. The tank plants at Omsk and Kharkov were used to produce the high cost tanks, first the T-64 and later the T-80.

Inside the T-72M1

One's first impression on climbing into the T-72M1 (Obiekt 172M-1) is how cramped it is inside, even compared with earlier Soviet tanks such as the T-62. American and European main battle tanks are luxurious in comparison. Neither the commander nor gunner can stand in the turret with the hatches closed because the autoloader cassette in the hull floor takes up so much space. It feels more like sitting inside an aircraft cockpit than inside a tank as far as space is concerned. The T-72 is definitely not for those prone to claustrophobia. Dimensionally, it is noticeably lower and shorter than the American M60A1 and British Chieftain tank and it is almost twenty tons lighter.

The T-72M1 is conventionally laid out with the driver located centrally in the hull, the commander in the right side of the turret, the gunner in the left side of the turret. The engine is mounted transversely in the hull with the engine accessories in the rear under the radiator.

The turret interior is dominated by the massive breech block of the D-81TM 125mm main gun, as well as its associated ammunition handling system. The gun uses a conventional sliding breech, combined with an elaborate autoloading system. The main ammunition reserve, consisting of twenty-two projectiles and twenty-two propellant cases, is stored in a rotating ammunition cassette on the floor of the tank. The projectiles are stored on the bottom layer, and the Zh40 propellant cases are on the top layer. The additional ammunition is stored in the hull around the turret: four projectiles and propellant cases in pockets in the right front fuel cells, two projectiles and Zh40 behind the commander's seat, two projectiles and one Zh40 immediately behind the gunner, three projectiles on racks on the left rear hull side, six projectiles on the rear firewall, and eight Zh40 propellants cases in cavities in the rear fuel tank on the floor behind the ammunition carousel. The only ammunition above the turret line are five propellant charges stowed near the gunner's and commander's station on later models of the T-72. The

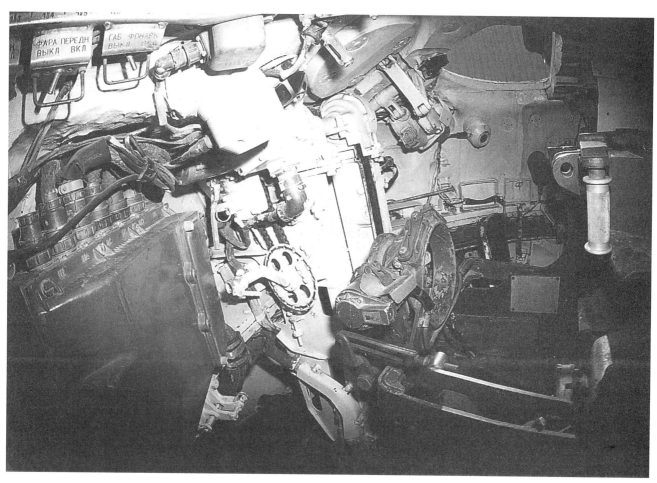

T-72M1 autoloader assembly.

7.62mm PKT co-axial machine gun is located to the right of the main gun, in the commander's side of the turret. Ammunition for the PKT is stored under the commander's seat in two ammunition boxes, with an additional six boxes stored in nooks and crannies in front of the commander and gunner.

The tank commander's station includes the vehicle radio. This is an R-173 in the case of the T-72M1 and operates in the FM mode. The R-173 operates at 30,000 to 76,000 kHz and has ten pre-selected frequencies. Unlike most Western tank radios, the R-173M operates via a throat mike that is part of the normal tanker's helmet. The same throat mike is also used with the tank's internal communication system. The tank commander's main sight is the bi-ocular TKN-3 sight, mounted in a fully traversable cupola. The tank commander can override the gunner's controls to traverse the turret, and he can bring the turret to his line-of-sight with the control handle on the TKN-3. The commander also has several small vision periscopes, though they provide a much more interrupted view than on Western tanks such as the M1 Abrams. Another significant difference between the T-72 and many Western tanks of the current generation is that the commander does not share an optical picture with the gunner. On tanks such as the M1 Abrams, there is an optical elbow between the gunner's main sight and the commander's station that helps the crewmen coordinate their actions and that allows the commander to verify the target being engaged by the gunner.

The gunner's station is even more cramped than the commander's station, mainly due to the presence of the bulky sighting system immediately in front of him. The gunner's sights consist of a TPN-1-49-23 active infrared night sight on the left and the primary TPD-K1 day sight with integral laser rangefinder immediately in front of him. Turret traverse is accomplished by a set of hand grips under the TPD-K1 sight, and a manual back-up is provided in case the electric drive is turned off or disabled. "Air conditioning" in the tank is provided by a small, unshielded plastic fan, located near the gunner's left knee. This may be adequate in Russian or European climates, but it would be of dubious comfort in the desert. To engage a target, the gunner begins by selecting the proper ammunition type on instructions from the tank commander. There are generally three ammunition options: sub-caliber (APFSDS), high-explosive-fragmentation (HE-Frag), and high explosive anti-tank (HEAT). Once selected, this puts the autoloader in motion. The autoloader cassette is fast, about seventy degrees per second in the horizontal axis, meaning that the ammunition is under the autoloader in less than three seconds. While this is taking place, the gunner aims the cross-hairs of the main TPD-K1 at the target and fires the laser rangefinder with a finger control. The range is displayed in the sight and then has to be manually inputted into the tank's analog ballistic computer.

Besides the range, the computer requires manual input of ballistic and meteorological corrections which are calculated from data available to the gunner before the engagement (barrel wear, charge temperature, barometric pressure, and ambient temperature). The only automatic data input is for vehicle movement. The T-72M1 fire control system does not incorporate corrections for cross-wind data or inherent system errors.

The 2E28M gun stabilization system is two-axes: a vertical stabilizer in the main sight and a horizontal stabilizer for the hull, supplemented by an accelerometer. The 2E28M is based on mid-1960s technology and does not incorporate dynamic cant data. In addition, the high shock levels from firing require that the system be recalibrated after five or six firings. This limits the tank's ability to fire on the move. This fire control system is similar in performance to that of the early 1970s generation of Western tanks such as the M60A1 RISE, Leopard-1A3, Chieftain Mk.5, or AMX-30. It is poorly suited to firing on the move because of the complexities of manual data input and inherent limits in the gun stabilization system. In the T-72, fire-on-the-move is accurate, even at short ranges, only when on level ground, at moderate speeds (up to twenty-five kilometers per hour) and against a target with small lead angles. Because of these limitations, all but the best crews will usually halt before firing.

While the gunner is aiming the main gun, the autoloader is preparing the ammunition for firing. This feature differs from most contemporary Western tanks (with the exception of the new French AMX LeClerc) which have a third turret crewman to load the gun. After the gunner selects the type of ammunition desired at the beginning of the engagement, an ammunition cassette in the floor rotates and stops under the autoloader hoist at the turret rear when the proper type of ammunition is located. The gun is automatically elevated into the proper loading position. The autoloader hoist then brings up the two-piece ammunition from the cassette storage, rams the projectile into the breech, and next, the Zh40 propellant charge. The gun then returns to the gunner's line of sight, and the weapon can be fired. The process will also eject the spent stub casing from the previous round out of the turret through a small port in the roof.

The entire process from selecting the round until the gun is ready to fire takes eight seconds. The system theoretically has a maximum rate of fire of eight rounds per minute. In the event that something goes wrong in the process, a manual crank is mounted on the autoloader hoist. The gun has a theoretical rate of fire of only two rounds per minute when loaded manually. The autoloader gives the T-72M1 a higher rate of fire on paper than earlier manually loaded guns such as on the T-62. But in practice, this has not proven to be particularly relevant in tank fighting. Manually loaded tanks such as the M1A1 Abrams have a theoretical rate of fire of only four rounds per minute, but in practice, a well-trained crew can get off three rounds in the first fifteen seconds of an engagement. Furthermore, the speed of target engagement is primarily dependent on the ability of the gunner to quickly acquire the target, perform the necessary calculations and aiming corrections, fire the gun and then switch to the next target. This is as much a function of training as technology. In both regards, NATO tanks have long enjoyed a significant advantage. In NATO tanks such as the M1 Abrams and Leopard-2, the gunner's engagement sequence is automated to a greater extent than on the T-72M, with many data being automatically entered into the computer, and the computer taking a greater role in ballistic corrections.

The D-81TM 125mm gun uses three primary types of ammunition: APFSDS, HEAT, and High-Explosive/Fragmentation (HE-Frag). A flechette anti-personnel round is available but seems to be fairly rare. A variety of APFSDS rounds have been developed for the 125mm gun. Several types were in service in the 1970s, including the homogeneous steel 3BM9, the tungsten carbide-cored 3BM12 and improved types such as the 3BM15 and 3BM17. The performance of the D-81 gun has been continually enhanced by ammunition improvements, with the Russians displaying depleted uranium BM-32 APFSDS rounds for the first time in 1992. Likewise, there were several HEAT projectiles which included the three-charge START-1 HEAT round — where the two outermost charges are precursor charges while the main charge resides in the middle of the round. The standard 3OF19 high explosive/fragmentation round has an explosive fill of 3.15 kilogram.

External armament consists of a 12.7mm NSVT heavy machine gun, code named Utes (Rock). Unlike the T-64, the machine gun mounting on the T-72M1 is entirely manual, not remote controlled. The machine gun is aimed using a K-10 reflex sight mounted in a small protective container above and to the right of the gun. Two additional boxes of 12.7mm machine gun ammunition are stowed externally on the turret side. In the front of the tank are twelve System 902B Tucha 81mm smoke dischargers. When fired together, this creates a smoke screen 300 by 300 meters lasting about two minutes. The T-72M1 also has the normal engine-mounted TDA smoke

T-72M1 V-46 diesel engine.

generating system. Crew protection is limited to a single AKS-74U or similar assault rifle, as well as grenades. Soviet tank crews are not regularly issued pistols.

The T-72M1 is provided with the usual PAZ nuclear protection system. This system detects the radiation wave from a nuclear blast and automatically carries out actions to minimize crew injury from the subsequent shock wave. The interior of T-72M1's crew compartment is protected with a layer of resin-impregnated lead antiradiation lining. This is found even on export tanks, for example, the Iraqi T-72M1s.

The driver sits centrally in the hull front, with a single large periscope mounted in the glacis plate, and two smaller periscopes mounted in the hatch over his head. This station is extremely cramped. There is a small belly escape hatch behind the seat, but it would take a very lithe gymnast to use it in combat. The driver's controls are traditional braking levers rather than the steering yokes found in most contemporary Western tanks. On either side of the driver are two large fuel cells which also have ammunition cavities.

The T-72M1 is fitted with a passive image-intensification night vision system for the gunner, fitted in the TPD-K1 system. The commander often is issued with an image intensification night sight, typically the PNV-57 biocular face-mask type. He uses these outside the tank. In the event of inadequate light, it depends on an active infrared system for night combat, based around the L-2AGM infrared searchlight on the turret. The driver can use a similar system connected to the driving lights, and the commander is provided with his own small infrared searchlight which is independent of the main gun Luna searchlight. This night fighting system is similar to those used by NATO in the 1970s. However, since the early 1980s, thermal imaging sights have been introduced on tanks like the M60A3, M1 and M1A1 Abrams, Leopard-2, Challenger, and AMX-30B2. Thermal sights have only recently been introduced on the latest models of the T-80 tank. The lack of modern night fighting sights is one of the main tactical drawbacks of Russian tanks compared with contemporary Western tanks.

The T-72M1 is powered by the V-46-6 multi-fuel, 12-cylinder supercharged diesel engine, developed by L.A. Vaisburd's design team in Chelyabinsk. By now, its power has been raised to 780 horsepower from the original output of 500 horsepower on the T-34 Model 1940. (The latest versions of the T-72 use yet another evolution, the V-84, with an output of 840 horsepower.) The V-46 is mounted in the traditional transverse fashion found on Soviet tanks since the T-44. It is located immediately behind the firewall that separates the engine compartment from the fighting compartment. Behind it is a large radiator for cooling the engine, with the vehicle transmission located under that. At the rear of the engine compartment is a large, circular fan which draws in air for the engine. The mechanical transmission has two planetary gearboxes with friction clutches and hydraulic steering. The T-72's transmission has a bad reputation in some armies because of the difficulty it presents when trying to recover a damaged tank.

In comparing the T-72 with contemporary Western tanks, it is important to remember how different the tanks

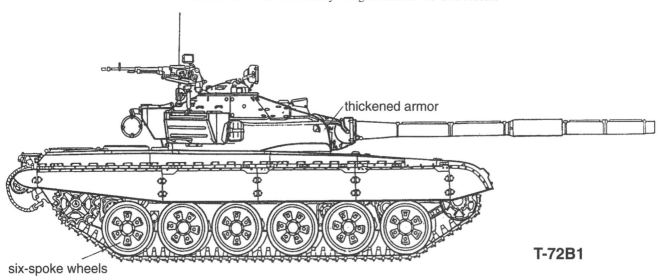

T-72B1

six-spoke wheels

are in terms of size and cost. The T-72 was designed as an economical way to replace the hordes of T-54, T-55, and T-62 tanks produced in the 1950s and 1960s. The designers accepted the many compromises in human engineering, firepower, night fighting and durability in order to stay within the limited goals of the program. The T-72 was not expected to challenge the new generation of NATO tanks; the more expensive and sophisticated T-80 was given at assignment. Forward deployed elements of the Soviet Army in Germany were equipped with the T-64B and T-80, not the T-72. By way of comparison, the T-72S has an export price of only $1.2 million; the T-80U, $2 million; the M1A1 Abrams, $3 million.

The T-72B Tank

By the mid-1970s, the Leningrad Kirov Plant (LKZ) tank design bureau (now under Nikolai Popov after Zhozef Kotin's retirement) was completing work on a

T-72B standard tank.

turbine powered challenger to the Kharkov T-64 as the Obiekt 219. This was accepted for production as the T-80 tank in 1976; the Kharkov bureau's improved T-64B appeared at this time as well. The T-80 was soon followed by an improved type with additional turret armor, the T-80B in 1978. The Uralvagon KB undertook a separate effort to bring the T-72 up to T-80B standards, at least in regards to armor protection. A new armor package was added to the turret front of this design, this early prototype was designated Obiekt 182. The new armor raised the vehicle weight by three tons, mandating an uprated diesel engine, the V-84. This new version entered production in 1985 as the T-72B Olkha (Obiekt 184) and T-72B1. This version introduced many small changes into the T-72 family, such as the substitution of new roadwheels with six indentations for the earlier pattern with eight indentations. The initial production batches had the System 902A smoke dischargers mounted on the turret front like the T-72A, but they soon were shifted over to the turret sides for reasons that will become apparent.

The much thicker turret frontal armor of the T-72B amd T-72B1 led to the unofficial US Army nickname "Super Dolly Parton" when it was first seen in 1986. The turret armor on the T-72B was the thickest and most effective ever mounted on a Soviet tank, surpassing even the T-80B. The primary aim of the new armor package was to defeat anti-tank missiles; thus it was tailored to defend best against HEAT warheads. It is the equivalent of 520 millimeter thick when faced by APFSDS kinetic energy penetrators, but an impressive 950 millimeters thick when attacked by HEAT shaped charge projectiles. These performance figures and characteristics are similar to those made possible by the "brow" armor developed by the NII Stali at this time for the T-55/T-62 upgrade program; it is possible that the T-72B's armor was based on the same configuration, but integral to the turret casting rather than as an appliqué.

Until the late 1980s, the T-72 tank was not capable of firing tube-launched antitank guided projectiles like the 9M112 Kobra fired by the T-64B and T-80. This was probably part of the effort to constrain the cost of the T-72. However, in the late 1980s, a new generation laser guided tank projectile family was developed to upgrade the T-55, T-62 as well as the T-72 and T-80. The basic element of this system is the 1K13 laser designator sight. This device is mounted over the gunner's station in lieu of the normal sight. The system selected for the T-72 is designated 9K120 Svir (the similar round on the T-80U is code-named Refleks). The projectile itself is called 9M119, and the entire ammunition round is called 3UBK14. It is a two-piece round and is stowed in the autoloader like any other type of ammunition. The new projectile considerably extends the range of the T-72, out to 5,000 meters. The Svir has an advanced warhead to

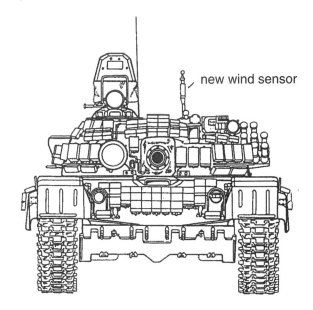

T-72S Shilden

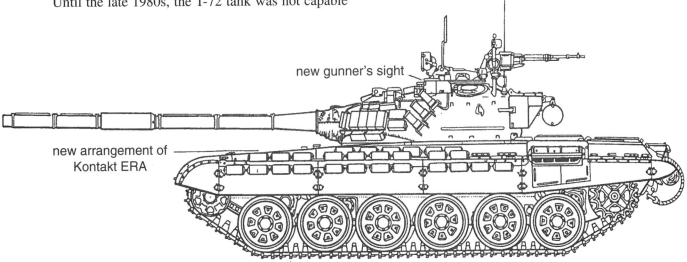

T-72S Shilden export tank.

Close-up of 1K13 gunner's sight and new wind sensor on T-72S.

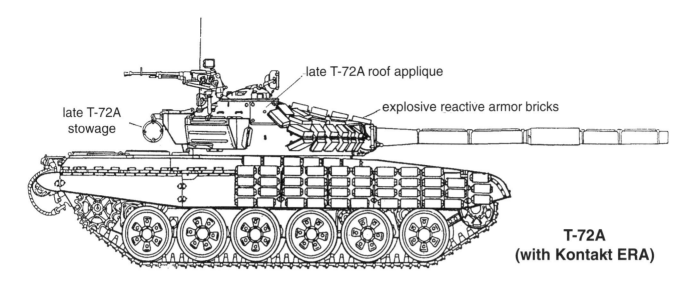

T-72A (with Kontakt ERA)

permit penetration of advanced tank armor; it is claimed to be capable of penetrating the M1A1 Abrams armor. The stated performance of the 4.2 kilograms shaped charge warhead is 700 millimeters penetration.

The laser guidance technique used with this projectile is of the beam-riding variety, not semiactive homing as is used with laser guided artillery projectiles. The 1K13 emits a laser beam that is optically rotated to form a narrow "funnel." The Svir missile automatically drops a small cover over the base of the projectile, which protects the optical port during the gun firing. The optical port senses the laser emissions. The laser signal is frequency modulated so that signal is varied from quadrant to quadrant. By monitoring the frequency, the missile's guidance system steers the Svir using small fins on the nose so that it remains in the center of the beam.

The propellant casing for the Svir is much smaller than the normal Zh40 because it was found that when a normal casing was used, it kicked up so much dust in front of the tank that it interfered with the laser signal. (On the earlier Kobra missile, the projectile had to be fired with the gun in elevation due to this problem.) The Svir fits into the ammunition cassette like any other round. Usually only four to six Svir are carried. This is due to the high cost of each round, about $45,000 on the export market. To put this in perspective, thirty rounds of Svir ammunition are equivalent in cost to an entire T-72 tank. As a result, the Svir is intended for specialized missions. Its main mission is to defend the tank against long-range anti-tank missile platforms. This would include ground-based missile-firing tank destroyers, as well as anti-tank helicopters fitted with anti-tank missiles.

The 9K120 Svir system was first mounted on the T-72B tank. The version without the Svir system is designated T-72B1. With the exception of the Svir fire controls, the T-72B and T-72B1 are otherwise identical.

The T-72S Shilden Export Tank

An export version was offered in 1987 to several countries, including the Warsaw Pact countries, as the T-72S Shilden (shield). Originally called T-72M1M (Obiekt 184), this is an export derivative of the T-72B. In the T-72S version, the tank is fitted with the 9K120 Svir missile system. There is also a T-72S1, which is probably without the Svir system. The T-72S has a different reactive armor package than the Soviet types with only 155 Kontakt ERA blocks instead of the usual 227. Some T-72S tanks have been fitted with a more advanced fire control system that includes a new wind sensor mounted on the rear roof of the tank.

T-72 Command Tanks

During the production of the T-72, command tank variants of most of these models were manufactured in parallel to the normal versions. The battalion and regimental command tanks are identified by the suffix "K" after their designation, such as T-72K, T-72AK, and T-72B1K, the "K" stands for Kommandniy. Regimental and battalion command tanks carry a ten-meter telescopic radio aerial in a small tube under the rear turret stowage bin. This mast antenna can be mounted only when the tank is stationary. The -K series of regimental and battalion command tanks are fitted with the R-130M radio system. The additional radio equipment as well as navigation equipment and an AB-1-P auxiliary electrical generator requires the deletion of six rounds of 125mm ammunition and 500 fewer rounds of 7.62mm machine gun ammunition.

A second family of command tanks also exists for company commanders, using the -K1 suffix such as T-72AK1. These tanks carry two R-123M or R-173 ra-

T-72B with Kontakt reactive armor.

dios as well as TNA-3 and GPK-59 navigation equipment. They do not carry the ten-meter command antenna and are externally similar to normal versions of the tank.

Reactive Armor T-72 Tanks

The armor protection of the T-72 was further enhanced by other improvements in defensive armor technology. In the 1982 Lebanon war, the Israeli Army first used Blazer explosive reactive armor (ERA) in combat. ERA such as Blazer is designed to enhance existing tank protection against HEAT shaped charge warheads especially those from antitank rockets and antitank missiles. The Syrians captured an M48 with Blazer that had been abandoned by an Israeli reserve tank unit following an attack by Mi-24 gunships. This was delivered to the main Soviet armor development research center, the NIIBT at Kubinka, for trials. The NII Stali (Scientific Research Institute for Steel) research center in Moscow had already been developing reactive armor in conjunction with NIIBT in Kubinka and VNII Transmash in Leningrad. It is called EDZ or DZ in Russian (elementy dinamcheskoi zashchity: dynamic protection elements) and explosive reactive armor (ERA) in the West.

Although the system had already been designed at the time of the 1982 Middle East war, there were no immediate plans to introduce it into service, the tank designers being unhappy that the added 1.5 metric tons of weight imposed only protected the tank against shaped charge warheads. However, they were overruled by the Soviet army, and on 15 January 1983, the "Act of the State Commission on the Adoption of Tanks with Explosive Reactive Armor" was signed. The first series manufactured armor, called Kontakt, was mounted on tanks in September 1983.[31]

In service, the Kontakt blocks are popularly called kosteki (dice). The Soviet Kontakt ERA differs from the Israeli Blazer type in a number of respects. When the HEAT warhead detonates against the Kontakt ERA brick, the hypervelocity jet of metal particles from the warhead penetrates the brick and detonates a thin sheet of high explosive. This explosion propels two steel plates located on either side of the explosive sheet. The outward facing plate is blown up into the penetrating jet, eroding the metal stream by forcing more and more metal plate into the path of the jet. At the same time, the other plate is propelled back towards the tank armor by the explosion, then rebounds off the tank's armor up into the remaining stream of the warhead jet, further eroding it. In this manner, the Kontakt ERA can substantially reduce

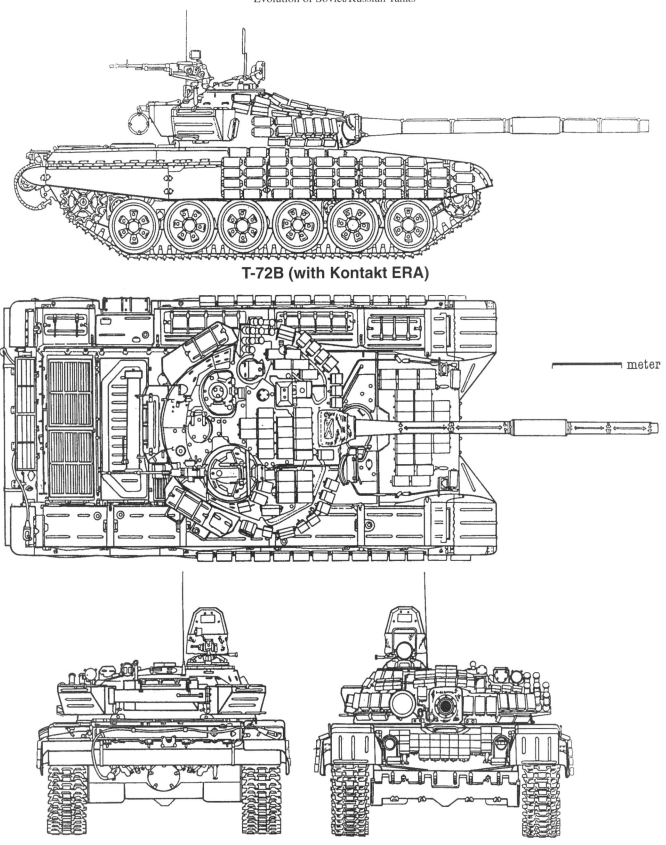

T-72B (with Kontakt ERA)

the penetration of the HEAT warhead.

Kontakt bricks were first spotted on Soviet tanks in Germany in December 1984. It was a very unpleasant surprise for NATO, which had come to rely very heavily on anti-tank missiles. The unexpected appearance of Kontakt ERA led to a crash program to develop tandem warheads and other technologies to defeat it.

The Kontakt ERA arrays began to be fitted on to the

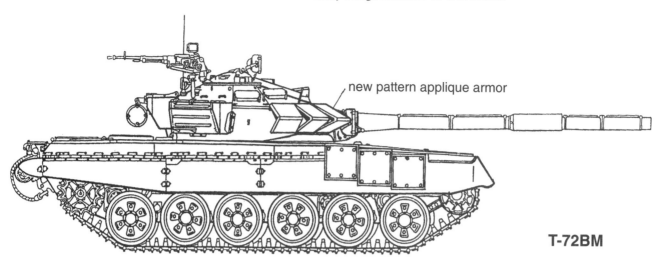

T-72BM

T-72A from 1987 to 1988 and later on the T-72B and T-72B1. The standard fit includes 227 bricks. Unlike the T-64B and T-80B tanks, which usually have the suffix "V" (vzryvnoi= explosive) added, such as T-64BV, to indicate Kontakt ERA the T-72 when fitted with Kontakt ERA, is not distinguished in this fashion. Controversy over Kontakt ERA reached a peak in August 1989 when a US Congressional delegation visiting the 24th "Iron" Guards Motor Rifle Division in Lvov was shown a T-72B1 with three stacks of Kontakt ERA instead of the usual one layer. It appears that the multilayer Kontakt ERA was a deliberate attempt at disinformation. The triple layer of Kontakt ERA is implausible given the dynamics of this generation of reactive armor.

The T-72BM (Obiekt 187) Tank

In the late 1980s, the NII Stali in Moscow, the advanced research institute of the steel industry, continued to develop more advanced versions of explosive reactive armor. The tank design bureaus had complained that the first generation ERA protected only against HEAT war-

T-72BM (Obiekt 184) standard tank.

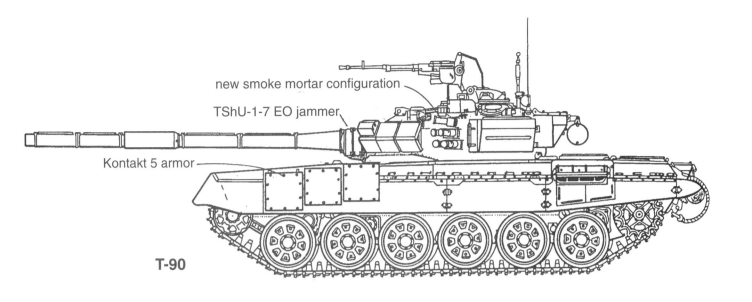

T-72BM (Obiekt 184) standard tank.

Close-up of Kontakt 5 reactive armor.

heads and was too heavy. The aim of the new program was to design appliqué armor that could degrade both kinetic energy penetrators and shaped charge penetrators.

This new Kontakt-5 applique armor was first mounted in 1985 on the T-80U. It increased the effective protection against APFSDS by about 200 millimeters, and against HEAT by 500 millimeters RHA equivalent. It was subsequently incorporated on an improved version of the T-72B designated Obiekt 184, and T-72BM when accepted into the Soviet army in 1989. This was the last version of the T-72 in production at the time of the Soviet Union's break-up in the early 1990s. The T-72BM is configured the same as the basic T-72B but has the new generation armor arrays. This consists of panels having a chevron cross-section on the turret front, large panels on the glacis plate and three square panels on the hull side, as well as small panels on the roof. This version, like the T-72B, is generally fitted with the 1K13 sight and so can fire the Svir guided projectile.

The T-90 (Obiekt 188) Tank

The T-90 is ostensibly the current choice of the Russian Army until a new generation tank becomes available in the next decade. Although not called a variant of the T-72, it is in fact a modestly upgraded version of the T-72B. The T-90 name was selected more for marketing than as an indication it is a new tank type.

The Russian Army decided to standardize on the T-90 tank in 1996 as a result of the Chechen war and the catastrophic decline of the Russian defense budget. This new tank offers little advantage beyond the tanks entering service in the final years of the Soviet Army such as the T-80U, and the decision to select the T-90 remains controversial. This action is a recognition of the limited potential of the current Russian defense budget to sustain the procurement of a more sophisticated and expensive new design. At the moment, the Russian arms export agency Rosvooruzhenie is focusing on the sale abroad of the less sophisticated T-72S Shilden tank and the more sophisticated T-80UK tank. Through 1996, the export policy on T-90 sales was that they would be permitted though not necessarily encouraged. It now appears from the IDEX-97 show that this policy has changed, and that T-90 is becoming a staple of the export business, much like T-80U.

A long-delayed next-generation tank is now in development, but production is unlikely both for technical and financial reasons over the next decade. As a result, the T-90 and its evolved versions are likely to be the most modern tanks in Russian service into the next century.

Origins of the T-90

The T-90 is not a new tank as its name would imply,

T-90 (Obiekt 188) standard tank.

but rather an upgraded version of the widely produced T-72 tank. In the final years of the Soviet Army, two tanks were still in procurement, the T-80U developed by the Spetsmash design bureau in Leningrad, and the T-72B developed by the "Vagonka" design bureau at the Uralvagon plant in Nizhni-Tagil. The T-80U was the more sophisticated of the two, with a superior fire control system and a more expensive turbine powerplant. This was evident in their export prices, the T-80U being offered for about $2 million, and the T-72 for about $1.2 million. By way of comparison, Russian assessments concluded that the T-80B was eleven to sixteen percent faster to acquire targets than the T-72B; it was twenty to forty-five percent more accurate in target engagements; and it had a thirty-seven percent better horsepower-to-weight ratio. There was considerable rivalry between the two design bureaus, with both offering upgraded versions for future Soviet Army requirements. Indeed, the T-72 and T-80 designs were nearly contemporaneous attempts to replace the T-64 tank, in one case with a focus on economy and in the other case with an emphasis on sophistication.

The imposition of the "defense sufficiency" doctrine in the late Gorbachev years, followed by the collapse of the USSR in 1991 had catastrophic effects on the Russian tank industry. During the 1980s, the Soviet Union had five tank plants, and by 1991 only three were active as production plummeted from 3,500 in 1987 to 1,000 in 1991. T-80U production in St. Petersburg at the Leningrad Kirov Plant ceased in 1990. The Kharkov tank plant in Ukraine, which in 1991 produced 800 T-80UDs, was cut off from the Russian Federation. The T-80 remained in production only at the "October Revolution" Transmash Plant in Omsk. Even at Omsk, production was slashed.

T-72 production continued only at the Ural Railcar Plant (Uralvagonzavod) in Nizhni-Tagil; T-72 production at the Chelyabinsk Tractor Plant (ChTZ) had ceased some years before.

In 1992, the Russian defense ministry made it clear that it could no longer afford to buy two main battle tanks simultaneously. Russian officials stated that they wished to cut production down to a single type, either the T-72 or the T-80. However, selecting one or the other tank meant that economic catastrophe would befall the losing city. Therefore, Russian officials continued to order both types in small amounts.

In 1992, the Russian Army ordered only twenty tanks, five T-80U tanks from Omsk and fifteen from Nizhni-Tagil. Production levels at Omsk and Nizhni-Tagil in 1992-93 were well beyond the puny state orders, though only a pale shadow of the mid-1980s. This additional production was not ordered by the Russian Army, but undertaken simply to keep the plants from closing. There was the hope that large export orders would emerge to save the plants and sop up this surplus. In 1992 the Omsk plant built 160 T-80U tanks even though only five were ordered by the Russian Army. Yet the anticipated export orders did not materialize.

As a result, Nizhni-Tagil has about 350 T-72S and T-90 tanks in its factory yards, and Omsk had 150 to 200 T-80U tanks which the Russian Army cannot afford to purchase and for which no export orders have been received. Some of these T-80U tanks were exported to Cyprus and South Korea in 1996. Lack of pay at the Nizhni-Tagil plant led to strikes in July 1995 during which the workers seized several of the idle tanks and drove them through the city to protest.

The T-90 Emerges

In 1988, Nizhni-Tagil upgraded the existing T-72B to the new T-72BM (Obiekt 187) configuration. The most significant change on this version was the substitution of the third generation Kontakt-5 appliqué armor, developed by NII Stali, for the earlier first generation Kontakt explosive reactive armor. NII Stali is Russia's primary research institute for armor development, and its primary developer of reactive armor. The Kontakt-5 had already been deployed on the rival T-80U, and the substitution of the this armor on the T-72B gave it equal or superior protection to the T-80U. Kontakt-5 offers protection against both HEAT and APFSDS, while the initial generations of Kontakt were valuable only against HEAT ammunition. The NII Stali claims that Kontakt reactive armor adds the equivalent of 500 to 700 millimeters of steel armor against HEAT warheads but nothing against APFSDS; Kontakt-5 offers the same advantage against HEAT as basic Kontakt, but also a further 250 to 280 millimeters against APFSDS. The T-72BM was produced in limited numbers, and saw combat during the conflict in the Caucasus in the early 1990s. It can be easily mistaken for the T-80U or T-80UD because of its use of the distinctive Kontakt-5 reactive armor.

The T-72 has traditionally been fitted with less sophisticated fire controls than the T-80, since it was intended only for second-line Soviet units and for export. In order to make the T-72 more competitive with the T-80, the Vagonka design bureau under general designer V. Potkin decided to adapt the more sophisticated T-80U fire control system to the T-72 chassis. The result was the T-72BU (Obiekt 188). Before the type entered service, the Soviet Union had collapsed. Around 1992, it was decided to rename the T-72BU as the T-90. This was apparently initiated by then-Defense Minister Pavel Grachev and personally approved by President Boris Yeltsin. The new name distanced the T-90 from the T-72A, which had performed poorly in the recently concluded 1991 Gulf War. The T-90 entered low-rate production around 1994, and by September 1995 a total of 107 were manufactured. They were first deployed with a regiment of the 21st Motor Rifle Division in the Siberian Military District in 1995. Although the T-90 was publicly displayed on Russian Tanker's Day in 1993 to 1996 at the Kubinka proving ground, it was not publically displayed until March of 1997 in the UAE and again in September at the Russian show in Omsk.

The Chechen War

The Russian defense ministry planned to choose between the T-80U and the T-90 sometime in 1994 but this selection was delayed due to the accelerating budget problems. The Russian Ministry of Defense was under pressure from regional industrial leaders who were concerned that the rejection of their tank would lead to further lay-offs and economic hardship in either Nizhni-Tagil or Omsk where the T-72/T-90 and T-80 were still in production. Undoubtedly, the war in Chechnya helped to accelerate the decision. The T-80, especially the earlier models, had a very bad reputation for high fuel consumption and low engine life. These problems were exacerbated by the improvised nature of the Chechnya deployments. When finally committed to combat in Grozny, tank regiments with the T-80BV suffered appalling losses. These losses were mainly due to poor Russian tactics and poor crew training, not technical shortcomings of the vehicles. However, the loss of so many of Russia's best tanks to a modestly equipped foe perplexed and angered senior Russian military leaders such as Minister of Defense Pavel Grachev.

The T-90 benefited from the Chechen war for several reasons. Its use of a more economical diesel engine gave it an advantage over the T-80U and its expensive turbine engine. It was not smeared in the press for poor performance in Chechnya since it was not deployed there, and its new marketing name distanced it from the T-72s which were also lost in large numbers in Chechnya. In January 1996, Colonel General Aleksandr Galkin, chief of Main Armor Directorate of the Russian MoD confirmed that a decision was made to gradually move to the T-90 as the single production tank of the Russian Armed Forces. The key expression in this decision was the phrase to "gradually move." In other words, T-80U production at Omsk is likely to continue at a low rate in spite of the T-90 selection to prevent undue economic hardship in Omsk, with the T-80U production intended primarily for export.

The selection of the T-90 has not met universal approval. In a September 1996 interview, General Galkin labeled the T-90 decision "a mistake" and he acknowledged that he still considers the T-80U to be a superior tank. The T-90 is overweight and underpowered, especially when compared to the very nimble T-80U. As mentioned below, there are a number of steps being undertaken to correct this problem. However, the continued production of the T-80U alongside T-90 production is likely to leave the issue of Russian Army standardization unsettled.

The Russian Army estimates that it needs to procure 300 new tanks annually to modernize its inventory, but in 1995, funds were available for only sixty to eighty tanks. In 1996, only fifty-eight tanks were ordered, and funding for these was to come from the 1997 budget. There were plans to fully equip one military district with

T-90 Turret Armor Protection*

Armor Configuration	Against APFSDS	Against HEAT
Basic armor (millimeters)	530	520
Kontakt 5 appliqué (millimeters)	250 to 280	500 to 700
Total	**780 to 810**	**1,020 to 1,220**

(armor thickness in millimeters of equivalent RHA)

the T-90 (probably in the Far East) in the 1996-97 budget. Besides production for Russian Army requirements, the Russian government projected that it would export 2,030 new production T-72 and T-80 tanks in 1994-97, which proved very optimistic.

T-90 Design

There has been considerable confusion over the T-90. It is almost identical in external appearance to the T-72BM, which shares the Kontakt-5 appliqué armor. Indeed, some Russian publications have mistakenly labeled photos of the T-72BM as the T-90. The major external distinguishing feature between the two types is the use of the Shtora tank protection system on the T-90 although there are other improvements as are detailed below.

Although it has been widely reported that the T-90 is fitted with a T-80U turret, this is clearly not the case. The T-90 is fitted with an improved derivative of the T-72BM turret. This confusion stems from the fact that the T-90 uses essentially the same fire control and gun system of the T-80U. The reason for the retention of the T-72B turret was that it was the best protected of contemporary Soviet tank turrets. The T-72B uses a NII Stali version of Chobham armor consisting of a basic armor shell the equivalent of 380 millimeter of RHA, with a 435 millimeter insert of alternating layers of aluminum and plastics and a controlled deformation section. The Kontakt-5 appliqué offers a thirty-four to fifty-seven percent increase in protection at a very modest weight cost of about three metric tons.

The heart of the changes in the T-90 are found in the weapon system. The T-90 employs the 1A45T fire control system, a derivative of the 1A45 used in the T-80U. The fire control system appears to be an analog and digital hybrid. It includes the new 1V528-1 digital ballistic computer, an upgrade from the 1V528 on the T-80U. The question of the gunner's fire controls have not been entirely settled. On the T-90 at IDEX-97, the sighting arrangement consisted of the TPN-4 night sight and the 1G46 day sight/laser range finder. There was no evidence of the small thermal imager video screen display found in the T-80UK at the show. An option is a thermal imaging night sight, and several alternatives are available including the older TPN4-49-23 Buran-PA used on the initial T-80U or the newer Agava-2 or TO1-KO-1 thermal imaging night sight. The Agava-2 is the newest of the Russian thermal imaging sights for tanks, and is manufactured by NPO Zenit in Krasnogorsk near Moscow for both current production T-80U and the T-90. The Agava-2 is mounted in place of the Buran-PA periscopic sight found on the T-80U. Unlike most Western integrated tank fire control systems, the Russian system divides the gunner's fire control systems into two blocks, the periscopic gunner's sight immediately in front of the gunner's hatch, and the integrated sight with laser range-finder/missile laser guidance system on the forward part of the turret. When using the thermal imager, the tank commander is provided with a small video display which provides the same image as seen by the gunner; this was not present on the IDEX-97 T-90.

Fire control inputs for the T-90 fire control system include twelve items: (1) fore/aft wind sensor, (2) cross-wind sensor, (3) barometric pressure, (4) barrel stiffness/droop, (5) ambient temperature, (6) propellant temperature, (7) kind of round, (8) number of rounds through barrel, (9) range of shot, (10) speed of target, (11) day or night, and (12) corrective bore sighting measurements. New ammunition types have been added to the ballistic computer's EPROM chip. The fire control system has a read-out to display if the system is functioning correctly, if not, manual settings can be input to physical controls

Comparative Fire Control Systems: T-72, T-80, and T-90

System Components	T-72BM	T-80U	T-90
Fire control system	1A40	1A45	1A45T
Gun stabilization	2E42-2	2E42	2E42-4
Gunner's fire control system	1K13-49	1A42	1A43
Ballistic computer	1V528	1V528	1V528-1
Wind sensor	Cross-wind	DVE-BS	DVE-BS
Guided missile	Svir	Refleks	Refleks

to compensate for system errors. The fire control system is set before battle and many of the inputs are not dynamic but set into the system before engagement. The dynamic inputs are numbers (1) through (7) and (9) through (11) with the static inputs being (8) and (12).

The commander is provided with a PNK-4S day/night sight with image intensification for the night channel.

Armament System

The T-90 uses the 2A46M-1 (also known by its bureau designation of D-81TM) 125mm smoothbore gun which is the same as in the T-80U. This gun was developed by the Spetstekhnika design bureau in Ekaterinburg (Sverdlovsk), and manufactured at the Motovilikha artillery plant in neighboring Perm. Motovilikha claims that the new 2A46M1 tank gun offers twenty to twenty-five percent increased accuracy compared to earlier models of the 2A46. There is also a new 2A46M-2 version that has a replaceable chromium barrel liner, an attempt to improve the notoriously short barrel life of Russian tank gun tubes.

One of the most important improvements in the new gun is its ability to fire the new generation of 125mm ammunition, developed by NIMI (Mechanical Engineering Research Institute) in Moscow. The current APFSDS round is the 3VBM17 which includes the 3BM42 tungsten carbide projectile; it has a stated armor penetration (at a range of 1,000 meters at sixty degrees) of over 250 millimeters. The standard production HEAT round is currently the 3VBK16 with the 3BK18M projectile; it has a stated penetration of 260 millimeters and uses a conventional copper cone. There are two new HEAT rounds, the 3VBK17 (3BK21B projectile) which uses a depleted uranium cone; and the triple charge 3BK29 projectile which uses a small precursor charge to strip off any reactive armor through the use of its precursor charge. The 3BK29 projectile is stated to have a penetration of over 350 millimeters plus any reactive armor layer. The latest high-explosive fragmentation round to enter widespread service is the 3VOF36 with the 3OF26 projectile.

The fire control system allows the T-90 to employ the 9K119 Refleks guided projectile; the earlier T-72 variants had been fitted with the related 9K120 Svir which has less range (four verses five kilometers) due to the laser fire control. Both use the same missile, the 9M119 (AT-11 Sniper). The Refleks is used in conjunction with the 9S515 missile control system, part of the T-90's 1A45T fire control system. The ammunition for the missile system is designated 3UBK14 and consists of the 9M119 missile, and the 9Kh949 reduced charge propellant casing with a spacer plug which seats the missile properly into the tank's gun. The 3UBK14 ammunition fits into the normal autoloader on the tank, like any other round of 125 mm ammunition. After launch, two sets of fins pop out, one for stability and the other for steering. The body of the 9M119 projectile contains an advanced 4.2 kilogram shaped-charge warhead with a penetration-to-diameter ratio of about 7:1, and a penetration advertised as 650 to 700 millimeters. The Refleks has undergone continual improvement.

After being fired, the Refleks missile drops a small cover over the tail of the missile which protects the rear-war pointing optical window. The T-90's fire control system's laser emitter creates a laser "funnel" with the missile riding in the center. The frequency of the beam is modulated in different sectors around the projected funnel so that if the missile deviates from the center, the guidance system onboard the missile interprets the signal and makes flight corrections to move it back into the center of the beam. The guidance system uses a timer so that the laser "funnel" is periodically altered in diameter so that, to the missile, it retains a near constant diameter. The Refleks is stated to have an eighty per cent probability of hit at 5,000 meters. The list price per round is $40,000 which limits the supply of the missile to an average of four per tank.

The Refleks is evidence of a different approach in Russian tank fire control design, opting for a lower cost and less sophisticated fire control and stabilization system adequate to 2,000 meters with conventional ammunition, supplemented by a small number of precision guided munitions for engagements over 4,000 meters. In contrast, most European and American designs of the current generation have excellent accuracy out to 4,000 meters.

The latest version is the 3UBK20 round which uses the improved 9M119M missile. This missile is fitted with a tandem shaped charge to deal with reactive armor. At least two new 125mm guided projectiles are being fielded, the 9M124 Agona and the 9M128.

Another external difference of the T-90 and T-72BM is the use of a modified version of the T-80 commander's cupola on the T-90. There are two significant differences between this cupola and that on the normal T-72B. In the first place, it uses a remotely controlled NSVT Utes 12.7mm machine gun with the PZU-7 sight and 1ETs29 stabilized fire control system. In contrast, the T-72 uses an ordinary manually aimed machine gun. Secondly, the new commander's cupola is fitted with a substantially improved vision system with the PKN-4S sight.

Defensive System

Externally, the most obvious difference on the T-90 is the Shtora electro-optical countermeasures system. This

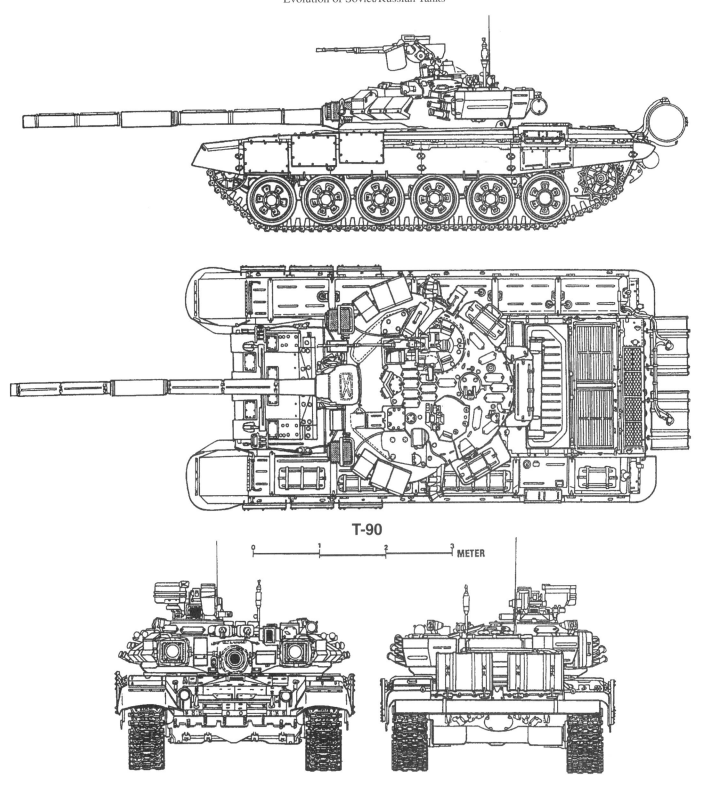

T-90

is an integrated defensive suite developed by VNII Transmash in St. Petersburg in cooperation with Elers-Elektron in Moscow which manufactures the system. The manufacturer claims that it reduces the hit probability of guided missiles such as TOW, Dragon, Maverick and Hellfire, as well as projectiles like Copperhead by four to five times, and reduces the hit probability of the HOT and Milan by three times. It also degrades engagements by enemy artillery or tank guns using laser rangefinders. The system has a secondary role in providing illumination for night vision systems. Several of the American missiles already have had counter-counter-measure systems introduced to minimize the effect of electro-optical dazzlers like that on the Shtora system prompted by the Iraqi use of dazzlers in the 1991 Gulf War.

The Shtora-1 system has four main elements: an

The turret top of a T-90 showing the roof armor protection and the forward-looking laser detectors.

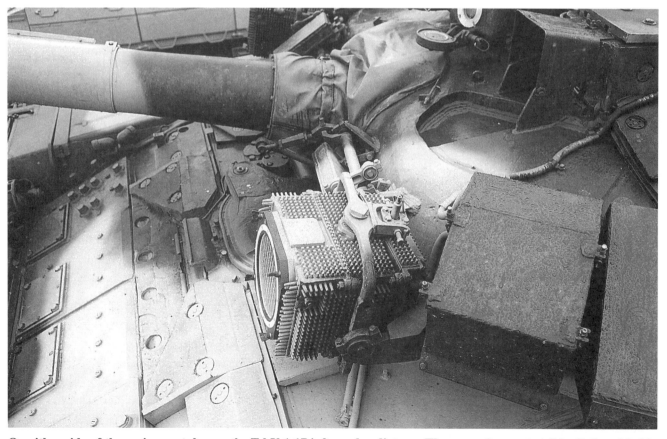
On either side of the main gun tube are the TshU-1-17 infra-red radiators. These are the most visibly distinguishable feature of the Shtora-1 countermeasure system.

The right side of the turret showing the remote-controlled NSVT Utes 12.7mm machine gun, the smoke dischargers, and the wind sensor just to the right of the radio antenna.

A close view of the turret Kontakt-5 armor.

T-90 Technical Data

Production plant	Uralvagonzavod, Nizhni-Tagil
Combat weight (metric tons)	50
Crew	3
Length (hull)	6.86
Length (overall)	9.53
Width (meters)	3.37
Height (meters)	2.23
Engine	V-84MS
Engine power (horsepower/kilowatt)	840 / 618
Fuel (integral) liters	1,000
Fuel (with external) liters	1,600
Maximum range/integral fuel (kilometers)	500
Maximum range/with external fuel (kilometers)	650
Ground pressure (kilogram per centimeter squared)	0.87
Fire control systems	1A45T
Gun stabilization	2E42-4 (electromechanical in azimuth, electro-hydraulic in elevation)
Gunner's sight	1A43
Gunner's thermal sight	TPN4-49-23 Buran-PA or Agava-2
Commander's day/night sight	PKN-4S
Ballistic computer	1V528-1 digital computer
Wind sensor	DVE-BS
Gun	2A46M-1
Main gun ammunition	43 rounds (22 in autoloader)
Guided projectile	9M119 Refleks (AT-11 Sniper)
Stowed guided projectiles	4
Secondary armament	12.7mm NSVT Utes machine gun
Co-Axial machine gun	7.62mm PKT machine gun
MG ammunition (12.7mm)	300
Radio	R-163-504 or R-173

electro-optical dazzler, laser warning detectors, anti-laser smoke grenade launchers, and system controls. There are two electro-optical interference emitters (dazzlers) fitted on both sides of the gun tube. This system functions like the US Army AN/VLQ-6 or VLQ-8, and is intended to bluff the missile tracker of standard wire guided anti-tank missiles by creating a pair of false images that mimic the tracking beacon on the rear of the missile. The system operates in the 0.7 to 2.5 micron range. This system uses the TShU-1-17 infrared radiator manufactured by the NPO Zenit in Moscow. The TShU-1-7 protects a sector four degrees in elevation and twenty degrees in azimuth. The light intensity is twenty millicandela and it draws one kilowatt of power.

The second element of the system is a set of four laser illumination sensors, fitted on the turret, which warn the crew of laser designation. These have a -5 to +25 degrees field of view in elevation and 90 degrees in azimuth. Data from the laser warning sensors is fed into a new microprocessor in the tank which provides information to the tank commander. The system can operate in an automatic mode, dispensing smoke grenades to defeat laser rangefinders or designators. Alternately, it can operate in a semi-automatic mode, with the tank commander deciding whether or not to trigger the smoke dischargers. The 3D17 smoke grenades are an evolutionary upgrade of the 3D6 grenades first used with the Type 902 Tucha system. They take about three seconds to bloom, and last twenty seconds. They provide obscuration in the 0.4 to 14.0 micron band covering the typical bands for thermal imaging systems, and bloom fifty to eighty meters from the tank.

Propulsion

The T-90 is powered by the V-84MS multi-fuel diesel engine. This engine is an upgrade of the V-84-1 used on the T-72BM, but has the same power output, 840 horsepower (618 kilowatt) even though the T-90 weighs more than two tons more. To compensate for the heavier weight, the torsion bars have been upgraded with 45KhNGMFA-3 steel which permits greater operating stress. As a result, the T-90 is somewhat more sluggish than either the T-72BM or the T-80U. It has a horsepower-to-weight ratio of 18.0 compared to 24.0 horsepower per ton on the M1A1 Abrams tank. The Chelyabinsk diesel engine plant has developed two alternative upgrades, the 950 horsepower V-92 and 1,100 V-96 diesels which may appear on future production variants to improve the T-90's sluggish performance. A T-90 prototype has also been built with a turbine engine. However, the turbine powered version raises the question of whether the Russian Army would have been better off staying with the T-80U which already has many of the bugs worked out of the turbine propulsion system.

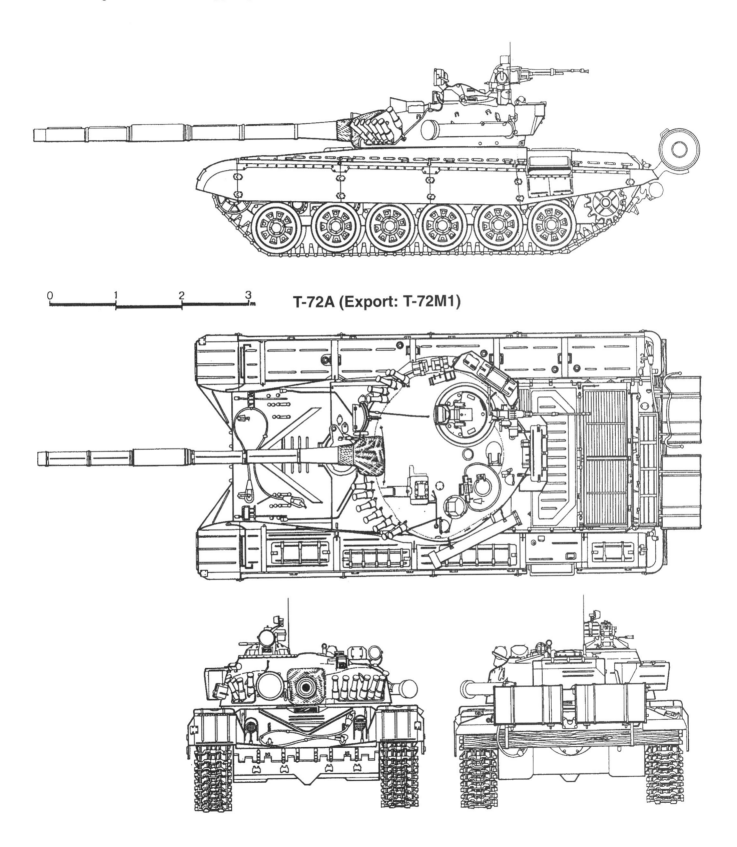

T-72A (Export: T-72M1)

T-90 Variants

Russian promotional material has described at least two versions of T-90 tank, the T-90E and the T-90S. Some descriptive literature states that the T-90S uses an analog, rather than a digital ballistic computer and the older R-173 radio. It is possible that this simply represents production variations from the earlier to the current configuration, and not a significantly different version. The T-90S displayed at IDEX-97 does appear to differ from vehicles described in Russian promotional literature in terms of some of its more expensive sub-systems such as the thermal imaging system.

The most recent production vehicles are fitted with an additional frame stowage bin on the rear of the turret. As mentioned earlier, work is underway on improved powerplants for the T-90 including both diesels and turbines.

Foreign T-72 Development

A secondary objective of the T-72 program was to develop a tank that would prove suitable for manufacture in the Warsaw Pact countries to replace the ubiquitous T-54A and T-55 tanks. In the mid-1970s, negotiations began with Poland and Czechoslovakia for production of the T-72. Both countries purchased small quantities of T-72 tanks, delivered in 1977, and concluded agreements to begin license manufacture of the T-72 in 1978.

Polish and Czechoslovak T-72 Production

The T-72M variant manufactured in Poland and Czechoslovakia had no direct equivalent in the Soviet Army. The T-72M turret was fitted with the new TPD-K1 laser rangefinder characteristic of the T-72A, but the turret armor was the thinner initial type of the basic T-72. Most T-72M used the "gill" flip-out side armor. Like the basic Soviet T-72, it is code-named Obiekt 172M in its manuals. Although this variant was designated T-72M in Poland and Czechoslovakia, it is also sometimes called T-72G in its Middle East export form for reasons that are not clear. Production in Poland was undertaken at the Bumar-Labedy plant in Gliwice which formerly produced the T-54A and T-55, while Czechoslovak production was undertaken at the ZTS (Czech = Zavod Turcanske Strojarne, Slovak = Zavody Tazkeho Strojastva) plant in Martin, Slovakia which previously manufactured the T-54A, T-55, and T-62.

During the late 1980s, many Polish and Czechoslovak T-72Ms, as well as those of their clients, underwent a gradual modernization program. Improvements from

T-72M standard tank (Polish construction).

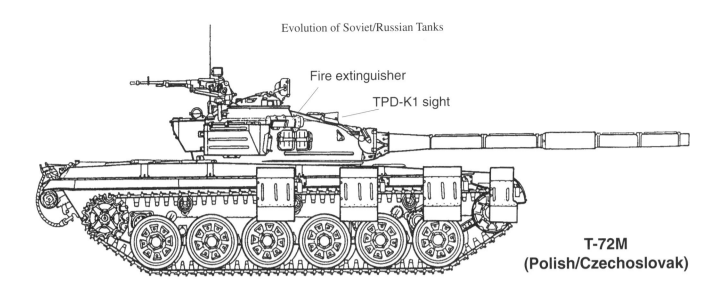

T-72M (Polish/Czechoslovak)

the Soviet T-72A program were gradually phased in including the substitution of full skirts for the gill armor, fitting the System 902B Tucha smoke grenade launchers and the addition of the seventeen millimeters applique armor panel on the glacis to boost protection up to T-72A standards.

In the mid-1980s, both plants converted to production of an equivalent of the Soviet T-72A (Obiekt 172M-1), designated T-72M1. (Soviet T-72A tanks that are exported are also called T-72M1.) This version had the improved turret armor and other improvements of the full T-72A/Obiekt 174 design. By 1991, Czechoslovakia had 897 T-72, T-72M, and T-72M1 in service and Poland had 757, nearly all locally produced. Poland produced 1,610 T-72s of all models through April 1993, of which about 900 were exported. Poland and Czechoslovakia sold T-72M and T-72M1 to most of the other Warsaw Pact armies so that by 1991 they had the following totals of T-72M and T-72M1: East Germany (549); Hungary (138), and Bulgaria (334).[32] Czechoslovakia and Poland produced about 1,700 T-72 for export to the developing world in the 1980s, with major clients including Syria, Libya, India, Iran, and Iraq. For example, Iraq was a major client for Polish T-72M and T-72M1, tanks and many of the tanks encountered in Operation Desert Storm were of Polish manufacture. Czechoslovak T-72M1 tanks were also apparently provided to the USSR.

Polish PT-91 Twardy

Polish T-72s have not been equipped with Soviet-pattern ERA. As a result in 1986, a design team under Adam Wisniewski at the WITU (Wojskowy Institut Technologie Uzbrojeniej = Military Institute of Armament Technology) in Zielonka developed a protective package designated ERAWA-1 and ERAWA-2.[33] The difference between the two versions is that the latter type uses a double layer of tiles. This system is noticeably thinner and lighter weight than earlier first-generation reactive armor packages, and may work in a modified fashion. The ERAWA Armor was also developed to reduce the radar signature of the tank against common battlefield surveillance radars, apparently through the use of a radar-absorbing material (RAM) on the surface. This armor can be retrofitted to older T-72s, but its main application appears to be oriented towards an upgraded

Polish PT-91 Twardy.

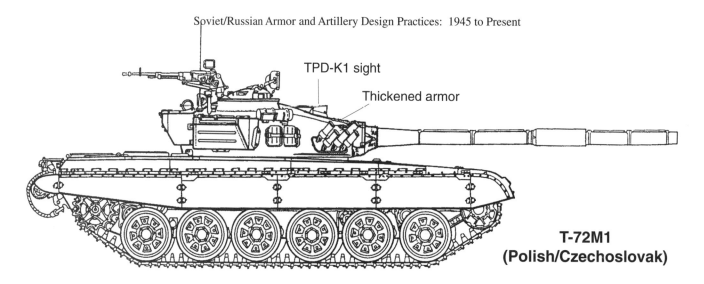

T-72M1 (Polish/Czechoslovak)

T-72 called the PT-91 (Polski Tank-91). On the Twardy, there are 394 tiles (compared with 227 Kontakt bricks on Russian T-72s). These include 108 on the turret, 118 on the hull and eighty-four on either hull side, totaling nine square meters of protected area.

The PT-91 Twardy (Hard) is a Polish attempt to field an equivalent of the T-80 at a lower cost. Poland acquired small numbers of T-80 tanks in the late 1980s and began negotiating license production rights to shift production from the T-72M1 to the T-80. This never transpired; instead Poland has continued to develop improved versions of the T-72M1. The PCE (Przemyslowe Centrum Optyki=Industrial Optics Center) in Warsaw developed the Drawa fire control system upgrade in cooperation with the IEK WAT (Electronic Systems Institute of the Military Technology Academy). This includes a new wind sensor, a new active/passive PCN-A gunner's night aiming sight, a new commander's POD-72 night sight, a new driver's PNK-72 Radomska night vision periscope, improved ballistic computer, and other features. The system gives the PT-91 an effective night range of 1,200 meters and boosts its overall night firing capability by 2.3 to 2.7 times according to Polish sources.[34] The package can also incorporate a TS 32D thermal imaging night vision system in lieu of the usual image intensification system, although Poland is studying other options apparently including a joint Polish-Israeli effort. The 2E28 stabilization system has been upgraded to 2E28M standards by its manufacturer, PZL Hydral, and a PDSU-1 display was added in the commander's station. The driver also has a digital USDK-1 diagnostic display which provides data on the powerplant. The Drawa system usually includes the OBRA laser warning system as well. The OBRA system is linked to two sets of smoke dischargers, twelve Tellur anti-laser smoke dischargers, and twelve standard Russian 902A Tucha smoke dischargers.

The normal V-46-6 engine, produced at the PZL Wola plant, has been modernized as the S-12U, boosting its power by ten percent to compensate for the increased weight of the tank. The Poles have also been discussing the addition of the Refleks guided anti-tank projectile to the PT-91 during negotiations with the Russians. Some of these upgrades can be retrofitted in a modular fashion to older T-72 tanks.

Tiger Fire Control System

In 1994, the Poles planned to use the new Tiger fire control system had been developed for T-72.[35] The Tiger fire control system was developed in South Africa under the leadership of the LIW Division of Denel who designed the G-5/G-6 artillery systems and all the Rooikat turrets. The two main additions of the system are the GS-72T, where the GS stands for gunner's sight, and the CS-65N, where the CS stands for commander's sight with a panoramic observation capability. The GS-72T gunner's sight contains an integral LE-72 Nd:YAG laser rangefinder and a TS-72 thermal imaging sight operating in two bands. It operates in x1 for target acquisition and in x8 for targeting and aiming. The commander's CS-65N is a day/night sight (with image intensifiaction for night observation). Unlike the normal fixed TKN-3 commander's sight in the standard T-72 cupola, the system has a rotating head to provide 360-degree coverage.

Slovak T-72M2 Moderna (Antares)

In 1993, the Slovak ZTS plant at Dubnica nad Vahom entered into a cooperative agreement with the Belgian electro-optics firm SABCA to develop an upgraded fire control and electronics package for the T-72M. This project was originally called Antares, but in 1994, it was renamed Moderna; the Slovak/Belgian electronics portion is named Lyra. The gunner's periscopic TPN-1-49-23 sight is replaced by the Vega thermal imaging sight, which links to the existing TPD-K1 laser sight. The thermal imaging sight has an optical elbow in the commander's station as well. The Vega Plus fire control package includes both the new sight as well as a new digital ballistic computer and various input sensors to improve gun performance.

A new commander's station is fitted with a panoramic French SFIM VS 580 stabilized sight with its own passive night channel. The 2E28M gun stabilization system is replaced by a new system to give the tank true fire on the move capability. The tank is also fitted with an intergated laser warning system, which has an automatic mode which can trigger countermeasures. A turret management system is fitted which provides readouts from the new systems, including the laser warning system. A new BAM frequency-hopping radio system with encryption is provided, incorporating an intercom with active noise reduction. There has also been a proposal to include a Sutrak air conditioner in the tank.

The original illustrations of the T-72 Antares released in 1993 showed the tank fitted with applique armor similar to the Kontakt-5 on the Russian T-80U or T-72BM/T-90. It is not clear from existing literature what the source of this armor package actually is; it may have simply been a paper proposal. In 1994 at the Brno IDET exhibition, the T-72M2 Moderna showed up instead with the Czech Dyna-72 reactive armor fitted. Another addition to the vehicle was a pair of Oerlikon-Contraves 20mm KAA cannon, one on either side of the rear of the turret, to provide air defense.

The T-72M2 Moderna was substantially reconfigured in 1996, substituting a single 2A42 30mm cannon mounted on the BMP-2 for the twin 20mm KAA-001 cannon, and being fitted with a different ERA array. In addition, the upgrade package includes a Slovak manufactured 2A46MS 125mm main gun. The new upgrade uses components from the Slovenian Fotona upgrade package including the LIRD-1 laser warner, and the Fotona TIGS (Thermal Imaging Gun Sight). This new configuration was first displayed at the IDET-96 exhibition in Trecin, Slovakia.

Czech Republic T-72M2

In 1993, the Army of the Czech Republic (ACR) announced its intention to modernize its T-72s under a program dubbed T-72M2. There were at least two efforts taking place, the T-72M2 effort by the ACR's own institutes and an industrial effort led by the RDP Group. Czech industry officials reached a decision on the program in 1996 and produced the first prototype in 1997.

The program has three main objectives: improve the tank firepower and night engagement capability, improve the armor protection, and improve the tank's mobility. The ACR program included an effort to update the fire control system to permit nightitme target identification at three kilometers and to extend the daytime engagement range to two kilometers. An effort was also underway to improve tank gun performance by adopting a new APFSDS round with performance comparable to current NATO rounds and to increase the performance of the HEAT projectile by sixty to seventy percent.

The new protective system could include the new Dyna-72 reactive armor package, which has already been experimentally fitted to several tanks by VOP 025 (Vojensky opravarensky podnik: Army Overhaul Facility) in Novy Jicin. A T-72M1 with the Dyna-72 ERA was displayed at the IDET-93 arms show in Brno; at the 1994 show, it was displayed on the Slovak T-72M2 Moderna upgrade. The package adds 1.5 metric tons to the weight of the tank. The requirement also called for the use of a laser warning system (SDIO= system detekce a indikace ozareni laserovymi). The added weight of the new systems was expected to bring the combat weight up to forty-six tons, leading to a requirement to increase the engine's horsepower and to improve the suspension.

The Czechs discussed the use of the SAVAN 15T gunner's sight for the program with the French SAGEM company which is already providing a version of the sight for the French AMX LeClerc tank. The majority of the T-72M2 program is being undertaken by Czech Army research institutes.

In a related effort, the Czech armaments consortium RDP is planned its own T-72 upgrade program called CZ-2000, and now referred to as T-72CZ. RDP has reached tenative agreements on this program with several French firms headed by SOFMA (Societe Francaise de Materiels d'Armemant). Included in the group were GIAT Industries, Thomson-CSF and SAGEM. SOFMA offered a new engine and transmission package, and SAGEM and Thomson-CSF offered fire control, night vision, and other electronics upgrades. The cooperative agreements stem from an April 1992 agreement between France and the Czech Republic over efforts to restructure to the Czech army and defense industries.[36]

In August 1995, the Czech government announced that the VOP-25 military repair depot in Novy Jicin would head the upgrade program, valued at six billion koruna ($220 million), for the modernization of the first 250 tanks. This upgrade is referred to as the T-72CZ or T-72M4. A total of nine companies were selected for the effort including: Officine Galileo (fire control); EL-Ops (Warsaw subsidiary of Israeli firm; laser warning receiver); NIMDA of Israel (powerpacks); Allison Gas Turbine (turbine APU); Racal Radio of the UK (radios); Deugra Ges. Fuer Brandschutzsystem in the FRG (fire suppression); Letecke Pristroje (navigation system); Meopta (night vision); and Synthesia (new ammunition).

At the IDEX-97 in the UAE, the first discussion of a

new entry into the Czech T-72 upgrade effort was announced, the T-72MP, being headed by the Czech industrial firm of PSP Bohemia, headquartered in Prague. This effort represents a more modest upgrade than the current T-72CZ (T-72M4) program, and would be suitable for the second group of Czech Army tank upgrades (referred to as T-72M3), or for other export clients. The effort is a tri-national affair involving PSP Bohemia on the Czech side, SAGEM from France, and the Morozov Design Bureau and associated Malyshev Plant from Kharkov from Ukraine. This is a unique program, and the first known instance of Ukrainian cooperation in tank upgrade programs.

Romanian T-72 Production

Romania purchased thirty T-72 from the USSR in 1979 and had plans underway in 1984 to license manufacture a version of the T-72 in Romania, powered by a French diesel engine. Designated TR-125, the Romanian T-72 variant had other changes as well, including seven roadwheels instead of the six on the normal T-72. Apparently, the Romanian program has not progressed beyond prototypes.

Yugoslav T-72 Production

The Polish/Czechoslovak development pattern was followed, with some differences, in Yugoslavia. The Yugoslav government purchased license production rights in 1979 and the first prototypes were completed in 1982. There was some delay in series production due to the decision to incorporate an indigenous fire control system into the vehicle. This included the DNNS-2 day/night gunner's sight, a wind sensor mast located on the turret roof above and behind the main gun, and a computerized fire control system. It was designated M-84 when it entered production at the end of 1984. In the late 1980s, the improved Obiekt 172M-1 turret was incorporated into the design, providing local fire control improvements and an uprated engine, resulting in the M-84A.

A total of 502 M-84 and M-84A tanks had been produced for the Yugoslav National Army by the time of the 1991 civil war. Additional production included at least one sample for the USSR. Kuwait ordered 200 M-84s in 1989, including fifteen command tanks and 15 ARVs. These vehicles had a variety of small improvements to make them more suitable for use in the desert, and they were designated M-84AB. About twelve had been delivered before the Iraqi invasion in 1990, and during the war, an additional shipment was made to the Kuwaiti 35th Fatah Brigade. A substantially modernized T-72 derivative, called the M95 or V-2001, entered development for the SDPR (Federal Directorate of Supply and Procurement) in 1990. This program was delayed by the Yugoslav civil war and the ensuing economic embargo and may be canceled. The main M-84 plant (Djuro Djakovic in Slavonski Brod, Croatia) is attempting to re-establish production of this tank; the Serbs are attempting to restart production at the "14 Oktobar" plant in Krusevac.

Croatian Upgrade Programs

Croatia has embarked on two tank programs involving components of the M-84: (1) a fire control upgrade to the base M-84A called M-84A4 Sniper; and (2) a new follow-on tank program called Degman.

The new M-84A4 is a substantial upgrade to the electronics of the base M-84A. The new fire control system is called OMEGA-84. This fire control system includes the SCS-84 stabilized sighting device (a three channel sight with day/night/laser channels), the DBR-84 ballistic computer, a servoelectronic unit, sensors for elevations, a gyro unit, and new control panels for the com-

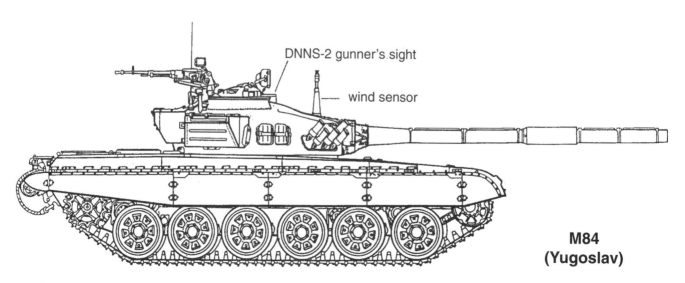

M84 (Yugoslav)

Romanian TR-125 (T-72 variant).

mander and gunner stations. Also included in the upgrade is an independent commander's day/night DNZF-2 second-generation image sighting system.

The Degman tank program is a new building program for a follow-on tank to the M-84A that will incorporate improvements in active/passive armor protection, mobility, and firepower. The Degman tank's base armor is greatly enhanced through the use of a multi-layer welded turret, new reactive armor panels, and a laser irradiation detector system (similar to the Russian Shtora system). In addition, according to Croatia promotional literature, work is underway on active protection armors, this may be a system similar in design to the Russian Drozd or Arena systems.

It is unclear given the break-up of much of the production base for the M-84 whether Croatia will be able to "go it alone" on their substantial upgrade to the M-84A or on a new-built production tank. However, given their recent wartime experiences, Croatia may feel it must develop an indigenous tank production capability and will earnestly strive to build one. It appears that the M-84A4 is a "stop-gap" measure allowing time for the development and eventual production of the indigenously designed Degman tank.

Indian T-72 Production

The first country outside Europe to begin producing the T-72 was India. In 1978, India began purchasing 500 T-72, T-72M and T-72M1 tanks directly from the USSR. At the same time, steps were taken to begin production of the T-72M1 at the Heavy Vehicles Factory at Avadi, totaling some 900 tanks. The T-72M1 is locally known as the Ajeya. Due to the slow pace of the indigenous

Yugoslav M-84 (Kuwaiti Army).

Croatian Degman

Chinese Type 90

Arjun next-generation tank program, in 1986 the Indian government approved a T-72 upgrade program, code-named Project Rhino. The main focus of the program is expected to be a fire control system upgrade, regarded by the Indians as the most serious defect of the T-72 design. This program has been seriously delayed by budget shortfalls, and at the time this study was prepared, the program had not progressed very far.

Chinese T-72 Efforts

China has been attempting to develop an equivalent to the T-72 for several years. This tank is called the Type 85 and was specifically designed to satisfy requirements of the Pakistani Army. A number of prototypes were displayed in Pakistan in 1992, and plans were underway to locally manufacture it under the name Khalid. This program was called into question in 1996 when it was announced that Pakistan was acquiring 300 T-80UD tanks from Ukraine. It was not clear if this decision was linked to a cancellation of the Khalid project, or if the T-80UDs are intended to provide a stopgap until the Khalid is available. There were reports in 1992 that China was negotiating with Russia for manufacturing rights to the T-72. North Korea is also reported to be working on a copy of the T-72. It is not certain if this is simply another offshoot of the Chinese Type 85, a license production from Russia, or an indigenous effort like many of China's most recent armored vehicles. China has developed a further evolution of the Type 85, called the Type 90.

Iraqi T-72M Assad Babil

Following the 1980-88 war with Iran, Iraq decided to begin license manufacture of the T-72M1 as the Assad Babil (Lion of Babylon). The first effort was directed towards manufacturing the 125mm gun tube. The service life of the tube was only 120 rounds, after which performance dropped markedly. The Iraqis frequently used tanks as mobile artillery, and barrel wear became a significant tactical problem. The Iraqis claimed to have assembled their first T-72M1s in 1989 from knock-down kits, but there is little evidence that any substantial number of vehicles were actually completed beyond prototypes. The Iraqis also began a modernization program. An electro-optical dazzler was provided by a foreign supplier, possibly China. This xenon strobe emits a beam that confuses the tracker of standard Western anti-tank missiles such as the TOW, Milan, and HOT. The tracker steers the missile by monitoring a flare at the rear of the missile; the dazzler mimics the flare and causes the missile to fly off course. Most of the T-72s and T-72M1s in Iraqi service were retrofitted with this device from 1990 to 1991.

Iraqi T-72M1 Assad Babil.

T-72 Variant Designations

Soviet	Warsaw Pact/Export	Command	STANAG (NATO)
T-72	T-72	T-72K	T-72
T-72	T-72M	T-72MK	SMT M1980
T-72A	T-72M1	T-72AK, T-72M1K	SMT M1980/1, M1981
T-72B1	T-72M1M	T-72B1K	SMT M1986
T-72B	T-72S	T-72BK	SMT M1988, FST-1
T-72BM			SMT M1990
T-90			SMT M1990

T-72 Comparative Data

	T-72	T-72A	T-72B/S	T-72BM
Prototype designation	Ob.172M	Ob.176	Ob.182/184	Ob. 187
Unloaded weight (metric tons)	38.6	38.9	41.9	41.9
Combat weight (metric tons)	41	41.5	44.5	44.5
Engine	V-46	V-46-6	V-84	V-84
Engine (horsepower)	780	780	840	840
Engine (kilowatts)	573	573	618	618
Fuel (integral/liters)	1,000	1,000	1,200	1,000
Fuel (with external/liters)	1,400	1,400	1,590	1,400
Maximum range/int. fuel (kilometers)	500	500	500	420
Max. range/with ext fuel (kilometers)	700	700	700	600
Gunner's sight	TPN1-49-23	TPN1-49	TPN1-49	1K13-49
Rangefinder	TPD2-49	TPD-K1	TPD-K1	TPD-K1
Rangefinder type	coincidence	laser	laser	laser
Main gun ammunition	39	44	46	45
Svir guided missile	no	no	yes	yes
Svir projectiles	0	0	4	4
MG ammunition (12.7mm)	300	300	300	300
MG ammunition (7.62mm)	2,000	2,000	2,000	1,000
Smoke dispensers	0	12	8	8
Gun stabilizer	2E28	2E28M		2E42
Turret armor versus HEAT (mm RHA)	500	560	950	?
Turret armor versus APFSDS (mm RHA)	410	500	520	?
Hull armor versus HEAT (mm RHA)	450	490	900	?
Hull armor versus APFSDS (mm RHA)	410	420	530	?
Radio	R-123M	R-173	R-173	R-173

T-72A Technical Data

Crew	3
Overall length (meters)	9.53
Hull length (meters)	6.86
Width (meters)	3.59
Height to turret roof (meters)	2.19
Ground clearance (meters)	0.49
Track contact (meters)	4.29
Track width (millimeters)	580
Maximum road speed (kilometers per hour)	60
Engine	V-46-6 12 cylinder, four-stroke, multi-fuel diesel, 780 horsepower (575 kW) @2,000 rpm
Fuel stowage (internal+external)	705 liters + 495 integral; +400 liters in two fuel drums
Fording (without preparation/with preparation)	1.2 meters / 5.0 meters
Slope (gradient/side slope)	30 degrees/25 degrees
Obstacle (vertical/trench)	0.85 meters/2.9 meters
Power to weight ratio (horsepower per ton)	19.8 (14 kW/T)
Ground pressure (kg/cm squared)	0.9
Main gun	2A46M (D-81TM) 125mm smoothbore
Rate of fire (autoloader)	8 rounds per minute (2 rounds per minute manual)
Gun stabilization	2E28M electro-hydraulic, two axis
Elevation	-6 to +14
Secondary armament	co-axial PKT 7.62mm machine gun
Anti-aircraft defense	12.7mm NSVT "Utes" machine gun
Smoke dischargers	Type 902A Tucha; 12 cover 300m^2 for 2 minutes
Gunner's night vision	TPN-1-49-23
Driver's day/night vision	TNPO-168 & TVNE-4B (active IR/passive II)
Commander's day/night vision	TNP-160 & TKN-3
Commander's searchlight	OU-3GKM with IR filter
Fire protection	Automatic freon system, 9 detectors
Crew self-defense	AK-74S assault rifle, 10 F-1 grenades
Unit price/with ammo and spares (1992 export)	$1,200,000m; $1,800,000

Turbine Tank: Leningrad's T-80 Tank

One of the most perplexing questions about post-war Soviet tank procurement was the decision in 1976 to begin production of the T-80 tank. The T-80 was yet another attempt to remedy flaws in the T-64 design; it substituted a gas-turbine engine for the diesel used in the T-64A. Yet instead of ending T-64 production and development in favor of the T-80, the Soviet Union simply added a third tank into production. This runs contrary to the popular Western stereotype of Soviet defense decision making which portrays the Soviet style as favoring standardization of a minimum of weapons types. The rationale for this decision is still not clear, though some suppositions will be presented here.

Development of the T-80 was undertaken by the design bureau of the Leningrad Kirov Plant (LKZ). The LKZ design bureau (called SKB-2 during the war) had been directed by General Zhozef Kotin but was shut off from further tank work by Khrushchev's decree in 1960 (ending heavy tank development) and directed to work in industrial tractor design. Some of the tank design staff went over to the VNII Transmash (All-Union Scientific Research Institute of the Transporation Machine Building Industry), also in Leningrad. The VNII Transmash, often nicknamed the "tank NII," suddenly became one of the two principal centers for advanced armored vehicle research (the other being the NIIBT at Kubinka). Following Brezhnev's coup d'etat against Khrushchev in 1964, the LKZ was allowed to reopen a tank design bureau, renamed the OKBT (Osoboye konsturktorskoye biuro-tankovoye: Special Tank Design Bureau), headed by Nikolai S. Popov.

Turbine Tank Propulsion

The Soviet Union was one of the pioneers of turbine power for tanks. This has not generally been appreciated in the West since most of the early programs have been classified until recently. As early as 1948 to 1949, the Turbine Group of the LKZ design bureau was commissioned by the Armored Force Directorate of the Soviet Army (GBTU) to undertake the development of a gas turbine for tanks. The program did not meet its requirements and was eventually rejected by the army. Again from 1955 to 1958, on the basis of a GBTU requirement, the LKZ design bureau began work on a new gas turbine for heavy tanks. A pair of gas turbine prototypes were mounted in the Obiekt 277 heavy tank (T-10 derivative) being developed by Zh. Kotin's team. This tank never reached the production stage, as in 1960, Khrushchev ordered an end to Soviet heavy tank production.

As mentioned previously in the section on Soviet heavy tank design, in 1956, the four tank design bureaus were ordered to begin work on turbine power plants for tanks. This program was considerably delayed by the lack of a suitable power plant. Automotive gas-turbine experiments began in 1958 by the NAMI (Nauchno-issledovatelskiy avtomobilniy i avtomotorniy institut: Scientific Research Automotive and Auto-Engine Institute) with the experimental NAMI-053 350 horsepower engine on a modified ZIL-127 bus.[37] The tank turbine program was revived in 1963 with the cooperation of the Leningrad Research-Production Association (LNPO) im. V. Klimov (formerly the State Aviation Plant (GAZ) Number 117) and the Glushenkov design bureau at the GAZ Number 166 imeni Baranov in Omsk.[38]

The Klimov design bureau was headed at the time by Sergey P. Isotov and had pioneered the use of gas-turbine engines on Soviet helicopters, notably the TV2-117A on the Mil Mi-8 (Hip) and later the Mil Mi-24 (Hind); the Glushenkov bureau had developed a 900 horsepower turbine, the GTD-3 for use on the Kamov Ka-25 (Hormone) naval helicopter. The Kharkov bureau adapted the two-shaft Glushenkov GTD-3TL to a T-64 chassis as the T-64T. In 1963, the Vagonka Design Bureau in Nizhni Tagil began similar efforts to adapt GTD-3T turbines to the Obiekt 167T tank (T-62 follow-on) and the Obiekt 150T (IT-1 tank destroyer). In addition, the Isakov design bureau at Chelyabinsk also mounted a turbine power plant in their Obiekt 775 low-profile missile-firing tank as the Obiekt 775T. The Klimov design bureau cooperated with the LKZ and VNII Transmash on an experimental turbine tank testbed, Obiekt 288, which used a chassis from the experimental Obiekt 287 Taifun missile tank destroyer. The tank testbed was powered by two GTD-350 gas turbine powerplants adapted from existing Mil Mi-2 (Hoplite) helicopter engines.

The early trials of these turbine-powered tanks was generally discouraging and led to an abandonment of any further efforts at three of the four plants. The conclusions of the design bureaus were summarized by the head of the Nizhni Tagil team in his memoirs. Although the turbine provided a significant increase in tank speed, this did not compensate for the increase in the per-kilometer fuel consumption as compared with diesel engines. This stands in contrast to the use of turbines in aircraft since the turbine is justified by the substantial increase in speed and power. In a turbine powered tank, the full speed potential of the turbine was impossible to realize due to the physical and dynamic loads on the crew from road travel. It is impossible to travel in a tank at a high-speed across a newly plowed field or on broken ground due to jolting. The turbine required a high degree of air purity and the air consumption in the turbine is significantly higher than in a diesel engine having the very same power. This re-

Obiekt 288 turbine test-bed.

Detail view of Obiekt 288 turbine engine deck.

Comparative Performance Data: T-64A, T-72A, and T-80

Type	T-64A	T-72A	T-80
Weight (metric tons)	38.5	41	42
Integral fuel (liters)*	815	1,200	1,840
Total fuel (liters)	1,145	1,600	2,240
Horsepower/weight (horsepower per ton)	18	19	24
Ground pressure (kilograms per cubic centimeter)	0.84	0.83	0.83
Maximum road speed (kilometers per hour)	60	65	70
Maximum road range, integral fuel (kilometers)	350	500	335
Maximum road range, total fuel (kilometers)	550	700	450
Fuel consumption (liters per kilometers)	2.08	2.4	5.49

*Integral fuel refers to fuel stowed on the tank not including disposable drums; it includes both the internal fuel tanks and the fuel panniers on the fenders.

quired a new air filtration system, which in the early Soviet prototypes was quite bulky.

The Soviet designers concluded that to guarantee the identical range for a tank with a turbine engine, it was necessary to carry 1.5 to 2 times more fuel than for a diesel engine, which would be unacceptable under combat conditions. Finally, the Soviet designers concluded that the cost of the gas turbine was somewhat higher than the cost of a diesel engine.

Although three of the four design bureaus were discouraged from any further work on turbine engines, the OKBT in Leningrad decided to continue their efforts with the Obiekt 288 testbed. This decision was probably affected by their lack of any other tank work since the cancellation of the heavy tanks. In addition, at this time the Soviet industrial ministries introduced the regional Sovnarkom system, which encouraged cooperative industrial programs within regions. The location of the OKBT tank design bureau and the Isotov design bureau in Leningrad (as well as the associated production plant at Kaluga) led to a natural combination to win local ministerial support. OKBT sponsored the development of a dedicated gas-turbine powerplant for tanks, the GTD-1000T (Gas turbine engine-1,000 horsepower for tanks) at the Klimov plant.

The new GTD-1000T engine was mounted on a modifed T-64 chassis, designated Obiekt 219. The OKBT apparently was as unhappy with the T-64 suspension as Nizhni Tagil, and so replaced it with a more conventional full hull-width torsion bar suspension with cast aluminum-alloy road-wheels. The prodigious fuel demands of the turbine required a significant redesign of the hull interior with no fewer than thirteen fuel cells including five external cells on the fenders. Leningrad, also unhappy with the Kharkov autoloader, adopted a new design that was being prepared for the T-64B with twenty-eight rounds in the magazine compared with thirty in the T-64A and twenty-two in the T-72A. Otherwise, the Obiekt 219 closely followed the T-64A design, having an identical fire control system. It was clearly designed as a premium tank like the T-64A, not an economy tank like the T-72. There is little information on the Obiekt 219's test program, but it was accepted for Soviet Army service in 1976, much earlier than has generally been appreciated in the West, and four years before series production of the M1 Abrams.

There has been virtually no unclassified discussion in the Russian defense press about the origins of the T-80 design requirements. While the Soviets were undoubtedly aware of US Army interest in turbine propulsion for their new tank (which would later emerge as the M1 Abrams), the evidence suggests that the OKBT was working in earnest on a turbine-powered tank well in advance of the start of the M1 program.

From the actual design of the T-80, it is apparent that neither advances in protection nor firepower superiority over the existing T-64 or T-72 were sought. Indeed, the actual design of the T-80 suggests that the only objective of the program was to increase the power. Given the very modest advance in actual road speed that this powerful new engine offered, it is possible that the Soviets were seeking some of the advantages also cited by turbine advocates in the United States, namely increased reliability and decreased maintenance. However, neither of these criteria have figured prominently in past Soviet design practice, and there is no evidence that these features were sought by the Soviet Army. Indeed, what is especially odd about this whole affair is that the Soviet Army would have sought a tank with a radical new power plant so shortly after its prolonged troubles with the T-64's far less revolutionary diesel engine. The Soviet Army had been forced into adopting the less sophisticated T-72 in 1972 to redeem the troublesome T-64; it is very hard to believe that few years later they would enthusiastically embrace yet another T-64 variant with a another potentially troublesome new engine. The close cooperation of the VNII Transmash and the OKTB may also have given the Leningrad tank proposal special clout.

The T-80 may simply have been another case where the industrial ministries successfully coaxed the Ministry of Defense into starting another production line to satisfy regional economic interests or bureaucratic interests and not based on army requirements.

There were significant problems in starting T-80 production. As the other design bureaus had discovered by their own turbine experiments, turbine propulsion for tanks is an expensive proposition compared with conventional diesel power. Besides the high production costs of the engine, the fuel consumption forced the designers to substantially increase the integral fuel load from 815 liters (T-64) and 1,200 liters (T-72) to 1,840 liters on the T-80. In spite of the large fuel capacity, the tank had markedly poorer range than the earlier tanks, 335 kilometers on roads compared with 500 kilometers for the T-72.[39] In addition, the GTD-1000T engine suffered from a very low engine service life of only 500 hours.[40] In addition, the high air intake demands of the engine led to serious complications in the design of the OPVT fording system, leading to a particularly large and cumbersome array of trunks and tubes.

The T-80B (Obiekt 219R) Tank

Production of the T-80 was very shortlived and may have only represented a pre-series batch for extended operational trials.[41] Although the reasons for this have not been explicitly described in Russian accounts, the reasons appear to be evident from changes introduced into the following Obiekt 219R (T-80B). The T-80B was called SMT M1983/1 by NATO when it was first spotted. In spite of the relatively high cost of the T-80, it was behind the T-64 in evolutionary development, being essentially equivalent in armor and firepower to the T-64A on which it had been based. In the meantime, however, the Kharkov design bureau had substantially modernized the T-64A as the T-64B with new armor, the new Kobra guided missile system, an upgraded fire control system and other improvements. The Obiekt 219R program appears to have been an attempt to catch up with the T-64B.

The T-80s of the initial series were characterized by lack of an anti-aircraft heavy machine gun, positioning of the L-2AG infrared searchlight on the left side of the main gun tube, and incorporation of a TPD-2-49 optical gunner's sight/rangefinder with one of its optical ports on the right turret side in front of the commander's station (like the T-64A). On the T-80B, the Luna L-2AG searchlight was switched over to the right side of the gun tube (as on the T-72), and the optical port for the rangefinder was deleted due to the addition of a laser rangefinder. The original T-80 used a TPD-2-49 optical gunner's sight with an optical rangefinder; the later batches of the T-80 were fitted with a TPDK-1 laser rangefinder. In the the T-80B, the improved 1A33 fire control system is used. The 1A33 system permits fire-on-the-move up to thirty kilometers per hour. The optical rangefinder sight was replaced by a 1G42 gunner's sight with an integral laser rangefinder for greater accuracy; the optical port in front of the commander's station was deleted. The 1G42 sight is stabilized in two-axes. The 1A33 complex also includes a 1V517 ballistic computer. Data input is broadened to include wind sensor information from a new sensor on the rear turret roof. The 2E42 stabilizer system was replaced by the 2E46M. The 9K112 Kobra guided missile system was incorporated into the tank to enhance its long-range firepower. [The T-80BK (Obiekt 630) command tank was not fitted with the Kobra system.] The armor of the T-80B was also improved, according to East European sources, but details are lacking.

After a production run of only about two years, yet another series of upgrades was introduced into the T-80B in 1980. The turret was unified with the T-64B to in-

T-80B (Obiekt 219R) standard tank.

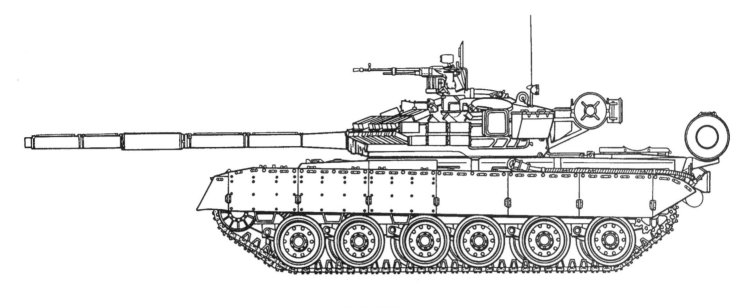

T-80BV

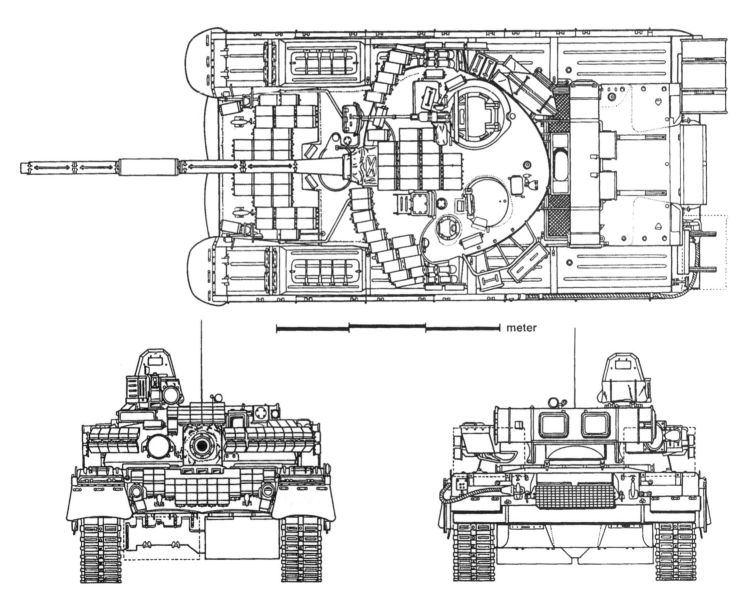

T-80BV (Obiekt 219RV).

clude as many common fire control components as possible and to reduce the logistics burden of supporting three different main battle tanks. In addition, the definitive production models of the T-80B introduced the improved GTD-1000TF (F = forsirovanniy, forced air) which boosted the output to 1,100 horsepower and doubled engine service life to 1,000 hours.

Eastern European sources indicate that in 1982, the turret armor was improved.[42] This probably refers to the substantial turret redesign seen on later production T-80Bs. This turret had a noticeably squarer rear section when viewed from above, and more vertical front panels, indicative of an internal turret armor change. In 1983, after complaints from troops about the poor range of the T-80, hardpoints were added to the engine deck to permit carrying an additional 200-liter fuel drum. In 1984 an additional plate of thirty millimeter armor was added on the upper glacis of the hull. Part of the T-80Bs were fitted with the 2A46-2 gun, a modified 2A46M-1 with an improved recoil system and fast change barrel that had been developed for the T-64B.

Production of the T-80BV (Obiekt 219RV) with the first generation Kontakt reactive armor began in 1985. This was identical to late production T-80Bs with the exception of the reactive armor applique. The T-80 and T-64 were given preference in this program, and many T-80Bs originally manufactured without reactive armor were rebuilt with the new feature at tank depots.

The T-80U Bereza Tank

During the early 1980s, the Shipunov KBP design bureau at Tula was engaged in the design of a new generation of tank-fired guided projectiles using a laser beam-rider guidance. These were ready for incorporation into new tanks from around 1985 to 1986 and resulted in the T-72B and the T-80U (Obiekt 219AS). (The "U" in T-80U stands for uluchsheniye, or improved.) Two slightly different versions of the 9K119 system were developed, the 9K120 Svir for the T-72B and the 9K119 Refleks for the T-80U. A small number of T-80Us were manufactured with the older Kobra system. They were designated as T-80U1 (Obiekt 219A).

The T-80U represented the third major variant of the T-80 family and introduced a host of other new improvements. This was the first T-80 variant to actually leapfrog ahead of the T-64, which by this point was finally ending its production. The most noticeable feature was the new Kontakt-5 generation of reactive armor, designed by NII Stali in Moscow. Unlike the first generation Kontakt reactive armor, which was designed primarily to degrade shaped-charge warheads, the Kontakt-5 applique armor had capabilities against both shaped charge and kinetic energy projectiles. According to a NII Stali, the Kontakt-5 package added the equivalent of 120 millimeters of steel armor against APFSDS projectiles and 500 millimeters steel equivalent against HEAT.[43] It also offered a modest weight reduction, from 1.5 metric tons with Kontakt to 1.3 metric tons with Kontakt-5. Anti-HEAT protection was improved by about twenty percent compared with the first-generation Kontakt ERA.[44] This armor had a distinctly different appearance from the reactive armor blocks. The Russians claim that this armor package, when combined with the integral T-80U armor, offers comparable or superior protection compared with the armor on the M1A1 Heavy Armor.

The T-80U also introduced the 1A42 fire control system with "doubled" fire controls. The gunner's controls consist of the Irtysh day optical sight with integral 1G46 laser rangefinder, located in the forward portion of the turret roof. Behind this is the Buran-PA stabilized day/night sight. This has combined night vision channels in-

T-80U (Obiekt 219AS).

Kontakt-5 armor.

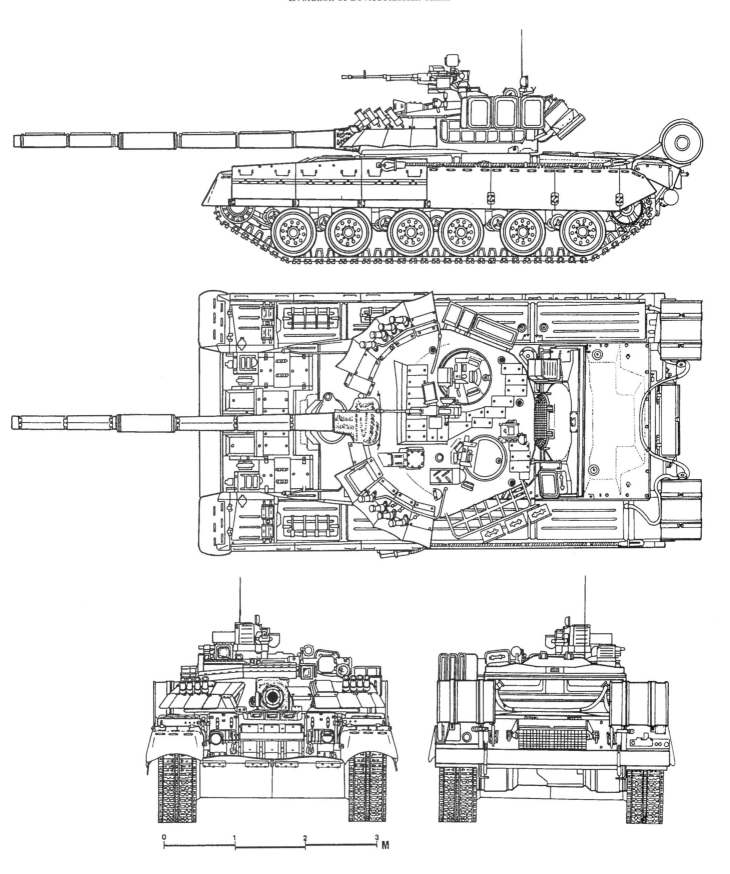

T-80UM Model 1993

9K119 Refleks missile.

cluding passive image intensification and active infrared viewing. The commander is provided with the PNK-4S day/night sight, which is stabilized in one axis. This gave the tank commander fire controls for the main gun except for the use of the Refleks missile. The T-80U tank was accepted for production in 1985 but does not appear to have been manufactured in any significant numbers until 1987.[45]

Upgraded T-80U Tanks

In 1990, production of the T-80 at the Kirov Plant in Leningrad (St. Petersburg) ceased. Currently, production is being undertaken at the Omsk Machine Construction Plant. There have been frequent upgrades to the T-80U. Around 1989, the T-80Us began to be manufactured with the uprated GTD-1250T turbine engine with 1,250 horsepower. In 1990, trials began of a T-80 powered by a 1,500 horsepower gas turbine.

A new rapid fording kit has been developed for the T-80U, designated the Brod-M system, to replace the clutter of tubes and adapter boxes previously used. The T-80U tank displayed at Abu Dhabi in 1993 had an improvised air filter added over the air intakes using portions of the Brod-M system, but with a simple particle filter added inside; this is not a standard feature on the tank since it prevents the turret from traversing fully. Beginning in 1992, the T-80U was upgraded with the new Agava M1 fire control system which incorporates a thermal imaging sight.[46] Some tanks publicly displayed in 1992 in Russia and Abu Dhabi had a different mounting for the heavy anti-aircraft machine gun off the commander's cupola. The remote control mounting was replaced by a simple pintle mounting. In some cases, the machine gun has added a small device about the size of a

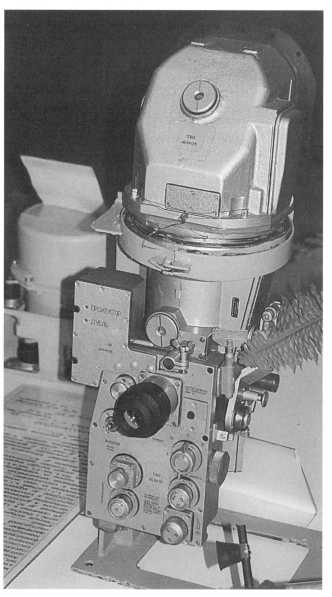

TPN-4 thermal imaging sight of the Agava system for T-80UM tank.

GTD-1250T turbine engine for T-80U tank.

T-80UM.

shoebox, which is the drive motor for the gun mount. It was placed outside to save space in the turret. These upgraded tanks do not have an official designation, but they have been referred to in Russian promotional literature both as the T-80U+ and as the T-80UM (Obiekt 219M).

Kharkov's T-80UD

With the completion of production of the T-64B at the Transport Machine Plant im. V. Malyshev in Kharkov, T-80 production began. Due to lingering distaste for turbine engines in parts of the Soviet Army, Kharkov embarked on a re-engining effort. The 6TD 1,000 horsepower turbocharged diesel engine (a six-cylinder version of the 5TD) had been developed for the short-lived T-64BM. This same engine was fitted into the T-80U under the code name Obiekt 478B. The prototypes of this tank, eventually called the T-80UD, were fitted with the original Kontakt-1 reactive armor, not the later generation Kontakt-5 armor usually associated with the T-80U. Surprisingly, in 1988, the Soviet Army permitted production of this tank to proceed, so that once again three different main tanks with three different engines were in production (T-80U with the GTD-1000TF turbine, T-72B with the V-84 diesel and T-80UD with the 6TD diesel). Some of the first production batches of these tanks were provided to the 4th Kantemirovets Tank Division im. Yuri Andropov, one of the Moscow show divisions. As a result, these tanks took part in the May 1990 Red Square parade celebrating the 45th anniversary of the victory over Germany in 1945.

The Ukrainian T-84 Tank

Production of the T-80UD in Kharkov was virtually paralyzed after the collapse of the USSR at the end of 1991 because about seventy percent of the components (hull, turret, gun, electronic components) were provided by other facilities located in Russia. In 1993, the Ukrainian government decided to place greater priority on restoring production capabilities at Kharkov. Under the current plan, an improved version of the T-80UD tank is being developed that will be built entirely by Ukrainian factories. About nine prototypes of the new tank, called the T-84, had been completed through the end of 1993.

T-80UD.

Reportedly, Pakistan examined a T-80UD and a T-84 in the Thar desert in the summer of 1993.[47] In 1997 Pakistan took delivery of the first two batches of T-80UDs, roughly 60-70 vehicles out of a total buy of 320 to 380.

The T-84 consists of a T-80UD chassis of the type developed and manufactured by Kharkov since 1986, fitted with a new turret. The turret had been the main roadblock in tank manufacture in Ukraine after the Soviet split-up in 1991, as the turret casting was provided from a plant in Russia. The new welded turret is designed to conform to the same shapes as the normal T-80UD turret, and the commander's and gunner's hatches appear to be in identical positions as are other major components. However, the external details of the turret differ considerably. The designers claim the new turret provides better protection than any other main battle tank built within the former Soviet Union. It is reported to be 1.1 to 1.4 times more effective against HEAT and 1.1 to 1.6 times better against APFSDS. The welded construction also offers greater internal volume. The precise composition of the turret was not disclosed, but was said to consist of multiple layers of steel and non-metallic armor similar to the M1A1 Abrams, six layers in the front and five on the sides. The turret of the T-84 is fitted with Kontakt-5 ERA. The smoke

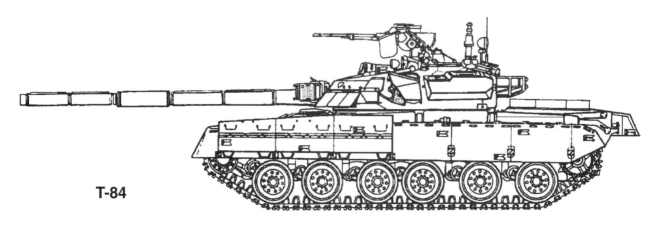

T-84

T-84 Technical Data

Crew	3
Combat weight (metric tons)	46
Length (meters)	9.72
Width (meters)	3.595
Height (meters)	2.215 to roof
Ground clearance (millimeters)	515
Power to weight ratio (horsepower per ton)	21.7
Ground pressure (kilograms per centimeter cubed)	0.93
Maximum speed (kilometers per hour)	60
Average cross-country speed (kilometers per hour)	35 to 40
Maximum road range (kilometers)	560
Maximum cross-country range (kilometers)	350 to 400
Gradient (degrees/percent)	32 / 63
Side slope (degrees/percent)	20 / 36
Vertical obstacle (meters)	1.0
Trench (meters)	2.85
Fording (meters)	1.8 meters
Engine	6TD 1,000 horsepower multi-fuel diesel
Fuel (liters)	1,300
Main gun	KBA-3 125mm smoothbore (2A46-2)
Gun elevation (degrees)	-5 to +13 (gun forward) or -2 to +16 (gun astern)
Ammunition stowage	45 (28 in autoloader)
Rate of fire (rounds per minute)	9
Gun stabilization (planes)	2

mortars on the turret side differ in appearance from the Russian T-80U. The same Tucha smoke mortars are used, but a thin metal cover has been added over them, reportedly to prevent the crew from being injured if standing outside the hatch when the mortars are discharged. A new stowage box has been fitted at the 8 o'clock to 4 o'clock positions, and the box exterior is armor to give stand-off protection.

The T-84 is fitted with a locally produced version of the 2A46-2 with fast change barrel called the KBA-3. Servicing of the recoil systems has been simplified by installing indicators on the recoil absorbers and recuperator. The autoloader system is the same as the T-80 and uses a two-way rotational conveyor with twenty-eight rounds of any of six different ammunition types, including APFSDS; HEAT; the 9M119 Refleks guided projectile; HE-fragmentation; and two types of HE munitions with time-fuzed detonation. The system is hydraulic and electromechanical with a constant loading angle. The loading speed is 7 to 12.5 seconds providing a rate of fire (per minute) of seven to nine APFSDS, HE-Frag or HEAT rounds, two to three guided missiles, or four delayed detonation HE-Frag projectiles.

The T-84 uses the same 1A45 fire control system as the T-80U and T-80UD. This includes the 1G46 Irtysh gunner's daytime sight and laser rangefinder; the Buran-E gunner's day/night stabilized periscopic sight, the commander's TKN-4S Agat day-night designation sight, the 9S515 semi-automatic laser guidance system for the 9M119 Refleks guided projectile, the 1V528 ballistic computer, and the 2E42 gun stabilization system. The Buran-E provides a maximum range of 1,200 meters at night. The T-84 brochure released at IDEX-95 mentions a thermal sight, but the tank featured at the show did not have the new Agava fire control system displayed on new production Russian T-80UM tanks like that at IDEX-95. The 1V528 fire control computer is electronic, digital/analog, and includes the upgrade features to permit use of new HE-Frag ammunition that can be airburst over targets by calculating tangent elevation and lateral lead and sending a laser signal to the projectile in flight. The fire control system automatically computes the effect of yaw, wind, tank and target speed as well as the angle of sight to ensure a ballistic performance for up to thirty types of projectiles. This includes meteorological considerations such as pressure; ambient temperature; shell temperature and wear on the barrel lining.

Unlike the Russian T-80UM which has been moving in the direction of simple pintle-mounted 12.7mm anti-aircraft machine guns, the T-84 retains the remotely fired 12.7mm NSVT machine gun, with the associated TKN-4S and PZU-7 sights. The PZU-7 sight vertically stabilized within -5 to +70 degrees or the gun can be controlled using the normal commander's TKN-4S sight in a

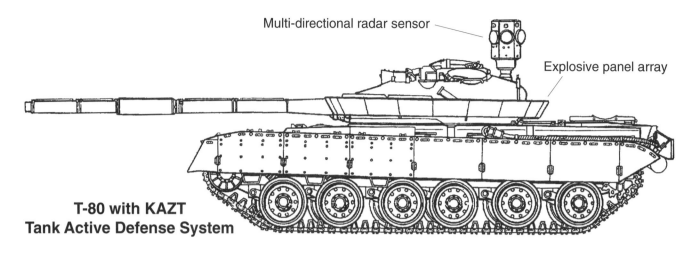

T-80 with KAZT Tank Active Defense System

vertical range of -5 to +20 degrees. NSVT machine gun ammunition totals 450 rounds.

The T-84 is fitted with the Shtora-1 countermeasures system including both the electro-optical jammers and the laser detectors. Ukrainian publications state that Shtora reduces the probability of destruction by 1.95 times. Other passive defenses include thermal protection of the hull roof plate over the motor and transmission compartments and protective coatings affording protection against radar location. The T-84 can be fitted with a blade mine-clearing device.

The T-84 can be fitted with one of three variants of the 6TD family of diesel engines. The power pack unit is compact (3.2 cubic meters) and is based around the 6TD engine and associated transmission. The 6TD is an evolutionary derivative of the 5TD used on the T-64 tank which added one more cylinder. It has been consistently improved during series production and has proved reliable under all conditions including dust, where the Ukrainians claim it enjoys a distinct advantage over gas turbine engines. The 1,000 horsepower 6TD-1 engine is a twin-stroke, multi-fuel, liquid-cooled, six-cylinder (120 millimeters) horizontal engine with fuel injection and 2 by 120 millimeters stroke opposed pistons with controlled exchange of gases, a supercharger and a gas turbine. The 1,200 horsepower 6TD-2 is a development of the 6TD-1 diesel engine and is fully interchangeable with it regarding fixture points and chassis mountings. The increases in engine power has been achieved by improving the blowing rate and fuel requirements. The 1,500 horsepower 6TD-3 is a modification of the 6TD-2, featuring intermediate cooling of supercharged air.

There are two ports to the engine air filtering system, the centrifugal pre-cleaners and the air cleaner casing. It enables the tank to be operated in hot and dusty conditions for up to 1,000 kilometers without a change of filters and to carry out combat under radioactive conditions. The transmission consists of a mechanical, epicycle train design with friction drive and power rating of 1,200 horsepower. It has seven forward gears and a reverse. It requires a major overhaul every 15,000 kilometers; weighs approximately 1,440 kilograms and has a capacity of 0.3 meters cubed.

The surface of the T-84 was fitted with several dozen small wedge-shaped devices about six inches long, and 1.5 inch wide. There were several welded to the hull side, and they appeared on various other locations including the stowage bin, gunner's sight and elsewhere. It is unclear if they are some sort of corner reflector, or attachment points for a camouflage system.

T-80UM-1 Bars (Snow Leopard)

In 1992, the Russians first revealed that they were working on a new active defensive system (first tested on the T-64) for the T-80 tank. In 1997, the Rusians first displayed the T-80UM-1 Bars in Omsk. This is the current production version of the T-80U fitted with the KAZT Arena active defense system, as well as other improvements such as the 2A46M4 gun.

The Arena KAZT (Kompleks aktivnoy zashchity tanka: Tank Active Defense System) is a new tank protective system developed by the Machine Design Bureau in Kolomna. The chief designer of the system is Nikolai Ivanovich Gushin.[48] The Arena KAZT is the second known Russian attempt to develop an active defense system. The first system, code-named Drozd (Thrush), was developed by the Shipunov design bureau at Tula and deployed on T-55AD tanks in the late 1980s. It was apparently too expensive and not effective enough to warrant its use beyond a small number of vehicles. The Drozd used a tube-fired munition as its missile kill mechanism, the Arena uses a sensor-fuzed explosive panel which sprays the incoming missile with pellets.

The KAZT system is designed to defend tanks against guided anti-tank missiles such as TOW and Hellfire, as

The Arena KAZT active tank defense system (on T-80B tank).

well as anti-tank rockets such as AT-4. Promotional material indicates that the system is effective against both ground-launched and helicopter-launched missiles. The system is fully automatic and reportedly doubles the survivability of the tank. Total added weight is 800 kilograms. The internal volume of the electronic equipment is thirty cubic centimeters, and system reaction time is 0.05 seconds.

The system employs a millimeter wave radar detection system, mounted on a stalk at the rear of the turret. Photographs and video footage of the system indicate that the detection system has at least six faces. There are two banks of sensors — a lower bank with circular antenna, and an upper bank. It is possible that the system employs passive infrared sensors for the initial detection of an incoming missile, activating the radar sensors when a likely target has been detected. This configuration would avoid the need to keep the active radar antenna emitting at all times. If not backed up by some form of passive sensor, the active radar emissions may make the tank more visible to hostile surveillance.

Missile detection takes place at fifty meters. The onboard computer then determines the velocity and direction of the missile attack, and selects which explosive panel it will detonate. At ten to five meters, the system has locked on and prepares to fire. The kill mechanism is a circular array of explosive panels around the turret. The panels are mounted around the turret of the tank, covering the entire area except for a small gap near the mantlet and main gun. The system launches a single panel (one of thrity-two) upward, which is detonated at a preset height by a trailing cord. The panel resembles a claymore mine, and on detonation sprays the missile below with metal fragments. The Arena KAZT coverage is about 340 degrees in azimuth and is -85 degrees to +25 degrees in elevation. In the event that a panel is blown off, the turret can be traversed to cover any gaps in the system. This would be accomplished manually, however, not by the system itself. It might be possible to link the control unit to the turret's traverse system to accomplish this, but this feature does not appear to be incorporated into the design as currently configured.

The Arena KAZT system was two to three years from completion when first revealed in May 1993 as the bureau was looking for partners or interested buyers. Bureau representatives estimated that the system cost was twenty percent of the cost of the tank. Given the announced export price of the T-80 as about $2 million, this would mean the system costs for export would be around $400,000. The demonstration system was mounted on a T-80 tank but can be configured for other tank types.

The Russian are vigorously marketing the Arena system throughout the world and appear to have had success in Europe. According to various media accounts at IDEX-95, the Russians have found partners in France's Thomson-CSF and Germany's Daimler-Benz Aerospace (these two companies have formed a joint venture company called TDA). According to the Russian developers

In 1997, the Russians showed a T-80U with Drozd.

Also shown at Omsk in 1997 was this T-80-based armored recovery vehicle concept.

T-80 Arena KAZT
Active Protection System

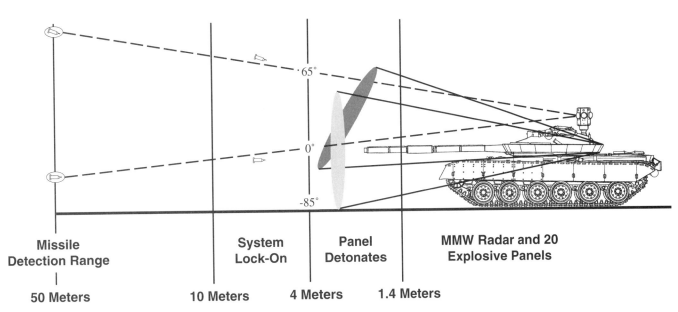

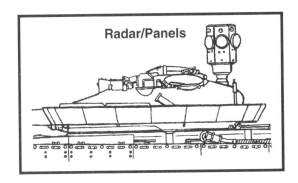

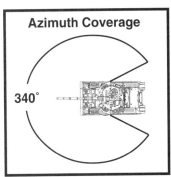

of Arena, the French-Germany combine now has exclusive rights to market Arena in Western Europe on tanks destined for export. The new joint partnership evolved after witnessing a successful firing demonstration of the system in Russia. The same IDEX-95 accounts stated that at least three prototype sets were developed and that the joint French, German, and Russian venture plans to show Arena fitted to a Western tank at a future arms show (possibly fitted to the French AMX-30 or LeClerc or the German Leopard 1/2)

Conversations with a member of the Kolomna design bureau in 1994 indicated that the system is capable of defeating missiles with overflight profiles (such as the Swedish RBS-56 Bill), but the system does not appear to be configured for dealing with missiles approaching at extreme diving flight profiles.

T-80UK

The version of the T-80U fitted with the current generation of passive Shtora-1 tank defenses is called T-80UK. It was first publically displyaed at the IDEX 1995 show in the United Arab Emirates. The T-80UK designation was intended to be applied to the usual command version of the T-80U, but Russian officials used the opportunity of the IDEX-95 show to display the command tank with the Shtora system as well. It is not clear how the Russians designate a basic T-80U with Shtora.

The Shtora-1 system has four main elements: electro-optical (EO) jammer, laser warning detectors, anti-laser smoke grenade launchers, and system controls. It is fitted with an electro-optical interference system (dazzler) fitted on the opposite side of the gun tube from the infrared searchlight. This system functions like the US Army AN/VLQ-6 or VLQ-8 and is intended to bluff the missile tracker of standard wire guided anti-tank missiles. The system operates in the 0.7 to 2.5 micron range. This system includes the TShU-1-17 radiator being promoted by the Zenit NPO in Moscow as an ATGM jammer.[49]

The second element of the system is a set of several

Russian T-80UK on the move in Abu Dhabi during a 1995 firepower demonstration.

Photographed in Abu Dhabi, this time in 1997, this T-80UK shows the turret roof.

STRUCTURE OF COMPLEX SHTORA-1

Protection techniques:

- light interference delivery to the visual channels of ATMC

- protected tank laser illumination screening with aerosol screens

EO interference station
Emitter, Modulator, Control panel
Weight — 80 kg

Aerosol screens distance laying system
Launcher, Grenade
Weight — 115 kg

Equipment for laser illumination indication
Precision head, Coarse head
Weight — 20 kg

Control system
Microprocessor, Manual screen laying panel, Control panel
Weight — 15 kg

Purpose:

- detection of the fact and direction of AT weapon laser target designation

- processing of information from equipment for indication and for delivery of instructions for aerosol screen laying;

- laser illumination signalling

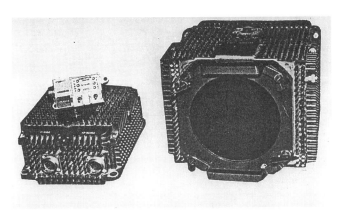

Shtora-1.

laser illumination sensors, fitted on the turret, which warn the crew of laser designation. These have a -5 to +25 degree field of view. Data from the laser warning sensors is fed into a new microprocessor in the tank which provides information to the tank commander. The system can operate in an automatic mode, dispensing smoke grenades to defeat laser rangefinders or designators. Alternately, it can operate in a semiautomatic mode, with the tank commander deciding whether or not to trigger the smoke dischargers. The smoke grenades are an evolutionary upgrade of the Type 902 Tucha system. They take about three seconds to bloom and last twenty seconds; they provide obscuration in the 0.4 to 14.0 micron band. In addition, according to Russian sources, further work continues on a follow-on system, Shtora-2, which will be able to degrade sensors and weapon guidance systems across the entire spectrum from millimeter wave to the optical wavelengths with multi-spectral obscurants and electronic jammers.

Russian officials have also indicated that there is some difference in the Shtora packages being produced for export (Shtora-1) versus those for the Russian Army (Shtora-1+). The most significant difference were reported to be the greater sensitivity of the laser warning sensors.

Current T-80 Production

In Russia, tank production has been trimmed back

Type 902 Tucha smoke mortars.

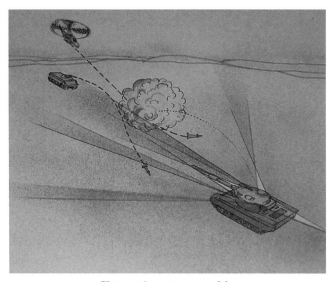

Shtora-1 system graphic.

considerably because of sharp defense budget cuts. T-80U production ceased in St. Petersburg in 1990 and is currently being undertaken at the Omsk October Revolution Machine Construction Plant alone. Production at Omsk has declined from 1,000 annually to only 160 in 1992, with only five tanks ordered by the Russian army and the rest for export.[50] Only one other tank plant is active, the Ural Railcar Plant (Uralvagonzavod) in Nizhni-Tagil, the home of the T-72/T-90. In 1992, the Russian defense ministry made it clear that it could no longer afford to buy two main battle tanks simultaneously. Russian officials stated that they wished to cut production down to a single type, either the T-72 or the T-80.

The changing strategic balance in the world has led Russia to offer the T-80 on the international arms market. India and Poland both were provided with small numbers of T-80 tanks in the mid-1980s for trials; both decided against buying the tank due to its high cost. The T-72S costs about $1.2 million, and the T-80U about $2.2 million; the Indians and Poles did not feel that the T-80 offered sufficient advantages for the price difference.

Morocco bought five T-80U tanks in the early 1990s, but the sale caused a scandal in Russia when it was learned that the tanks were destined for Western intelligence services. With the relaxing political climate in Europe, in 1992 Russia directly sold another T-80U to Britain, and other T-80s were provided to Sweden to take part in the

T-80U in Abu Dhabi showing air filter improvised from Brod wading trunk.

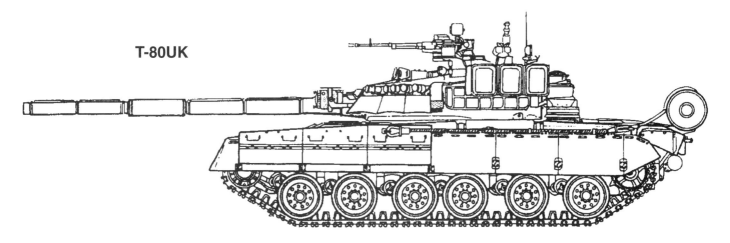

T-80UK

tank competition there. Russia has made small sales of the T-80U to Oman for trials and is currently trying to interest Dubai in the United Arab Emirates in the tank as a less expensive alternative to Western tanks like the American M1A2 and French LeClerc.[52] Although more expensive than the T-72, the T-80U is still a bargain compared with the European and American tanks. There have been repeated reports that Syria has a standing order for the T-80.[52] Reports that China has purchased the T-80 have been vigorously denied by Russian authorities.[53] Further, Cyprus displayed in a military parade the first batch of T-80s in October 1996 of a total buy of roughly forty T-80Us from Russia. Russia has also deliveried T-80Us to South Korea. In June 1996, Pakistan announced that it was planning the purchase of 330 T-80UD tanks from Ukraine at a cost of $650 million and received the the first two deliveries in 1997.

The T-80's Performance in the Chechen War

Then Russian Defense Minister Pavel Grachev during a scientific-technical conference on the future of Russian armored vehicles, held at Kubinka in February 1995, stated ". . . that the (T-80 main battle tank) turned out to be a junk heap on the battlefields of Chechnya."[54] The tank's armor plating proved to be too thin, resulting in ninety-eight percent of the kills of T-80 being caused by RPG man-portable weapons, according to Russian sources.

Many of the RPGs were utilized by Chechen rebels firing down on the vehicles as they passed by large multi-story buildings in the capital of Grozny. Such hits on a T-80 or for that fact, any Russian main battle tank, would result in catastrophic damage. The two primarily hit locations on the T-80 were the turret roof and the engine deck. Both of these locations are very lightly armored and are also not covered with explosive reactive armor. When a tank was hit in these locations it almost always resulted in a catastrophic kill due in part to secondary ammunition detonations. These secondary detonations resulted from the explosion of ammunition located in the unarmored autoloader of the T-80's 125mm gun. One Russian soldier was quoted as saying that the lesson he learned from the war in Chechnya is that, ". . . a T-80 turret can be tossed into the air over 300 meters."[55]

These survivability issues have prompted Grachev to insist that all T-80s, located in the North Caucasian region, be fitted with additional armor on the outside and

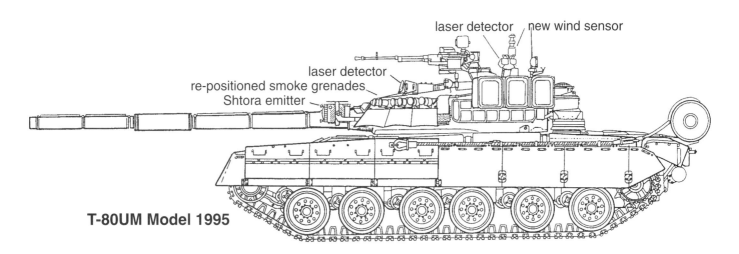

T-80UM Model 1995

that new designs be fitted into the interior to protect ammunition storage areas in less than three weeks. In addition to more appliqué armor, these vulnerable areas would be covered with explosive reactive armor, however, Russian tank specialists pointed out that the placement of explosive reactive armor in these areas was problematic.[56]

The main reason for this concern resides with the second major deficiency of the T-80 in the Chechen war, poor turbine-engine performance. The turbine engine is a voracious consumer of fuel. Grachev ordered tank designers and engineers to convert the T-80 from aviation kerosene to diesel fuel and to improve the performance of the engine to fight for eight continuous hours or travel 400 miles on one refueling.

Grachev's remarks reflected his amateurish understanding of tank design. A former paratrooper officer, he used the alleged technical flaws in Russian tank designs as a scapegoat for his own incompetent leadership in Chechnya. No contemporary main battle tank is designed for 360 degree protection, nor is any current tank designed to fight in an urban environment with untrained crews, no machine gun ammunition, and no infantry support.

T-80 Comparative Technical Data

	T-80B	T-80U
Crew	3	3
Weight (metric tons)	42.5	46.0
Length (meters)	9.65	9.65
Width (meters)	3.58	3.6
Height (meters)	2.2	2.2
Clearance (millimeters)	451	446
Main gun	125mm 2A46M	125mm 2A46M-1 (or 2A46-2)
Elevation (degrees)	-7 + 20	-4 + 15
Rate of fire (rounds per minute)	6 to 8	6 to 8
Effective range (meters)	4,000	5,000
Autoloader magazine (rounds)	28	28
Main gun ammunition	38	45
Missile system	9K112-1 Kobra	9K119 Refleks
Gunner's missile controls	GTN-12 (9S461-1)	(9S515)
Gun stabilization system	2E36M	2E42
Fire control system	1A33	1A45
Ballistic computer	1V517	1V528
Co-axial machine gun	7.62mm PKT	7.62mm PKT
Anti-aircraft machine gun	12.7mm NSVT (Utes)	12.7mm NSVT (Utes)
Anti-aircraft machine gun sight	ZPU-5	K-10T
7.62mm ammunition	1,250	1,250
12.7mm ammunition	500	450
Gunner's ranging sight	1G42 laser	1G46 laser
Driver's periscopes (day/night)	TNPO-160/TVNE-4B	TNPO-160/TVNE-4B
Commander's sight (day/night)	TKN-3	PKN-4S
Main infrared searchlight	Luna L-4A	Luna L-4A
Commander's infrared searchlight	OU-3GK	OU-3GKU
Engine	GTD-1000TF	GTD-1250
Power (horsepower)	1,100	1,250
Power-to-weight (horsepower per ton)	25.8	27.2
Auxiliary power unit	no	GTD-18
Integral fuel (liters)	1,840	1,770
Ground pressure (kilogram per square centimeter)	0.865	0.93
Gradient/side slope (degrees)	32/30	32/30
Ditch crossing (meters)	2.85	2.85
Unprepared fording (meters)	1.8	1.2
Prepared fording (meters)	5	6
Maximum road speed (kilometers per hour)	70	70
Maximum range on integral fuel (kilometers)	325	325
Maximum road range (kilometers)	370	400
Internal communications	R-124 TPU	R-174
Radio	R-123M	R-173

Future Tank Design: T-80 Low-Profile Testbed

Russia has built a number of experimental prototypes to test new design concepts (e.g., T-62 with a 125mm gun). It appears that Russia continues to field such experimental prototypes and testbeds at the Scientific and Research Institute for Armor and Technology.[57]

One recent experimental testbed appears to be a T-80 chassis with an externally mounted long-barreled gun with a main gun caliber between 135 to 152mm.[58] Such a gun caliber range reflects similar exploration by the West. Russian tank gun designers have always been interested in fielding a larger caliber gun than the West. As illustrated in this study the decision to up-gun the T-64 main battle tank with the 125mm was spurred in part, by the fielding of 120mm gun on the British Chieftain.

Several Western European countries to include the United States have looked into utilizing larger gun calibers on current and future armor designs (e.g., the German Leopard 2 main battle tank follow-on is looking at the possibility of utilizing a 140mm main gun). A main gun in the 135 to 152mm range would have a considerable increase in muzzle velocity, and projectile weight, which would result in a sizable increase in armor penetration. A larger gun also means fewer rounds could be carried by a main battle tank. The Low-Profile T-80 has a three-man crew who are seated in a side-by-side configuration. The crew consists of a commander, gunner, and driver. The turret of the vehicle is a low-profile turret-shape reminiscent of the General Dynamics Land Systems modified M1 Tank Testbed. The sensors for the gun are located on either side of the turret. The small turret presents a considerably reduced target area and would in theory be harder to target. The tank would be required due to the crew layout, weight, and size of the projectiles to utilize an autoloader.

It is uncertain at this time whether this vehicle is a testbed of a future main battle tank. It illustrates that Russian tank designers are still exploring new and innovative designs to field on their future main battle tank. However, it is unclear at this time whether the Russian Ministry of Defense will be able to devote the kind of resources required to field a next generation main battle tank design for the foreseeable future.

Black Eagle

In September 1997, the Omsk tank plant demonstrated a mock-up of a future tank called Black Eagle. This version of the T-80U has a new turret with a new generation armor, new fire controls, and a very large turret bustle incorporating compartmentalized ammunition stowage and a new autoloader. The first prototype is supposed to be completed in 1998, and appears to be an effort to reinvigorate export interest in Russian tanks rather than a solution to Russia's long term tank needs.

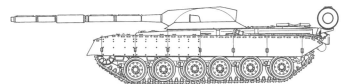

T-80 Low-Profile Turret Test-Bed

The Black Eagle concept as displayed in Omsk, Russia, September 1997.

Future Trends in Russian Tank Development

In 1966, the minister of defense of the USSR ordered the T-64 into service. In 1972 Marshal Grechko accepted the T-72 and in 1976, Minister Ustinov gave the "thumbs up" to T-80 series production. Over a period of ten years, the USSR armed forces accepted three expensive lines of combat vehicles virtually identical in their basic tactical and technical characteristics and combat capabilities, but different in their design features and absolutely different in operation from each other . . . The fact of the matter was that neither of these "new" tanks had anything which made them superior in characteristics to the T-64; they remained tanks of second post-war generation . . . In comparison with us, both the USA and Germany already have third generation tanks, and they are conducting intensive work on projected vehicles that will become the fifth and sixth generation tanks. Us? We are stumbling on with the second generation, and this stumbling means that we won't be able to get back on our feet.

Colonel (Professor) Boris A. Didusev,
Scientific Research Institute for Tank Technology, Kubinka, Russia

Until recently, the Western appreciation of Soviet tank development have rested on two pillars: technological imperatives and tactical/doctrinal imperatives. While these two factors no doubt play a crucial role in many aspects of Soviet armored vehicle policy, they alone are not sufficient to explain many of the key decisions taken in the past. For example, neither by themselves satisfactorily explains the proliferation of main battle tank designs in the 1970s (T-64, T-72, and T-80). As has been suggested in this study, other factors, such as the political influence of design bureaus, internal rivalries within the military industrial complex, regional industrial policies, and other internal factors have played a significant role in the selection of particular armored vehicle designs. It can certainly be argued that tactical/doctrinal influences are strongest in setting overall procurement goals, while technological and bureaucratic factors have greater influence on the actual technological characteristics of armored vehicle designs.

Future trends in armored vehicle design are likely to be influenced by all these factors as well as by factors not previously important in Soviet acquisition decisions. With the collapse of the USSR, the future configuration of the Russian army is far less predictable. The orientation of these forces could change dramatically over the next decade for a variety of reasons.

The armored vehicle industry still could continue to stagnate due to funding constraints imposed by a parsimonious Duma, an event unthinkable until the 1993 budget debates. Significant curtailment of research and development (R&D) funding and procurement could alter the nature of the armored vehicle design and acquisition process. Until recently, Soviet armored vehicle design was predicated entirely on Soviet requirements; little consideration was given to export requirements outside the Warsaw Pact. Under the new budgetary realities, Russia could be moving in the direction of countries such as France, which exports a very large fraction of its total armored vehicle output and is therefore, very sensitive to market demands. Indeed, we are now seeing the first signs of Russian developments inspired primarily by export demands.

Future Armor Developments

Although the Russian Army would like to modernize its tank force through standardization of existing designs and the adoption of a new generation tank, this has proven impossible for the past five years due to staggering budget problems. At its peak, the Soviet Union had five tank plants: Leningrad, Chelyabinsk, Kharkov, Omsk, and Nizhni-Tagil. Of these, the Kharkov plant is now in Ukraine; the Russian plants in Leningrad and Chelyabinsk no longer produce tanks. Only two plants remain active: Nizhni-Tagil and Omsk. In the early 1990s, Russian Army orders for tanks fell precipitously due to a lack of funds. However, the tank plants continued to produce vehicles in order to keep workers employed. As a result, there were over 500 tanks manufactured in the mid-1990s which remain in manufacturer's yards without orders from either the Russian Army or export clients. The Russian Army now estimates that it needs to procure 300 new tanks annually simply to make up for attrition and maintain a modest rate of modernization. But in 1995, funds were available for only sixty to eighty tanks. A total of fifty-eight tanks have been ordered in 1996, but the funding is so inadequate that these were not paid for until 1997.

Modernization of the Russian tank force has been stymied both by funding problems and a lack of consensus about a future tank. In the mid-1980s, the Kharkov tank design bureau developed a radical, new-generation tank called the Obiekt 477 Molot (Hammer) to replace the existing generation of T-72 and T-80 tanks. Yet the project was deemed too complicated and too costly and was canceled. At the time of the Soviet collapse, work was underway in Leningrad and Nizhni-Tagil on test-beds for new generation tanks, but these were too immature for production. The 1991 Gulf War cast some doubt on Soviet tank design due to the poor performance of the

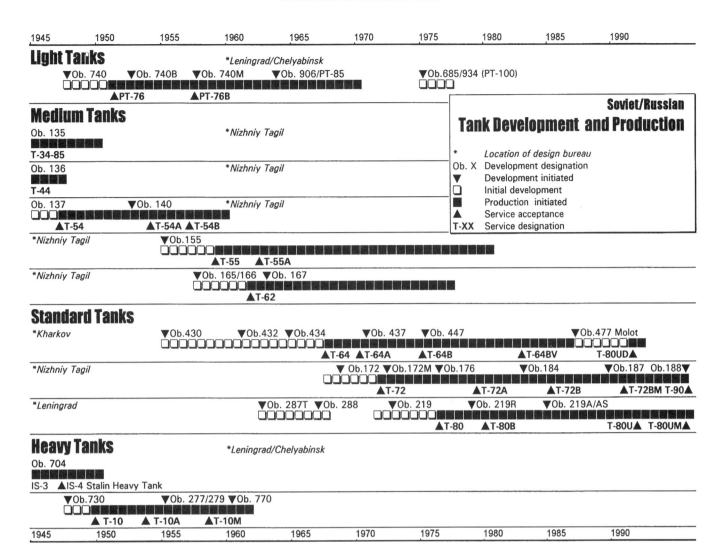

CFE Counts for Armor (European Russia): 1990 to 1997

	PT-76	T-54	T-55	T-62	T-64	T-72	T-80	T-90	Total
1990	602	1,593	3,130	2,021	3,982	5,092	4,876	0	21,296
1991	602	1,593	3,130	2,021	3,982	5,092	4,907	0	21,327
1992	483	539	1,266	948	1,038	2,293	3,254	0	9,821
1993	397	515	871	948	705	1,923	3,031	0	8,390
1994	282	394	637	688	625	2,144	3,004	1	7,775
1995	126	94	308	761	449	1,938	3,144	2	6,822
1996	25	5	38	25	161	1,848	3,311	2	5,515
1997	5	1	65	97	186	1,980	3,210	2	5,546

T-72 tank. Russian tank experts dismiss this, and some go so far as to say that the many photographs of T-72s with their turrets blown off, were American propaganda efforts to undermine Russian export sales. Such paranoid delusions are not universally shared, and some Russian tank specialists are concerned that Russian tank design went seriously off-course in the 1970s and has still not recovered.

The lack of a new generation tank design, the shortage of funds for developing a new design, and the general shortage of funds for production has forced the Russian tank force to fall back on less lofty ambitions in its near term program. As a short-term goal, the Russian Army would like to standardize on a single tank type to ease the logistics burden of such a varied inventory. Of its current tank force west of the Urals, forty-seven percent are T-80 and twenty-nine percent are T-72 tanks, so the choice boiled down to one of these types. This was further reinforced by the fact that of the two remaining tank plants, the Omsk plant manufactures the T-80U, and the Nizhni-Tagil plant manufactures the T-72S and its modernized derivative, the T-90. Although a decision was planned for 1994, this was put off for several years due to continuing collapse in state funding. In December 1994, the Chechen war intervened.

Coming two months after the 31 December 1994 debacle in Grozny, on 20 February 1995, then-Minister of Defense Marshal Pavel Grachev addressed a conference on the future of tank technology in Russia at the Kubinka armored vehicle center. Grachev's comments on existing Russian armored vehicles were scathing. The T-80 was singled out for particular censure due to alleged problems with its armored protection, survivability, fuel system and reliability. Grachev's remarks soon found their way into the Russian press. They proved particularly embarrassing to the Russian defense industry, which was only weeks away from marketing many of the same weapons at the IDEX-95 exhibition in the United Arab Emirates. Over the course of the next several weeks, the Russian Army attempted to put a more favorable spin on Grachev's remarks — with little effect. Coming on the heels of the disastrous performance of Russian weapons in the 1991 Gulf War, Grachev's remarks rang true.

The stinging criticism of Russian weapons, and its armored vehicles in particular, were due in no small measure due to the embarrassing defeat of the Russian Army during its 31 December 1994 attack on Grozny. During the battle, the 131st Motorized Rifle Brigade was virtually wiped out, and about seventy percent of the 200 tanks involved in the attack were knocked out. For the next few days, Russian and world media were filled with lurid imagery of charred Russian corpses littered around burned and smashed armored vehicles. For the Russian public, long shielded from such painfully realistic journalism, the effect was especially shocking. In the wake of such a disaster, the attempts to portray Russian tanks in a more favorable light by senior Russian tank officers seemed like a half-hearted whitewash. The chief of the Main Armor Directorate, Colonel General Aleksandr Galkin, tried to rebut the more extreme charges but he was widely ignored.

Yet Galkin's remarks were probably closer to the truth than the sensationalism in the Russian press. While the attempts to depict the T-80U as the world's best tank can be discounted as the usual sort of marketing hype regularly heard at major arms shows, Galkin laid out many plausible arguments to exonerate the quality of Russian weapons. Like many other senior Russian commanders, he was hamstrung in discussing the painful problems that lay at the heart of the Grozny debacle. To have done so would have put him at odds with the Minister of Defense, and would have forced him to criticize senior officers in charge of the Grozny operations.

The Chechen war may force the Russian Army to examine its ideas about armored vehicles. Instead of confronting NATO in the central European plains, the Russian Army is now more likely to face irregular forces in urban or other fine-grained terrain where conventional tanks and infantry fighting vehicles are not well suited. This raises the question of whether a new category of vehicle might be needed.

But money, not technical/tactical requirements, is likely to be the driving force over the next decade. At the moment, Russian Army officers appear more interested in simply keeping the existing tank park functional with parts, fuel and attrition replacements. Modernization seems like a distant dream.

Russian officers clearly appreciate that the current generation of tanks are not ideal. But the Russian Army's main problem is not the technological sophistication of its tanks. Its three modern tanks, the T-64, T-72, and T-80, all have similar combat features but all have substantially different logistics requirements. The three different types were foisted on the army due to industrial politics rather than requirements. This legacy continues to haunt the Russian Army. Since 1992, Russian officers have stated that they would like to standardize on a single tank type, but this has proven impossible.

In 1996, the Russian Army selected the T-90 as its new main battle tank even though many, including the General Galkin feels the T-80U is a superior design. In a recent interview, General Galkin told the Russian press that the decision to proceed with the T-90 was a mistake. There were plans to fully equip one military district with the T-90 (probably in the Far East) in the 1996-97 budget, but this seems wildly optimistic at the moment.

It seems likely that a modernized version of the T-90 with a new engine will be fielded in the next few years. Russian designers are currently examining upgraded versions of the T-90 with both improved diesel engines and gas turbines. Standardization on a single type remains an ideal, rather than a practical program.

However, should funding become available, will the Chechen experience affect future developments? At the moment, the answer is "maybe." Work is currently underway on a new generation tank by at least three design bureaus in Nizhni-Tagil, Omsk, and St. Petersburg. The Russian Army expects to field prototypes by 1998. Few details of the design have been released, though most sources agree it will be armed with a 135mm, 140mm, or 152mm gun. Russian tank designers categorically deny the reports of a "T-95" tank, pointing out that such a designation will not be given until the tank passes state trials and is accepted for service. The new tank is expected to be heavier than existing Russian tanks, but even at fifty or sixty metric tons, it will probably be smaller and potentially lighter than contemporary Western tanks.

What remains to be seen is whether the Russian Army can afford to produce the new design. The Russian Army takes great pride in its tank force, and efforts at modernization are likely to continue no matter how penurious the budget.

Future Russian Main Battle Tank Development: Three Scenarios

Scenario One: Stagnation

Russian tank design may stagnate for several completely different reasons. Russian economic collapse or regional disintegration remains a real possibility. Even if this does not occur, the weakening of R&D funding due to budget disagreements between the Russian Duma, the central government and the armed forces could cause the tank development efforts at Omsk and Nizhni-Tagil to wither. This tendency could be accelerated by doctrinal trends in the Russian Army. For instance, a policy favoring the Mobile Force could lead to increased funding for light deployable forces and their equipment needs. Russia has lost a great deal of its airlift to the break-away republics. If the air mobile and air transport forces are given priority, this could suck funding out of heavy armored vehicle development and procurement. In addition, a reorientation of the Russian Army away from conventional confrontations with NATO and toward the capability to fight low intensity border wars would also lead to a shift of funding away from heavy forces. In Afghanistan, the newer T-64, T-72, and T-80 tanks were simply unnecessary, and attention turned instead toward giving the infantry heavy firepower support with light armored vehicles. Finally, the Russians could begin to enjoy export success with the existing tanks. The T-80/T-90 enjoy significant cost advantages over the M1A2, LeClerc, Leopard-2, and buyer hesitance due to the poor performance of the T-72 in Desert Storm may eventually wane with time. If export becomes a significant fraction of total Russian tank production, there might be a reluctance to tamper with a good thing, particularly if a new design would raise the cost to the point where Russian products are no longer competitive.

Scenario Two: Rationalization

Although then Defense Minister Grachev was pushing for the Mobile Force, the heavy mechanized force has powerful advocates throughout the Russian Army. The minimal objective of this constituency will be to rebuild the Russian Army's armored force so that it can win in any confrontation with neighboring Ukraine or China. In terms of main battle tanks, the most serious problem confronting the Russian Army is not the technological level of its best equipment. The T-80U and T-72B are clearly superior to anything in the Chinese inventory and equal to anything in the Ukrainian inventory. However, the vagaries of Soviet force dispositions have left the Russian Army with a hodge-podge of tank types. Only about twenty-five percent of the Russian tank force is made up of the better T-72B and T-80B/U tanks.[59] The immediate aim of the Russian armored force, if it can get the funding, will be to standardize the tank production on one existing type and begin to manufacture tanks in quantities sufficient to rationalize the force. This type could incorporate improvements over the existing T-80U and T-72B, such as the T-90 with the Shtora electronic self-defense suite. But it is unlikely to be a new main battle tank given the budget and economic realities.

Scenario Three: Revolution

Although Russian analysts were quick to publicly dismiss Operation Desert Storm as a barometer of the state of Soviet tank design, it is quite evident to many that the T-64, T-72, and T-80 has reached the end of its evolutionary potential. Russian tank design remains strong in the three traditional areas of armor, firepower and mobility. No matter how good Russian tanks appear on paper when comparing these features with foreign designs, Russian tanks come off second best in actual confrontations because they have serious deficits in other features. This could act as an incentive to press for a new tank design. Two situations could occur that would help foster such a program. On the one hand, a "Resurgent Russia" scenario could mean a return to pre-1991 priorities, and greater influence of the military and military industrial sectors. On the other hand, Russian tanks could continue to meet strong export resistance to unfavorable views fostered by the Desert Storm example. The defense industrial lobby could argue that the industry needs a new design to compete better with the LeClerc and M1A2 in the international market. This would simply be an extension of the current defense industrial lobby's argument that defense conversion needs funding, which is available only through hard currency export earnings.

Assuming that Russia decides to pursue a new main battle tank, how will past design trends affect future requirements? Recent Russian writing on tank design offers some hints of some of the lessons that would shape a future tank design:

Russian tank designs are poorly configured for crew performance. Excessively small interior volumes lead to premature crew exhaustion, asphyxiation and poor performance.[60] This is partly due to a traditional lack of concern over the human element in armored vehicle design, and partly forced on Russian designers due to the small size of the current main battle tanks. This can be remedied by allowing the size of a future main battle tanks to increase, or by reducing the amount of material stowed under armor (such as fuel and ammunition).

Russian tanks have mediocre sensors. The normal daylight vision optics are barely adequate, and the commander station design is poorly suited to allowing the commander to ride with his head out to observe terrain. This has been a traditional failing of Soviet/Russian tanks since World War 2, and may rest as much with training and tactics as technology. The thermal imager equipped M1A1s has made it clear that attention has to be paid to the ability of Russian tanks to acquire and engage enemy vehicles at greater ranges than possible with current daylight/passive image intensification night sights. This is likely to lead to wider adoption of thermal sights, or development of more advanced sights such as millimeter wave which could assist the crew in rapid target recognition. The Russians are already known to be working both approaches, with thermal imager-based fire control systems such as Agava and with automated millimeter wave systems such as Volya.[61]

Russian tanks are inordinately vulnerable to catastrophic internal ammunition fires. The small internal volume of Russian tanks means that ammunition compartmentalization is impossible. Russian designs have succeeded in moving nearly all ammunition below the turret line, but penetration by kinetic energy (KE) rounds is so destructive that it renders this design ineffective. There are three possible solutions. The Russians could move to a larger tank which would allow compartmentalization and decrease the number of rounds stowed onboard, or they could move to a radical innovation in gun design with less combustible propellant. Soviet tanks in the past have had as few as twenty-three rounds of ammunition (the IS-2M Stalin heavy tank), but the Russians would probably be reluctant to go below thirty rounds. This study has argued that the Soviet decision to concentrate on light main battle tanks (MBT) was a fluke caused by Khrushchev's rash and ill-considered cancellation of further Soviet heavy tank design in 1960. However, the Russian tendency towards main battle tanks smaller than their Western counterparts has become ingrained in Russian design practices. This may lead the Russians to examine solutions other than a significantly larger tank to cure this problem.

Russian tanks require improved main gun performance against new armor concepts and technologies. The Russian's may look to radical innovation in tank gun technology, such as a liquid propellant gun to improve their main gun performance. In addition, liquid propellants are potentially much less vulnerable to catastrophic fires than conventional propellant and they have the added advantage of permitting bulk storage that makes better use of limited internal stowage which reduces internal ammunition fires. Indeed, the Russians might attempt something even more extreme, such as stowing the propellant externally to free up limited internal hull volume, much as they do now with externally stowed fuel on current tanks. Such a direction would be in keeping with traditional Russian tendencies toward radical innovations in tank and armored vehicle gun designs. The Russians have traditionally set the pace in world tank gun development since the 1930s.[62] One reason why the Russians have managed to set the pace is that they have followed the old engineering maxim "Perfection is the enemy of excellence." They have been willing to adopt new technologies even if the bugs were not completely worked out. The US Army developed smoothbore guns and autoloaders in the 1950s but did not adopt them. The Germans developed reactive armor before the Russians but did not adopt it. The Russians may be willing to press ahead with radical gun solutions, and their technology in this field has almost always been apace with the West.

It is not clear what impact such a weapon would have on gun caliber. The increase in gun caliber over the last two decades has been predicated on the need to increase the gun tube pressures through increased projectile charges in order to increase the velocity of the APFSDS projectile. Some have suggested that some configurations of liquid propellant guns could effectively increase the velocity of the projectile. More efficient timed burning of the projectile as it moves down the gun tube would obviate the need for another increase in gun tube diameter to 140mm or more. However, failure to field a radical design could lead to the adoption of a larger gun bore to allow for increases in round weight and projectile velocity to penetrate current and future armor schemes.

Russian tank protection is adequate. Although Desert Storm showed the vulnerability of Soviet tanks to the standard NATO 120mm gun, the T-72B and T-80 tank have substantially better armor protection. This is unlikely to lead to complacency among Russian tank designers, who are no doubt aware of US and European work on 140mm tank guns. In addition, as a result of lessons learned from the Chechen war, improvements will be made to the ability of MBT's to fight in built-up urban areas. The Russians have examined radical alternatives in tank protection, such as the Drozd and the new Arena system, but both of these are tailored toward defeating shaped charge warheads rather than the KE threat. A future radical Russian MBT would probably incorporate a variety of defensive technologies to minimize its vulnerability. This is likely to include a new generation of reactive or active protection to minimize vulnerability to ATGMs and rockets, and an enhanced traditional armor, probably similar in protective quality to the M1A1HA tank. An alternative would be to disengage from the defensive armor treadmill. On the assumption that offensive tank gun and KE penetrator technology will continue to overcome all conceivable defensive technologies for the foreseeable future, the Russians may simply allow the armor protection to stagnate near current levels. The current levels, such as the M1A1HA (M1A1 with Depleted Uranium Armor) level, protects the MBT frontally against the vast majority of world threat systems. Furthermore, the end of the Cold War maybe the occasion to end the tank gun race and as a result, limit the growth of tank armor.

Russian tank mobility is adequate. The Russian T-80U has the world's highest weight-power ratio. Indeed, it is hard to see how any additional power will benefit the tank. However, higher weight-power ratios only contribute to higher speeds when combined with improvements in running gear. The end result is a MBT with greater enhancements to its mobility and agility. Higher speeds are a questionable proposition in combat since terrain conditions and human physiology place an outer limit on speeds that current tanks have already reached. Future Russian tank power plant concerns are more likely to focus on logistics than enhanced capabilities. This would include concerns over engine volume, engine fuel consumption, engine maintainability, and engine life expectancy.

2
ANTI-ARMOR DEVELOPMENTS

This section examines the development of Soviet and Russian anti-armor missiles (anti-tank guided missiles: ATGM). It begins with an overview of the development of Soviet anti-armor systems since World War 2. The primary focus here is on precision guided anti-armor weapons, not unguided anti-armor weapons such as rocket launchers, towed anti-tank guns, or anti-armor mines.[1] Anti-armor missile development is treated here in three main sections: heavy, light, and guided projectiles. Heavy ATGMs are those generally carried on vehicles or helicopters. Light ATGMs are those designed to be man-portable for infantry use. Guided projectiles are those launched out of armored vehicle gun barrels, primarily on main battle tanks (MBT) and the BMP-3 infantry fighting vehicle (IFV). This section concludes with a brief survey of precision guided artillery munitions with potential anti-armor uses, and an overview of likely development trends in this sector over the next few years.

World War 2 Lessons

At the outset of World War 2, the Red Army had four primary types of anti-armor weapons: anti-tank rifles, anti-tank guns, anti-tank grenades, and anti-tank mines.

Anti-tank rifles were the most common type of anti-tank weapon, with 216 anti-tank rifles in each rifle division. This type of anti-tank weapon had first been used by the German army in World War 1, and was common in European armies of the 1930s, including Germany, Poland, Italy, and Sweden. The Red Army was confronted by anti-tank rifles both in Poland in 1939 and Finland from 1939 to 1940. Indeed, it was the effectiveness of Finnish Lahti 20mm anti-tank rifle in the 1940 Winter War that led to the Red Army's adoption of such weapons. Each infantry battalion had two anti-tank rifle platoons (four anti-tank rifles each) and each infantry regiment had an additional anti-tank rifle company (two platoons). There were two types of anti-tank rifles in Soviet service, the PTRD Model 1941 (protivo-tankovie ruzhya Degtaryeva: Degtaryev anti-tank rifle) and PTRS Model 1941 (Simonov anti-tank rifle). The PTRS was the more complicated of the two, having a five-round magazine; the Degtaryev was a simple, bolt-action, single-shot weapon designed for ease of manufacture. Both were 14.5mm weapons, using the new BS-41 projectile. Both had identical anti-armor performance: 40 millimeters of rolled homogenous armor (RHA) at 300 meters and 25 millimeters of RHA at 500 meters. They were both very large and cumbersome weapons, the PTRS weighing 44.7 pounds, the PTRD weighing 35.2 pounds. There were 17,688 PTRD manufactured through September 1941 and a further 184,800 in 1942. The more complicated PTRS was manufactured in much smaller numbers: seventy-seven in 1941 and 63,308 in 1942.

These anti-tank rifles were generally crewed by two men. They were only marginally effective even during the first year of fighting, and quickly became obsolete as German tanks became more thickly armored to deal with the threat posed by the Soviet T-34 and KV tanks. For example, the standard German Pz.Kpfw. III Ausf. J tank of 1941 had fifty millimeters of protection on the mantlet and frontal armor (side armor was thirty millimeters). So the anti-tank rifles could only penetrate the dwindling number of light tanks, or penetrate the medium tanks at close-range from the side. Later tanks, such as the Panther were nearly invulnerable to such weapons. Nevertheless, such rifles remained in service through the end of the war due to the lack of an effective substitute. They were still useful against some targets including log bunkers even if their anti-tank capabilities were minimal. The Red Army also issued the infantry with the RPG-40 Model 1940 anti-tank grenade, but this was largely ineffective against any modern tank.

The Red Army did not follow the course of most other major armies in the European theatre which deployed various forms of rocket grenade launchers by 1943.[2] The Red Army had been one of the earliest developers of rocket weapons, including a 65mm rocket launcher very similar to the later US Army 2.35 inch "bazooka" rocket launcher. The Soviet weapon was experimentally developed in 1931, but proved ineffective against armor since it relied on a high-explosive blast warhead rather than a shaped-charge warhead. Likewise, the Soviets developed a number of recoilless rifles that were actually deployed in modest numbers in the late 1930s.[3] Without a shaped-charge warhead, they too were ineffective against tanks and their development atrophied before the outbreak of the war.

The Germans began introducing the Raketenpanzerbusche 42 (RP 42) in late 1942, which fired an 88mm anti-tank rocket. This was a crew served weapon much like the American bazooka, and did not become common until the fall 1943 campaigns. The Germans also deployed a less complicated rocket-propelled grenade, the Panzerfaust, in the autumn of 1943. Unlike the RP 42 or bazooka, this weapon was disposable once fired. It was so inexpensive that it could be issued *en masse* to the infantry, a distinct departure in anti-tank tactics from previous weapons.

In the absence of novel anti-armor technologies, and especially the shaped-charge, the Soviets made a number of attempts to develop improvised anti-tank weapons. The most unusual weapon deployed for the anti-armor role was the ampulomet. This was a crude mortar which fired a ball-shaped vial containing jellied gasoline to a range of 250 meters. It weighted twenty-eight kilograms, and was crewed by three men. It could be fired at a rate of about eight rounds per minute, but was neither very accurate nor very lethal when fighting tanks. It was used during the desperate days of late 1941 and early 1942, but soon passed from the scene due to its ineffectiveness. Considerable attention was paid to the use of man-portable flame-throwers as an anti-tank weapon, and in 1943, the Red Army even formed separate motorized anti-tank flamethrower battalions to this end.

Although the Red Army was undoubtedly aware of German anti-tank rocket grenade launchers from contact in combat, the response of the Soviet military industries was almost non-existent. There is some reason to believe that the Soviets may have begun manufacturing copies of the Panzerfaust in 1945. By the final months of the war, large stockpiles of German Panzerfausten had been captured, and were used on a wide scale by Soviet infantry to make up for deficiencies in their own anti-armor equipment.

Russian anti-tank artillery was deficient for much of the war. The standard anti-tank weapon at the outset of the war was the 45mm Model 1937 anti-tank gun. This was a license copy of the standard German Rheinmetall 37mm anti-tank used by the Wehrmacht during the first years of World War 2. The Soviet weapon had its bore increased to enable it to fire both Armor Piercing (AP) and High Explosive (HE) projectiles so that it could be used as an infantry gun.[4] There were forty-eight per rifle division: one anti-tank platoon per rifle battalion (two guns); one anti-tank battery with six guns per rifle regiment, and a divisional anti-tank battalion with twelve guns. The 45mm anti-tank gun was essentially the same as the weapon in Soviet tanks of the 1930s and could penetrate 42mm of armor at 500 meters. The Red Army had planned to replace it with the very effective 57mm ZIS-2 anti-tank gun, but production was curtailed in 1941 after only a few hundred had been built due to controversy over the need for even larger caliber weapons.

The 45mm anti-tank gun began to lose its effectiveness against German armor by mid-1942. As a stop-gap, the 45mm Model 1942 anti-tank gun was introduced which had a lengthened tube to increase anti-armor performance (armor penetration of 95 millimeters of RHA at 300 meters). This became the standard divisional anti-tank weapon due to the growing obsolescence of the anti-tank rifle. However, it was far from satisfactory. In a large defensive battle near Siauliai in August 1944, four 45mm anti-tank guns were lost for every German armored vehicle knocked out. This weapon could be supplemented by other divisional weapons, such as the 76mm ZIS-3 divisional gun, a weapon widely misconstrued by Western historians as an anti-tank gun.

With the appearance of the German Tiger I tank on the Leningrad Front in January 1943, the ZIS-2 57mm anti-tank gun was revived and put back into production. This could penetrate 100mm of armor at 500 meters using the normal AP round, and 145mm using a new sub-caliber projectile. This weapon was generally deployed in special anti-tank units, such as army- and front-level anti-tank brigades. It had superior performance to comparable Allied anti-tank guns of the period such as the British 6 pounder and its American copy, the 57mm M1 anti-tank gun; it was also comparable to the standard German anti-tank gun of the period, the 75mm PaK 40. It could penetrate most German tanks of the period at ranges under 500 meters except for the Tiger I heavy tank. The appearance of the Tiger led to the development of the heaviest Allied anti-tank gun of the war, the BS-3 100mm field gun. This weapon bore no immediate relation to the 100mm D-10 developed for the SU-100 tank destroyer, though they shared common ammunition. It was produced in very modest numbers during the war, and its use was confined to special anti-tank brigades attached at front- and army-level.

Technically, the Red Army was not as well equipped for anti-tank defense as their German opponents. Infantry anti-tank weapons were an area of special weakness, but divisional anti-tank weapons also were inadequate in the later years of the war. While the 57mm and later 100mm guns could have solved this problem, they were not produced in sufficient quantities to equip many rifle divisions. This equipment deficiency was not particularly manifest during several of the more prominent armored clashes of the war such as the Kursk-Orel battles in the summer of 1943, since the Red Army could concentrate large quantities of non-divisional anti-armor assets in key sectors where major tank assaults were expected. Nevertheless, Soviet infantry never posed the threat to armor in close-grained terrain that German infantry posed to Allied armor during the war.

Post-War Anti-Armor Development

The Soviet Army began a broad effort to improve its anti-armor capabilities in the wake of World War 2. The lack of infantry anti-tank weapons was addressed by Design Bureau 3 (KB-3) in Sofrino, later called the GSKB-30. This design center was organized immediately after the war, and accumulated a number of German weapons engineers familiar with the Panzerfaust and other infantry anti-armor weapons. The first product of this design center was the RPG-2. Although closely linked to the wartime German Panzerfaust in warhead and propellant technology, the RPG-2 was not disposal, but rather reloadable like the American bazooka. However, the novel design made it a much smaller and more compact weapon than its American counterpart, the 2.35 inch bazooka. Production of this weapon began in 1949 to replace the archaic anti-tank rifles. Each rifle platoon received three RPG-2 (one per squad); there were 243 per division (compared to 216 anti-tank rifles in a 1943 rifle division). By the early 1950s, the 45mm anti-tank gun was replaced by the ZIS-2 57mm anti-tank gun, with three per rifle battalion.

As in many other armies of the period, the Soviet Army began to explore the use of recoilless rifles as a direct fire weapon with significant anti-armor capability. The KB-3 at Sofrino developed two related weapons, the light-weight 82mm B-10 and the towed 107mm B-11. The B-10 was issued on a scale of three per rifle battalion, supplementing the 57mm anti-tank gun. The qualitative improvements in anti-tank weapons extended upwards in the division. Rifle regiments replaced their 76mm ZIS-3 divisional guns with six 85mm divisional gun which had significantly better anti-armor capabilities. In addition, the regiments received six 107mm B-11 anti-tank recoilless rifles. These anti-armor improvements were evolutionary rather than revolutionary.

Tank armor continued to increase, so that by the mid-1950s, NATO main battle tanks had an effective frontal protection equivalent to about 300 millimeters of steel armor; the penetration capability of the heaviest towed anti-tank guns of the period such as the Soviet 100mm gun firing armor piercing was about 250 millimeters of RHA. Of the changes, potentially the most important was the addition of RPGs to the infantry platoons. By the 1960s, each infantry squad would be armed with an anti-armor weapon thereby substantially increasing the depth of anti-armor protection for the infantry. This was the beginning of a trend, first started by the Germans in World War 2, that considerably increased the number of infantry anti-tank weapons that made it increasingly risky for tanks to encounter enemy infantry at close ranges.

The most revolutionary anti-armor technology to appear in the 1950s was the anti-tank guided missile. The Germans had developed the X-7 wire-guided anti-tank missile in 1945, but it was available too late to see combat use. The ATGM offered several intriguing possibilities. To begin with, ATGMs held the promise of substantially better accuracy at long range than conventional anti-tank guns. Soviet wartime experience had concluded that it took eight to ten rounds from an anti-tank gun to knock out a tank at 1,000 meters.[5] Missiles, due to their lack of recoil, could be fired from lighter weapons platforms, making them smaller and more mobile (and therefore less vulnerable to counterfire) than towed artillery. Missiles were not limited to the diameter of a gun tube, and so their shaped charge warhead could be considerably larger than comparable anti-tank guns, thereby offering better anti-armor penetration and better behind-armor effects.

France pioneered ATGMs. In 1946, the French government directed its Arsenal de l'Aeronautique at Chatillon-sous-Bagneux to begin work on guided missile technology using the experience of captured German missile engineers. This involved the further development of the German X-4 air-to-air missile into the AA-10, and the development of the X-7 anti-tank missile into the SS.10. The SS.10 was the world's first series produced anti-tank missile and entered production in 1955 by Nord Aviation (now Aerospatiale). It was exported to several countries including the US, and was the first ATGM used in combat (Israel versus Egypt in 1956).

The appearance of the SS.10 in France, quickly followed by the SS.11 and the Entac, was followed with interest in the USSR. In 1956, a dedicated program began in the Soviet Union to deploy comparable weapons. The 1957 program can be traced to several factors. From the tactical side, the "revolution in military affairs" brought about by tactical nuclear weapons sounded the death nell for conventional towed artillery in the view of many tacticians. Towed anti-tank guns were especially vulnerable to radiation, since the crew was completely exposed. There was no easy way to enclose the crew short of mounting the weapon in a heavy armored vehicle. A missile-armed vehicle could perform the same function, but the lack of recoil and light weight of the launcher would entail a much lighter and less expensive chassis than a gun-armed tank destroyer. Of equal importance, Nikita Khrushchev was becoming infatuated with missiles as a revolutionary alternative to conventional weapons.

Khrushchev was convinced that the Soviet Army had to be trimmed back after its large expansion in the early 1950s. He sought a leaner and more modern Soviet Army as a more cost effective alternative to the outdated Stalinist force of the early post-war period. Khrushchev's missile panacea paralleled the "jeune ecole" movements earlier in the century in many navies (including the Russian) which saw motor torpedo boats and submarines as a less costly and more effective alternative to battleships and other large warships. Khrushchev saw anti-tank missile launchers as an alternative to heavy tanks, and tactical ballistic missiles as an alternative to heavy artillery. There were many technical reasons to doubt the efficacy of this approach, but in the heavily centralized post-Stalin system, these arguments carried little weight. Khrushchev intended not only to revolutionize the army, but to modernize the defense industrial base at the same time. Some bureaus, closely associated with conventional weapons, were simply closed. The artillery design bureaus were heavily hit, most notably Grabin's bureau which had developed many of the most successful World War 2 anti-armor weapons including the ZIS-2 57mm anti-tank gun, the ZIS-3 76mm divisional gun, and F-34 76mm tank gun. Other bureaus were allowed to remain open, but were redirected to the new "hi-tech" fields of missile technology.

The development of anti-tank guided missiles did not completely end development of other anti-armor weapons. In the towed artillery field, the 85mm gun was replaced by the smooth-bore 100mm T-12 and MT-12.[6] The B-10 82mm recoilless rifles were replaced by the SPG-9 82mm recoilless rifle. The RPG-2 was replaced by the RPG-7, and a host of evolutionary outgrowths. However, the arrival of ATGMs placed a cap on the procurement of other technologies, particularly at the higher end such as anti-tank guns.

Early Soviet ATGM Development

The earliest work on guided anti-tank missiles was begun by KB-1 (Design Bureau-1), a large collection of design efforts under NKVD/NKGB direction in Moscow. This bureau was responsible for integrating German defense technology into the Soviet system, particularly in high technology fields such as missile guidance, radars, avionics and computers. One of the senior engineers, A. A. Raspletin, was heavily involved in early missile guidance research. Work on anti-tank guided missiles began in earnest in the mid-1950s in response to developments in the West. The Council of Ministers directed the precision machinery industry to take on this new work, the economic sector already involved in small arms and light weeapons development and manufacture. As in most other Soviet weapons systems, the advanced research was conducted by Scientific Research Institutes.

The principal research institute of the small arms industry was NII-61 in Klimovsk, later known as TsNII-Tochmash (Central Scientific Research Institute for the Precision Machinery Industry). Several small arms bureaus were directed to begin ATGM research including the pistol design bureau headed by Nikolai Makarov; the mortar design bureau in Kolomna headed by Boris Shavyrin; and the aircraft cannon bureau OKB-16 in Moscow headed by Aleksandr Nudelman.[7] Of the three bureaus, only the latter two were successful, and Makarov's design remains unknown. These new weapons entered production in the late 1950s, and their development is recounted in more detail below.

Heavy ATGM Systems
AT-1 Snapper

The Special Mortar Design Bureau (SKB Gladkostvolnoi artillerii) in Kolomna, headed by Boris Shavyrin was one of the first bureaus assigned to the task of developing guided anti-tank missiles. This system was developed in parallel to the Nudelman bureau's Falanga, and an unknown missile by the Makarov pistol bureau. The system was designated 2K15, and the missile was designated 3M6 Shmel (Bumblebee). The Shmel was patterned after similar French missiles of the period, notably the Nord Aviation SS.10, though it was significantly larger.[8] Guidance was by manual-command-to-line-of-sight (MCLOS) with the command signals passed from the operator to the missile over a thin trailing wire link. Flight control was unconventional, using a set of vibrating spoilers. The missile was too large to be man-portable, so was mounted on vehicle launchers.

There were two standard launcher configurations for the Shmel, the 2P26 and 2P27 tank destroyers. The 2P26 "Baby Carriage" consisted of four Shmel launch rails mounted on the rear of a modified GAZ-69 light truck. The missile operator had a small joystick control and could guide the missile either from a launch station on the truck, or could remotely launch the missiles from a short distance away using a portable control and wire reel. The main advantage of this system was that it was very cheap; the main disadvantage was that it had no armored cover. This led to the development of a corresponding

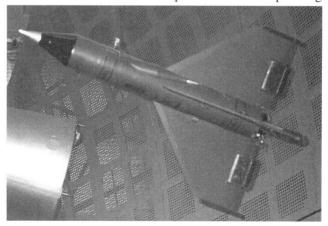

3M6 Shmel (AT-1 Snapper).

2P26 tank destroyer (GAZ-69 with AT-1).

9P27 tank destroyer (BRDM-1 with AT-1).

armored version, the 2P27, mounted on a modified BRDM-1 armored vehicle chassis. The 2P27 mounted three missile launchers in a rear compartment on the BRDM-1 and the missiles were protected under an armored cover until they were deployed for launch.

The Shmel entered Soviet service in 1958 and was first publicly displayed in 1963. They were intended to supplement more conventional anti-tank weapons like the T-12 100mm anti-tank gun, or gun armed tank destroyers such as the SU-122. While the accuracy of conventional guns dropped dramatically at ranges beyond 1,500 meters, the accuracy of the Shmel remained high as far as 2,500 meters. The new wheeled tank destroyers were typically deployed in anti-tank batteries attached to mo-

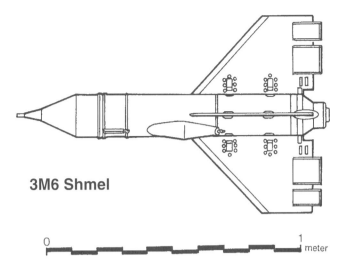

3M6 Shmel

torized rifle regiments. Each battery had three platoons with three launch vehicles and a single command BRDM each, totalling nine 2P26 or 2P27 tank destroyers and four BRDM-1, or BRDM-2 command vehicles.

The Shmel was still fairly primitive. The Shmel used manual command-to-line-of-sight (MCLOS) guidance which required a very experienced operator to obtain the desired level of accuracy. The missile was very slow, and took thirty seconds to fly to its maximum range of 2,500 meters (compared to about three seconds for an anti-tank gun round). During this thirty seconds, the tanks could take evasive action, or try to supress the missile by firing on the gunner. The missile was so large that only three could be carried in the BRDM-1/2P27.

Rate of fire was also extremely slow since only one missile could be controlled at a time from each launch vehicle. The MCLOS guidance method required considerable dexterity and skill on the part of the operator and required a high level of continual training. Although such missiles had a high theoretical probability of kill, in actuality, their performance was signficiantly less. One of the few engagements in which both MCLOS and the later improved SACLOS missiles were used was the 1972 North Vietnamese offensive; this provides some sense of the comparative performance. US forces used both the MCLOS SS.11, which was similar in technology to the Shmel, as well as the new SACLOS guided BGM-71 TOW. During the fighting, the SS.11 displayed a probability of hit of under ten percent while the TOW demonstrated over fifty percent.

The first combat employment of the Shmel was by the Egyptian Army in the 1967 war, which used the 2P26 vehicles. Few were used in combat, and there was only one tank kill attributed to the system. During the fighting, its probability of hit was under twenty-five percent due to the difficulties inherent in any MCLOS guidance system. The acknowledged shortcomings of this system

led to the deployment of a second generation of guided anti-tank missiles beginning in the 1960s.

AT-2 Swatter

The 2K8 Falanga radio-command guided weapon was developed by the Nudelman OKB-16, a design bureau previously associated with aircraft cannon development. It was developed as a heavy ATGM for both ground-based launchers and helicopters. The original version entered service about the same time as the Shmel in 1962. It was described in one Russian source as "notable for its complexity and low reliability." The original Fleyta (Flute) missile (AT-2a) employed manual-command-to-line-of-sight guidance (MCLOS) with a radio command link.

As in the case of the Shmel, the Falanga system was too large to be used in a man-portable role, and so was mounted on a light armored vehicle chassis, the 2P32 on the BRDM-1. The Falanga offered several advantages over the Shmel: it was more compact, so a 2P32 Falanga could carry four launchers instead of the three on the 2P27 Shmel. It was deployed in the same type of anti-tank batteries as the 2P27 Shmel, with nine 2P32s per battery.

The early version of the Falanga was plagued with technical problems. The original missile was judged to have inadequate range, and an improved missile, the 9M17 Skorpion (AT-2b), was developed as a successor for the improved 3K11 Falanga system. It used the same MCLOS radio guidance system, but with an increase in range from 2.5 to 3.5 kilometers. Externally, both missiles were very similar in appearance but could be distinguished by small details such as the forward canard fin shape. The standard production version was the improved 9M17M Skorpion-M and it was accepted for service in 1968.

In the mid-1970s, the Falanga system was substantially improved by conversion to a semi-automatic-command-to-line-of-sight (SACLOS) command system. In

3M11 Falanga (AT-2 Swatter).

9P110 tank destroyer (BRDM-1 with AT-3).

9P122 tank destroyer (BRDM-2 with AT-3).

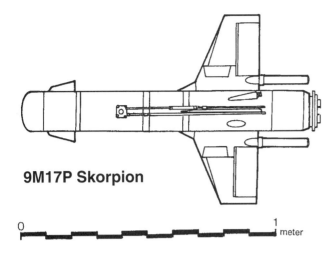

9M17P Skorpion

this system, the operator merely had to keep the crosshairs on the target, and the guidance system determined the proper flight adjustments. This substantially improved hit probability and at the same time decreased the inordinate training requirements of the earlier MCLOS guidance system. The new missile for this system was the 9M17P Skorpion-P (P= poluavtomaticheskiy, semi-automatic) and it was accepted for service in 1969. The final version was a product-improved SACLOS type, the 9M17MP which had a more powerful engine, and an improved guidance signal lamp.

By the time that the newer SACLOS 9M17P missile arrived, the BRDM-1 vehicle was out of production. As a result, it was mounted on the BRDM-2 as the 9P124, which was accepted for service in 1972. The older 2P32 vehicles were later upgraded to handle the improved 9P17M missile. Neither of these types was fielded in significant numbers, as the Soviet Army was generally shifting to the use of the medium range ATGMs such as the 9M14 Malyutka (AT-3) and the later SACLOS Fagot/Konkurs (AT-4/-5).

The Falanga system was the first Soviet ATGM to be deployed from helicopters. It was fitted in small numbers to an armed version of the Mil Mi-4A helicopter, the Mi-4AV. The helicopter-based version of the Falanga system required two new systems, the light weight 2P32M/K-4V launcher system, and the Raduga-F guidance antenna package. The most widely used version of the helicopter-based Falanga system was on the Mil Mi-24 Hind helicopter. It was used as the standard anti-tank system on the initial Mi-24 Hind-A and on the more common Mi-24D and export Mi-25 Hind D series.

The missile is aimed from the weapons officer's station in the nose of the Mi-24D, using a telescopic sight (varies from 3.3 to 10 power) mounted on the right side of the cockpit. Guidance signals are transmitted via the Raduga-F antenna, located under the left side of the forward nose section. The missile tracker is integrated into the electro-optical sighting system mounted under the right side of the nose. These helicopters were initially fitted with the Falanga-M MCLOS system, but this was replaced by the Falanga-P using the 9M17P Skorpion-P missile, and the later Falanga-MP system with the 9M17MP missile. This system is still in widespread use around the world with Mi-24 Hind operators, and has been extensively used in combat including the 1982 Lebanon war, the Iran-Iraq war, Angola, and Afghanistan.

The assault version of the Mil Mi-8 Hip transport helicopter, the Mi-8TB (Hip-E), required some modification to the system to integrate it into the helicopter since it lacks the electro-optical weapons aiming system fitted to the Mi-24. This led to the Falanga-DB system which uses the 9M17DB Skorpion-DB missile and a separate gyro-stabilized missile tracker assembly and guidance antenna. This version was used mainly by the Soviet Union.

3M7 Drakon

In 1956, the four main tank design bureaus were ordered to begin work on a missile armed tank. This took different forms, with the Isakov bureau developing a tube-fired weapon, and the Kartsev and Kotin bureaus adapting rail-launched missiles to armored vehicles. These programs were delayed until the early 1960s when the missiles matured to the point that engineering development of the associated vehicles could begin.

The Kartsev design bureau at Nizhni Tagil began work on a missile-armed derivative of the T-62, designated the Obiekt 150. The Obiekt 150 was based on a standard T-62 hull with a modified turret. The turret contained a missile launcher which popped up from a roof hatch. A magazine of twelve to fifteen missiles was contained within the turret, along with the associated launcher guidance system. The 3M7 Drakon (Dragon) missile was developed by the A.E.Nudelman OKB-16 design bureau in Moscow, and the missile guidance system by A.A.Raspletin's KB-1 design bureau.

A small test series of Obiekt 150 were ordered as the IT-1 (Istrebitel tankov= Tank destroyer). In 1964, Khrushchev had been removed from power and so the pressure to adopt missile tanks disappeared. This concept was extremely unpopular in the Soviet military which resented Khrushchev's intrusion into the minutest details of tactics and technology. In 1965, the IT-1s were used to form two tank destroyer battalions, one manned by tankers, the other by artillery troops to test the concept. One battalion served with a motor rifle division in the Carpathian Military District, the other in the Byelorussian MD. Small scale production of the vehicle took place in 1968 to 1970. However, the IT-1 was not well liked in service and the vehicles were converted to recovery vehicles. A turbine powered version was also developed, the Obiekt 150T for further trials. So far as is known, no other use of the Drakon missile was envisioned.

The Kotin bureau's counterpart to the Obiekt 150 was the Obiekt 287, based on a modified T-64 chassis. This was armed with an unusual weapon system developed by the Tula KPB (Instrument Industry Design Bureau). In place of a conventional turret, the Obiekt 287 had two small fixed cupolas armed with the 73mm 2A28 Grom low-pressure gun. Between these gun positions was a hatch for a pop-up missile launcher, armed with the new Taifun (Typhoon) anti-tank missile. Both the guns and missile launcher were serviced by an autoloader in the turret basket below, with fifteen stowed missiles and thirty-two gun rounds. The Taifun was apparently radio command guided, but details are almost entirely lacking. In any event, the Obiekt 287 was not accepted for service, and the Taifun missile never entered series production.

AT-5 Spandrel

Development of a new generation of wire-guided anti-tank missiles was begun in 1958 to 1962 by the Tula Instrument Design Bureau (Konstruktoskogo biuro priborostroeniya). The general designer of the facility was Vasily Gryazev, and the anti-tank missile program was headed by chief designer Arkadiy G. Shipunov who had transferred to the facility from the anti-tank missile design group at the TsNII Tochmash (Central Scientific Research Institute for Precision Machinery institute) in Podolsk. The Tula KBP was previously involved in aircraft automatic cannon design, and the decision to begin anti-tank missile work there was another manifestation of Khrushchev's plans for accelerating the development of advanced technology.

KBP was assigned to develop two different missiles: the light man-portable 9K111 system and a heavier, longer ranged 9K113 system for arming tank destroyers and other armored vehicles. The SACLOS guidance system was developed by K. Zvyagin and V. Kurnosov of the TsNII Tochmash. This program was a direct counterpart to the Franco-German ATGM effort that eventually resulted in the Milan and HOT missiles, and there has long been suspicion that design similarities between the systems was more than coincidental. The heavier, longer-range 9K113

9M113 Konkurs (AT-5 Spandrel).

Konkurs (Konkurs= Contest; US/NATO name: AT-5 Spandrel) was a direct counterpart to the Euromissile HOT. The 9M111 missile was accepted for service first, in 1970, followed by the 9M113 in 1974.

The 9M113 missile is delivered in a glass-reinforced plastic launch tube/container. At the rear of the container is a gas-generator to eject the missile. Once clear of the tube, a sustainer engine ignites, venting its exhaust from two ventral ports. The larger container forward of the gas-generator and behind the engine is a spool of trailing wire for the command link and an infrared lamp for missile tracking. Missile stability is provided by four gas-inflated thin metal fins which curl around the missile body while in the launch cannister. The basic 9M113 Konkurs uses a 2.7 kilograms 9N131 shaped charge warhead with a penetration of about 600 millimeters of RHA. The improved 9M113M Konkurs-M uses a tandem warhead with an extensible stand-off probe for dealing with explosive reactive armor and has an advertised penetra-

9M113M Konkurs-M (AT-5b Spandrel).

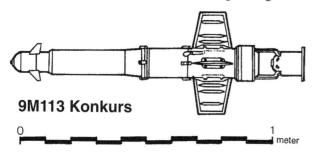

9M113 Konkurs

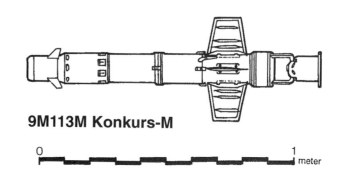

9M113M Konkurs-M

9P148 tank destroyer (BRDM-3 with AT-5 Spandrel).

9M111 Fagot and 9M113 Konkurs missiles on 9P148 launcher.

tion of 750 to 800 millimeters of RHA.

The 9K113 Konkurs missile was intended to be used as secondary armament on armored infantry vehicles such as the BMP-2 and BMD-2. These use a fixed 9S428 guidance controller and launch system on the vehicle roof with a 9Sh119M1 optical fire control system integrated into the gunner's BPK-1-42 fire control system. Four missiles are carried internally and are reloaded via the gunner's hatch by traversing the launcher and folding it vertically to accept another missile tube. The launch system can also fire the smaller 9M111 Fagot missile.

The other major role for the 9K113 Konkurs is to arm wheeled tank destroyers based on the BRDM-2 (GAZ-41-08) chassis with a five-rail launcher on the roof. It has a crew of two, the driver and gunner/commander. The five missiles can be reloaded using an automated system which takes twenty-five seconds. There are two versions of this tank destroyer. The initial 9P137 was designed to carry and launch the 9M113 Konkurs exclusively; it carries a total of fourteen 9M113 missiles internally. The 9P148 tank destroyer can fire both the smaller 9M111-2 Fagot missile, and the 9M113 Konkurs. The usual mixture is ten of each type of missile, for a total of twenty missiles. Both these vehicles have an electro-optical tracker assembly on the right superstructure side. Although the basic guidance method is SACLOS, there is a warning system incorporated in the tracker to alert the gunner when optical jamming is taking place, and there is an elementary manual system override for this eventuality.

The main production centers for the 9M113 Konkurs in Russia are the Degtaryev Machinery Plant in Kovrov, the Shchit Machinery Plant in Izhevsk and the Tulskiy Armaments Plant in Tula. Export prices for these systems as of 1992 were $13,000 for the 9M113 Konkurs missile and $135,000 for the improved 9P135M firing post. License production of the 9M113 Konkurs is currently being undertaken in Bulgaria by the Vazov Engineering Plant in Sopot, in India by Bharat Dynamics, and in Germany on the basis of an existing East German government agreement with the USSR.

AT-6 Spiral

The 9K114 Shturm missile system has no immediate

9M114 Shturm/Koken (AT-6 Spiral).

Western counterpart, though it is somewhat similar to the US Army's AGM-114 Hellfire in role. It was developed by the Kolomna Machine Design Bureau under S. P. Nepobidimy, the reorganized bureau formerly headed by Boris Shavyrin which had developed the Shmel (AT-1) and Malyutka (AT-3). It was primarily intended as a helicopter-launched anti-armor missile, and was originally called the AS-8 under the US designation system, later being reclassified as the AT-6 Spiral. The 9K114 system is called Shturm-V (Shturm= storm; V= vertoletniy, helicopter) in the helicopter version and Shturm-S (S= sukhoputniy, ground) in the ground-launched version.

The 9K114 Shturm was developed as the first supersonic Soviet ATGM. The slow speeds of the Falanga meant that the helicopter had to remain exposed too long while engaging hostile targets. Like the Falanga that preceded it, the Shturm uses radio-command guidance, rather than the wire guidance or laser guidance used on most Western ATGMs. It was first deployed on the Mi-24V Hind E in 1978. Like many contemporary Russian tactical missiles, the 9M114 Kokon (Cocoon) missile is transported and launched from a glass-reinforced plastic storage/launcher tube. The missile uses a Soyuz NPO solid-rocket with a small booster stage at the tail of the missile to propel it out of its launcher tube. The missile guidance system on the Mi-24V Hind E is configured much the same as the Falanga system on the Mi-24D, with a new podded Raduga-Sh antenna replacing the Raduga-F under the left nose of the helicopter and a new 8-power telescopic sight for the weapons operator integrated into the new electro-optical system/tracker. The

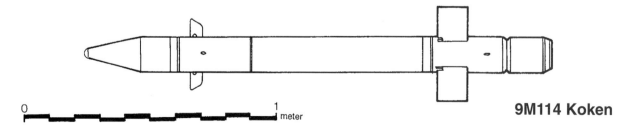

9M114 Koken

AT-6 Shturm-S.

radio command system operates on five fixed frequencies in the millimeter wave band with two coding systems to minimize the risk of electronic jamming.

The Shturm was first employed in Afghanistan. There were guidance problems using the missile against mujihadeen defensive positions in caves and rocky areas according to some Russian accounts. The shaped charge warhead had dissappointing performance against such targets, leading to the development of a special enhanced blast warhead as mentioned below.

There are at least four versions of the 9M114 missile. The basic 9M114 employs a conventional shaped-charge anti-armor warhead capable of penetrating up to 600 millimeters of RHA. This missile was used in Afghanistan, but the warhead was only marginally effective against dug-in guerilla forces or against reinforced buildings and caves. As a result, a special bunker-buster warhead was developed, using an enhanced blast warhead similar to the fuel-air explosive warhead used with the RPO.[9] This warhead uses a dispersed metallic powder for the blast, not a liquid.

In recent years, two additional variants have been fielded with an enhanced anti-armor warhead capable of penetrating tanks protected with reactive armor, presumably a tandem shaped-charge type. These variants differ in the length of their rocket engines, one with a range of six kilometers and the other with a range of seven kilometers. These are referred to here for convenience as the 9M114M1 and 9M114M2 on the accompanying data chart, but these designations are not known. As mentioned in the section below, the relation of these variants and the 9M120 Ataka is not clear (they may in fact be the Ataka). The 9M114 Kokon missile is manufactured by the Shchit Mechanical Plant in Izhevsk. Its 1992 export price was $50,000.

In 1990, the 9K114 was adapted to the ground role, being mounted on an automated launcher on the MT-LB light armored vehicle as the 9P149 tank destroyer. This vehicle was developed by the design bureau at the Saratov Machinery Plant headed by Viktor Petrov; the MT-LB itself was developed and manufactured at the Kharkov Tractor Plant. The 9P149 tank destroyer is currently being marketed for export, but it was not adopted by the Russian Army. The export price for the 9P149 vehicle was stated to be $1 million in 1992, a not inconsiderable price in view of the $1.6 million price of a fully equipped T-72S tank. The launcher assembly is being offered both in the 9P149 form, and as a modular unit that can be used as a static pillbox, or mounted on various armored vehicle hulls including surplus T-55/T-62 medium tanks or light armored vehicles. The system is entirely automated, with the launcher assembly protected under armor until placed into use. The autoloader assembly holds twelve missiles and the rate of fire is three to four missiles per minute.

In recent years, it the Shturm-V system has been

9M114 Shturm-V (AT-6 Spiral on Mi-28).

mounted on other attack helicopters including the Kamov Ka-29 Hormone naval helicopter. Besides its use as an anti-armor system, it is also employed in a secondary role as an air-to-air missile against enemy helicopters and slow flying aircraft.

AT-9

The 9M120 Ataka (AT-9) is a new derivative of the 9M114 Shturm (AT-6 Spiral) and also developed by the Kolomna design bureau. The Ataka missile has two significant differences from the earlier type. The missile fuselage is longer, with a lengthened solid rocket engine for greater range (six versus five kilometers). In addition, a larger range of warheads have been developed. It is not clear what the relationship of this missile is to the upgraded 9M114 missiles mentioned in Russian sources in 1992. It is possible that these are in fact the 9M114 variants and have simply been renamed, or they may in fact be further evolutions.

There are at least three different versions of the Ataka distinguished by warhead types. The anti-tank version uses a stand-off probe to assist in penetrating advanced and reactive armors. The bureau's advertisement claims that the 9M114 will penetrate not less than 850 millimeters of homgenous steel armor, while the 9M120 is capable of penetrating 800 millimeters of RHA, after having stripped away reactive armor. As in the case of the 9M114, there is also a special blast warhead using fuel air explosives for attacking soft targets, bunkers, ships, and other targets.

The third and most novel warhead type is a special anti-helicopter warhead. This type has an expanding rod warhead and is detonated by a proximity fuze. This is the only known case of an anti-tank missile specifically configured for the anti-helicopter role. Most helicopter defense missiles have relied on manportable, infra-red homing air defense missiles (in the Russian case, the Igla/SA-16).

The 9M120 Ataka missile has been developed both

9M120 Ataka-V (AT-9).

for helicopter and vehicle platforms. The Ataka-V (V=vertolet: helicopter) has been fitted to the naval Ka-29TB marine attack helicopter, the Mi-24 Hind and the Mi-28 Havoc. The Ataka-V system on the Mil Mi-28 Havoc carries double the number of missiles as the earlier Shturm system. It is not clear if the platforms have to be modified to fire the new missile, but if so, the similarity between the two missiles probably require only minimal modification. In the ground role, the Ataka-S (S= sukhohoputniy: ground) appears to be interchangeable with the 9M114 on the 9P149 tank destroyer on the MT-LB chassis.

AT-14

The Kornet (AT-14) is the latest product of Tula's KBM design bureau. The missile is obviously based on the bureau's earlier 9M131 Metis-2 and resembles an enlarged version of this missile. However, there are several significant differences. The Kornet is considerably larger than the Metis-2. The reason for this is that the roles are entirely different. The Kornet is clearly intended as a heavy anti-tank missile for use from vehicles. It is intended to replace the Konkurs (AT-5) not the manportable Metis or Metis-2. Secondly, the Kornet uses

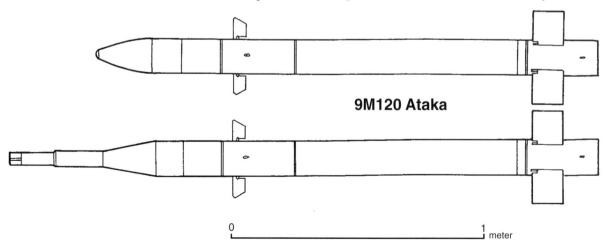

Anti-Armor Developments

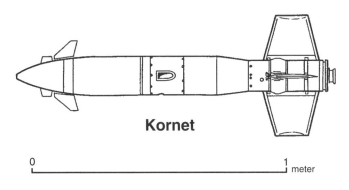

Kornet

9M133 Kornet (AT-14) launcher on UAZ-469.

laser beam riding guidance, not wire guidance as has been the case with all previous KBM anti-tank missiles.

In its basic configuration, the Kornet is launched from a simple tripod firing post. The most significant innovation on the firing post is that it has been fitted with a thermal imaging sight from the outset, providing it with the ability to engage targets under a broader variety of battlefield conditions. The electronics package consists of the thermal imaging sight on the right side of the launcher, and the guidance assembly on the left. An optical elbow has been provided, allowing the operator to remain prone to minimize exposure to enemy fire. The launcher is relatively heavy at nineteen kilograms (forty-two pounds) and the missile in its container weighs a further twenty-seven kilograms (sixty pounds). It does not appear that the Kornet is intended as a truly man-portable system. Rather, it will be carried on vehicles, and dismounted when necessary. One of the standard configurations is to mount the launcher and missiles on a light truck, such as the UAZ-469. This is the same configuration developed for the Konkurs, with a special reinforced mounting over the rear bed of the vehicle, and special ammunition stowage racks configured into the vehicle. Such a vehicle will have obvious appeal to light forces, since it can be more readily deployed than previous armored tank destroyers such as those based on the BRDM-2 chassis.

A dedicated tank destroyer vehicle for the Kornet was also developed, the 9P162 based on the BMP-3 universal chassis. The first information about the 9P162 Kornet missile tank destroyer became public in mid-1995 when the vehicle was displayed on Russian television and was displayed for the first time at the IDEX-97 show. The Kornet (AT-14) missile itself debuted in 1994 and Russian designers have stated it was designed to defeat the U.S. M1A1 Abrams.

The 9P162 Konkurs missile tank destroyer vehicle was developed by the Volsk Mechanical Plant in the Saratov region of Russia. This bureau designed the ear-

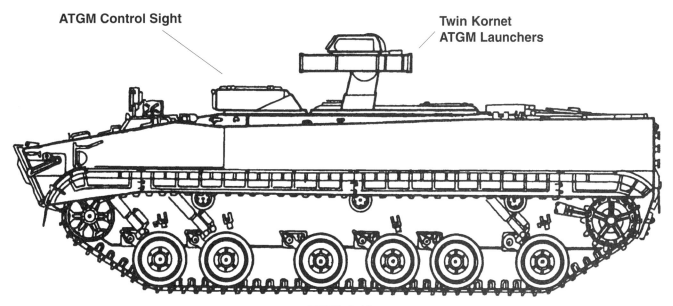

BMP-3 Kornet
Self-propelled Anti-Tank Missile Complex

The ground mount for the Kornet is seen here next to the vehicle-mounted version

The top deck of the BMP-3 Kornet.

9M133 Kornet launcher (AT-14).

9M133 Kornet (AT-14).

lier 9P149 Shturm (AT-6) tank destroyer on the MT-LB chassis, a type which does not appear to have ever won any orders from either the Russian Army or export clients. The 9P162 Kornet vehicle is one of two missile-armed tank destroyers on the BMP-3 chassis, the other being the 9P157-2, armed with the Krizantema (AT-X-15) anti-tank missile. These appear to be competitive types, the Kornet having been developed by the Tula KBP design bureau while the Krizantema was developed by the rival Kolomna KBM. The Krizantema tank destroyer has not yet been publicly displayed at any arms show, although photos and brochures about the system have been released.

The Kornet missile tank destroyer vehicle consists of a standard BMP-3 chassis, with the turret replaced by a new missile armament system. This is a traversable platform with two elevating launchers. The launchers consist of a retractable device which grasps the missile launch canister from above. Each of these launchers is reloaded through a rectangular door immediately below the launcher assembly. Each launcher is fed by a revolving drum containing six Kornet missiles. According to design representatives, one drum is generally armed with the anti-tank version of the missile, while the other drum is armed with the high-explosive (FAE) fire support version. The vehicle contains a total of sixteen missiles: twelve in the revolving drums under the launcher, and four additional missiles stowed in racks on the floor of the vehicle under the turret. These can be reloaded into the revolver racks by the crew from under armor protection. The vehicle is fairly spacious inside, and it would appear to be possible to carry more missile canisters than the official total of sixteen. The tunnel at the rear of the vehicle over the engine does not appear to be in use, although it is not readily accessible by the crew from under armor protection.

Russian descriptive literature indicates that the crew is two. However, from an inspection of the vehicle, there are in fact three seats: the gunner under the guidance/sighting cupola in the left front corner of the hull, the driver in the center and a third crew position to the right front of the hull. This third position may be occupied by the gunner when the vehicle is in transit, but more likely it is the location for a vehicle commander. Rather surprisingly, the vehicle has no self-protection weapon such as a hull machine gun. This is presumably due to the range of its weapons which would keep it significantly behind the forward edge of battle.

The missile fire controls are located under a traversable sighting unit located in the left forward corner of the hull. This contains the laser guidance for the missile. The limits on the traverse of this sight of about 240 degrees due to hull obstructions towards the rear limits the engagement envelope of the missile system. Since 1996, the Russians have advertised the 1PN80 Kornet-TP thermal imaging system for the Kornet system. The thermal imaging sight has a detection range of five kilometers, an identification rage of 3.5 kilometers and was developed by the State Institute for Optical Applications (GIPO). The missile gunner also has a mode control panel below the main missile control panel, but fixed to the vehicle bulkhead, and a radio below this attached to the floor. The missile gunner does not have a hatch in this area, but apparently accesses the hull through a large hatch on the roof forward of the two o'clock position of the turret. This hatch is also used for reloading the missile cassettes.

The Kornet missile follows the general layout of the Metis-2 missile but has four rear fins instead of three. The engine configuration of this missile appears to be significantly different than previous Russian ATGMs, due to the guidance method. The engine appears to consist of boost-sustain elements. The booster ejects the missile from the launch tube. There are ten exhaust ports at the base of the missile. The rocket sustainer is configured like many recent Russian ATGMs, with a pair of lateral exhaust ports. The missile's rearward-pointing laser receiver is mounted at the base of the missile, surrounded by the booster exhausts. The guidance method is apparently the same technique used with the KBMs tank-fired guided projectiles such as Bastion and Svir/Refleks. The launcher emits a modulated cone of light beams and the missile receiver keeps the missile placed near the center of the cone. There have been reports that the missile executes a terminal maneuver, popping up and descending at a steep angle at the target. The guidance method for such a maneuver is not clear.

The Kornet missile comes in two versions, with a shaped-charge anti-tank warhead or with a fuel air explosive (thermobaric) enhanced blast warhead for attacking bunkers and other soft targets. This dual warhead option has become common on recent Russian ATGMs. It can be traced back to the Afghan experience, where anti-tank missiles were widely used to attack guerillas hiding in rock formations. It provides an interesting illustration of the Russian view of the ATGM system as a precision support weapon, not limited to the anti-tank role, but suitable for use against a wide range of pin-point targets. The reported total penetration of the Kornet anti-tank warhead is 1,200 millimeters or 980 millimeters beyond explosive reactive armor. This makes the Kornet warhead the most potent of any existing Russian ATGMs.

The Kornet has a relatively high speed of 240 meters per second and an effective range of 5,500 meters. The increase in range over previous missiles of this size has been made possible by the use of laser guidance, since the missile does not outrun the spool of guidance wire as occurs in earlier types such as Konkurs. On the other hand, laser guidance has two potential shortcomings. The laser beam can be interrupted by obscurants such as dust, battlefield smoke or deliberate enemy countermeasures. It is not clear what type of laser is used in the guidance system from open literature. Also, the laser beam can presumably be detected by laser warning receivers on enemy vehicles, prompting countermeasures. It is not clear what techniques, if any, the designers have employed to reduce the vulnerability of the system to countermeasures.

AT-X-15

The 9M127 Krizantema (Chrysanthemum, AT-X-15) was developed by the KBM design bureau in Kolomna. The Krizantema is configured as a heavy anti-tank missile system, designed specifically to deal with contemporary MBTs (specifically the M1A2 and the uparmored Leopard 2 (Improved)) and has one of the largest warheads in its class. Krizantema is fitted with an advanced anti-armor shaped charge warhead of 150mm diameter capable of penetrating at least 800 millimeters of armor.

The Krizantema uses a new mixed guidance system. The target can be acquired and tracked by either an electro-optical sight or a millimeter wave radar sensor with the radar sensor being the primary guidance option.

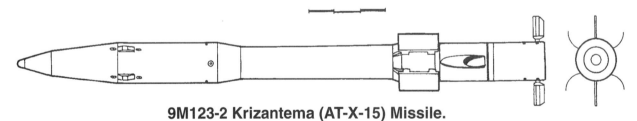

9M123-2 Krizantema (AT-X-15) Missile.

BMP-3 Missile Tank Destroyers Comparative Technical Data

System Description	Kornet	Krizantema
Missile design bureau	KBP, Tula	KBM, Kolomna
Launcher system designation	9P162	9P157-2
Missile rate of fire (rounds per minute)	2	2
Maximum range (kilometers)	5.5	6
Missile supply in vehicle	16	15
Missile flight speed (meters per second)	225	400
Missile launch tube length (meters)	1.2	2.75
Time from march to first launch (minutes)	1.5	3

Anti-Armor Developments

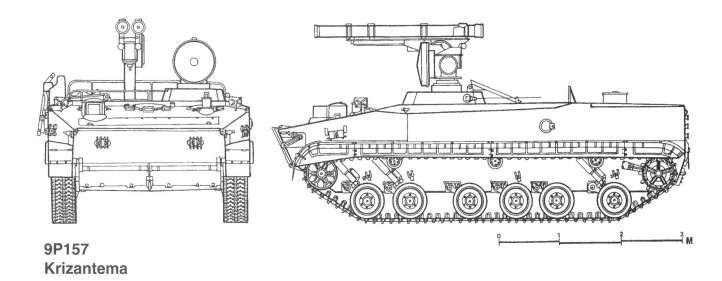

9P157 Krizantema

The missile is automatically guided using command guidance from the radar, or using semi-automatic laser beam riding. It has a range of five kilometers, a speed of 170 meters per second. The Krizantema is launched from a standard transporter/launcher canister. The standard platform for the Krizantema is the 9P157-2 tank destroyer which is based on the BMP-3 infantry vehicle chassis.

AT-16

The 9K121 Vikhr (Whirlwind: AT-16) resembles a lengthened 9M114 Kokon/Shturm, but employs semi-active laser guidance. The missile is about 1.5 meters long, and the launch cannister appears to be fitted with a small optical port to allow the laser seeker to lock-on before launch. The missile has a maximum effective range of about eight kilometers and weighs about sixty kilograms in its launch cannister. It has been demonstrated on the Kamov Ka-50 Oboroten (Werewolf, NATO: Hokum) attack helicopter and can also be launched from fixed wing aircraft such as the Sukhoi Su-25T Frogfoot. When mounted on aircraft, it uses an APU-8 launcher which permits eight missile tubes to be fit to a single hard-point. The launcher array on the Kamov Ka-50 Hokum has a different configuration with six missiles per hardpoint.

9M127 Vikhr

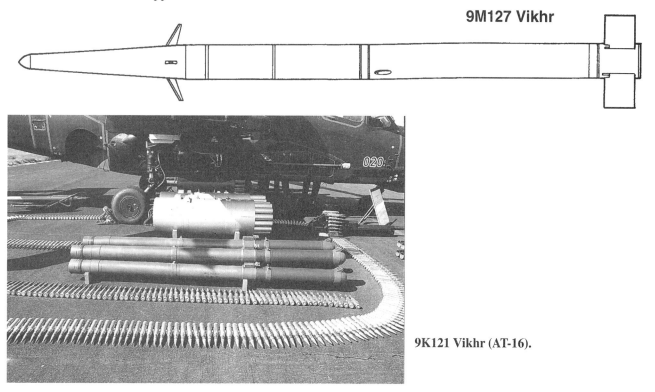

9K121 Vikhr (AT-16).

187

Russian Vehicle/Helicopter Mounted ATGM: Comparative Technical Data***

Missile name	Shmel	Skorpion	Konkurs	Kokon	Kokon-M*	Kokon-M*
Russian System	2K15	3K11P	9K113	9K114	9K114	9K114
Russian Missile	3M6	9P17P	9M113	9M114	9M114M1*	9M114M2
System name		Falanga-P		Shturm		
US Designator	AT-1	AT-2c	AT-5a	AT-6a	AT-6b	AT-6c
NATO Codename	Snapper	Swatter	Spandrel	Spiral		
IOC	1958	1962-68	1974	1978	1992	1992
Guidance	MCLOS	SACLOS	SACLOS	SACLOS	SACLOS	SACLOS
Command link	radio	radio	wire	radio	radio	radio
Missile length (millimeters)	1,150	1,165	1,150**	1,625	1,745*	2,045*
Missile diameter (millimeters)	136	140	135	130	130	130
Missile span (millimeters)	750	695	468	360	360	360
Missile weight (kilograms)	24.0	29.0	14.6	31.4	33.5	40
Warhead weight (kilograms)	5.4	5.4	3.0	5.3	7.4	7.4
Penetration (millimeters RHA)	300	650	650	600	950	950
Average speed (meters/sec)	120	150	208	345	345	345
Minimum range (meters)	600	1,000	75	400	400	400
Maximum range (meters)	2,000	4,000	4,000	5,000	6,000	7,000
Cannister length (millimeters)	n/a	n/a	1,260	1,830	1,950*	2,250*
Cannister weight (kilograms)	n/a	n/a	25.2	46.5	49.5	57.0

*Data not confirmed **Missile length including gas generator ***n/a=not applicable

New Russian Vehicle/Helicopter Mounted ATGM: Comparative Technical Data*

Missile name	Ataka	Kornet	Krizantema	Vikhr
Russian System	9K120	9K129	n/a	9K121
Russian Missile	9M120	9M133	9M123-2	9M127
System name	Shturm	n/a	n/a	n/a
US Designator	AT-9	AT-14	AT-X-15	AT-16
NATO Codename	n/a	n/a	n/a	n/a
IOC	1993	1994	1995	1993
Guidance	SACLOS	SACLOS	SACLOS	Beam-Rider
Command link	radio	laser	laser	laser
Missile length (millimeters)	n/a	1,200	n/a	1,500
Missile diameter (millimeters)	130	152	350	130
Missile span (millimeters)	750	n/a	n/a	n/a
Missile weight (kilograms)	n/a	n/a	n/a	45.0
Warhead weight (kilograms)	n/a	n/a	n/a	n/a
Penetration (millimeters RHA)	800	1,200	800	1,000
Average speed (meters/second)	400	420	170	400
Minimum range (meters)	400	100	500	n/a
Maximum range (meters)	6,000	5,500	5,000	8,000
Cannister length (millimeters)	n/a	1,200	n/a	n/a
Cannister weight (kilograms)	n/a	27.0	n/a	60.0

*n/a=data not available

Russian Anti-Tank Missile Launchers

Designation (Russian)	Missile (US/NATO)	Missile (Russian)	Launcher Chassis
2P26	AT-1 Snapper	3M6 Shmel	GAZ-69
2P27	AT-1 Snapper	3M6 Shmel	BRDM-1
2P32	AT-2 Swatter	3M11 Fleyta	BRDM-1
9P110	AT-3 Sagger	9M14 Malyutka	BRDM-1
9P111	AT-3 Sagger	9M14M Malyutka	Firing Post
9P122	AT-3 Sagger	9M14M Malyutka	BRDM-2
9P124	AT-2 Swatter	9M17P Skorpion-P	BRDM-2
9P133	AT-3c Sagger	9M14P Malyutka	BRDM-2
9P135	AT-4/AT-5	9M111/9M113	Firing Post
9P137	AT-5 Spandrel	9M113 Konkurs	BRDM-3
9P148	AT-5/AT-4	9M111-2 /9M113	BRDM-3
9P149	AT-6 Spiral	9M114 Kokon	MT-LB
9P151	AT-7 Saxhorn	9M115 Metis	Firing Post

Manportable ATGM Systems

AT-3 Sagger

The 9K11 Malyutka was the Soviet Union's first manportable anti-tank guided missile and entered service in 1963. It was developed by the Kolomna Machine Design Bureau under S. P. Nepobidimy, the reorganized bureau formerly headed by Boris Shavyrin which had developed the Shmel (AT-1). The Malyutka was patterned after Western ATGMs of the late 1950s such as the French Entac and the Swiss Cobra. The original version of the 9M14 missile employed manual-command-to-line-of-sight guidance (MCLOS) with the command signals being sent along a trailing wire. It was followed shortly afterwards by the improved 9M14M Malyutka-M, which was the most common production version of the missile.

The manpack version was generally operated by a three-man team. The first load, carried by the gunner, consists of the 9S415 joystick control unit, and the attached 9Sh16 periscopic sight. The second and third team members each carry a fiberglass "suitcase" each containing a single 9M14 Malyutka missile. The 9P111 suitcase serves as the launch-pad for the missile and contains a small launcher rail which is fitted to the top of the box. It takes about five minutes to deploy the launch equipment. The missiles can be set up at distances up to fifteen meters from the 9S415 joystick controller, the link being a simple cable. The gunner can control the missile by eye if firing at targets under 1,000 meters, and he uses the 8 power 9Sh16 periscopic sight for longer ranges. A single 9S415 joystick unit can control up to four 9P111 launchers in consecutive order.

The manportable 9K11 Malyutka is deployed as part of the anti-tank platoon of Soviet motor rifle battalions. Each platoon has two Malyutka sections, each with two teams for a total of four launcher/control stations. One assistant gunner in each team also serves as an RPG-7 gunner. This is necessary since the Malytuka has a 500 meters dead-zone since the missile is impossible to control for a few seconds after launch. The platoon also has two SPG-9 recoilless anti-tank weapons.

The 9K11 Malyutka system was also widely deployed on armored vehicles, both as secondary armament on armored infantry vehicles and as primary armament on tank destroyers. The 9K11 Malyutka was an integral part of the armament of the BMP-1 infantry combat vehicle and BMD-1 airborne fighting vehicle. These vehicles, which share nearly identical armament systems and turrets, carry three missiles internally and can carry a fourth on the external launch rail. The guidance optics for the missile are built in to the gunner's 1PN22 sight, and the joystick controller is folded under the gunner's seat in the turret when not in use. The 9K11 Malyutka was used as secondary armament on many other armored vehicles outside the USSR including the Yugoslav BVP M80, Romanian TAB-77, and Polish SKOT-2A.

The 9K11 Malyutka has also been fitted to helicopters including the Mi-8TVK Hip F and some export

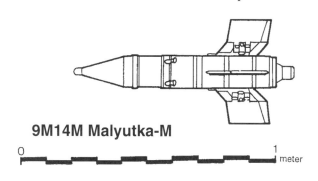

9M14M Malyutka-M

9M14 Malyutka (AT-3 Sagger).

Mi-25 Hinds. However, the longer-ranged Falanga (AT-2) system was a far more common helicopter ATGM system in Soviet service. Other countries fitted the 9K11 Malyutka system on helicopters, notably the Yugoslav Gama system for the Aerospatiale SA.342 helicopter.

The 9M14 Malyutka served as the basis for the most common BRDM tank destroyer versions. The first of these was the 9P110, based on the BRDM-1 which carried six missiles on launcher rails and a further eight reloads. This type was shortlived owing to the advent of the improved BRDM-2 armored vehicle. The initial version of the 9K11 Malyutka system on the BRDM-2 (GAZ-41-06) was the 9P122, also fitted with six launch rails and carrying eight reload missiles. This version was equipped to fire the 9M14M Malyutka-M (AT-3b) missile. The later 9P133 version has been modified to fire the improved 9M14P and 9M14P1 Malyutka P (AT-3c) missiles which use SACLOS guidance. The 9P133 has an enlarged optical sight system and can carry sixteen missiles of either the 9M14M or 9M14P types; six on launch rails and ten reloads. These BRDM tank destroyers serve in the antitank battalions of Soviet/Russian motor rifle divisions; each battalion has an ATGM battery with nine of these vehicles.

The effectiveness of the Malyutka anti-tank missile, like many first-generation MCLOS missiles is the subject of some controversy. Early unclassified assessments estimated that these missiles had a probability of a hit (Ph) of between sixty-seven to ninety percent at ranges over 2,000 meters, and a sixty-five percent probability of kill (Pk) if the tank was hit. These were idealized estimates which assumed a highly trained crew, lack of enemy countermeasures and other factors which were extremely unlikely in the real world. To become proficient, a Malyutka operator had to fire at least 2,300 simulated rounds on a missile trainer, and had to regularly fire fifty to sixty simulated rounds per week on a trainer. Beyond these technical skills, an operator had to have a particularly stout heart when actually using such weapons in combat, as they took about thirty seconds from launch to target impact, with the operator vulnerable to enemy supressive fire the whole time of the launch-hit interlude.

The manportable Malyutka was used by North Vietnamese forces beginning in 1972 with some success against ARVN armor units. The use of both manportable and vehicle launched Malyutka's in the 1973 Arab-Israeli war came as a shock to many observers, the first

9K11M Malyutka firing post and guidance box.

time that ATGMs were used on a large scale. The press interpreted Egyptian successes with the Malyutka against Israeli tank formations in the opening days of the war as a signal of the demise of the tank, akin to the demise of the armored knight following the advent of the longbow at Crecy. More sober observers felt that the initial Israeli tank losses were due as much to Israeli overconfidence and poor tank-infantry cooperation. During the later phases of the war, the successes of the Malyutka teams disappeared as Israeli tankers came to understand their vulnerability to counteraction during the launch-hit interlude. The 1973 war demonstrated that the actual probability-of-hit of one of these missiles was dramatically lower than the idealized figures of sixty-five to ninety percent, generally averaging twenty-five percent or less, and often as low as two percent. The Egyptian Army used Malyutka missiles in massive numbers. On average, each launcher expended two units of fire (twenty missiles) or about 2,000 missile per division during the war. One division fired 460 missiles on one day alone. Soviet accounts claim that the Malyutka was responsible for 800 Israeli tank losses.[10]

The Soviet Army was already aware of the shortcomings of the MCLOS guidance system, deploying a semi-automatic-command-to-line-of-sight system for the Malyutka in 1969, called the 9M14P Malyutka-P (P=poluavtomaticheskiy, semi-automatic). This was called the AT-3c Sagger C under the US/NATO reporting system. This version introduced an improved warhead which boosted penetration from 400 to 460 millimeters of RHA. A slightly improved version was fielded in the early 1970s, the 9M14P1 shortly before production began to shift to next-generation systems like the 9M111 Fagot.

In recent years, there have been at least two upgrade programs for the Malyutka warhead. The improved

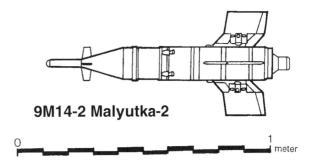

9M14-2 Malyutka-2

9M14MP1 warhead used a new stand-off probe on the nose, increasing armor penetration from 460 to 520 millimeters of RHA. This was followed in 1992 by the 9M14-2 Malyutka-2 which has an enhanced warhead and also a new solid-fuel rocket propellant which gives it higher average flight speed (130 versus 115 meters per second). The Malyutka-2 warhead is 3.5 versus 2.5 kilograms on earlier types, and increases penetration from 460 to 800 millimeters of RHA through the use of a new generation shaped charge.

The 9M14 Malyutka is probably the most widely produced ATGM of all times with Soviet production in the 1960s and early 1970s averaging 25,000 missiles annually according to published DIA estimates. The unit cost was extremely low, about 500 rubles per missile or about $500 in then-year dollars. Besides manufacture in the USSR, the 9M14M was built under license by the Vazov Engineering Plant in Bulgaria, in Romania, and by Krusik in Yugolavia and in several Warsaw Pact countries. Unlicensed copies of the 9M14M are still being manufactured by China (Red Arrow 73), Iran (RAAD), North Korea, and Taiwan (Kuen Wu 1).

AT-4 Spigot

Development of a new generation of wire-guided anti-tank missiles was begun in 1962 by the Tula Machinery

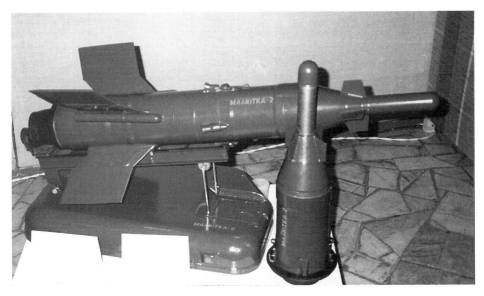

9M14-2 Malyutka-2 (AT-3 Sagger).

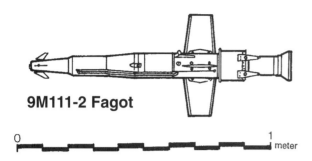
9M111-2 Fagot

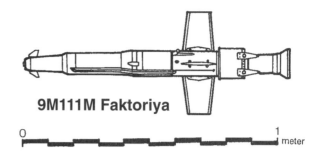
9M111M Faktoriya

Design Bureau (Tula KBP), headed by Arkadiy Shipunov. The aim of this program was to develop a new generation of SACLOS guided ATGMs suitable both for the man-portable role (AT-4 Spigot) and the heavier vehicle-mounted tank destroyer role (AT-5 Spandrel). Both missiles share similar technology, differing only in size and capability. The lighter of the pair, the 9K111 Fagot (Fagot=Basoon; US/NATO name: AT-4 Spigot) was a direct counterpart of the Euromissile Milan. The associated 9P135 firing post is very similar to the firing post used by the Milan which has led to some suspicion that espionage played a role in the program. However, there are several distinct technical differences between the Fagot/Konkurs and the Milan/HOT. All these missiles employ a gas-generator to boost the missile out of the launch tube. But the Milan/HOT system expels the launch tube backward as a counterweight to reduce the recoil on launch, while the Russian approach expels the gas-generator backwards as a counterweight. The 9K111 Fagot system was first deployed in 1973, the larger 9K113 Konkurs followed in 1974.

The basic 9M111 Fagot uses a 1.8 kilograms 9N122 shaped charge warhead with penetration of 400 millimeters of RHA. The 9K111 Fagot has gone through two evolutionary steps. The 9M111-2 (AT-4b) has greater effective range through the use of an improved sustainer motor and lengthened guidance wire. It is fitted with an improved warhead with penetration raised to 460 millimeters of RHA. The most recent model, the 9K111M, has been renamed Faktoriya (Trading post, AT-4c). This variant uses a tandem warhead for better penetration of explosive reactive armor.

The 9P135 firing post is mounted on a simple tripod, with the gunner laying prone for firing. A 9S451 guidance controller box is fitted to the tripod, with an attachment for the missile launch tube immediately above. The 9Sh119 optical sight is attached to the left side of the launcher. The launcher system weighs 22.5 kilograms and has a rate of fire of three rounds per minute. The 9P135 is limited to the 9M111 missile; 9P135M series

9P135M launcher with 9M111 Fagot missile (AT-4 Spigot).

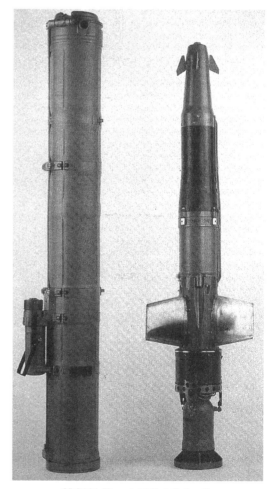
9M111 Fagot (AT-4 Spigot).

can fire either the 9M111 Fagot or the heavier 9M113 Konkurs/Faktoriya. There have been at least two upgrades to the system, the 9P135M1 and 9P135M2. An improved 9P135M3 night-capable firing post was introduced in the early 1990s. This -M3 upgrade adds a thirteen kilogram TPVP thermal imaging night sight, which clips on above the 9Sh119 sight. Using this sight, the 9P135M3 firing post has an effective range of 2.5 kilometers at night. TsNII Tochmach has also been developing and upgrade for the 9S451M1 guidance control box which would harden the missiles against electro-optical dazzler countermeasures like the US Army VLQ-6 and VLQ-8, or the Russian Shtora system. The new guidance tracker increases the weight of the complete firing post from 22.5 to 23.5 kilograms.

The 9K111 is a battalion-level anti-tank system with the anti-tank companies of Russian motor rifle battalions in place of the earlier 9K11 Malyutka system. The anti-tank company deploys two or three 9K111 sections, each with two 9P135 launchers, for a total of four-to-six 9P135 launchers. Each ATGM team consists of a gunner, carrying the launcher and tripod in a back-pack; and two assistants, each carrying two 9M111 missiles. Unlike the 9K11 Malyutka, the team does not carry an RPG-16 since the 9K111 has a very small dead-zone in front of the launcher. The standard firing unit for a team is eight 9M111 missiles, the additional missiles being carried on the section's BTR transporter. The 9P135 launcher can also fire the larger 9M113 Konkurs missiles, but generally they employ the lighter 9M111 Fagot.

The 9K111 missile was used to upgrade older BMP-1 and BMD-1 armored infantry transporters by replacing the obsolete 9K11 Malyutka missile. The modified versions of the vehicles are designated BMP-1P and BMD-1P, the P indicating poluavtomaticheskiy (semi-automatic guidance). The conversion involved the removal of the launch rail and guidance system for the 9K11. A small attachment post is welded to the turret roof for mounting the 9P135 firing post. To fire the missile, the upper portion of the 9P135 firing post minus the tripod base is removed from the troop compartment and mounted on the roof along with a missile. The gunner is exposed when firing the missile, since he is obliged to use the guidance system in identical fashion to the normal infantry version.

The main production centers for the 9M111 Fagot in Russia are the Degtaryev Machinery Plant in Kovrov, the Shchit Machinery Plant in Izhevsk and the Tulskiy Armaments Plant in Tula. Export prices for these systems as of 1992 were $10,000 for the 9M111M Fagot-M missile, $85,000 for the basic 9P135 firing post, and $115,000 for the 9P135 firing post. License production of the 9M111 Fagot is currently being undertaken in Bulgaria by the Vazov Engineering Plant in Sopot, and in Germany on the basis of an existing East German government agreement with the USSR.

AT-7 Saxhorn

The 9K115 Metis (Mongrel) is a light-weight infantry anti-tank missile somewhat similar to the US Army MGM-52 Dragon or Aerospatiale Eryx. It was developed by the Tula KBP, headed by Arkadiy Shipunov, and introduced in 1979 to supplement the larger 9K111 Fagot at company level. The lighter weight of the system is due to a less sophisticated firing post and a lighter missile. The 9P151 Metis tripod firing post is only 10.2 kilograms compared to the twenty-three kilograms for the 9P135M firing post of the 9K111 Fagot. In terms of the missiles, the 9M115 Metis missile is only about half the weight of the 9M111 Fagot, with a weight of four versus eight kilograms. The difference in weight is due to the smaller amount of rocket propellant in the 9M115 Metis missile. This is most evident in the range differ-

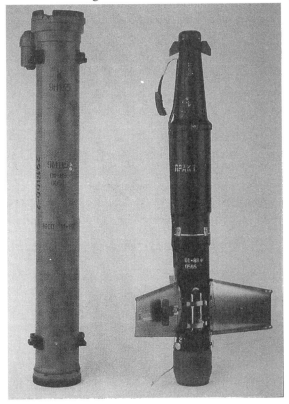

9M115 Metis (AT-7 Saxhorn).

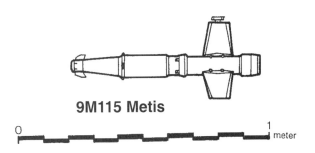

9M115 Metis

9P151 launcher with 9M115 Metis (AT-7 Saxhorn).

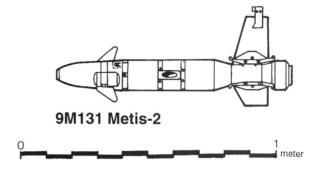

9M131 Metis-2

9P151 launcher with 9M131 Metis-2 (AT-13).

9M131 Metis-2 (AT-13).

ence, with the Metis being credited with a maximum range of 1,000 meters, while the Fagot has a range of 2,500 meters.

The Metis has a variety of intriguing features. The 9S816 guidance system is powered by a thermal battery attached to front of the launch tube prior to launch. The missile contains no battery; necessary power is provided via the guidance wire at relatively high voltage. The system can be fired from the shoulder, as well as from the tripod, but this requires a great deal more skill. Like the French Eryx, the Metis can be fired from within an enclosed space such as a building, though it requires at least six meters of clear space behind and an internal volume of 100 meters cubed. The missile is launched from the cannister by a booster stage, not a gas generator as is found in the Fagot/Konkurs missiles also designed by the Tula KBP bureau.

The significant size and weight difference between the two missiles is due to their different roles. The AT-4 Spigot is deployed in Soviet BTR-equipped motor rifle battalions in a dedicated anti-armor platoon. (BMP battalions have organic anti-tank defense on their vehicles.) The AT-7 Saxhorn has been added to the inventory to deepen Soviet anti-tank defensive capabilities; it does not replace an existing weapon, but adds new capability. It is being deployed at a lower level than the AT-4 Spigot, in the motor rifle companies, with three launchers per company (for a total of nine in the battalion). The AT-7 Saxhorn is operated by a two man team. The gunner carries the 9P151 firing post and one 9M115 missile, and the assistant gunner carries three additional 9M115 missile cannisters. The missile's short minimum range of forty meters, makes it more suitable for urban or close terrain fighting than the AT-4, which has a minimum range of seventy meters. The missile can engage targets moving at speeds up to sixty kilometers per hour. The basic 9M115 Metis has a unitary shaped charge warhead with penetration of 460 millimeters of RHA.

The 9K115 has undergone one major upgrade program, with the new missile redesignated 9M131 Metis-2. This missile can be fired from the usual 9P151 firing post, but the missile is substantially larger and heavier, and closer to the French Eryx in perofrmance. The improved 9M131 Metis-2 has two alternative warheads: a 4.6 kilograms tandem shaped charge with precursor charge for penetrating ERA with a penetration of 800 to 900 millimeters of RHA, and a 4.95 kilograms fuel-air explosive warhead for attacking bunkers and similar targets.

The Metis is manufactured in Russia by many of the same plants as the 9M111 Fagot and 9M113 Konkurs, of the former Ministry of Precision Machinery. It is also license produced by the Vazov Engineering Plant in Sopot,

Russian Manportable ATGMs: Comparative Technical Data

Russian Name	Malyutka-M	Malyutka-2	Fagot	Faktoriya	Metis	Metis-2	Avtonomiya
Russian System	9K11M	9K11-2	9K111	9K111M	9K115	9K115-2	n/a
Russian Missile	9M14M	9M14-2	9M111	9M111M	9M115	9M131	n/a
US Designator	AT-3b	AT-3d	AT-4a	AT-4b	AT-7a	AT-13	n/a
NATO Codename	Sagger	Sagger	Spigot	Spigot	Saxhorn	n/a	n/a
IOC	1963	1990	1975	1991	1979	1990	1993
Guidance	MCLOS	SACLOS	SACLOS	SACLOS	SACLOS	SACLOS	n/a
Missile length (mm)	860	985	1,030*	1,030*	740	910	1050
Missile diameter (mm)	125	125	120	120	94	130	125
Missile span (mm)	393	393	369	369	300	400	n/a
Missile weight (kg)	10.9	12.5	12.5	12.9	5.5	13.0	10.0
Warhead weight (kg)	2.6	3.5	2.5	2.5	2.5	4.6	n/a
Penetration (mm RHA)	400	800	400	460	460	900	n/a
Average speed (m/s)	115	130	186	180	223	200	n/a
Minimum range (m)	500	500	70	75	40	80	n/a
Maximum range (m)	3,000	3,000	2,000	2,500	1,000	1,500	500
Firing post	9P111	9P111	9P135	9P135	9P151	9P151	n/a
Firing post weight (kg)	8.1	8.1	22.5	22.5	10.0	10.0	n/a
Cannister length (m)	-	-	1.1	1.1	0.78	0.98	n/a
Cannister weight (kg)	-	-	13.0	13.4	6.3	13.8	n/a

*Length with gas generator, missile is 875mm without gas generator **n/a=data not available

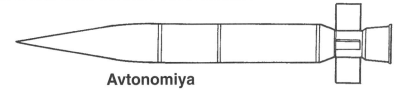

Avtonomiya

Bulgaria. Export prices at the end of 1992 were: $13,500 for the 9M131 missile and $70,000 for the 9P151 firing post.

Avtonomiya

In 1994, the Russians displayed a new light manportable anti-tank missile for the first time, called Avtonomiya (Autonomous). The missile is the first ATGM developed by the TsNII Tochmash (Central Scientific Research Institute for Precision Machinery) in Klimovsk which is Russia's primary center for small arms development. The missile has a diameter of 125 millimeters and a length of 1,050 millimeters. The configuration is similar to the Fagot/Konkurs (AT-4/AT-5) with a small gas generator at the base of the missile to eject it from the tube, followed by a sustainer rocket engine in the main fuselage of the missile. The guidance method of the missile was not disclosed but as the name implies, is autonomous. In view of the fact that the missile obviously lacks an imaging seeker, the most likely method of guidance would be some form of inertial platform akin to that being developed by the US Marine Corps for the Predator light anti-tank missile. TsNII Tochmash has been working on inertial platforms for small artillery forward observer navigation aids (such as Riga), and this guidance package is probably a spin off of that work.

The Avtonomiya has an effective range of only 500 meters, so the main purpose of the guidance unit is to prevent the missile from veering from the aiming point during its short flight. The warhead on the missile is described as being capable of defeating tanks with reactive armor. In comparison to the Marine Corps Predator, the Avtonomiya is longer, narrower, and significantly heavier (10 versus 6.4 kilograms).

The launcher system for the Avtonomiya is very simple consisting of a lightweight tripod and a small daylight optical sight. The missile is delivered in a typical launch container, protected at either end by large plastic foam discs to prevent damage in transit. The total system weight including the tripod is nineteen kilograms, of which the missile weighs ten kilograms. This makes it about double the weight of the US Marine Corps Predator, although other performance characteristics appear to be similar.

Anti-Tank Missile Sights

A Russian night set developer, Elers-Electron has developed a new family of thermal sights for the standard range of Russian infantry anti-tank missiles. At least three add-on thermal sights are now available, the Trakt (Fagot/AT-4); Mulat-115 (Fagot/Konkurs; Metis-M), and Sokol (Metis-M). These use a CdHgTe array nitrogen-cooled to seventy-seven degrees Kelvin sensitive in the 7.5 to 12.5 micron infrared band.

Guided Tank Projectiles

Guided anti-tank missiles fired through tank guns or other similar weapons are a rarity. Although several countries have developed these weapons, such as the French ACRA, only two armies have fielded such weapons: the US Army with its 152mm MGM-51 Shilellagh, and the Soviet/Russian army. This type of weapon has generally been abandoned over the past decade in Europe and the United States, with Russia remaining the only country actively pursuing this technology. In Russia, these missiles are called TUR (Tankovaya upravlenaya raketa: Tank guided missiles).

Early Soviet TUR

Development of ATGMs began in the USSR in 1956. V. A. Malyshev, who had led the Soviet tank industry in World War 2, was delegated by Khrushchev to impose the "new thinking" on the ground weapons designers. As deputy to the chairman of the Council of Ministers handling major military programs, and the former head of the Soviet nuclear weapons program from 1945 to 1956, he wielded considerable authority. On 31 January 1956, he held a meeting with the heads of the four tank design bureaus (Morozov from Kharkov, Kartsev from Nizhni Tagil, Kotin from Leningrad and Isakov from Chelyabinsk). He asked if any of the designers had thought about incorporating missiles into their future tank designs. The responses were negative, with Morozov in particular sticking with a gun armament. Malyshev chaired a second meeting of the designers in the Kremlin in August 1956 along with some of the new missile designers. Malyshev chastised the group for their failure to take his hint from January, and told the designers to prepare a rough outline of a future missile-armed tank. When the group reconvened two days later, the commander of the Soviet tank force, P. P. Poluboyarov gave a presentation which did not reject missiles, but stressed the versatility of tank guns and their developmental potential. Malyshev responded menacingly "Are you trying to drag us backward? Do you want the sailor's fate to befall the tank forces."[11] There was still stubborn resistance to the missile idea, and General Lieutenant I.A. Lebedev, head of the GBTU, laid out a proposal for the future tank-fired missile that essentially required it to be a tube-fired device identical to a conventional projectile except for the guidance. Malyshev rejected this position and insisted that the design teams begin to work on missile armed tanks using more conventional ATGMs as well as tube-fired missiles. The missile designers present generally felt that tube fired missiles would be a much greater challenge than more conventional missiles. This meeting led to the development of the Drakon and Taifun missiles mentioned above in this section, as well as the first generation tube-fired ATGM, the Rubin.

Development of the Rubin was tied to the radical new Obiekt 775 tank that was being developed by the Isakov KB design bureau in Chelyabinsk as a competitor to the T-64 tank. The Obiekt 775 was armed with a special 125mm smooth-bore weapon that could fire either high explosive rocket projectiles, or a guided projectile. The guided 125mm projectile was designated the Rubin (Ruby) and used radio-command guidance developed by A. A. Raspletin at KB-1 in Moscow. Since it's speed was inadequate for kinetic energy penetration, it relied on a shaped charge warhead. The unguided 125mm rocket projectile was designated Bur (Drill). The Obiekt 775 carried twenty-four Rubin missiles and forty-eight Bur rockets. In the end, the revolutionary armament system proved a failure. It's shaped charge warhead was ineffective and there was concern that the radio-command system could be jammed.

In spite of these setbacks, development efforts on guided tank projectiles continued through the 1960s. In contrast to contemporary American designs, such as the Shillelagh missile, the Soviet designers favored radio command guidance rather than optical (infra-red) command guidance. A revival of interest in guided tank projectiles took place in the 1970s for two main reasons. On the one hand, the US Army was introducing its own guided anti-tank missile, the MGM-51 Shillelagh, on the M60A2 tank and on the M551 Sheridan scout tank. The American program was an attempt to develop the capability to accurately hit enemy tanks at ranges of over 2,000 meters, far beyond the capability of conventional tank projectiles of the period.

The second motivation was the Soviet recognition of the anti-tank missile and attack helicopter threat. Anti-tank missiles such as TOW and HOT, mounted on armored vehicles, could outrange contemporary tank guns with a higher degree of accuracy at these ranges. In addition, the US Army was beginning to add the TOW anti-tank missile to AH-1 Cobra gunships, and army aviation was touting the combination as the most lethal threat to tank forces to date. The Soviets took this threat seriously, no doubt connected with their own experiences with their new Mi-24 attack helicopter armed with the Falanga missile system. A guided projectile offered to extend the effective range of tank guns, as well as provide a limited amount of self-defense against the threat of attack helicopters.

The 9K112 Kobra TUR

Development of a second generation tube-fired 125mm projectile began in the mid 1970s, by the KB Tochmass under the designation 9K112 Kobra (US/NATO: AT-8 Songster). Russian sources have not provided many details of its development. The 9K112-1 Kobra missile system was first mounted on a new version of the T-64 tank, the T-64B in 1980. Technical details of the 9M112 Kobra missile are incomplete since the Kobra was restricted from export. The 9M112 Kobra missile consists of two elements, a head and tail section. The 9M43 head section contains the 9M129 shaped-charge warhead, and the propellant for the missile's 9D129 sustainer motor. The rocket sustainer motor has a pair of lateral exhaust vents near the center of the projectile. The 9B447 tail section contains flight controls and battery. At the base of the projectile are a rearward pointing command-link antenna and a light source used by the tank's guidance system to locate and track the projectile in flight. The two sections are stored in the autoloader like a conventional two-piece 125mm round.[12]

Command guidance of the projectile is handled by the tank's 1A33 fire control system. There are three firing modes for the 9M112 Kobra missile. In the normal mode, the projectile is fired with the barrel elevated to minimize the amount of dust kicked up in front of the tank. Dust can interfere with the system's optical tracking of the missile. The 1G21 rangefinder/sight located in front of the gunner's station acquires the missile from its rearward pointing modulated light source. The guidance system in the tank sends out signals via the GTN-12 (9S461-1) radio command antenna located in an armored box in front of the commander's cupola. This sends guidance commands to the missile so that it will conform to the gunner's line-of-sight. The Kobra system uses semi-automatic SACLOS guidance. The gunner keeps the target centered in the cross-hairs of his sight, and the guidance system onboard the tank sends appropriate command signals to the missile until impact. Russian sources claim that the probability of hit at 4,000 meters is about eighty percent in the normal mode.

The second firing mode for the Kobra system is a shaped-trajectory in conditions when the missile's flight close to the ground could kick up dust and thereby interfere with the command link. In this mode, the missile flies a higher ballistic flight path, but this requires that the gunner input accurate range data. The third mode is an emergency mode when the target is under 1,000 meters and the missile is already loaded. This mode allows the gun to be fired without superelevation, and the command link is activated almost immediately after the missile is only about eighty to one hundred meters from the tube. Probability of hits in this mode are reduced due to the higher risks of interruption in the command link between the tank and the missile.

When introduced into service in the early 1980s, the T-64Bs were generally deployed in independent tank regiments, used in support of normal T-64A units. There were suggestions by General Yuri Potapov, an influential tank commander, that a new deployment scheme be adopted, with a special platoon of four T-64B tanks being added to each T-64A tank battalion. This scheme was vetoed by General P. P. Poluboyarov, head of the armored force directorate at the time, as unwieldy. The later T-80B was also equipped with the Kobra system and it was deployed in the same fashion as normal tanks. The standard combat load for the T-64B and T-80 tanks were four Kobra missiles each.

The Soviet interest in the Kobra system stands in contrast to Western developments at the time. By the early 1980s, Western tank designers had abandoned guided tank projectiles due to their high cost compared to conventional tank ammunition. The new generation of tank fire control systems, combining digital computers with advanced sensor input, offered long-range accuracy equal or superior to guided tank projectiles in the usual 100 to 2,500 meters combat ranges and were more convenient to use. Although anti-tank missiles had a nominal advantage at ranges over 3,000 meters, such engagements represented only a very small fraction of likely combat encounters. The Soviets clung to the concept since the role of guided projectiles was viewed differently than in NATO. The Kobra provided some ability to defeat tanks at extended ranges. But more importantly, it provided the capability to engage long-range anti-armor platforms such as anti-tank missile launcher vehicles and attack helicopters firing ATGMs.

100mm Laser Guided TUR

The Kobra was a sophisticated and expensive system, confined to front-line tank units such as the Group of Soviet Forces Germany, and not exported. In the late 1970s, work began at the KBP in Tula on a third generation tank-fired projectile, but using laser guidance rather than radio command guidance. Development of the laser guidance was undertaken by Igor Aristarkhov, and the weapon itself by Petr Komonov. Laser guidance was probably selected since there was long-standing concern, even since the Rubin projectile of 1962, that radio command guidance links could be electronically jammed.

The new program took two directions: a new missile and guidance system suitable for retrofitting on older T-55 and T-62 tanks to enhance their long range firepower,

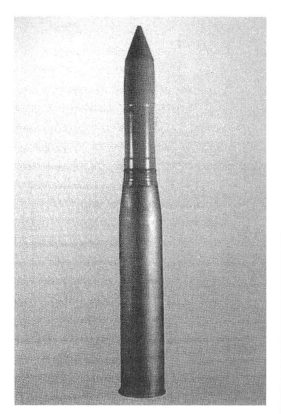

3UBK10 100mm round with 9M117 Bastion missile (AT-10 Stabber).

3UBK10-1 100mm Guided Projectile

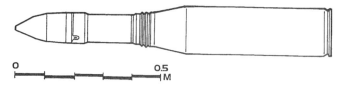

3UBK12 100mm Guided Projectile

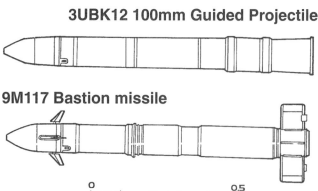

9M117 Bastion missile

3UBK12 100mm round above, 9M117 missile (AT-10 Stabber) below.

and a second program to develop a missile to replace the Kobra on main battle tanks with 125mm guns.

The 100mm laser guided projectile was accepted for service in 1983. Designated 9K116 Bastion (US/NATO: AT-10 Stabber), the system consisted of the new 9K117 missile and an associated 1K13 fire control sight. The new ammunition, designated 3UBK10-1, consists of the 9M117 Bastion missile and a propellant sleeve which fits over the rear end of the projectile. The 3UBK10 resembles a normal round of 100mm ammunition and is handled in the same fashion. The 9M117 Bastion missile follows the layout of the Kobra in many respects, with a nose-mounted shaped-charge warhead, a center-body sustainer rocket motor, and a rear section containing the flight controls and battery. There is a small set of steerable fins on the nose, and a set of pop-out fins at the rear for greater stability. The rocket ignites 1.5 seconds after launch and burns for six seconds.

After being fired, the Bastion missile drops a small cover over the tail of the missile which protects the rearward pointing optical window at the base of the missile.

The 1K13 laser fire control system directs a coded beam. The laser emitter creates a laser "funnel" with the missile riding in the center. The frequency of the beam is modulated in different sectors around the projected funnel so that if the missile deviates from the center, the guidance system onboard the missile interprets the signal and makes flight corrections to move it back into the center of the beam. The guidance system uses a timer so that the laser "funnel" is periodically altered in diameter so that, to the missile, it retains a near constant diameter. The Bastion is stated to have an eighty percent probability of hit at 4,000 meters. Armor penetration is stated to be 550 to 600 millimeters of RHA, and the projectile has an effective envelope of 100 to 5,000 meters. Flight time to 4,000 meters is about twelve seconds. The projectile has a maximum endurance of twenty-six to forty-one seconds at which point it self-detonates. The 9K116 Bastion system was fitted to modernized T-55 tanks including the T-55AM1 and T-55AM2. Further details of the tanks fitted with this system are contained in the section dealing with T-55 tank development.

There were three other versions of the 100mm 9M117 missile developed at the same time. The 9K116-1 Sheksna (AT-12) system is essentially the same as the 9K116 Bastion, except designed to fit the T-62 tank. It uses the same 9M117 missile, but the 3UBK10-1 round is adapted to fit the T-62's 2A20 (U-5TS) 115mm smoothbore gun. The

Anti-Armor Developments

tion of a guided projectile for the BMP-3 was probably based on the convenience of this arrangement compared to the traditional use of an externally mounted wire-guided missile as on the BMP-2. Although the 3UBK14 ammunition is about twice as expensive as the 9M113 Konkurs antitank missile on the BMP-2, the tube-fired ammunition is more convenient to use. It also removes the need for an expensive and vulnerable external missile launch system.

100mm TUR Improvements and 105/120mm Western Offers

Russian officials of the Tulamashzavod revealed a newly developed tandem-shaped charge fitted to their 100mm TUR family. The improved Bastion missile, designated 9M117M, incorporates a new tandem shaped charge. This tandem-shaped charge consists of a precursor and a main shaped-charge. The new 9M117M missile is twenty-two centimeters longer than 9M117 at 1.106 meters.

Tula's promotional literature, claims that 9M117M can defeat reactive armor and can penetrate 550 millimeters of RHA. Tula's other brochures claim that the single shaped-charged 9M117 is only capable of penetrating 275 millimeters of RHA protected by ERA. The capabilities for this system are identical for the other three members of the 100mm TUR family. The designation for the new Bastion round with missile is 3UBK10M for the 100mm anti-tank MT-12 gun, 3UBK10M-1 for the T-55, and 3UBK10M-3 for the BMP-3.

According to Tula's promotional literature Bastion, Sheksna, Kastet, and Basnya round's could also be fitted to fire out of a Western 105mm or 120mm main tank gun by modifying the round's propellant case.

These new 100mm guided projectiles provide considerable enhancement for the outdated T-55, T-62, and Western 105/120mm tanks in terms of range, accuracy and armor penetration.

125mm Laser Guided TUR

In parallel to the development of the 100mm 9M117 missile, the KBP at Tula also developed the 9M117 125mm laser guided projectile (AT-11 Sniper) for the new T-72B and T-80U tanks which was accepted for service in 1984. The 9K119 Refleks system was designed for the T-80U tank and works in conjuction with the 1A45 fire control system, the 1G42 laser sight, and the 9S515 missile control system. The 9K120 Svir system was developed for the T-72B tank and works in conjunction with the 1A40-1 fire control system, and the gunner's 1K13-49 laser guidance/fire control sight. Both systems use the same missile but the Svir is credited with a range of 4,000 meters and the Refleks with a range of 5,000 meters.

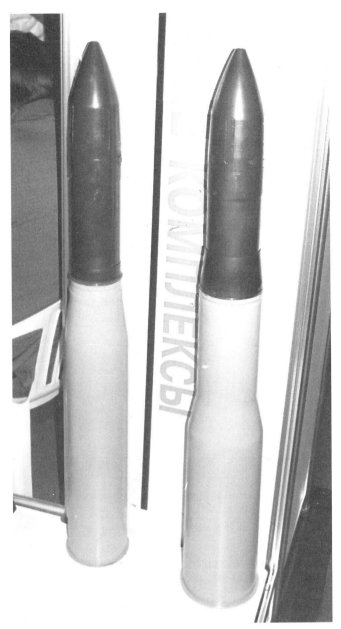

3UBK10-1 115mm round with 9M117-1Sheksna (AT-12) guided missile (to right; 100mm Bastion to left).

guidance system is identical. The Kastet is another derivative adapted to the MT-12 towed 100mm tank gun. In order to employ the Kastet, the MT-12 must have a special laser designator mounted nearby which provides the laser guidance in place of the 1K13 sight/fire controls used with the tank-fired missile.

The fourth version of this missile system is the 3UBK12 projectile, designed for firing from the rifled 2A70 100mm gun on the BMP-3 infantry fighting vehicle. The same 9M117 missile as the two earlier types is fired but it uses a significantly different propellant casing because of the ballistic differences of the BMP-3's 2A70 100mm gun. The BMP-3 has a special compartment for six 3UBK12 rounds, necessary since the ammunition will not fit in the normal autoloader. The selec-

199

9M119 Svir/Refleks 125mm guided projectile (AT-11 Sniper).

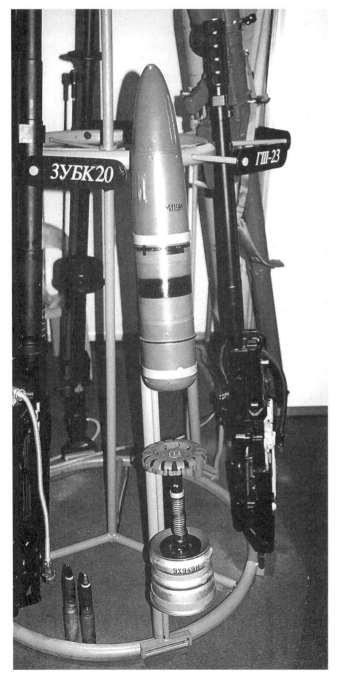

9M119 Svir/Refleks with 9Kh949 propellant charge below.

Rear of 9M119 missile showing optical guidance port.

The difference is due to the power of the laser emitters.

The 9K119 Refleks/9K120 Svir 125mm guided weapon system is significantly different in appearance from the 9M117 Bastion, but operates in a similar fashion. The difference is attributable to the need to design a round capable of fitting in the standard 125mm autoloaders used on these tanks, so the projectile/missile is stubbier. The ammunition for the system is designated 3UBK14 and consists of the 9M119 missile, a 9Kh949 reduced charge propellant casing with a spacer plug which seats the missile properly into the main section of the tank's gun. The reduced propellant charge is used since it was found that the normal Zh40 or Zh52 charge created such an enormous dust cloud in front of the tank that it was difficult for the laser emitter to transmit to the projectile. The 3UBK14 ammunition fits into the normal autoloader on the tank, like any other round of 125mm ammunition. After launch, two sets of fins pop out, one for stability and the other for steering. The body of the 9M119 projectile contains an advanced 4.2 kilograms shaped-charge warhead with a penetration-to-diameter ratio of about seven to one, and a penetration advertised as 650 to 700 millimeters of RHA.

The internal configuration of the 9M119 missile con-

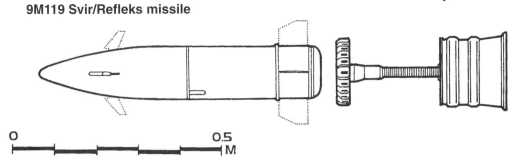

9M119 Svir/Refleks missile

3UBK14 125mm Guided Projectile

sists of flight controls and battery in the nose, with the rocket sustainer engine in the center, exhusting through lateral ports. The shaped-charge warhead is located in the rear section of the missile, with the optical tracking equipment behind it. The guidance technique for the 9M119 missile is the same as for the 9M117 100mm missiles. The list price per round is about $40 000.

Trends in Russian Guided Projectiles

The new generation of guided projectiles give Russian armored vehicles significant long-range firepower. Of special significance is that this is the first generation of guided tank projectiles being offered for export. In the case of the 100mm and 115mm Bastion and Sheksna series, they offer a considerable enhancement for the out-

125mm Guided Round Comparative Technical Data

Missile Name	Refleks	Svir
US designator	AT-11	AT-11
NATO codename	Sniper	Sniper
System designation	9K119	9K120
Missile designation	9M119	9M119
Ammunition designation	3UBK14	3UBK12
Caliber	125mm	125mm
Platform	T-80U	T-72B
Weight of round (kilograms)	28	24.3
Missile weight (kilograms)	17.2	17.2
Armor penetration (mm)	700	700
Maximum range (kilometers)	5	4

Tula's 100-115mm TUR Family: Comparative Technical Data

Missile Name	Kastet	Bastion	Basnya	Sheksna
US designator	AT-10	AT-10	AT-10	AT-12
NATO codename	Stabber	Stabber	Stabber	
System designation	9K-116	9K116-1	9K116-3	9K116-2
Missile designation	9M117/ 9M117M	9M117/ 9M117M	9M117/ 9M117M	9M117-1/ 9M117M-1
Ammunition designation	3UBK10/ 3UBK10M	3UBK10-1/ 3UBK10M-1	3UBK10-3/ 3UBK10M-3	3UBK10-2/ 3UBK10-2M
Fuze:	9E256	9E256	9E256	9E256
Seeker head:	9H136M	9H136M	9H136M	9H136M
Brusting charge:	OKFOL-3.5	OKFOL-3.5	OKFOL-3.5	OKFOL-3.5
Caliber	100mm	100mm	100mm	115mm
Platform	MT-12	T-55	BMP-3	T-62
Weight of round (kilograms)	24.5	24.5	24.5	24.5
Shell case (kilograms)	5.5	5.5	5.5	5.5
Missile weight (kilograms)	18.4	18.4	17.6	18.4
Armor penetration (millimeters)	275/550	275/550	275/550	275/550
Round length (meters)	1.098	1.098	1.106	1.098
Shell case length (meters)	0.695	0.695	1.106	0.695
Missile length (meters)	1.084/1.106	1.084/1.106	1.084/1.106	1.084/1.106
Missile wing span (meters)	0.255	0.255	0.255	0.255
Min/Max. range (kilometers)	0.10 to 40	0.10 to 40	0.10 to 40	10 to 40
Missile time of flight (seconds)	13	13	13	13
Muzzle velocity (meters/second)	370	370	370	370

dated T-55 and T-62 tanks in terms of range, accuracy and armor penetration. They might also be attractive to many of the forty-five countries using these tanks, since the necessary upgrade is very modest, consisting mainly of the addition of the 1K13 BOM sight unit which can be installed at depot level since no major structural change is required. The main drawback to the export of these weapons is likely to be price. The 125mm round has a base price of $40,000 and the 100mm round starts at $25,000. Therefore, a standard combat load of six rounds would cost from $150,000 to $240,000. This exceeds the initial purchase price of T-55 and T-62 tanks. While Russian export agencies undoubtedly offer substantial discounts, it is worth noting that the asking price of a T-72S tank is equivalent to only thirty rounds of guided ammunition.

This highlights the most persistent problem of guided tank projectiles, their exorbitant cost. This was the reason that this approach was abandoned in the West, including the US Army which fielded the Shillelagh guided missile in the 1970s. For the cost of a small load of guided projectiles, a tank can be modernized with an improved fire control system to greatly improve the accuracy of the tank's main gun. Combined with new generations of APFSDS ammunition, the rejuvenated tank can have comparable accuracy to a tank firing the new guided ammunition at normal engagement ranges. And the conventional ammunition is only about one to two percent the cost of the guided ammunition.

Admittedly, the fire control upgrade is unlikely to give these older tanks comparable accuracy over 2,000 meters. But tank engagements over 2,000 meters would have been uncommon in Western Europe where most engagements would be at 500 meters due to sighting limitations. This capability would only be effective in an open desert battlefield where engagements at 2,000 meters would be commonplace.

In view of the cost factors, why did the Russians opt for guided projectiles? As this chapter has suggested, these weapons are not intended to replace conventional APFSDS projectiles in tank fighting, rather they are for specialized applications. The three main missions for these projectiles are: defense of tank units against enemy tank destroyers firing long-range anti-tank missiles, self-defense of tank units in the field against the missile-firing attack helicopter threat, and long-range engagement of enemy tanks. The economics of the situation are unlikely to permit tank units to carry any significant numbers of these guided projectiles, and training will be circumscribed by cost as well. Nor are the Russians unaware of the advantages offered by fire control improvements to increase the effectiveness of conventional ammunition: the T-64B, T-80, and T-90 are already fitted with wind sensors and Russian fire control systems have improved steadily in the late 1980s.

Russian "Competent" Tank Projectiles

The main drawback to guided tank projectiles is their high unit costs. The Russians are trying to skirt around this problem by fielding alternative types of "semi-smart" or "competent" unguided projectiles for this role. As suggested here, there are two particular targets for these guided tank projectiles: anti-tank helicopters and long range anti-tank missile teams.

To defeat helicopters, the Russian NIMI (Research Institute of the Machine Industry) has developed a pair of proximity fuzed 125mm projectiles, similar to what the US Army has attempted with its 120mm M830A1 MP AT round. The main advantage of such a round is that the proximity fuze provides a limited amount of aim-miss distance leeway. A direct hit on the helicopter is not necessary, as the proximity fuze will detonate the round in the event of a near miss. The NIMI anti-helicopter projectiles come in two varieties, a dual-purpose HEAT round, similar to the M830A1 with a conventional shaped-charge warhead and proximity fuze. The second round is designed specifically for this role, and uses a High Explosive-Fragmentation (HE-Frag) warhead tailored for the anti-helicopter role.[13]

Although entrenched enemy anti-tank teams can be attacked in a conventional fashion using high explosive projectiles, NIMI has attempted to enhance the effect with a 125mm semi-smart high explosive-fragmentation round for the T-72B and T-80U tanks.[14] The idea behind this round is that the lethal effect of the projectile is enhanced by detonating the round over the target, in traditional artillery fashion. The problem with using traditional artillery techniques against small pin-point targets such as these teams is that they are insufficiently precise.

The new projectile takes advantage of the laser rangefinder/guidance system now found on a growing number of Russian tanks. The tank gunner lases to the target to get a precise range. The gunner aims slightly over the target, and when the anti-ATGM round is fired, the tank's ballistic computer times the sequence in relation to the range to target. When the projectile flight time is reached, the fire control signal emits a laser signal to an aft-pointing sensor in the base of the projectile which detonates the round. A Russian presentation on this approach suggested that the technique increased the effective destructive power of the round by three times, and doubled the target kill probability compared to normal firing tactics.

Indirect Fire Anti-Armor Projectiles

Russia has been very active in recent years in the

development of guided, indirect-fire artillery projectiles. While these are not exclusively anti-armor munitions, they are the nonetheless often promoted for this role. In nearly all the cases cited here, these weapons use semi-active laser homing like the US Army's M712 Copperhead 155mm projectile. Laser designation is provided by a forward observer, often located on an armored artillery reconnaissance vehicle such as the PRP-4 or by a dismounted team with a small laser designator such as the 1A35 fire control system with 1D15 or 1D22 tripod-mounted laser designator. Two design bureaus have been active in this area. The NII Ametekh (Automation and Mechanical Technology Scientific Research Institute) Scientific Technical Company under Vladimir Vishnevskiy and the Tula KBP were assigned to competitively develop a 152mm semi-active laser guided projectile in the mid-1980s, probably in response to the US Army's Copperhead. In recent years, the Tula KBP appears to have been the more active of the two design centers, offering a much wider range of PGMs on the international export market.

Most Russian semi-active laser guided projectiles (SAL-GP) have been developed by the Tula KBP, but Ametekh had several parallel programs displayed for the first time at IDEX-97. The Ametekh SAL-GP use a common guidance infrastructure, called RCIC. Laser target designation is provided by the 1D15, 1D20, or 1D22 laser system, generally mounted on a command-observation post vehicle. The command post vehicle is fitted with the 1A35E portion of the 1A35 command net, which provides a radio and data link back to the artillery position, which uses the 1A35K link. The communication link between the 1A35E (1A35I) and 1A35K is either a R-107M (R-108M) radio or a TA-57 field telephone via P-274M cables.

Gran 120mm SAL-GMP

The Gran is a Tula semi-active laser guided projectile, intended for Russian 120mm mortars. It weighs twenty-five kilograms with a 5.1 kilograms HE fill, and has an effective range of 7.5 kilometers.

Kitolov 120/122mm SAL-GP

In 1993, the Russian Tula Machine Design Bureau began marketing its Kitolov semi-active laser guided projectile. This projectile is designed for standard Russian artillery like the D-30 towed howitzer or the 2S1 122mm self-propelled howitzer. In addition, a 120mm version of this round can be fired by the 2S31 Vena Self-Propelled Gun-Mortar system. The projectile has an effective range of twelve kilometers, and carries a 5.5 kilograms HE warhead.

2K24 (3OF38) Santimetr 152mm SAL-GP

This is a second semi-active laser guided projectile, but developed by Vladimir Vishnevskiy of the Ametekh (instead of the Tula bureau responsible for most of the other SAL-GP described here). The 3F5 round has a maximum effective range of twelve kilometers (limited mainly by the laser designator). The projectile weighs 49.5 kilograms and has a 8.5 kilograms HE fill. This particular round was a competitor to the Krasnopol, and so far as is known, it has not yet been exported. At IDEX-97, Ametekh indicated that it could also provide this round in 155mm caliber, known as Santimetr-1. Further, the company indicated that this round could also be adapted to be fired from the Russian M-160 160mm mortar.

9K25 Krasnopol 152mm SAL-GP

The 9K25 Krasnopol is the Russian equivalent of the US Army's M712 Copperhead semi-active laser guided projectile. It was developed in the early 1980s, and can be fired from any of the current generation Russian 152mm guns or howitzers. Like the Copperhead, it depends on laser designation from an independent laser designator.

Germes 155mm SAL-GP

In 1993, the Russians began to display a new laser guided projectile for the Germes (Hermes) 155mm gun/mortar system on a modified BMD-3 armored vehicle.[15] The rocket-powered projectile uses semi-automatic laser guidance. It was developed by the Tula Machine Design Bureau, and appears to be aimed primarily for export. So far as is known, none have been deployed with the Russian Army, and the system is probably still in development.

1K113 Smelchak 240mm SAL-GMP

The Smelchak is another product of the Ametekh Institute and is the world's largest SAL-GP. It is a 240mm semi-active laser guided mortar bomb, and is designed to be fired from the M-240 240mm towed mortar. The projectile was first tested in combat in Afghanistan in the late 1980s. It has an effective range of 3.6 to 9.2 kilometers. The projectile weighs a whopping 134.2 kilograms (295 pounds) and has a thirty-two kilograms HE fill.

Amtekh's Modular 110mm SAL-GP Family

The latest development from Ametech is a family of SAL-GP submunitions. The idea behind this is to develop a modular 110mm diameter SAL-GP that can be adapted to a wide range of small caliber artillery pieces simply by fitting different sabots. These include the Ugrosa-1 for the 122mm Grad multiple rocket launcher, the Firn-1 for 130mm M-46 artillery pieces and the Alfred for 155mm artillery. Ametech indicates that all of these can be adapted to other weapons. For example, the Firn-1 can be adapted to 120mm mortars and 122mm howitzers, and the Alfred can be adapted to 152mm howitzers. These are all stated to have an accuracy of between 0.8 to 1.8 meters.

The US Army has generally abandoned further development of semi-active laser guided artillery projectiles. Although these munitions offer pinpoint precision, they are very awkward to use in combat due to the complicated command and control involved in coordinating the forward observers, fire direction center and artillery piece. During Desert Storm, very few rounds of Copperhead were used. European firms have likewise dropped nearly all work on semi-active laser guided artillery projectiles. Russian interest remains high, but this may parallel early US Army optimism about the potential of this technology prior to the disillusionment caused by the tactical difficulties of employing these weapons.

European and American interest has generally turned to sensor fuzed munitions which deliver one or more anti-armor submunitions by an artillery cargo round. The submunition then floats downward, senses the tank or other target using an infrared (IR), imaging infrared (IIR), millimeter wave (MMW), or dual-mode sensor, and then attacks the armored vehicle, usually using an explosively formed penetrator. The first of these munitions was the US Army SADARM, designed to be delivered by either a M898 155mm projectile or by MLRS rocket. Comparable programs are underway in Europe including the joint Swedish/French BONUS, and the German SmART 155. The US SADARM program has seen extensive program slippage and cost escalation due to the novelty of the technology, and no munition of this type has been successfully fielded to date. In view of the fact that SADARM development started a decade ago, it seems likely that the Russians have either considered or begun development of such a munition. It is worth noting that in 1993, the Bazalt NPO advertised a RBK-500 tactical anti-armor submunitions dispenser, similar to the US Air Force Sensor Fuzed Weapon (SFW). The Russian SPBE-D submunition contained in this dispenser is more similar in configuration to the SADARM submunition than the SFW submunition. Video tapes showing the submunition being tested, and the fact that it is being offered for export suggests that series production will probably occur in the near future if it has not already begun.

Precision Multiple Rocket Launcher System Warheads

The Splav and Bazalt State Research and Production Enterprises are both offering a variety of new payloads for several Russian MLRS models. This is partly an effort to accommodate its new customers like Kuwait which has purchase the 300mm Smerch MLRS and to keep pace with foreign competitors offerings. In addition to the 300mm MLRS, Russia is also offering to sell unique payloads for the 220mm URAGAN and 122mm GRAD (see Appendix A: Multiple Rocket Launch System).

One of the more intriguing sub-munitions offered by Bazalt and Splav is the Universal Sub-Munition (USM) also referred to as Grad-M after the 122mm Grad MLRS system. The USM is designed to attack the top of armored vehicles where their armor is the thinnest. The USM is equipped with both a millimeter wave and infrared sensors which positions the explosively formed projectile (EFP) to penetrate 60mm of RHA. The USM can be fitted to the Russian 300mm Smerch (twenty per round), the 220mm Uragan (five per round) and the 122mm Grad (one per round). In addition to Russian MLRS, up to two USM can be fitted to the U.S. MLRS.

Other guided sub-munitions includ the anti-armor Motiv-3M which is a larger predecessors of the USM. This system was designed by Bazalt for the 300mm Smerch and is also called the SPBE-D when delivered by the RBK-500 air delivered bomb. The Motiv-3M can penetrate up to 100mm of RHA. Bazalt also designs a wide-range of sub-munitions for air-delivered bombs. These sub-munitions could also be offered, like the Motiv-3M/SPBE-D, as payloads for Russian MLRS. These sub-

Ametekh RCIC SAL-GP

System Configuration	Ugrosa-1	Firn-1	Alfred
Caliber	122mm	130mm	155mm
Associated weapon	BM-21 Grad MRL	M-46 gun	NATO 155mm howitzers
Submunition caliber (millimeters)	110	110	110
Projectile weight (kilograms)	67	33.4	46
Munition weight (kilograms)	16	12	16
Projectile length (meters)	3.0	0.67	0.9
Munition length (millimeters)	740	580	740
Warhead	Tandem HEAT	6 kg HE-FRAG	Tandem HEAT

munitions included the anti-armor PTAB, the anti-material/anti-armor OFAB-50 and anti-material/anti-personnel OAB-2.5.

Further additions to the Russian MLRS inventory are evident by the recent offering of a miniature UAV fitted into the 300mm Smerch MLRS. This UAV system would allow the Smerch MLRS to perform various missions such as reconnaissance, laser-spotting, and battle damage assessment. The UAV could be equipped with a variety of sensors but is advertised with a camera as its primary sensor. The UAV would send its data back in real-time to the MLRS system. The 300mm Smerch rocket could launch the system out to seventy kilometers and the UAV can be maintained on station for up to thirty minutes.

Other unique payloads for the 122mm GRAD include rounds for smoke, illumination, and jamming. The 122mm smoke round designated 9M43 can provide a continuous smoke screen that is impenetrable by IR and laser devices operating in the electromagnetic spectrum from 04. to 14.0 microns. The smoke lasts for up to 320 seconds and covers a frontage of one kilometers to a depth of 800 meters. The illumination round for the 122mm is designated 9M42 and provides illumination to a radius of 500 meters. The final round is a jammer round designated 9M519-1-7. The 9M519-1-7 employs a R-032 jammer (also referred to as Liliya-2) which jams both FM and RF communication channels. A Russian advertising brochure stated that the system uses seven rockets which together cover the communication bands from 1.5 to 120 MHz. The jamming duration is given as sixty minutes, and the rockets have a range of 4.5 to 18.3 kilometers. This system is being developed jointly by Splav and the Electron Progress company of Bulgaria. Representatives from the Bulgarian company have claimed that the 152mm artillery jammer projectile was used by the Iraqis to good effect against Coalition communication systems during Desert Storm.

Future Trends in Russian Guided Anti-Armor Projectiles

Russia currently produces a wider range of ATGMs than any single country offering everything from the lightweight Metis up to the heavy Vikhr aviation anti-tank missiles. Since the 1960s, the USSR has generally kept pace in nearly all fields of direct-fire anti-armor guided technology. It has pioneered certain categories of anti-armor technology, notably tube-fired tank guided projectiles and enhanced blast warheads to give anti-armor missiles greater versatility when dealing with non-armored targets.

However, there are several areas where little or no development has been pursued or publicly announced. There are three possible explanations for these apparent omissions: technological barriers; technological approaches of little interest; or continued secrecy. Several of these technology approaches are important, since the advent of reactive armor, compound armors with depleted uranium layers, and other advanced armors raise questions about the viability of direct-attack approaches using conventional guidance. Several of the new technologies make possible top-attack, fly-over attack, and other techniques to attack weakly protected portions of the tank.

To date, the Russians have not displayed a fly-over attack ATGM similar to Swedish RBS.56 BILL or US Army BGM-71F TOW-2B. This is a strange omission given the Russian recognition of the level of protection against shaped-charge attack embodied on newer Western tanks such as the M1A1, Leopard 2, and Challenger. It is possible that such approaches have been taken with newer Russian ATGMs that have not been publicly shown. The use of small sensor-fuzed submunitions on the Russian Air Force's new SPBE tactical munitions dispenser indicates that Russia is manufacturing miniaturized target detection devices for such applications.

There has been little announced interest in fiber-optic guidance. Several countries have active fiber-optic guided missiles including the US (EFOG-M); France/Germany (Polyphem); Japan (XATM-4); Italy (MACAM), and Brazil (FOG). The lack of Russian activity in this sector may be due to general problems in fiber optic technology; Russia is not in the forefront of civil fiber-optic communications technology and may not have benefitted from spin-offs out of the civilian sector.

There is no open evidence of Russian interest in hypersonic missiles comparable to the US Army LOSAT program. This approach has some technical appeal since hypersonic missiles use kinetic energy for penetration, and therefore skirt around the problems confronting all existing ATGMs which rely on shaped-charge warheads. There is no evidence of any Russian lag in hypersonic research, and Russia's work in laser beam-riding technology suggests that the guidance portion of this technology would not be a major impediment. This may simply be a case where there is little interest in this approach, or little confidence it will turn into a viable weapon. The Russians may remain confident that shaped charge warhead technology is not yet exhausted and can overcome defensive technologies as it has done over the past decade.

The Soviet Union generally lagged behind in thermal imaging sensors and in related low-cost imaging infrared missile (IIR) seekers. The pattern seen in other countries has been the adoption of large, high cost Forward Looking Infrared (FLIRs) devices on helicopters and tanks, followed by high cost air-to-surface IIR seekers, followed by low cost IIR ATGMs. The Soviet Union has been about a decade behind the US in this area, only beginning to field the Agava thermal imager on the T-80U in the early 1990s (The US Army adopted the M-60A3 TTS in the late 1980s). Quite surprisingly, the new Russian attack helicopter, the Mil Mi-28 Havoc, lacks a thermal imaging night sight, and Mi-28N ("N" equates to night capable) version did not reach prototype stage until early 1997. There has been no evidence of an imaging-infrared air-to-surface missile comparable to the US Air Force AGM-65D IIR Maverick, low-rate production of which started in 1983. This lag in thermal imaging technology has been attributed to difficulties in mass-production of key sensor array technologies in Russia. These problems are likely to delay Russian development of missiles using guidance technology similar to the US Army's new Javelin ATGM which is expected to reached service in 1997.

Russia may try to circumvent weaknesses in thermal imaging technology by using alternative approaches such as millimeter wave guidance, like that being attempted on the US Army's AGM-114 Longbow Hellfire. Millimeter wave offer near-imaging capabilites for precision attack, and may not pose the industrial challenges of FLIR. A model of the Mi-28N shown in Moscow in 1994 showed a mast-mounted sensor, but it is not clear if this will entail a millimeter wave sensor like the Longbow.

Russia has generally shown more interest in guided anti-armor projectiles than the US or European armies. There have been no descriptions of Russian programs similar to the US Army's 120mm STAFF, which is the first new US guided tank projectile in many years. The STAFF uses a sensor to detect when the projectile passes over the target, and then fuzes a downward pointing warhead to attack the thinner upper surfaces of the tank. Another guided 120mm round, the X-Rod, is also being developed which if successful would be the first guided Kinetic Energy (KE) penetrator. Given Russian interest in guided tank projectiles, it would not be surprising to find that Russia is developing comparable munitions.

Russian interest, and willingness to invest, in advanced anti-armor technology will be conditioned on Russian perceptions of the combat utility and export potential of these systems. The Gulf War probably reinforced Russian interest in anti-tank missile technology. Helicopter-fired missiles were a particularly lethal system, evident from the destruction of fleeing elements of the Hammurabi Republican Guards Division in March 1991 in the 24th Infantry Division sector. In a single attack, an Apache battalion effectively rendered a mechanized brigade ineffective in less than an hour. Combined with Soviet experiences in the use of helicopter gunships in Afghanistan and the new importance given to air-mobile units under the current Russian Mobile Force concept, continued development of this field of weapons is assured.

Ground fired anti-tank missiles were used in large numbers during Operation Desert Storm, including about 3,000 TOW missiles and smaller numbers of HOT and Milans. Their performance was generally satisfactory from published reports. Although detailed unclassified figures are lacking, it is evident that the primary tank killing weapon in most large armored engagements was the 120mm tank gun, but ATGMs gave light armored vehicle significant anti-armor capability. The Iraqis made minimal use of newer generations of Russian ATGMs, so there was probably little evidence of "lessons-learned" on this score.

Russian success in exporting ATGMs will also effect future developments in this area if Russian defense budget cuts continue. Russian ATGMs were not tarred by any specific accusations of poor performance of the type leveled against Russian tanks. While this hardly constitutes a ringing endorsement, it also suggests that there will probably not be any strong resistance against acquiring such weapons by most potential Russian clients. What remains to be seen is whether the Russians will be able to survive in a very competitive export market which has a wider range of competitors than large systems such as fighter aircraft and tanks. Russia's main selling point in the short term will be the relatively low costs of their systems compared to American and European systems, even if the Russian systems do not yet match some Western systems in some areas of technology such as top-attack.

3
ARMORED INFANTRY VEHICLES

Soviet/Russian Armored Infantry Vehicle Development — An Overview

Unlike the United States, Britain, and Germany, the Soviet Union was a relative late-comer to infantry mechanization. In the view of their German opponents, the Soviets' failure to mechanize portions of its infantry in World War 2 was a major impediment to their offensive combined arms tactics. The Soviet Union did not begin manufacturing armored infantry vehicles until after the Second World War. The reason for the slow Soviet progress in this area during the war was mainly industrial production limitations rather than doctrinal shortcomings. The production capacity of the automotive industry was focused on higher priority weapons, including light tanks and assault guns.

Soviet development of armored infantry vehicles accelerated after World War 2 to make up for this shortcoming. Early post-war Soviet armored transporters were technically conservative and not as well designed as their NATO counterparts. The BTR-152 was little more than an armored truck and was inferior to wartime German and American half-tracks; the tracked BTR-50 was adequate for transporting troops, but access from the vehicle by the infantry was poor.

Following Stalin's death in 1953, doctrinal development flourished, and more attention was paid to the implications of the presence of tactical nuclear weapons on the battlefield. The watershed event in post-war Soviet history was the decision in 1957 to convert the Soviet Army's 150 plus rifle divisions into motor rifle divisions (mechanized infantry divisions in US terms). This implied the need for a massive increase in armored infantry vehicles. The Soviet Union continued to follow a dual track procurement approach: a tracked and sophisticated infantry vehicle for the tank divisions, and a less costly wheeled vehicle for the motor rifle divisions. The doctrinal ferment led to the decision to adopt the world's first true infantry fighting vehicle, the BMP, to satisfy the high-end requirement. The low end requirement was satisfied by the BTR-60. The mechanization process was slowed by Khrushchev's lack of enthusiasm for the expensive BMP, and procurement was not possible until after the Brezhnev coup in 1964.

For the subsequent two decades, Soviet infantry vehicle development was an evolutionary path on the basis of the BMP/BTR-60 pattern. The BMP-1 evolved into the BMP-2, while the BTR-60 evolved through the BTR-70 and BTR-80. There is little primary source material in Russian on the decisions involved in the design of Soviet infantry vehicles, and the associated design bureaus have not generally exploited perestroika and the post-1991 reform movements to trumpet their accomplishments. Nevertheless, the release of other information has made it possible to depict some of the options considered during the 1970s and 1980s.

The most recent Soviet innovation in armored infantry vehicles, the BMP-3, remains an enigma. Several suggestions are made here for the possible rationale for its design. But it remains a unique vehicle in its class. There have been strong hints from Russian sources that it is not an entirely successful design, and a more satisfactory vehicle could appear by the end of the decade.

This section is divided into three main parts: wheeled infantry transporter, tracked infantry transporters, and airborne armored vehicles.

Early Development Trends

The Red Army was very slow to grasp the need for armored infantry vehicles to support tanks. Experiments were conducted in the 1930's with armored infantry vehicles, but none were ever accepted for quantity production.[1] This is all the more surprising, as Soviet tactical developments in armored warfare in this period were impressive. Nearly all the other major European armies were mechanizing elements of their infantry to cooperate with tanks, even some of the smaller armies such as Poland and Italy. The problem became extremely evident in the 1940 Russo-Finnish war which led to the adoption of improvised armored sledges that were towed behind attacking tanks, carrying prone infantrymen. Beyond a few more experimental infantry vehicles, no further attempts were made to begin mechanization of the infantry as was occuring in France, Britain, Germany, and the US at the time.

During World War 2, the need for armored infantry vehicles became even more manifest. Tanks, unaccompanied by infantry, were vunerable to close-in attack by enemy infantry. Tanks were also fairly blind, and often unable to identify and suppress anti-tank gun positions. Foot infantry could not keep up with tanks, and infantry mounted in trucks were unable to follow tanks in rough terrain, snow, and many other conditions given the backward nature of the Soviet road network. Although the Soviets had witnessed the considerable success of German panzergrenadier tactics using armored infantry halftracks, they were unable to emulate these tactics due to industrial constraints. The Soviet automotive industry was hard pressed to provide sufficient vehicles and light tanks, and could not afford the diversion of resources to armored infantry vehicles. As a result, during World War 2, the Soviets produced no armored infantry vehicles aside from a few experimental models, while at the same time, nearly one-third of German, American, and British armored vehicle production consisted of armored infantry vehicles. The Red Army did receive British and American armored infantry vehicles through Lend-Lease, such as the Universal Carrier (Bren Carrier), US armored halftracks, and US M3A1 scout cars. However, these were never available in large enough numbers to develop motor rifle regiments, and so they were used as scout vehicles or as armored transporters for unit commanders.

Soviet limitations in this field led to the development of improvised tank-infantry tactics, notably the "tank desant" tactic. Tank desant troops were trained to ride tanks into combat, clutching to hand-holds on the turret and hull sides. This tactic was extremely costly on infantry troops, since they were completely unprotected. Such tactics were often successful, but came at a high price. This concept remained in the post-war Soviet Army, and Soviet medium tanks through the T-62 had large turret hand-holds specifically to facilitate this tactic. Shortcomings in infantry mechanization was one of the singular failures in Soviet tactical doctrine during World War 2. Many German officers felt that Soviet shortcomings in this field were a major contributing factor to German defensive successes against Soviet mechanized offensives in the 1943 to 1945 period.

Early Post-war Development

One of the central debates in the development of armored infantry vehicles was whether to base the vehicles on tracked or wheeled chassis. Both Germany and the United States adopted a compromise (on their wartime vehicles), the half-track. While offering better mobility in poor terrain conditions, there was the general consensus by the end of the war that half-track technology was a dead end. It did not offer the mobility of fully tracked vehicles, or the high road speed of wheeled vehicles. In addition, it was more expensive than comparable wheeled configurations, and had higher maintenance demands. The only army to opt for the fully-tracked approach was Britain, which had adopted the Universal Carrier as its standard mechanized infantry vehicle before the war. This was a relatively small vehicle employing light tank components, and typically carrying up to six men. It's armor protection was very modest, even compared to German and American half-tracks. By 1944, the British Army was being forced into improvised solutions, notably the conversion of obsolete Canadian Ram tanks into the Ram Kangaroo. The Kangaroo was a Ram tank minus the turret and turret basket. The infantry section simply rode in the hull where the turret had been located, and the conversion was so spartan that not even bench seats were fitted. This improvised solution proved more useful than the Universal Carrier in infantry assaults against well defended German positions. The US Army introduced a small number of M39 Armored Utility Vehicles towards the end of the war, which were M18 76mm tank destroyers with the turret replaced by a cargo compartment. These were not specifically intended as infantry carriers, but there was some use in this role.

By the end of the war, both Britain and the United States were leaning towards the use of fully-tracked infantry vehicles. In addition, there was a growing consensus over the need for full armor protection. Wartime vehicles, including the half-tracked types, did not have overhead armor protection. This made the infantry squad inside vulnerable to overhead artillery airbursts as well as small arms fire and grenades. Improvised solutions such as the fitting of overhead armor was impossible due to the automotive limits of the chassis.

Post War Soviet Infantry Vehicle Development

The Soviet Army, lacking extensive wartime experience with mechanized infantry tactics, took a broad look at infantry vehicle technology. By the late 1940s, a wide range of development programs were underway, examining both wheeled and tracked armored infantry vehicles. This study begins by examining wheeled infantry vehicles.

Wheeled Infantry Vehicles

The BTR-152 Armored Transporter
Following the war, attention began to be paid to the requirement for infantry vehicles. In May 1946, the ZiS automotive plant in Moscow completed the first proto-

BTR-152 armored transporter.

BTR-152A fire support vehicle.

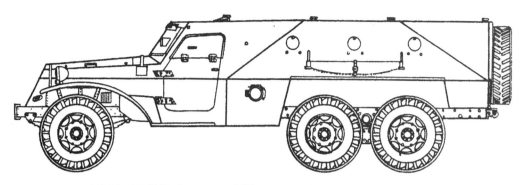

BTR-152V2 Armored Transporter

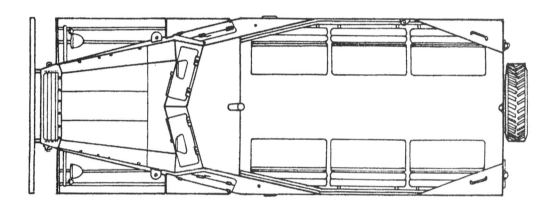

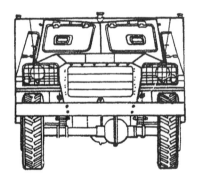
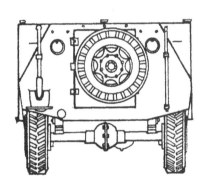

type of its new 2-1/2 ton army truck design, the ZiS-151, which was copied from Lend-Lease American International Harvester K designs.[2] A new design bureau, headed by B.M. Fitterman, was assigned the task of developing an armored transporter (bronetransporter or BTR) called izdeliye 140, or Project 140. The izdeliye 140 used a modified version of the ZiS-151 chassis, the ZiS-123, to carry an infantry squad of seventeen men (plus two crew). It was a simple armored truck, with armor up to 13mm but no overhead protection for the crew. The first two prototypes were completed in May 1947 and were sent to state trials. The new vehicle was accepted for limited production in 1949 as the BTR-152 with sufficient numbers manufactured later in the year for operational trials.

Following the troop trials, the BTR-152 was oficially accepted for Soviet Army service on 24 March 1950, and entered regular service later in 1950. It was the first Soviet designed armored infantry transporter to reach series production.

In comparison to armored infantry transporters in service elsewhere at the time, the BTR-152 was a regressive design. It was open-topped, exposing its squad to overhead artillery airbursts, grenades, and small arms fire. The chassis was not significantly different from the normal truck chassis, except being encumbered with five tons of armor.

As a result, it was a very sluggish vehicle with no real cross-country ability. Indeed, it was inferior in many

BTR-152 command vehicle.

respects to wartime infantry transporters such as the American M3 half-track or German Sd.Kfz. 251 half-track in terms of mobility. By way of contrast, the US Army of the time was adopting the M75 armored infantry transporter which was tracked and fully armored. The BTR-152 was selected by the Soviet Army for its low cost. It was an easy vehicle for the ZiS Automotive Plant to manufacture, and it served as a worthwhile first-generation infantry vehicle with which the Soviet Army could experiment with motorized infantry tactics.

Concurrently with the izdeliye 140, a related infantry support vehicle was also developed, the BTR-152A. This vehicle was equipped with a new twin ZTPU-2 heavy machine gun mount fitted with the new Vladimirov KPV 14.5mm heavy machine guns. This vehicle was intended to provide fire support for mechanized infantry both against low-flying aircraft and ground targets. The BTR-152A was not accepted for service until 1952, about two years after the basic BTR-152.

In 1951, design work on the BTR-152 series was transferred to two of Fitterman's subordinates in the ZiS design bureau, V. F. Rodionov and N. I. Orlov. Work was begun on a PUA (punkt upravleniya artilleriyei-Artillery Command Post) version. This was completely armored, and had raised sides to permit artillery command officers to stand within the rear hull and to erect maps. This was accepted by the Soviet Army in 1952 as the BTR-152B. At the time, the ZiS Automotive Plant in Moscow was developing a tire pressure regulation system on the basis of American systems received in the Lend Lease program. These systems could lower tire pressure from the driver's station, thereby increasing vehicle traction in poor soil conditions or snow. Wheeled vehicles like the BTR-152 had ground pressures at least five times higher than comparable tracked vehicles. The tire regulation system theoretically could reduce this ground pressure to levels closer to that of tracked vehicles, but the vehicle could not be used at high speeds in this mode for fear of damaging the tires. This system was first used on the ZiL-157 truck and was adapted to the BTR-152 in 1954 as the BTR-152V.[3] The first twenty pre-series BTR-152V were used during the 1954 Byelorussian army maneuvers, where they won the personal endorsement of Marshal G. Zhukov who was impressed by the cross-country performance improvements and who pushed to adapt this system to all infantry transporters.

The BTR-152V had very noticeable external air lines running from the hull to the wheel hubs, and also had a winch added to the front bumper; it was accepted for Army use in October 1955 and entered quantity production at the end of 1955. In 1956, this system was improved using internal air lines. This improved system, along with infrared night driving lights resulted in the BTR-152V1 which entered series production in 1957. Earlier versions of the BTR-152 were rebuilt with the new tire regulation system, rebuilt BTR-152 becoming the BTR-152V2 and rebuilt BTR-152V becoming the BTR-152V3.

Development also proceeded on further support and command vehicles. An improved version of the BTR-

BTR-152V armored transporter.

BTR-152V1 armored transporter.

152A fire support vehicle, based on the BTR-152V chassis, was accepted for production in 1955 as the BTR-152D. Another version based on the BTR-152V, using a quad ZTPU-4 KPV 14.5mm heavy machine gun system was also produced in small numbers as the BTR-152E but did not see extensive troop use. Two further command versions of the BTR-152V were developed, the BTR-152I being a counterpart to the earlier BTR-152B artillery command post, and the BTR-152S being a new command post vehicle, designed for motor rifle unit commanders.

In 1959, shortly before production ceased, the BTR-152V1 was modified with overhead protective armor, designated BTR-152K (K= Krytkiy, closed). Some earlier vehicles were later rebuilt in this configuration. Production of the BTR-152 ceased between 1959 and 1960 in favor of the newer BTR-60. Total production was about 15,000 vehicles of all types. It was exported to most of the Warsaw Pact countries, and later was exported to the Middle East, serving with the Egyptian Army in the 1956 and 1967 wars. The BTR-152 disappeared as a troop transporter in the Soviet army in the late 1960s and early 1970s as the newer BTR-60 became available. However, many were retained and rebuilt as armored ambulances, command vehicles, mobile radio stations, and combat engineer mining vehicles, thereby serving well into the 1980's. Some are still in service today in the Third World.

The BTR-60

Experience with the BTR-152 provided the Ground Forces with critical experience in armored infantry tactics. With the decision between 1956 and 1957 to begin to convert all rifle and mechanized divisions into motor rifle divisions, a new vehicle was needed to make this conversion possible. The principal shortcoming of the BTR-152 was its poor cross-country ability, and in 1959, a new Technical-Tactical Requirement was issued for the development of an armored transporter with improved mobility, including amphibious capability. At the same time, work began on a more expensive and sophisticated tracked carrier which would eventually emerge as the BMP. The BMP was intended to serve in tank divisions, while the new BTR was intended to mechanize the more numerous motorized rifle divisions. Tracked chassis, although offering better mobility in poor ground conditions, are more expensive to produce and more expensive to operate and maintain than wheeled systems.

The requirement for the new izdeliye 49 was issued to at least two design teams, Dedkov's design bureau at GAZ in Gorkiy, and the Rodionov/Orlov team at ZiL. The Dedkov team had already developed a four-axle, all-terrain chassis in the winter of 1956, the izdeliye 62B which formed a basis for their work on the new izdeliye 49.[4] The requirement called for the use of a gasoline engine (since suitable diesel engines were still at a premium in the USSR) and the use of an existing engine. The Dedkov team selected the GAZ-49 engine which they had employed already in their earlier BRDM scout vehicle design (the GAZ-49 was a derivative of a wartime US Dodge engine used in 3/4 ton trucks). The Dedkov team proposed several novel technical features in the design, but most of these were rejected in the state trials in favor of simpler automotive components which would be less trouble to manufacture. The ZiL team developed a 6 x 6 design, designated the ZiL-153. The hull shape was similar to the Gorkiy design. The prototypes were submitted for state trials in 1960, and the Dedkov entry was selected. It was authorized under the designation BTR-60P.

The initial version of the BTR-60, the BTR-60P, had an open roof. It is unclear why such an archaic feature was retained in view of the general trend in the Ground Forces towards sealed vehicles which could be fitted with

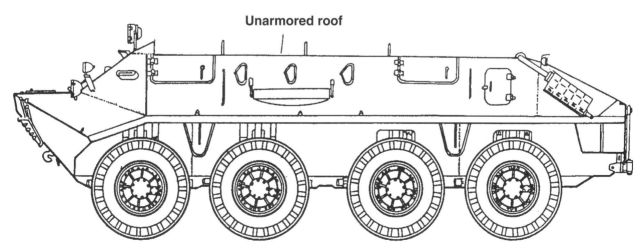

BTR-60P Model 1960 Armored Transporter

ZiL-153 (BTR-60 competitor).

BTR-60P armored transporter.

nuclear protective equipment. In any event, the BTR-60P was shortlived, being series produced only from 1960 to 1963. In 1963, it was supplanted by the BTR-60PA, with an armored roof which remained in production through 1966. On this version, the troop size was reduced from sixteen (fourteen + two crew) to fourteen (twelve + two crew). The main shortcoming of the BTR-60PA was that the squad machine gun could not be fired from within the protective shell of the vehicle. There was also some criticism that the armored transporters were becoming increasingly "toothless" and could not deal with hostile infantry transporters on the battlefield. For example, the German HS.30 was armed with a 20mm gun turret. As a result, in 1964, work was undertaken on an improved version mounting the 14.5mm KPVT heavy machine gun in a small turret. The first production model of this vehicle appeared in 1965 as the BTR-60PAI, but it was quickly replaced by a slightly improved type, the BTR-60PB which had an improved sighting system for the gun. The BTR-60PB became the standard production model of the family, and was in series production from 1966 through 1976. The turret adopted for the BTR-60PB was the same as that later appearing on the BRDM-2 in 1966. The seating in this vehicle was more complicated than the simple bench seats of previous models. The infantry squad was seated mainly on bench seats along the outer edge of the hull, on a single bench seat along the rear engine compartment bulkhead, and in two seats facing forward immediately behind the driver's and commander's seats.

The turmoil in BTR-60 development highlighted the tactical ferment in the Soviet Ground Forces at the time.

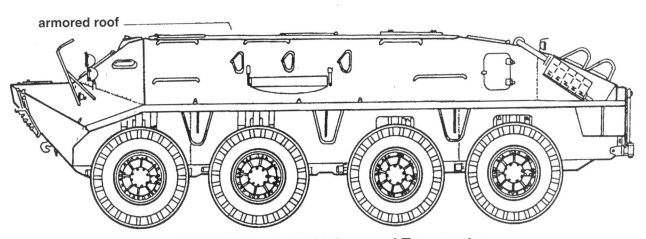

BTR-60PA Model 1963 Armored Transporter

BTR-60PA armored transporter.

There was still some controversy over the tactical requirements for an armored transporter for the new motor rifle divisions. The BTR-60P represented the last trace of a doctrine which viewed armored transporters as nothing more than taxis to bring the infantry near the battleline. The BTR-60PB, which appeared only four years later, represented the first glimmers of a new view that an infantry transporter had to be survivable on a nuclear battlefield and that the troops had to be able to fight from within the confines of the vehicle. The BTR-60PB was not an infantry fighting vehicle in the current sense of the word, since only six of its eight squad members could fire their weapons from within the armored vehicle, but it was a step in that direction. The BTR-60PB has a normal

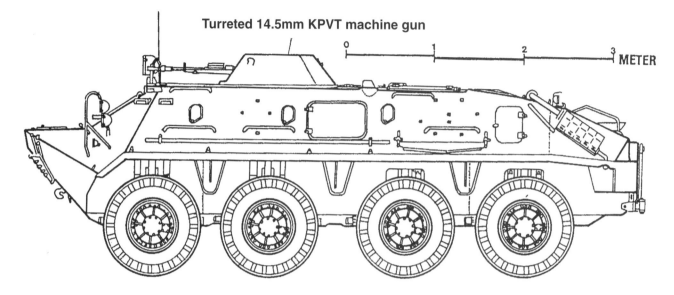

BTR-60PB Model 1966 Armored Transporter

BTR-60PB armored transporter.

complement of ten: two crew (driver and machine gunner) and an eight man squad. In comparison to NATO designs of the day like the US M113 or the British FV432, the BTR-60PB was not impressive. The design turmoil between the open-topped BTR-60P and the closed BTR-60PA and BTR-60PB had resulted in a vehicle poorly suited to crew entrance and exit. The doors on the BTR-60PB were small and badly situated. A squad disembarking under hostile fire was extremely likely to suffer casualties. In contrast, the NATO designs had large rear doors which permitted easy entrance and exit, and a modicum of protection for the squad if exiting under fire. The only advantage offered by the BTR-60PB was the turret machine gun station. In contrast, the machine gun on the American M113 was completely exposed. The BTR-60PB was automotively inferior to the M113. The BTR-60 series uses two ninety horsepower GAZ-49B, each of which powers one side of four wheels. The selection of this peculiar configuration was due to the shortage of production capacity for larger truck engines at the time. Such a configuration demands the use of two starters, two transmission trains, four transmission boxes, and a general doubling of engine and powertrain subcomponents. While inexpensive to produce, the vehicle powertrain is difficult to service, and problems with one of the two engines can immobilise the vehicle. Engine synchronization was a frequent problem, as were other powertrain failures. Due to the hull shape, the fire-prone gasoline engine, and its automotive shortcomings, the BTR-60PB was popularly called the "wheeled coffin" (kolesniy grob) amongst Russian motor riflemen. In spite of its shortcomings, it was a cheap method to mechanize the bulk of the Soviet Army.

The BTR-60 was designed to be amphibious without preparation. It uses a hydrojet system similar to that employed on the PT-76 light tank. This offers good swimming ability in rivers so long as the current is not above eight kilometers per hour. The BTR-60 has a water swimming speed of ten kilometers per hour. The BTR-60 series is faster and better handling in water than NATO systems like the track propulsion used on the M113, but the BTR-60PB has problems exiting rivers. The water propulsion system can only be operated when the normal transmission for the wheels is in neutral, so there can be troubles in shallow water with soft river bottoms since only one propulsion method can be used at a time.

The BTR-60 series was in production from 1960 to about 1976. Total production of all models is believed to have reached about 25,000 vehicles. Final production models, called the BTR-60PZ, had a number of small improvements over the earlier BTR-60PB. These features include a new 1PZ-2 roof-mounted periscopic sight for the KPVT heavy machine gun and higher maximum elevation for the machine gun. The BTR-60PB's with the Limited Contingent of Soviet Forces-Afghanistan were modified with an AGS-17 30mm grenade launcher mounted in place of the normal KPVT machine gun. There were a number of further experiments with the BTR's armament.

A related BTR-60 variant is the MEP, which is apparently intended for escort duty. This vehicle is essentially similar to the normal infantry versions, but has a TKB-0149 turret which allows the turret occupant to stand up in the turret via a new roof hatch. The turret is armed

BTR-60 MEP armored transporter.

BTR-60PU command vehicle (early configuration).

BTR-60PU-12 air defense command vehicle.

only with a 7.62mm PKT machine gun. The new turret adds a day/night vision sight for surveillance up to 1,500 meters. The MEP based on the BTR-60PB has an additional box on the right-center side of undetermined role.

As has been the case with so many Soviet light armored vehicles, the BTR-60 fostered a very large family of derivative types. Besides the basic troop carrying varieties, a number of specialized command vehicles have also been fielded, which are generically called BTR-60PU (upravleniya-command). Infantry command vehicles are primarily intended as a radio vehicle for battalion and regimental command staffs. The earliest versions of these command vehicles were usually simple adaptations of the turretless BTR-60's with additional whip antennas. The antenna fit and internal arrangements are tailored to meet the requirements of the user. A typical BTR-60 command vehicle, used by a motor rifle battalion chief of staff, could be fitted with an R-130 high frequency (AM) transceiver to communicate with regimental HQ on the regimental AM radio net; an R-107 very-high frequency (FM) amplifier equipped transceiver to communicate on the regimental FM command net; another R-130 to operate on the battalion AM command net; an R-123 very-high frequency (FM) transceiver to communicate on the main battalion net used by all battalion vehicles; and finally an R-311 high frequency AM receiver to listen in on the regimental or divisional CBR and air warning network. In contrast to this staff vehicle, the commander of a motor rifle battlion would have a BTR-60 command vehicle with an R-107, R-123, R-126, and R-130 transceiver. A company commmander's vehicle would have a similar fit, minus the R-130.

Another common type is the BTR-60PU-12, used as a command post for air defense units; the similar BTR-60 VVS is used by air force units for forward air control. These are fitted with an additional generator in a rectangular container in lieu of the turret, and a telescoping antenna (NATO code name: Top Ball) to improve transmission. The basic BTR-60/R-145 radio is fitted with a R-145 radio inside along with a clothesline antenna running around the hull, at least one additional generator in a rectangular container in lieu of the turret, and a telescoping Top Ball antenna. The BTR-60/R975 Forward Air Control (FAC) versions of the BTR-60PB include a type with a large generator on the rear of the hull and a plexiglass window replacing the armament in the turret. There is another FAC version of the BTR-60, which has a new, higher turret than the normal BTR-60PB, but without armament.

The Romanian BTR-60 Variants

Romania was the only other country to license manufacture the BTR-60PB. In the late 1960s, the Romanian Socialist Army (ASR) approached the Military Industrial

BTR-60 Variants (as listed in CFE documents)

Designation	Mission/Role
BTR-60 PAU	Bulgarian artillery command vehicle
BTR-60 PU-12	Air defense command vehicle
BTR-60 VVS	Air force air controler vehicle
BTR-60 P-238BT	Field telephone communications vehicle
BTR-60 P-239BT	Field telephone communications vehicle
BTR-60 P-240BT	Field telephone communications vehicle
BTR-60 P-241BT	Field telephone communications vehicle
BTR-60 LBGS	Field telephone cable laying vehicle
BTR-60 R-137B	Turreted radio vehicle (minus 14.5mm hmg armament)
BTR-60 R-140BM	Turreted radio vehicle (minus 14.5mm hmg armament)
BTR-60 R-145	Turretless radio vehicle, clothesline antenna
BTR-60 R-145BM	Artillery communications
BTR-60 R-156BTR	Turretless signals vehicle, inverted "U" clothesline antenna
BTR-60 R-161B	HF Radio vehicle
BTR-60 R-409BM	Turretless radio vehicle
BTR-60 R-975	Turreted forward air controller vehicle
BTR-60 R-975MI	Turretless forward air controller vehicle
MEP	Infantry command post
1V18	Artillery battery command observation post
1V19	Artillery battalion command observation post
BTR-60 Z-351BR	Power generator (*zaryadnaya stanitsya*)
MTP-2	Turretless maintenance vehicle

Committee of COMECON about the possibility of license producing the BTR-60PB. The Soviets consented, probably feeling that this was preferable to yet another co-development scheme between Romania and the heretic Yugoslavs. The initial, turretless model, called TAB-71, entered production in 1970, and was basically a BTR-60P with two Romanian 140 horsepower engines in place of the normal ninety horsepower engines. This version was not produced in significant numbers. This was soon succeeded by a version of the BTR-60PB with a modified turret and uprated engines, designated TAB-71M (sometimes also called TAB-72). The only external difference between a BTR-60PB and the TAB-71M is the protruding gun sight on the left turret side of the Romanian vehicle. Total production of the TAB-71 was 1,878 vehicles through 1990. A number of subvariants of the TAB-71 were manufactured. These included three command versions: TAB-71AR1 450, TAB-71AR1 451, and TAB-71AR1 452; the TAB-71AR 82mm mortar version (sometimes called TAB-73); and two maintenance and recovery versions, the TERA-71 and TERA-71L.

The OT-64 SKOT

In 1959, Czechoslovakia and Poland discussed co-producing a new armored infantry vehicle in lieu of purchasing the BTR-60PB from the USSR, called the SKOT (Czech: Stredni kolovy obrneny transporter; Polish: Sredni kolowy opancerzony transporter; Medium Wheeled Armored Transporter). The BTR-60A was viewed skeptically by the Polish and Czechoslovak armies due to its open top and poor crew access/egress. This was an unusual case where the Soviet Union permitted the Warsaw Pact countries to develop their own military equipment, violating the usual insistence on standardization. Responsibility for the work was divided, with Czechoslovakia responsible for the chassis and automotive components, and Poland for the armored body and armament. The first prototype was completed in 1961, and production began around 1963 with chassis production at the Tatra Automotive Factory in Koprivnice, Czechoslovakia and armored hull work and assembly at the FSC (Fabryka Samochodow Ciezarowych) in Lublin, Poland. The initial production version, called OT-64 SKOT-1A, corresponded to the Soviet BTR-60PA and had no armament. The principal difference between the OT-64 SKOT and the BTR-60P was that the SKOT could carry a larger number of troops, was configured for more satisfactory troop entrance and exit, and incorporated a CBR protective system from the outset. The OT-64 has two large rear doors, which speeds troop access, and reduced their vulnerability to small arms fire. As in the case of the BTR-60, the OT-64 SKOT underwent continual evolution and many subvariants were developed. Production continued until 1990 and 10,300 were manufactured.

The BTR-70 Armored Transporter

If the BMP was the most revolutionary armored vehicle developed in the Soviet Union since the Second World War, the BTR-70 was one of the least radical. The BTR-70 is a straight evolutionary development of the BTR-60PB, also designed by the Dedkov design bureau in Gorkiy/Arzamas.[5]

Development of the BTR-70 began in 1971 to cure some of the problems with the BTR-60. The major aim of the redesign was to make the BTR series more compatible with the BMP in terms of tactical training. Towards this end, the Gorki team designed the Obiekt 50 (also known as GAZ-50). This mated a modernized BTR hull with a BMP-1 turret. The Obiekt 50 had reconfigured hatch and seating arrangements, permitting the infantry squad to fight from inside like a BMP, and to more safely exit the vehicle from lower in the hull. Rather than opting for a complete reconfiguration such as that offered by the Czechoslovak/Polish OT-64 SKOT, the Dedkov OKB opted for a simpler change. Hatches were provided for entrance and exit in the lower hull between the second and third wheel stations. The new design also offered protection advantages over the BTR-60PB by improving the hull shape. Internally, the Obiekt 50 offered a reconfigured seating arrangement more in keeping with the mounted infantry tactics being used in BMP motor rifle regiments. In the center, a bench seat was located down the middle of the vehicle for six riflemen sitting back to back, each with their own firing port. Two more riflemen behind the driver and vehicle commander face forward, and have side firing ports as well. The squad size was the same, two crew plus eight squad members.

There was some debate in the Soviet army about the armament for the new armored transporter, probably due to the cost involved in adding the more sophisticated BMP-1 turret. The design bureau prepared another prototype with the standard BTR-60PB 14.5mm heavy machine gun turret, designated Obiekt 60. In the end, the Soviet Army decided to stick with the simple 14.5mm machine gun turret, probably on cost grounds. Series production of the Obiekt 60 vehicle began in 1972 as the BTR-70.

The BTR-70 retained many features of the BTR-60PB design, including the twin engine arrangement. The BTR-70 is powered by two fuel-injected ZMZ-4905 115 horsepower diesel engines which provide the vehicle with 230 horsepower. In contrast to the BTR-60 powertrain, on the BTR-70 the torque from the right engine powers the first and third roadwheels on both sides, and the the left engine the second and fourth, allowing the vehicle to

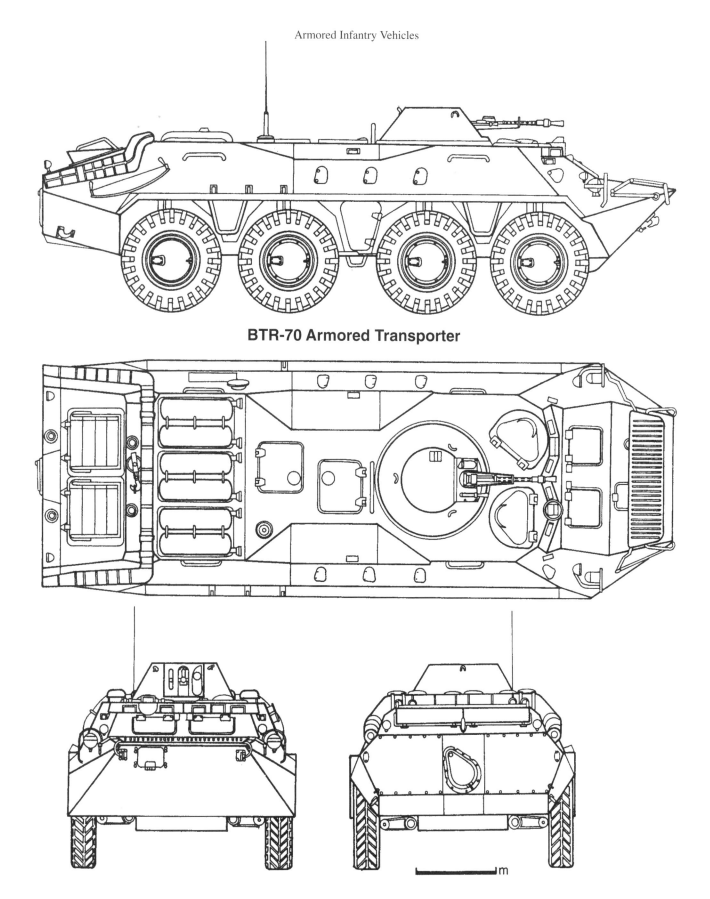

BTR-70 Armored Transporter

Obiekt 50 armored transporter.

BTR-70 armored transporter.

continue moving even if one engine fails. The retention of this configuration was due to successes in overcoming some of the technical problems associated with the BTR-60's tandem configuration as well as continued shortages of large gasoline and diesel engines due to demands from other sectors of the military economy.

The BTR-70 was a clear example of the Soviet predilection towards a high-low mix in armored vehicle procurement. In the Soviet Ground Forces in the late 1970s, the BMP was issued to the motor rifle regiments in tank divisions, and to one of the three motor rifle regiments in forward-deployed motor rifle divisions. The remaining two motor rifle regiments in the motor rifle divisions were equipped with the BTR-70 (or the older BTR-60PB). The BTR-70 allowed these regiments to emulate BMP mounted and dismounted tactics, but the BTR-70 did not offer the armored protection, mobility or firepower of its more expensive tracked counterpart.

The BTR-70 offers only an incremental advance over the earlier BTR-60PB, and not a major increase in combat capability, as occurred when the Soviets switched production from the BTR-152 to the BTR-60 series in 1960. Funding went instead to permit wider introduction of the more expensive BMP beyond its use in tank divisions.

During production, many small changes were gradually introduced into the basic BTR-70. The early production vehicles were fitted with the same wheels as the BTR-60PB. The post-1982 production types were fitted with a new wheel design and some minor hull changes, and the engine power was increased from 230 to 240 horsepower. During the Afghanistan War, there were extensive field modifications to the BTR-70 including addition of AGS-17 grenade launchers to enhance firepower, and addition of appliqué side armor to compensate for the vehicle's inadequate protection against heavy machine guns. There was also at least one standardized design for a turret mounted auxiliary 30mm grenade launcher, somewhat like the type later developed for the BMP-1PG upgrade. Like the BTR-60PB and BMP, many specialized versions of the BTR-70 are also in service as noted below.

The basic BTR-70 is fitted with the R-123M radio transceiver, while the platoon commander's BTR-70K vehicle is fitted with the R-123M and an R-126 radio transceiver. Company and battalion commander's BTR-70K vehicles are fitted with four radios, the R-105, R-123, R-126, and R-107 at company level and the R-104, R-123, R-126, and R-107 at battalion level.

The East German NVA modified some of their BTR-70 into chemical scout vehicles, using features from the normal BRDM-RKh; this does not appear to have been a standard Soviet practice.[6]

The MEP is a modified version of the BTR-70 used as an escort vehicle or mobile command post. This vehicle is essentially similar to the normal infantry versions, but has a TKB-0149 turret which allows the turret occupant to stand up in the turret via a new roof hatch. The turret is armed only with a 7.62mm PKT machine gun.

BTR-70 MEP armored transporter.

BTR-80 armored transporter.

The new turret adds a day/night vision sight for surveillance up to 1,500 meters. The MEP/BTR-70 has a new stowage box located on the right-front side.

In the late 1970s, the Romanians acquired license production rights for the newer BTR-70 which entered production around 1977 as the TAB-77; production through 1990 for the Romanian ASR was only 154 vehicles. The Romanian version is essentially similar to the Soviet version, but has launchers for the 9M14M Malyutka missile on the turret sides to give the vehicle some anti-tank protection. A locally developed, shortened version of the TAB-77, with four instead of eight

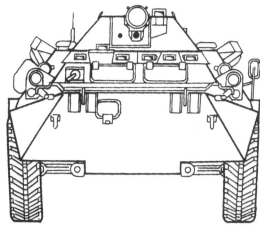

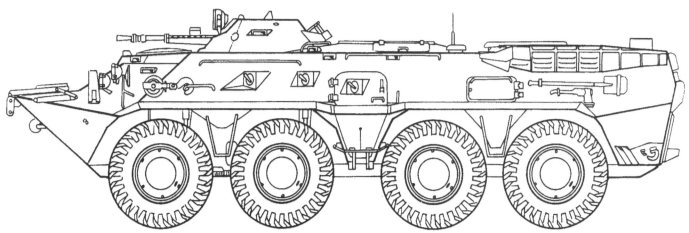

BTR-80 Armored Transporter

This BTR-80 displays an add-on armor package to the hull.

Looking forward into the driver's compartment of the BTR-80.

BTR-80A armored transporter.

wheels, entered production around the same time as the TAB-C. This appears to be intended as a scout vehicle like the BRDM-2 rather than a troop carrier. Variants of the TAB-77 include the TAB-77 PCOMA artillery command post; TAB-79 A-POMA artillery observation post; TAB-79AR 82mm mortar carrier; TAB RCH-84 CBR scout vehicle; TAB-77A R1 451/M command vehicle; and TERA-77L maintenance/repair vehicle.

The BTR-80 Armored Transporter

The BTR-70 displayed one of the shortcomings of reliance on conservative evolutionary development of armored vehicles. The design clearly did not cure all of the BTR-60PB's problems, which became evident during early fighting in Afghanistan in 1980-81. As a result, in 1982, a further evolution of the BTR-70 undertaken, the GAZ-5903 by the design team at AMZ (under the direction of Aleksandr Masyagin since Dedkov's retirement).[7] The evolutionary leap between the BTR-70 and BTR-80 was very modest.

There were three main changes. The twin engine configuration was dropped in favor of a simpler single 260 horsepower diesel engine with a less complicated power train. The side doors were improved to allow easier troop entrance and exit. The turret was redesigned to permit elevation of the KPVT heavy machine gun to sixty degrees, compared to thirty degrees on the earlier vehicles, allowing the weapon to be used against troops in hilly terrain (like Afghanistan) and aerial targets like helicopters. There have been a number of other detail design improvements, including the addition of Type 902 Tucha smoke mortars behind the turret, and improved firing

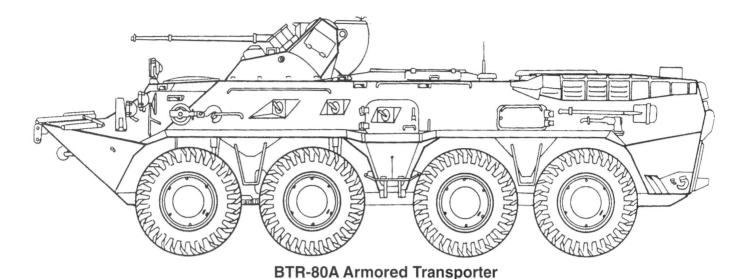

BTR-80A Armored Transporter

ports. The GAZ-5903 was accepted for production in 1984 as the BTR-80, replacing the BTR-70.

The BTR-80A

The BTR-80 has undergone gradual evolution since entering production. A fire in 1993 destroyed the plant manufacturing the BTR-80's original engine. As a result, in 1994 a new engine option was developed, using either the KAMAZ 7403 260 horsepower truck engine or the YaMZ-238-M2 240 horsepower engine.

Besides the engine option, in 1993, the Tula design bureau developed a modular weapons mount based around the 2A72 30mm cannon. This is the weapon used on the BMP-3, and is essentially similar to the 2A42 30mm cannon used on the BMP-2, but is simpler and has only a single rate of fire to reduce the problem of excessive fumes that plagued the earlier weapon. To minimize its size, the weapon is mounted externally along with a co-axial 7.62mm PKT machine gun. This weapons option was first advertised in 1993 at international arms shows. In 1994 at the Nizhni Novgorod arms show, it was revealed that the modular weapons station/BTR-80 combination has been standardized by the Russian Army as the BTR-80A which implies that it has been accepted for Russian service. At the moment, it is not clear if this is intended to replace the normal BTR-80, or supplement it with a more potent weapon.

As in the case of the BTR-60, the BTR-80 has formed the basis for a number of support vehicles and derivative types. The basic GAZ-5903 chassis is used for the armored personnel carriers; and enlarged GAZ-59032 series is used for mobile command posts.

2S23 Nona

The 2S23 Nona-SVK is a self-propelled 120mm gun/mortar vehicle armed with the 2A60 weapon in a large

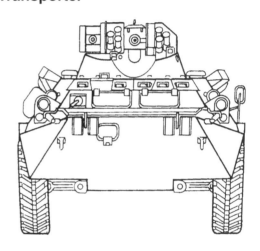

turret. Further details on this system are found in the self-propelled artillery chapter.

BREM-K

The BREM-K (Bronirovannaya remontno-evakuatsionnaya mashina, Armored Repair-Recovery Vehicle, also called BREM-80 and GAZ-59033) is a specialized recovery vehicle for light armored vehicles that was developed in 1988 and went into limited production in 1993. It is designed to tow stuck vehicles (maximum towing force of six metric tons), and has a nose mounted spade to secure the vehicle during this process. There is a small jib crane for mounting on the hull roof, and an "A" frame crane that is erected on the vehicle nose for removing powerpacks and other equipment. A stowage rack on the hull roof has a capacity of 0.8 metric tons. It is also fitted with an electric welding generator.[8]

BTR-80KSh

Russian sources have referred to this command version of the BTR-80, based on the GAZ-59032 chassis

2S23 Nona-SVK 120mm self-propelled weapon.

BREM-K recovery vehicle.

with enlarged superstructure. This is presumably intended to replace the many BTR-60PU variants still in service.

BTR-80 PU-12M

This is the latest variant of the PU-12 air defense command and control vehicle previously built on the BTR-60P chassis. The new vehicle, first unveiled at the 1994 Nizhni-Novgorod arms show, is based on the enlarged GAZ-59032 chassis. The vehicle is intended to provide command and control for divisional air defense systems including the Strela-1M (SA-9 Gaskin), Strela 10SV (SA-13 Gopher); ZSU-23-4 Shilka, and Osa-AK (SA-8 Gecko). The BTR-80 PU-12M6 is provided with digitial data transmission and reception systems (234 468) baud reception). It receives data from the local air defense radar net and other command posts, and can accept data from up to ninety-nine targets. The on-board system can track five to seven targets. It accepts radar data from only one source at a time. Data output to the air defense systems or back to the radars and higher command posts are via five radios and five telephone lines. It has an effective radio range of about twenty-five kilometers and a field cable range of fifteen kilometers. The vehicle has its own navigation and terrain orientation equipment, as well as its own APU for power generation.[9]

BTR-80 RKhM

This is a NBC scout version of the BTR-80, which basically parallels the older BRDM-2 RKh. It is fitted with NBC sensors, and has a flag marker dispenser on the hull rear to mark contaminated areas.[10]

PKNP Kushetka-B

The Kushetka-B PNKP (Podvizhniy komandno-nablyudatelniy punkt: Mobile command-observation station) is a BTR-80 variant that serves as part of the OBAK Battery Automated Fire-control System for the new 2S19 152mm self-propelled gun. This system was developed by the Radiopribor NPO in Zaporozhe, Ukraine in cooperation with VNII Signal in Kovrov. The Kushetka-B is based on the GAZ-59032 chassis, which uses the basic BTR-80 suspension but has an enlarged hull superstructure. The Kushetka-B is forward deployed to conduct fire reconnaissance and observation. It is fitted with a laser designation/ranging system for use with precision guided munitions such as the Krasnopol semi-active laser guided projectile. Communications equipment includes two R-171M UHF radios, one R-163-50U UHF radio, one R-163-10V UHF radio, one R-163UP receiver, one R-134M HF radio, data and telephone encyphering equipment. The crew of the vehicle is five. The Kushetka-B is used in conjunction with a battery fire control station based on the Ural 43203 truck. The fire control system was developed by the VNII-Signal in Kovrov.

GAZ-59037

This is an unarmored "civilianized" version of the BTR-80 intended for operations in rough terrain such as emergency rescue, geological work, construction, etc. It dispenses with the turret and some of the body armor. Several other civilianized BTR-80 variants have also been developed such as the ML-102 logging vehicle which uses a modified BTR-80 chassis with a completely unarmored hull.

The BTR-90 Armored Transporter

In 1994, the Russian Army publicly displayed the new (as yet unofficially designated) BTR-90 armored transporter for the first time. This vehicle was apparently developed by the Arzamas AMZ plant under the designation Obiekt 51 and GAZ-5923. The chassis is a new 8 x 8 design, and is significantly larger than that used with the previous BTR-60/-70/-80 types. The armament system is basically the entire turret package from the BMP-2. The level of armor protection of the vehicle is not yet known, and technical details are still unavailable. In recent accounts of Russian armored development in

GAZ-59037.

BTR-80RKhM NBC reconnaissance vehicle.

Kushetka-B.

BTR-90 Armored Transporter

BTR-90 armored transporter.

the Russian press, there have been acknowledgements that Middle East customers are very interested in wheeled AFVs. It is not clear at the moment if the BTR-90 is intended for Russian Army service or primarily to satisfy export demands.

The BTR-90's hull was built by the Chelyabinsk Plant and utilizes a welded hull construction capable of withstanding the fire of up to a 14.5mm heavy machine gun. The powerplant for this vehicle utilizes a new diesel engine developed at Chelyabinsk rather than Kamaz truck plant which suffered a fire at the plant in 1994.

The BTR-90 has a crew of three (commander, gunner, and driver) and carriers up to ten fully loaded combat infantry men. The armament of the BTR-90 is a 2A42 30mm cannon, a co-axial 7.62 PKT machine gun, and a roof-mounted 9K111-1 Konkurs ATGM (AT-5 Spandrel). In addition, this vehicle may also potentially utilize the new Russian Kornet ATGM in the production version. There also appears to be an effort to improve the main gun armament through the replacement of the 30mm for

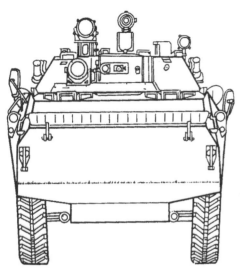

a larger 45mm cannon mount.

The BTR-90 is in many respects highly suitable for the export market, due to its excellent road mobility, firepower, and relative low cost. The BTR-90 should be considerably cheaper to produce than the BMP-3 and should be attractive to cash-starved third world militaries. However, for Russian service, the future of all wheeled-vehicles is in doubt with the exception of the reconnaissance role, due in part to their poor off road performance, according to statements by Russian General V.L. Bryev.

Tracked Armored Infantry Vehicles

The first post-war Soviet tracked carrier to be devel-

Comparative Armor Data-Wheeled Armored Infantry Transporters

ARMOR (millimeters)	BTR-152V1	BTR-60PB	SKOT-2AP	BTR-70	BTR-80	BTR-80A	BTR-90
Turret front	n/a	7	7	7	7	7	n/a
Turret sides	n/a	7	7	7	7	7	n/a
Hull bow	11-15	9-11	10	10	10	10	n/a
Hull sides	8-9	7-9	9	7-9	7-9	7-9	n/a
Hull rear	7	7	7	7	7	7	n/a

n/a = data not available

Comparative Technical Data-Wheeled Armored Infantry Transporters

	BTR-152V1	BTR-60PB	SKOT-2AP	BTR-70	BTR-80	BTR-80A	BTR-90
Factory designation	Obj. 140	Obj. 49	n/a	Obj. 50	Gaz-5903	Gaz-59034	Gaz-5923
Crew+squad	2+17	2+9	2+10	2+8	2+8	3+7	3+10
Weight (tons/combat load)	9.8	10.3	14.8	11.5	13.6	14.5	17.0
Length (mm)	6,830	7,220	7,440	7,535	7,650	7,650	n/a
Width (mm)	2,320	2,820	2,550	2,800	2,900	2,900	n/a
Height (mm)	2,000	2,410	2,060	2,235	2,350	2,410	n/a
Ground clearance (mm)	280	470	400	475	475	475	n/a
Ground pressure (kg/cm^3)	3.7	2.5	2.5	2.8	3.09	3.09	n/a
Main armament	SGMB	KPVT	KPVT	KPVT	KPVT	2A72	2A42
Machine gun caliber (mm)	7.62	14.5	14.5	14.5	14.5	30	30
Depression/elevation (°)	-6+23	-5+30	-4+89	-5+30	-4+60	-5+70	n/a
Ammunition stowed	1,250	500	500	500	500	n/a	n/a
Co-axial armament	n/a	PKT	PKT	PKT	PKT	PKT	PKT
Gun caliber (mm)	n/a	7.62	7.62	7.62	7.62	7.62	7.62
Ammunition stowed	n/a	2,000	2,000	2,000	2,000	2,000	n/a
Engine designation	ZiL-123	GAZ-49B	T-928-14	ZMZ-4905	KamAZ-7493	KamAZ-7403	n/a
Type gasoline	gasoline	diesel	diesel	diesel	diesel	diesel	diesel
Horsepower	110	2 x 90	180	2 x 115	260	240 or 260	n/a
Fuel stowed (liters)	300	290	330	300	300	300	n/a
Max. road range (km)	650	550	750	450	600-800	600-800	n/a
Max. road speed (km/h)	65	80	95	80	80-90	80-90	90+
Water speed (km/h)	n/a	n/a	n/a	n/a	10	10	n/a
Ditch crossing (m)	0.7	2.0	2.0	2.0	2.0	2.0	n/a
Vertical obstacle (m)	0.6	0.4	0.5	0.5	0.5	0.5	n/a
Radio	10RT-12	R-123	R-113	R-123M	R-173	R-173	n/a

CFE Counts MT-Lb and BTR-50 to BTR-80 (European Russia)

Year	MT-Lb	BTR-50	BTR-60	BTR-70	BTR-80	Total
1990	1,300	7	4,191	3,936	1,130	10,564
1991	1,300	0	4,233	4,070	1,149	10,752
1992	1,226	0	2,747	1,830	1,287	7,090
1993	986	0	2,534	1,712	865	6,097
1994	895	10	1,014	1,598	965	4,482
1995	392	15	97	1,144	959	2,607
1996	448	0	67	1,031	939	2,485
1997	649	0	56	1,038	1,099	2,842

oped was a 1947 effort by Col. A. F. Kravtsev at the VRZ Number 2 tank rebuilding plant near Moscow in conjunction with the Moscow Engineering Academy. Kravtsev developed a tracked armored transporter, designated the K-75, based on elements of the obsolete T-70 light tank. Although the front of the vehicle was lightly armored, the rear compartment had only side armor protection. It was powered by a YAZ-206 140 horsepower engine, and weighed 7.5 tons. It could theoretically carry up to seventeen troops, but without much equipment as the cargo compartment was quite small. Only a single prototype was completed, and no series production ever took place. In 1949, the design bureau at the Uraltransmash Plant Number 50 in Sverdlovsk headed by Lev I. Gorlitskiy was assigned to develop a new family of medium self-propelled guns. A new chassis was developed for this requirement, called the izdeliye 100. Besides the self-propelled guns which emerged from this project (the SU-100P self-propelled gun, SU-152P 152mm self-propelled gun, and SU-152G 152mm self-propelled howitzer), a tracked armored transporter was also developed, the izdeliye 112 (also sometimes called the BTR-112). The izdeliye 112 armored transporter program has never been extensively discussed in any open Russian source, and it is not clear if the program was based on a Soviet Army requirement, or if it was a local initiative of the design team attempting to convince the Army of the desirability of a common family of vehicles on a shared chassis. In either case, the izdeliye 112 was the first significant tracked infantry transporter developed in the USSR. It weighed 18.2 metric tons, and its 400 horsepower engine gave it a road speed of sixty-five kilometers per hour. Armament consisted of a 14.5mm KPV-44 heavy machine gun in a partially armored turret, and a

K-75 armored transporter.

BTR-112 (izd. 112) armored transporter.

pintle-mounted SG-43 7.62mm machine gun. The crew compartment was located at the rear of the vehicle and could accomodate up to twenty-five lightly equipped infantrymen. It would appear that the prototypes did not have an armored roof for the rear compartment. Ultimately, it was not selected for production, probably due to the high costs it would have represented compared to other alternatives such as the BTR-50 and BTR-152.[11] However, the vehicle design was resurrected in 1963 and formed the basis for a wide range of new specialized vehicles including the 2P24 (TEL for the SA-4 Ganef), 1S32 (Pat Hand radar vehicle for SA-4), 2S3 Akatsiya 152mm howitzer, 2S4 Tyulpan 240mm self-propelled mortar, and 2S5 Giatsint 152mm self-propelled gun.

The BTR-50 Tracked Infantry Vehicle

In 1949-50, work began on a new light amphibious tank, which would later emerge as the PT-76. At least three different designs were considered, Kravtsev's K-90, Kotin's PT-76, and an unknown type from Astrov's design bureau. Parallel to the light tanks, the design bureaus were also assigned to develop a related amphibious infantry transporter. Kravtsev's team prepared the K-78, based on the K-90 light tank. Balzhiy's BNK design bureau in Chelyabinsk developed a transporter patterned on their PT-76 light tank, the Obiekt 750.[12] With the Kotin PT-76 design winning the light tank competition, its related Obiekt 750 armored transporter, later named the BTR-50P, was the obvious choice.[13] The rationale for production of a full-tracked armored transporter in parallel to the wheeled BTR-152 was that the tank divisions would need a vehicle capable of keeping up with the tanks in all types of terrain, an impossibility with the wheeled BTR-152. The BTR-50P was accepted for service by the Soviet Army in 1954 and entered production that year.

The BTR-50P was a straightforward adaption of the PT-76; in place of the turret, a simple, open box superstructure was added on the hull front. The resulting vehicle could be used to transport twenty troops on simple bench seats, or could be used to transport up to two tons of equipment. It could be fitted with two folding ramps on the rear so that a GAZ-69 jeep, 85mm anti-tank gun, or B-11 107mm recoilless rifle could be rolled up onto the rear deck for transport over water obstacles. Shortly after production began in 1954, an improved version was developed, the BTR-50PA, which had a mounting for the 14.5mm KPVT heavy machine gun added over the left front bay. This type later superceded the BTR-50P in production. As in the case of the parallel BTR-152 program, a fire support version was developed by mounting a ZPTU-2 dual 14.5mm KPVT heavy machine gun turret in the cargo bay of a BTR-50P. Experiments were also conducted with the quad ZPTU-4 gun mounting on the BTR-50P, but none of these fire support types was accepted for production.

K-78 amphibious armored transporter.

BTR-50P armored transporter.

BTR-50PK armored transporter.

Due to doctrinal discussions in the mid-1950s over use of tactical nuclear weapons, it was prudent to improve the BTR-50P to fight on the nuclear battlefield. An armored roof was added to the design along with a PAZ air filtration system on the Obiekt 750PK prototype. It entered series production as the BTR-50PK in 1958 and became the standard production model of the family.

The BTR-50P series was not entirely satisfactory as an armored infantry vehicle. It was certainly more mobile than the BTR-152. However, the design was poorly configured for troop entrance and exit. The troops had to climb up the sides, which was a time-consuming nuisance when field equipped. Exit was equally slow, and far more dangerous if under hostile fire since the troops had to clamber over the roof. In addition, the Soviets were coming to realize that armored infantry vehicles should be tailored to squad size. The BTR-50P carried a bit less than two squads, or twenty troops. Since Soviet motor rifle companies were triadic (three platoons of three squads each), there was inevitably a bit of confusion mixing squads from different platoons together in the same transporter. As in most NATO countries of the period, the 1960s saw a general shift away from large armored troop carriers to slightly smaller carriers suitable for a single squad. The crucial step (in the Soviet case) occured in the 1960s when the motor rifle battalion's armored transporter companies were abolished in favor of each motor rifle company having its own armored infantry transporters. This neccessitated a carrier tailored to squad size.

During the development of the BTR-50P, it was decided to develop corresponding command vehicles. The first of these was the BTR-50PN which entered production in 1958 alongside the BTR-50PK. The more common command type was designated the BTR-50PU. This command version resembled the BTR-50PK, but had a modified armored roof with oval doors and additional vents to dissipate the heat from added electronics and generators. The original version, based on the BTR-50P, had a single bay on the left side of the hull glacis for the driver. The subsequent, and more common production model on the BTR-50PK chassis, had two bays. The radio fit of the BTR-50PU varied depending on its role. Variants were built with assortments of the R-112, R-113, R-105, or R-105U; R-403BM radio relay (half-set); R-311 radio receiver, P-193A field telephone system. The boxy hull of the BTR-50P made it very suitable for this role, and these command vehicles remained in service long after the demise of the BTR-50P infantry carriers. Indeed, some of the BTR-50P infantry carriers were rebuilt in the command vehicle role. Later improved versions of the BTR-50PU family included the BTR-50PUM

BTR-50PU armored command vehicle.

and the BTR-50PUM-1. The BTR-50 also served as the basis for a number of specialized armored vehicles including the UR-67 minefield breaching vehicle and the MTP-1 technical repair vehicle.

Due to the shortcomings in the BTR-50PK design, its production ended in the mid-1960's. However, the BTR-50PU command vehicle had proved a very useful vehicle, and production continued until 1972-73. Total BTR-50 production was much lower than BTR-152 production, amounting to about 6,500 vehicles of all types. The BTR-50 was produced in too small a number to significantly affect the mechanization of the Ground Forces. It was never available in sufficient numbers to entirely equip the motor rifle regiments of all the tank divisions.

The OT-62 TOPAS

In the late 1950s, the Czechoslovak People's Army (CSLA) was encouraged by the Soviets to emulate the Soviet reorganization and equipment of its infantry forces. The CSLA decided to adopt the Soviet BTR-50P for the motor rifle regiments of its tank divisions, but sought license production rights rather than outright purchase. Negotiations were concluded in 1958, but the CSLA decided to develop an improved version rather than a straight copy. The V-6 engine was improved as the PV-6 by turbocharging which boosted its output, and a new transmission was developed, derived from the wartime German AK 7-200 used on the Panther tank. The body was redesigned with side hatches to permit easier access and exit from the vehicle. Although not readily apparent, the hull superstructure was redesigned as well. This was accepted in 1962 as the OT-62 TOPAS (Transporter Obrneny Pasovy: Tracked Armored Transporter) and entered production in 1963 at ZTS Detva in Slovakia. Several versions of the TOPAS were manufactured. The original version closely resembled the BTR-50PK, and carried no special armament. This was followed by the OT-62 TOPAS-2A which carried a small machine gun cupola over the right bay, also armed with a Tarasnice 82mm recoilless rifle. Two command versions were also manufactured, generically called TOPAS-VS or TOPAS-Velitelske Stanoviste; one version as a simple company command vehicle, the other as a more elaborate command/staff vehicle with additional radio masts and generators. At least four of the later type have been identified, varying in radio fits: OT-62R-2; OT-62R-2M; OT-62 R-3MT; and OT-62 R-4MT. Other specialized versions included the DTP-62 engineer vehicle and the WPT Maintenance vehicle.

The Polish People's Army (LWP) decided to acquire the OT-62 TOPAS in lieu of the BTR-50PK. The basic production model was ordered, with an aim towards modi-

OT-62 TOPAS armored transporter.

fying it to satisfy Polish requirements. The basic TOPAS was used in some units, but later production deliveries were shipped to LWP workshops where they were modified with the WAT turret developed by the Military Technical Academy for the OT-64 SKOT-2AP. This turret was armed with a 14.5mm KPVT machine gun with a special high elevation trunnion for secondary use in an anti-aircraft role. This reduced the crew from eighteen on the TOPAS to fifteen on the TOPAS-2AP. The Poles also converted some of the TOPAS-2AP into a special carrier for mortar squads, carrying two 82mm mortars on external mountings, as well as fitting ammunition racks inside the hull. This version, the TOPAS-2AP (mozdzierz) carries a three man crew and two four-man mortar squads. The Poles also developed a light recovery vehicle on the basis of the TOPAS, designated WPT-TOPAS. TOPAS production continued into the early 1970s, and about 3,500 were manufactured. Besides use by the CSLA and LWP, the TOPAS was exported to the Middle East, Africa, and India where it has been used in combat.

The BMP Infantry Combat Vehicle

The first Soviet infantry vehicle to be developed from the outset with the needs of the nuclear battlefield in mind was the BMP. The Soviet BMP was the world's first infantry combat vehicle and one of the most significant innovations in infantry tactics in the latter half of the 20th Century. The BMP represented a significant watershed in mechanized infantry tactics, providing the infantry squad with unprecedented firepower, mobility, and protection. In subsequent years, other armies followed the Soviet pattern with vehicles like the US Army's M2 Bradley, the German Marder, and the British Warrior.

Mechanized infantry tactics up to the late 1950s used the armored transporters as battlefield taxis. They brought their infantry squads to the battleline, where the riflemen disembarked and fought on foot. This tactic was deficient on the nuclear battlefield, where the effects of radioactive or chemical weapons contamination could incapacitate or kill exposed infantry. Armored vehicles provided the solution: the armor plate not only protected against bullets, but provided a significant measure of protection against most forms of radiation. By reconfiguring the infantry vehicles, the infantry could fight from within the protective confines of the armor. So was born the the concept of the infantry combat vehicle, (Boyevaya mashina pyekhota or BMP). The infantry could fight from within the BMP in contaminated areas, or dismount and fight in a traditional fashion on the conventional battlefield.

The BMP requirement was examined by the Main Administration of the Armored Force (GBTU) in the late 1950s. The requirement stressed the need for high vehicle speed, good armament, and the ability of all members of the squad to fight from within the confines of the vehicle. The armament issue was controversial. The original conception envisioned a 23mm cannon, which was felt adequate to defeat comparable infantry vehicles like the US Army M-59 or the German HS.30. This weapon was probably influenced by the German use of 20mm cannon on the HS.30 armored infantry vehicle, which was apparently viewed by the Soviets as the most potent of the existing armored infantry transporters.[14]

However, given Khrushchev's penchant for missile weapons, some discussion began on arming the vehicle with a weapon capable of defending the BMP against tanks. The Gryazev/Shipunov bureau in Tula (officially called the KB Priborostroyeniye: Design bureau of the Instrument Industry), which developed the 23mm aircraft cannon being considered for the BMP, offered a novel solution, the 2A28 Grom low pressure gun, developed by V. Silin. It is not clear if the Grom was developed as a local initiative, or on the basis of a GRAU requirement.[15]

The Grom was a relative of the SPG-9 recoilless rifle being developed as an infantry weapon. It fired a rocket-propelled grenade, related to the common infantry RPG-7 anti-tank rocket launcher. In contrast to conventional recoilless rifles, it used a very small charge to eject the RPG out of the tube. The charge was so small that there was no need to vent the gases backward to reduce the recoil. This enabled the weapon to be used from within the confines of an armored vehicle without the problems attendent to conventional recoilless rifles. The Shipunov design bureau integrated this weapon with the new 9M14 Malyutka (AT-3 Sagger) wire guided ATGM developed by the Nepobidimy design bureau in Kolomna. The 73mm Grom weapon covered short-range targets out to 700 meters, while the Malyutka covered longer ranged targets from 700 to 3,000 meters. This combination was an ingenious and unique engineering solution for light armored vehicles. Many other armies had attempted to develop light armored vehicles armed with recoilless rifles, but they had often proven to be awkward to employ due to the need for the crew to exit the vehicle to reload the weapon. The Grom provided a practical means to use rocket projectiles without the disadvantages attendent with recoilless rifles and is yet another example of the Soviet's traditional interest in novel technological solutions in armored vehicle armament.

The Grom was selected as the basic armament package for the BMP prior to the requirements being issued to several design bureaus. The armor protection requirement called for defense against the NATO 20mm cannon (20 x 139 millimeter ammunition) in the frontal quadrant, the ammunition for the German HS.30 and the later Marder. There was no conensus on the precise configuration of the BMP, including the issue of whether it should be tracked or wheeled. There was evidently some resistance within the Ground Forces over adopting a sophisticated tracked vehicle in infantry units due to the maintenance and training burden. Therefore, wheeled configurations were still a possibility. The GBTU decided to competitively develop the BMP in order to explore a wide range of alternative configurations. These included the Gavalov KB in Volgograd, (best known for their later BMD-1 airborne assault vehicle); the Isakov KB, (the heavy tank design team in Chelyabinsk, formerly under M.F. Balzhiy), and smaller design teams connected with the automotive plants in Rubtsovsk and Briansk. All four competitors used the same Grom armament and turret package as developed by the Gryazev/Shipunov design bureau. The technical-tactical requirement was issued to the bureaus in 1959 or 1960.

The Briansk Obiekt 1200 was a wheeled BMP, selected because of the extensive experience of the Briansk automobile plant (BAZ) with heavy cross-country trucks (such as the contemporary BAZ-135 used as the basis for the Luna-M (FROG-7) artillery rocket launcher vehicle). This design resembled a more sophisticated alternative to the BTR-60. The most unusual design came from the Rubtsovsk Machine Building Plant (RMZ: Rubtsovskiy mashinostroitelniy zavod).

The Rubtsovsk Obiekt 19 was a mixed wheel-cum-track design reminiscent of pre-World War 2 Austrian Saurer designs. This type of suspension had never proved practical since it tended to be more complicated than either tracked or wheeled configurations, and not as mobile as either. Unlike the Austrian Saurer which used the tracked suspension as the primary type, with the wheeled suspension as the secondary configuration, the Obiekt 19 used the wheeled configuration as its primary mode. The conventional wheeled 4 x 4 suspension was used for road operations or in less strenuous cross-country conditions, and its secondary track assembly located between the wheels could be lowered to the ground to improve traction when confronted by rough terrain. Both of the Rubtsovsk and Briansk designs shared a common configuration problem. The engines in both vehicles were mounted in the rear of the vehicle, which inevitably led to awkward troop access through roof hatches behind the turret.

The most serious contenders for the BMP requirement came from the Gavalov and Isakov design bureaus. The Gavalov design bureau had been formed in the 1950s

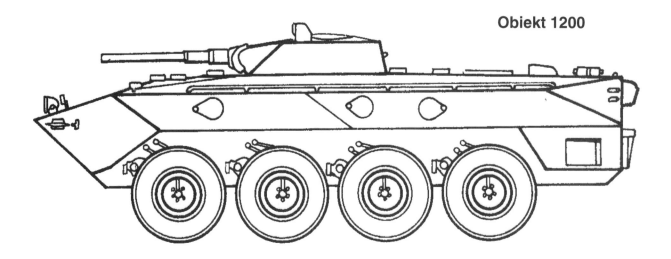

Obiekt 1200

when production of the PT-76 light tank and the associated BTR-50P infantry transporter was started at the Volgograd Tractor Plant. The bureau was formed to continue the evolution of the PT-76 and associated light armored vehicles. The Gavalov design bureau offered two designs using components from the PT-76, the wheel-cum-track Obiekt 911 and tracked Obiekt 914. The Obiekt 911 had a conventional tracked suspension, but underneath the hull were four retractable wheels.

When operating on the road, these wheels could be lowered to propel the vehicle to high speeds. The Obiekt 914 was a conventional tracked design. Like the PT-76 amphibious tank from which they were derived, both designs had a rear-mounted engine. Crew exit was through the roof but there was a small alley at the rear of the hull which made climbing on and off the vehicle easier than on the Obiekt 19 or Obiekt 1200.

The fourth BMP contender was the production of the Isakov design bureau in Chelyabinsk. This bureau was a descendent of the "New Design Bureau" created in 1947 at Chelyabinsk under M. F. Balzhiy to pursue heavy tank designs in competition with General Zh. Kotin's OKBT in Leningrad. The Isakov design bureau was well regarded at the time, having been responsible for the excellent Obiekt 770 heavy tank design, and the radical two-man Obiekt 775 missile tank. The advantage that Isakov

Obiekt 1200 (BMP competitor).

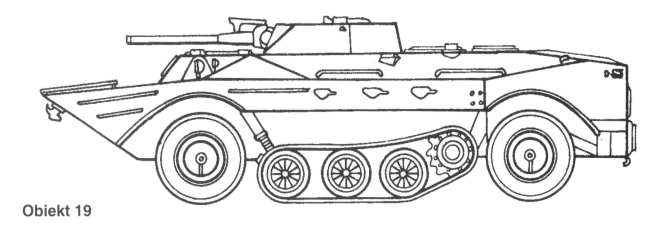

Obiekt 19

Obiekt 19 (BMP competitor).

enjoyed was that, unlike Gavalov, his design team could start from a clean slate not tied to an existing armored vehicle design. Their contender for the BMP competition was designated as Obiekt 765. The Obiekt 765 Korshun, (Kite bird) was a conventional, fully tracked design. Its most significant difference compared to the other contenders was the use of a front mounted engine which permitted a rear mounted troop compartment. It is not at all clear why the Isakov design was the only contender with such an obvious layout, as it permitted the use of rear doors for easier troop exit. It is possible that the other design bureaus placed little emphasis on crew access/egress, since the whole point of the design was to permit combat from the mounted position. They may have felt that rear doors would complicate the amphibious characteristics sought in the design. It is possible that there was a fifth contender for the BMP design. In a recent article, a Russian author mentions a prototype armed with a conventional 76mm gun, but further details are lacking.[16]

Obiekt 911 (wheel/track BMP competitor).

Obiekt 914 (BMP competitor).

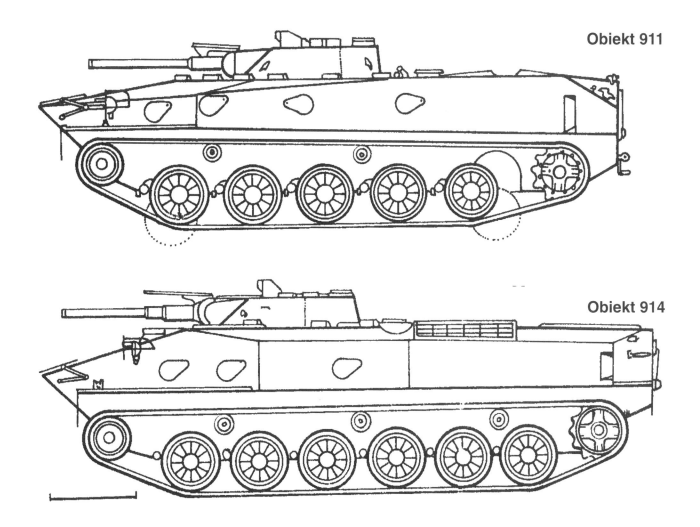

Obiekt 911

Obiekt 914

The prototypes were ready for trials in 1964 which were conducted at the main proving ground at Kubinka and at Rzhev, both near Moscow. The decision on the BMP was delayed due to Khrushchev's disfavor with the adoption of an expensive infantry vehicle. He told the army that "if there is a projectile capable of defeating the BMP armor, then it would be much more reasonable to keep transporting the motor rifle troops in trucks!"[17] As a result, the BMP program went into limbo for a time. Following Khrushchev's ouster in the Brezhnev coup in the autumn of 1964, the BMP was revived. The final decision to proceed to production was made in 1966 when the Isakov Obiekt 765 was accepted for the Soviet Army's BMP requirement. The reasons can be surmised.

The wheeled Obiekt 1200 was too similar to the less expensive BTR-60PB, though better armed and armored; its armor was also considered deficient.[18] The performance advantages of the wheel-cum-track Gavalov Obiekt 914 and the Rubtsovsk Obiekt 19 did not prove themselves, especially in light of the added complexity implied by such designs. The real choice fell to the more conventional Gavalov Obiekt 914 and the Isakov Obiekt 765. The Isakov Obiekt 765 was the more practical design with a better layout of the troop compartment, and the Obiekt 914 "had a number of serious defects" according to a recent Russian account.[19] In 1966, the first production of the Obiekt 765 at Chelyabinsk began to permit further operational trials. Because of the substantial numbers required and Chelyabinsk's commitment to tank production, the Soviet government decided to convert the Kurgan Machine Building Plant (Kurganskiy mashinostroitelniy zavod) to BMP production, and series production began there in 1968.[20] It was later extended to Rubtsovsk as well. Eventually, a design bureau under A. Blagonravov was also set up at Kurgan, a continued reflection of the "doubling" policy in tank design mentioned earlier in the section on tank design in the 1950s.

The BMP was originally called BMP-765 due to its izdeliye/obietkt number, or simply BMP. Initially, it was

243

BMP Model 1970 infantry fighting vehicle.

called BMP-76PB by NATO when first spotted in 1966 for operational trials. The operational trials of the BMP were protracted due to continuing bugs in the design. The suspension had to be strengthened, especially in the front, due to the speed of the vehicle. The swimming vanes on the fenders had to be modified, and the internal venting system, which expelled fumes when the crew fired their AKM assault rifles, had to be improved. In addition, the Ground Forces decided that besides protecting against nuclear contaminants, the BMP would also have a chemical protection filter system. There were at least four subvariants of the BMP between 1966 and 1969 as these difficulties were ironed out. Small production batches of the improved type, designated BMP-1 (or BMP-A by NATO) were first produced in 1968, but series production did not begin until 1970. The BMP-1 incorporated the new chemical filter system to the left of the turret. A twenty centimeter extension was added to the bow to shift the center-of-gravity aft, and to prevent the BMP from "submarining" in the water while swimming. Also, a new swimming air intake was added with a low, erectable snorkel, to prevent water from flooding the air intake as had been the case on the earlier BMP. The BMP-1 incorporated the improved 9M14M missile.

The BMP-1 Design

The BMP is a marked departure in Soviet armored vehicle design. Contrary to the usual Western stereotypes of Soviet vehicles, the BMP is elegantly engineered, and surprisingly complicated. It does follow typical Soviet design practices in other respects, especially in its compact size and low silhouette. It is much lower to the ground than Western infantry vehicles such as the German Marder, British Warrior, or American M2 Bradley. This smaller size has meant lower weight and lower cost, but at the price of habitability under prolonged battlefield conditions.

The BMP is divided into three main sections: the powerplant in the hull front, the turret in the center and the troop compartment in the vehicle rear. The powertrain compartment at the front of the hull is very spacious by Soviet standards. The reason for this was to counteract the weight of the forward compartment to provide sufficient bouyancy for the BMP when swimming. The transmission is a five-speed manual type with five forward and one reverse gear. The master clutch is a hydraulically or air-pressure operated twin disc. The steering clutch is a planetary type using seven dry discs, and transmits power to the final drives. Front mounted powertrains can be problem in service since the front drive sprocket is vulnerable to damage if the vehicle hits an obstruction. On the BMP, this has been minimized by locating the drive sprocket far enough to the rear to avoid such damage.

The engine and engine cooling system are mounted in the center and right side of the hull, behind the powertrain compartment. The engine is in the center of the hull, with engine accessories to the right, and the main radiator overhead and to the rear. The engine is a 300 horsepower six-cylinder water-cooled UTD-20 diesel (also called 5D20) with an injection pump fuel system, developed and manufactured by Transmash in Barnaul.

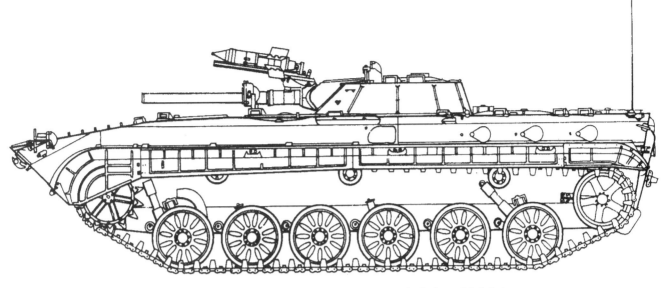

BMP-1 Model 1970 Infantry Fighting Vehicle

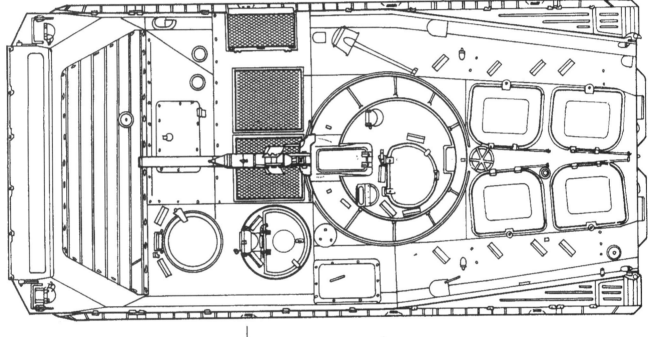

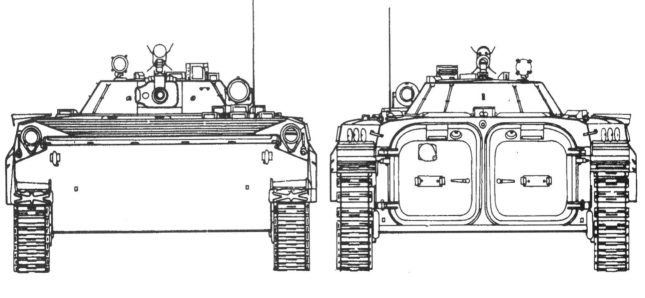

BMP-1 infantry fighting vehicle.

Air filtration is provided by cyclone filters. The vehicle exhaust is vented through a port on the extreme right side of the hull roof.

The driver is stationed on the far left side of the front hull, immediately to the left of the engine compartment. The driver can enter the vehicle through a roof hatch immediately overhead his station, or through a narrow alley-way that runs alongside the left side of the turret. The driver can adjust his seat to ride with his head outside the hatch when outside of the combat area. When operating with the hatch closed, the driver can view through three TNPO-170A vision periscopes. Night vision is provided by substituting the TVNE-1PA active infrared night vision metascope in place of the front day periscope, and active infrared lighting is provided by the vehicle's FG-125 infrared driving lights or the several infrared searchlights elsewhere on the vehicle. When the vehicle is swimming, the TNP-350B periscope is substituted for the usual TNPO-170A allowing the driver to see over the bow plane.

The BMP was the first Soviet combat vehicle to introduce a simplified steering system instead of the traditional tractor-style levers. The T-bar steering yoke operates much like a steering wheel in an automobile. The other driving controls are conventional. There is a high/low range lever gear shift lever on the left of the T-bar control. The vehicle is normally operated in the high range; the low range is used in rough terrain or steep hills and provides more power to the tracks at the expense of speed. For navigational aid in driving, the vehicle is fitted with a GPK-59 electrical gyrocompass.

There is a single seat immediately behind the driver for the squad commander. The commander's position is fitted with a rotating hatch and his main observation device is a TKN-3B (which is basically the same sight used by the commander in the T-62 tank). This is a day-night sight, with a metascope to view active infrared images in the night mode. Immediately above the sight is an OU-3GK infrared searchlight, which the commander uses at night in conjunction with the TKN-3B sight. This has an effective range of about 400 meters (1,300 feet).

The infantry squad configuration selected for BMP was unique at the time, though it has since become commonplace. Eight members of the squad were seated in the rear, back-to-back, facing outwards. Six firing ports and associated periscopes were positioned on either side, and two more towards the rear. The seats are mounted against a center compartment which contains the vehicle batteries and the main 330 liter fuel cell. The stowage containers under the seat are for vehicle tools. Each infantryman is provided with a firing port, the forward most on the BMP-1 being designed for the PKM squad light machine gun, the others for AKM assault rifles. This layout clearly reflected the experiences of Soviet motor rifle units with the inferior layout of the BTR-50 and BTR-60. The layout allowed the entire squad to fire its weapons from within the confines of the vehicle without exposure to an outside (and potentially contaminated) environment. The internal compartment was protected by a nuclear protection (PAZ) system relying on a detector, filtration system, and atmospheric overpressure to keep out airborne contaminants.

To fire his weapon from the port, the infantryman opens the port hatch using an internal control, clips a sleeve over his AKM-S assault rifle and places the sleeve into the firing port. The sleeve is designed to prevent outside contamination from entering the BMP if the vehicle is operating in a chemically or radioactively contaminated environment. After mounting the assault rifle in the firing port, the infantryman snaps a shell-deflector/gas evacuator over the shell ejection port of the rifle. This prevents the spent casings from hitting and injuring adjacent squad members. The deflector is attached to a small hose, connected to the vehicle air filtration system, which sucks away any fumes from firing the rifle and exhausts them outside the vehicle. There is a periscope at every station for the infantrymen, and each firing port has a small aiming viewport of bullet-resistent glass. There are several innovative features in this arrangement, including the use of heated periscopes which remove frost and condensation. The utility of these firing ports is open to question, since it is impossible to aim the rifles with any accuracy when the vehicle is moving since the rifle is rigidly mounted and subject to the bouncing and rolling motions of the vehicle hull. These firing ports are intended for general suppressive fire, not precision marksmanship.

When the vehicle is traveling, the infantry squad in the BMP can stow their assault rifles in front of themselves using clips. There is also a set of straps provided for stowing small rucksacks. Stowage inside the BMP is inadequate by US Army standards. Although the BMP troop compartment is better arranged than previous Soviet infantry vehicles, it is still woefully inadequate by Western standards. The interior is much too small for taller soldiers and equipment stowage facilities are inadequate for field conditions. A US Army study of a captured Syrian BMP-1 from the 1973 war concluded that with hatches closed, the height of the rear compartment would only accomodate men up to the 25th percentile if wearing light tropical battledress, and only up to the 15th percentile if wearing arctic battledress. In terms of shoulder width, the bench seats are so narrow that only men up to the thirty-fifth percentile fit properly. Admittedly, US average heights are somewhat greater than the Soviet average, but the study provides some clear evidence of the degree of congestion in the compartment. By way of comparison, the design norm of US Army vehicles such as the M2 Bradley is to accomodate a ninety-fifth percentile soldier (a height of six feet five inches).

In the 1960s when the BMP was designed, the average Soviet infantryman was seldom issued a rucksack of the size issued in most Western armies, and never issued a sleeping bag, so the problem was not very apparent in training. During the Afghanistan war, the inadequacies of this spartan simplicity became evident when operating under prolonged field conditions. Inevitably, Soviet BMP squads shifted their equipment into improvised external stowage racks or boxes. This often meant that the roof hatches were permanently closed due to the gear stowed on them, and the turret weapons were limited in their rearward arc of fire since the stowed gear obstructed traverse to the rear. Eventually, a rear-turret stowage bin was developed and attached by field depots, but even this proved too small.[21] In some other armies, the lack of stowage in the BMP led them to adopt the practice of reducing the number of troops carried in the BMP, with the additional space being used for stowage. In the East German NVA, the rear compartment of the BMP-1 was usually limited to seven or eight infantry instead of the intended nine.

The BMP can carry its own air defense weapon. Above the central fuel cell/backrest is a stowage clip for a 9M32 Strela 2 (SA-7 Grail), 9M36 Strela 3 (SA-14 Gremlin), or 9M313 Igla-1 (SA-16 Gimlet) manportable air defense missile. The scale of issue of these weapons has changed through time. Originally, the usual issue was about one per two vehicles; today the practice is for each vehicle to carry a launcher gripstock and one or two missiles.

Exit from the BMP is either through the two rear doors or through the four overhead roof hatches. The rear doors hinge from the center and open sideways. They double as fuel cells, being connected to the main cell between the troop seats. This arrangement is a major improvement over the poor exit arrangement on previous Soviet infantry transporters such as the BTR-50 or BTR-60PB. The BMP-1 is protected against radioactive and chemical/biological contamination. The vehicle can be hermetically sealed by closing and locking all hatches and it uses an air filtration system bolstered by an air overpressure system to keep out contaminants. The vehicle carries a GO-27 radiation and chemical agent detector which can be operated in automatic or manual modes. Two TDP chemical decontamination kits are carried in the vehicle.

There are two methods for deploying protective smoke. There is a built-in TDA thermal smoke emitter in the engine compartment that creates smoke by injecting a small amount of diesel fuel onto the engine manifold. On later BMP-1Ps and on all BMP-2s, this has been supplemented by an array of System 902V Tucha smoke grenade launchers. On the BMP-1P upgrade, these smoke dischargers are mounted on the rear edge of the turret roof. These fire the 3D6 81mm smoke grenade about 200 to 300 meters in front of the vehicle. These create a smoke screen about eighty meters wide and lasts from

1.7 to 2.4 minutes depending on the ground wind.

The BMP-1 Armament

The turret in the BMP-1 is occupied by the gunner, who sits to left of the gun breech. The BMP-1's main weapon is the 2A28 Grom 73mm low-pressure gun, serviced by an autoloader. The forty round ammunition reserve for the autoloader is located on the right side of the turret floor. When the Grom gun was originally developed, there was only one type of ammunition available for it, the PG-15V with a shaped-charge anti-armor warhead. A related high explosive projectile did not become available until 1974. The PG-15V is similar to the PG-7 rocket grenade fired from the RPG-7 rocket grenade launcher, but with a small propellant casing added at the base of the rocket which ejects it out the gun barrel. After leaving the gun tube, the rocket engine is ignited, propelling the PG-15V projectile to the target. The PG-15V round has an initial muzzle velocity of 400 meters per second increasing once the rocket engine ignites, and it has a maximum range of 1,300 meters. Soviet sources credit it with an effective range of 700 meters, but operators have found it to be limited to about 500 meters (1,640 feet). It is armed with a 0.322 kilogram hexogen shaped charge warhead. This warhead has a nominal penetration of 350 millimeters (fourteen inches) of steel armor, but tests showed it to have an average penetration of 280 millimeters (eleven inches). This means that it can penetrate the thickest front armor of standard NATO tanks of the 1970s including the US M60A1, British Chieftain, or German Leopard 1. It cannot penetrate the frontal armor of contemporary tanks such as the M-1A1 Abrams, Leopard 2, or Challenger.

The ballistic trajectory of the PG-15V is flat out to 800 meters, making it easy to aim in still air conditions. Accuracy is degraded by the need to use fins for stabilization. In the case of a cross-wind, the PG-9 projectile tends to fly into the wind. The 2A28 gun has a seventy percent hit probability against a tank at a 500 meter range, and a fifty percent hit probability at 800 meters when firing from a static position in still air. The 2A28 Grom gun has very poor accuracy on the move since the gun is not stabilized. The autoloader used with the 2A28 Grom gives it a rate of fire of six to eight rounds per minute. Like most Soviet autoloaders, it requires the gun to be depressed to reload, meaning that the gunner loses any aiming advantage from the first round, as the barrel is not returned very precisely after loading. Night firing is possible with the gunner's 1PN22M1 sight which can be used with the turret's active infrared searchlight, or in a passive mode using an image-intensification channel on later production vehicles that depends on ambient starlight and moonlight.

In the mid-1970s, the Soviet Army introduced the delayed OG-15V projectile. This is a high explosive projectile designed for use against troops or field fortifications. The high explosive fill has been doubled to 0.73 kilograms of TNT, substantially enhancing its blast effect compared to the anti-armor PG-15V. The autoloader has a switch to select the type of ammunition.

The autoloader has proven to be one of the less popular features of the BMP-1. It is dangerous to the gunner if he carelessly leaves the guard off, since it can easily catch on baggy clothing. It works poorly if knocked out of alignment and can easily jam if old or not properly maintained. In several armies, the autoloader was simply dismantled. Some Iraqi BMP-1s had the autoloader removed, and this has occured in other armies such as Finland. The rounds are very easy to manually load and some armies found that they could be manually reloaded more quickly than with the autoloader.

To supplement the 2A28 gun at ranges beyond 700 meters, the BMP-1 is fitted with a launcher for the 9M14M Malyutka (AT-3 Sagger). The 9S415 launch rail is fitted immediately above the main gun tube. There is a small hatch behind the launch rail for reloading the missile. The 9M14M is a manual command-to-line-of-sight missile comparable to 1950s vintage missiles like the French SS-11. The gunner has a small joystick controller which he uses to steer the missile. After launching the missile, the gunner must keep track of the relative position of the missile and target, and steer the missile manually using the joystick. This requires repeated training to maintain any proficiency. Under the stress of combat conditions, this system has proven to be very inaccurate. Reloading the missile is difficult from under armor. The missile is loaded out on the launch rails with its fins folded. After it is secured to the launch rail, the gunner must then fold out the fins using a small stick. This is not a particularly convenient feature under combat conditions. A total of four 9M14M missiles are stowed internally in the BMP-1, two to the right of the gunner in the turret, and two in the hull space to the right of the turret.

The shortcomings of the 9M14M led to the BMP-1P in the late 1970s. This version has a small mounting lug added to the turret roof to the right of the gunner's hatch. To use the missile, the gunner must mount the 9P135 firing post on the lug, from its storage position in the hull. He then loads a 9M111 Fagot (AT-4 Spigot) missile from stowage racks in the hull. This launcher works in exactly the same fashion as the normal infantry firing post, so the gunner is exposed to enemy fire while employing the system. The 9M111 Fagot, and the larger 9M113 Konkurs (AT-5 Spandrel), both use semi-automatic command-to-line-of-sight guidance. After the missile is fired, the gunner keeps the cross-hairs of the firing post on the target, and the guidance system automatically corrects the missile flight path over a thin wire link. This system is far more accurate in combat conditions, with hit probabilities well over fifty percent at normal combat ranges. In the early 1990s, a further upgrade to the BMP-1P was developed, the BMP-1PG. This is essentially similar to the BMP-1P upgrade, but also has a 30mm AGS-17 Plamya grenade launcher added on the rear left side of the turret to provide additional firepower.

Many of the other features of the BMP-1 were novel. It was designed with a new type of track, similar to that of the T-64. Since the BMP-1 was expected to keep up with the T-64, it was designed for high speed. It was the first Soviet tracked armored vehicle to use a simple driving yoke steering system instead of the antiquated clutch and brake steering system previously found on all Soviet armored vehicles. It was also designed to be amphibious. Unlike some of the other contenders which used a hydrojet system, the BMP-1 uses the track propulsion system similar to that used on American APCs. The BMP-1 was the most novel and radical departure in Soviet armored vehicle design since World War 2, and it was the world's first true infantry fighting vehicle, preceding the German Marder I by about a year.

The decision to adopt the BMP in the Ground Forces provoked a vigorous debate. The BMP was an extremely expensive vehicle, and many tank officers questioned whether it was prudent to spend so much money for a vehicle. In the end, the BMP was still very lightly armored and lightly armed compared to a tank, when cheaper wheeled BTR-60s could be obtained in larger numbers. The debate was also sparked by continuing doctrinal development in the Ground Forces. By the early 1970s, the Ground Forces had begun to shed their fixation with nuclear warfighting. The USSR was approaching strategic weapons parity with the USA. Nuclear parity implied that a European war might be confined to purely conventional weapons, with both sides fearing the provocative consequences of the use of tactical nuclear weapons. Even if a confrontation with NATO did go nuclear, there would be many occasions when the battlefield would not be contaminated, and traditional tactics would prevail. With this in mind, attention again shifted to conventional battlefield tactics and doctrine.

In the eyes of many Soviet tacticians, the BMP-1 was not entirely suited to the conventional battlefield. On a nuclear battlefield, NATO anti-tank guided missile and rocket teams would be severely inhibited by the contaminated enviroment. The BMP-1 could reign freely on the battlefield at the head of combined tank-motor rifle groups. On a conventional battlefield, there would be a profusion of anti-tank teams. The BMP-1, with its light armor, was especially vunerable to the wide range of infantry anti-armor weapons available to NATO ranging from hand-held weapons like the M72 LAW or the Carl Gustav, up to crew-served anti-tank guided missiles. The Soviet debate questioned how the BMP could be employed in these different conditions, and concluded that new tactics were required.

The new tactics accepted the use of mounted infantry BMP actions where there was little resistance, such as during the break-out phase of offensive operations, or during pursuit of a disorganized enemy force. When resistance was stiff, the BMP-1 would be used as a part of a tank-infantry team with the infantry dismounted. A unit of tanks would be placed in a wave in the vanguard, since they were better able to absorb the blow of anti-armor defenses. Infantry would follow 200 meters behind the tanks to help route out hostile infantry anti-armor teams. The BMPs would follow no more than 300 to 400 meters behind the infantry, providing fire support for the tanks, and preparing to move forward to pick up the infantry once the opposition was overcome.

The BMP in the 1973 Middle East War

The viability of the BMP on the conventional battlefield was first tested in the 1973 Middle East war. Egypt received its first batch of about 80 BMP-1s in July-August 1973 and after hasty training, they were put into service with the 4th Armored Division and two other units in September 1973. A second batch of about 150 BMP-1s arrived from August to September 1973 and they were rushed into service in time to take part in the October fighting. The standard organization was forty BMPs per mechanized infantry battalion. Because of its amphibious qualities and combat capabilities, the Egyptians used a small number of BMPs, supported by BRDM-2 scout vehicles, for the initial water crossing of the Suez Canal. The general Egyptian evaluation of the BMP was that it was a very good vehicle, and its high speed and maneuverability were appreciated. On the other hand, it was poorly ventilated and it became unbearably hot inside when the vehicle hatches were closed. Inevitably, some of the hatches had to be left open. The Egyptians also found that the infantry compartment in the rear was far too cramped for a full squad of eight men, and usually six or fewer were carried. The Egyptians were especially happy with its performance in the northern canal area around the Kantara salt marshes due to its very low ground pressure, but units in the south were not as happy with it due to its tactical shortcomings in combat.

The Syrians also received the BMP-1 for the first time in 1973, and had received 150 to 170 by 1973, of which about 100 were committed to combat (the remainder were kept with Assad's presidential guard units). The general impression of the BMP was that it was very fast and

Syrian BMP-1 infantry fighting vehicle.

nimble. The Syrians found that the 2A28 Grom gun was effective only at close ranges, no greater than 500 meters. The 9M14M Malyutka missile system was almost wholly ineffective due to the difficulty in accurately steering it from the confines of the vehicle. The Syrian assessment after the war was that "the BMP was like a Mercedes when we really need a simple Ford."

The Israelis destroyed or captured forty to sixty Egyptian BMPs and about fifty to sixty Syrian BMPs. About half of the Syrian BMPs were abandoned due to mechanical problems. The Israelis were especially impressed by the performance of the Egyptian BMPs around Kantara where they could pass over the salt flats where Egyptian and Israeli tanks bogged down. This vindicated the view of the tracked vehicles advocates, demonstrating the exceptional performance of tracked vehicles over wheeled vehicles in terrain extremes. In the southern canal area around Wadi Mabouk, the BMPs were very roughly handled, and the Israel infantry found that they could be knocked out by .50 caliber machine gun fire against the turret rear or the vehicle's side. The BMP-1 proved very vulnerable to Israeli infantry 106mm recoilless rifles. As the Egyptians had noted, the BMPs had to be operated with the hatches open due to the heat, and so the Israeli infantry was often able to disable vehicles be firing into open hatches if they had positions on hilly ground.

In reality, the 1973 war was not an entirely fair test of the BMP. Neither the Egyptian nor Syrian armies had the BMP in service long enought to adequately train with it. This was most clearly the case with the Syrians who suffered extensive losses due to the unfamiliarity of the BMP crews with the maintenance needs of the new vehicle. In addition, the tactics employed by the Syrians and Egyptians, though based on the Soviet model, were not entirely the same. Although the 1973 Middle East War did not provide any conclusive evidence about the viability of infantry combat vehicle tactics, the war did display some of the technical shortcomings of the BMP design. The Soviet GBTU sent technical teams to Syria after the war to study these lessons, and the US Army also studied the lessons of the war. These experiences confirmed problems with the BMP that had become evident in the USSR during field exercises.

The gun on the BMP is less than six feet off the ground, so that it will hit its own dismounted infantry advancing in front of it. This forced the adoption of tactics where each infantry squad would allow a fifty meter zone of fire between it and the squads on either side to permit the BMP to fire its weapons. Such a tactic is easier to propose in peacetime training than to carry out in the confusion of a modern battlefield. Furthermore, the use of the BMP to the rear of the tanks and infantry often precluded use of its main weapon, the 2A28 low pressure gun. The 2A28 had an effective range of about 700 meters, and the Egyptians and Syrians found that it was actually effective only to 500 meters. The Soviet assault wave would form up about 1,000 meters before the forward edge of the enemy lines, placing the BMP as much as 1,500 meters away from its targets. The BMPs' low pressure gun would only come in range of hostile targets as the tanks approached a scant 200 meter from the forward edge of the enemy lines. The PG-15 round's tendency to shuttlecock in a headwind made it dangerous to employ in the fire support role in windy conditions.

The supplementary 9M14M Malyutka (AT-3 Sagger) missile launcher was of little consolation. As previously discussed, the missile was inaccurate and slow to reload. The effective rate of fire was about one missile a minute and the gunner's preoccupation with reloading kept him from using other turret weapons.

Upgrading the BMP

Assessments of the performance of the BMP-1 in the Middle East war as well as experiences in peacetime expercises led to the decision to significantly revamp the BMP design. Work was already underway on improved versions even before the war. The first major new direction was taken with the Obiekt 680, developed in 1972 by the new design bureau headed by A. Blagonravov at Kurgan before the Middle East fighting. This vehicle was inspired by the German Marder 1 which had entered production shortly after the BMP-1. The armament configuration on the Marder 1 was substantially different than on the BMP-1, an externally mounted 20mm autocannon.

The Obiekt 680 was a technology testbed to examine the concept of a reduced profile turret for the BMP, with the new Shipunov 2A42 30mm autocannon instead of the 73mm 2A28 Grom low-pressure gun. Two other innovations were incorporaed into the design. The turret diameter was enlarged to accomodate two men, the squad commander being moved from the forward hull to the turret. To further enhance the firepower of the BMP, a 7.62mm PKT machine gun barbette was added in place of the former squad commander station immediately behind the driver. This also was probably inspired by the Marder, which employed a similar remote-control machine gun barbette at the rear of the vehicle for self-defense. The Obiekt 680 was never acccepted for service, but proved a useful demonstration of alternative approaches in armament.

The lessons of the Middle East war, combined with assessments of foreign infantry combat vehicle designs such as the Marder, led to the 1974 BMP replacement program. The first stage of the program was a modest BMP-1 upgrade to solve the most serious shortcomings, called BMP-1P. The 9M14M Malyutka anti-tank missile system was widely regarded as a failure due to the difficulty steering it using the primitive MCLOS (manual-command-to-line-of-sight) system. In the meantime, the KBM (Gryazev/Shipunov) design bureau in Tula had developed a new generation system of anti-tank missiles, the 9K111 Fagot and 9K113 Konkurs (NATO: AT-4 Spigot and AT-5 Spandrel). These missile used semi-automatic command-to-line-of-sight (SACLOS) guidance, like the

Obiekt 680 infantry fighting vehicle.

BMP-1P with external 9P135 ATGM launcher.

American TOW, or the new Euromissile Milan and HOT. Both Fagot and Konkurs could be fired from the same 9P135 launcher, their principal difference being that the Konkurs was a larger, heavier missile with greater range. The BMP-1P was fitted with a mounting lug on the turret roof for the 9P135 launcher for these missiles. This system was awkward to use, since the gunner had to fire the missile from outside the protective confines of the armor. Its main advantage was that it offered substantially better accuracy than the flawed Malyutka which had temporary filled a tactical void. Other changes were later added to the BMP-1P, including an array of 902V Tucha smoke grenade launchers at the rear of the turret. The BMP-1P replaced the BMP-1 on production lines in the late-1970s. In addition, BMP-1s being returned to depots for their periodic captial rebuilding were also upgraded to BMP-1P standards in the 1980s.

The BMP-2 Yozh

Although the BMP-1P was adequate as a stop-gap program, more substantial improvements were required. Between 1974 and 1977, two parallel efforts were begun, the Obiekt 675/681 by Blagonravov's design bureau at Kurgan using the basic BMP-1 hull, and the Obiekt 768/769 by Isakov's design bureau at Chelyabinsk, using a new lengthened chassis. All four of these designs used two-man turrets. There was a general consensus that the one-man BMP-1 turret was a mistake and that the squad commander should ride in the turret, not in the hull. This was done for two reasons. On the BMP-1, the commander's station is fitted with an infrared searchlight. This acted as an obstruction to the turret armament, and the gun had to be elevated to clear this obstruction if engaging targets to the left front corner of the vehicle, creating a weapon dead-zone in the front left corner. The squad commander also had very poor vision in his hull location. The turret over his right shoulder prevented him from any vision in that direction. By moving the commander into the turret, the obstruction was removed, and the commander gained a better 360 degree view.

It is interesting to note that the US Army came to the same conclusion at this time. Its original infantry fighting vehicle, the XM723 MICV, had the same crew configuration as the BMP-1. However, before production began, it was reconfigured with a two-man turret and entered service in the late 1980s as the M2 Bradley IFV. The main drawback of adding a two man turret is that it tends to take up a disproportionate amount of hull room,

Armored Infantry Vehicles

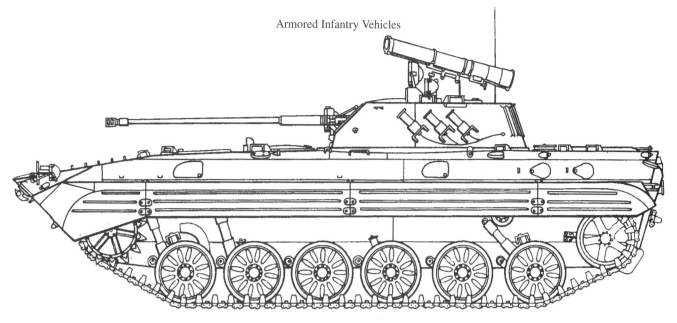

BMP-2 Infantry Fighting Vehicle

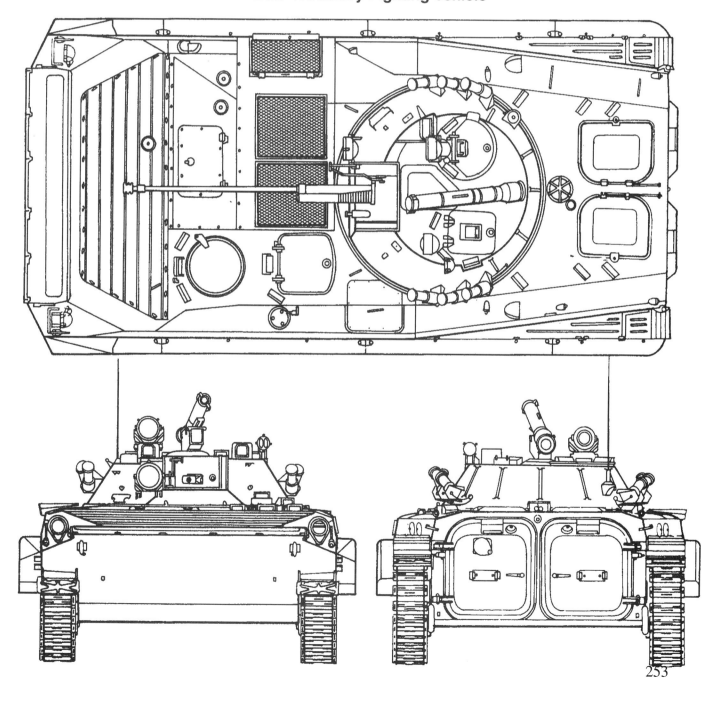

Obiekt 769 with lengthened BMP with Zarnitsa 73mm gun.

forcing a reduction in squad size. However, infantry squad sizes have been shrinking in most European armies since World War 1 due to the increasing firepower in the rifle squads. The lengthened hull alternatives, the Obiekt 768 and 769, were apparently attempts to deal with this problem by increasing hull volume.

The reason for two alternative versions of each hull concept was indecision over the future armament of the BMP. The GRAU (Main Missile and Artillery Directorate) insisted that the BMP remain armed with an anti-tank weapon, not the type of 30mm weapon examined on the Obiekt 680.[22]

As a result, the new, long-barreled Zarnitsa (Lightning) 73mm low-pressure gun was developed which offered better range than the 2A28 Grom even though it fired the same projectile.[23] This weapon was demonstrated on both hull configurations (the Obiekt 681 on the BMP-1 hull and the Obiekt 768 on the lengthened BMP hull). The other two vehicles, the Obiekt 675 on the BMP-1 hull and the Obiekt 769 on the lengthened BMP hull, were armed with the 30mm 2A42 autocannon first demonstrated on the Obiekt 680 of 1972 but in a more conventional, full-height turret. Although lacking GRAU support, the 30mm cannon alternative was supported by the commander of the armored forces, Marshal Amazasp Babzhanian. All four of the vehicles used the new 9Sh119 missile launcher that could fire either the Fagot or Konkurs missile. But unlike the BMP-1P, the gunner remained safely inside the turret when firing the missile.

In spite of trials favoring the Obiekt 675 with its 30mm cannon, there was still considerable controversy over the selection of the future BMP armament. For a time, the Soviet army considered producing both the Obiekt 681 (73mm Zarnitsa gun) and Obiekt 675 (30mm cannon). Combat lessons from the Afghanistan war helped to decide the fate of the program. General of the Army Mikhail Zaitsev, at the time commander of the Group of Soviet Forces-Germany, called the Central Committee and urged them to adopt the Obiekt 675. He had become familiar with the design during operational trials in the Byelorussian Military District (where he had formerly been the commander). Zaitsev argued that the high elevation of the 30mm cannon made it particularly suitable for use in Afghanistan, where the limited elevation of the 73mm Grom gun had been a hindrance. Production of an initial batch of twenty-five was completed in April 1980, and in August 1980, the Obiekt 675 was officially accepted for service as the BMP-2.[24] The lengthened hull types were probably rejected since they would have required more extensive retooling of the main BMP plants at Kurgan and Rubtsovsk.

The main advantage of the 2A42 30mm gun over the 73mm low pressure guns was that it offered far better

Obiekt 768 lengthened BMP with 30mm cannon.

range (2,000 to 4,000 meters) which made fire support of the lead waves of tanks more practical and was well suited to dealing with the new menace of missile-firing attack helicopters like the US Army AH-1S/W Cobra and AH-64 Apache, as well as land-based anti-tank missile teams. This decision displayed a pragmatic reevaluation of the utility of low pressure gun systems, which seem to offer a useful anti-armor capability, but which have significant range and accuracy shortcomings when used to provide overwatch fire support. Although the 30mm gun cannot penetrate the frontal armor of a main battle tank, the weapon is highly effective against the profusion of light armored vehicles on the battlefield, including infantry fighting vehicles and APCs, and is very lethal against infantry and unarmored targets. In addition, it can disable main battle tanks by damaging fire control optics and track. The AT-4 and AT-5 anti-tank missile systems on the BMP-2 offer some measure of self-defense against tanks.

The BMP-2 Armament

The BMP-2 turret is considerably larger than the BMP-1 turret, and houses two crewmen, the squad commander to the right and the gunner to the left. The main armament is a 2A42 automatic cannon manufactured at the Tula Machinery Plant. The gun has a selectable rate of fire, a slow rate of 200 to 300 rounds per minute, and a fast rate of 550 rounds per minute. The main problem of the fast rate of fire is that it tends to fill up the turret with fumes. The gun is fed from two ammunition trays located at the rear base of the turret floor, the usual storage being 160 rounds of armor piercing and 340 rounds of high explosive/incendiary. One of the main problems encountered by the BMP-2 in service was the poor design of the ammunition loading system. Even if the ammunition was ready and prepared in belts, it takes about 1.5 to 2 hours to reload the magazines. There are two types of high explosive rounds that can be selected, the 3UOF8 high-explosive-incendiary (HEI) or 3UOR6 high-explosive-tracer (HE-T) which are often intermixed. There are also two types of armor-piercing ammunition, the 3UBR6 armor-piercing tracer (AP-T) with an initial muzzle velocity of 970 meters per second and penetration of 20mm of armor at sixty degress at 700 meters, and a new armor-piercing discarding sabot-tracer (APDS-T) with a muzzle velocity of 1,120 meters per second and penetration of twenty-five millimeters at sixty degrees at 1,500 meters. This is more than adequate to pierce the armor or contemporary infantry vehicles such as the M2 Bradley and Marder 1, but the AP-T may not penetrate the reinforced armor of the M2A3 Bradley or Marder 1A4 which were specifically designed to resist this weapon.

The 2A42 gun employs an electro-mechanical two-plane stabilization system which gives it good accuracy on the move at typical vehicle speeds up to thirty-five kilometers per hour (twenty-one miles per hour). The gun can be aimed from either the gunner or commander's station, though it is usually aimed by the gunner. It has an effective range of about 2,000 meters with AP ammunition and 4,000 meters with HE ammunition. The gun has been given an unusually high maximum elevation (seventy-four degrees) in order to make it more suitable for use against anti-armor helicopters, and the commander has a 1PZ-3 anti-aircraft sight specifically for this role.

The gunner's BPK-1-42 sight is suitable for night operations, using a passive image-intensification channel with an effective range of 650 meters, or on very dark nights using an active infrared searchlight, with an effective range of 350 meters. The squad commander is also provided with a TKN-3B day-night vision periscope, but the night vision channel is limited to active infrared. Illumination for the active infrared night sights for the gunner comes from the FG-126 searchlight mounted co-axially with the 30mm gun on the right side of the turret front while the commander's night sight can be supported by the OU-3GA2 searchlight on the cupola.

The 2A42 30mm cannon is supplemented by a 9Sh119M1 missile launcher for the 9K111 Fagot and the 9K113 Konkurs (NATO: AT-4 Spigot and AT-5 Spandrel). The missile rail is located on the roof of the vehicle, and four rounds are stowed internally. As mentioned earlier, the 9M113 is a SACLOS wire-guided missile, and is similar in performance to the American TOW or the Euromissile HOT. As in the case of these Western missiles, there are improved versions of the Konkurs, the 9M113M Konkurs-M which has a precursor charge to defeat reactive armor. The infantry squad also is equipped with a portable, dismountable 9P135M firing post for this missile.

The BMP in Combat with Soviet Forces

The BMP-1 was first used in combat by Soviet troops in Afghanistan in 1979. Three drawbacks of the BMP-1 became immediately apparent: the thin side armor, inadequate elevation of the gun, and poor crew habitability. The armor of the vehicle was designed to withstand frontal attack from heavy machine guns under conventional conditions of mechanized combat. In Afghanistan, the BMPs were often ambushed from the side by mujihadeen firing from very close ranges. At close ranges, the 16mm side armor could be penetrated by the 12.7mm DShK heavy machine gun. This led to a depot rebuilt starting in 1982, called the BMP-1D (D= dorobotanaya: upgrade), which added a layer of ten millimeters spaced armor around the hull side. This made the BMP-1D less vulnerable to heavy machine gun fire, though still vulnerable to RPG-7 attack. The habitability problem contin-

BMP-2 infantry fighting vehicle.

BMP-2D infantry fighting vehicle in Afghanistan.

ued to plague the BMP through its service in Afghanistan and was solved by field improvisations involving the permanent attachment of additional stowage bins and water containers on the upper hull roof, in spite of the obstruction that this created for the gun.

There was no immediate solution to the gun elevation problem. Due to the mountainous terrain of Afghanistan, the elevation of the 73mm Grom gun was inadequate since it could not be elevated high enough to hit some targets on high mountain slopes. The Soviet Army decided to dispatch the newer BMP-2 to Afghanistan in 1982, which had a high elevation mounting for the gun already due to the need to combat attack helicopters. The 30mm gun was a superior alternative to the 73mm 2A28 Grom in any event, as the targets were invariably guerilla troops, not armored targets. As a result, the BMP-2 became the preferred variant in service, largely replacing the BMP-1 by 1987. As in the case of the BMP-1, most BMP-2s in Afghanistan were eventually uparmored at depot level, becoming designated BMP-2D in this configuration. This depot refit usually included the addition of a large stowage bin on the turret rear, which somewhat improved the stowage problem.

Although the BMP-2 was considered much superior to the BMP-1 during the war, the gun system was too slow to reload, and at maximum rate of fire it tended to overwhelm the turret crew with fumes. As a result, the gun was generally fired only at the lower rate. This problem led to the development of the later 2A72 30mm cannon (now used on the BMP-3 and BTR-80A). The 30mm gun was also considered inadequate when dealing with entrenched infantry armed with RPG anti-tank rocket launchers and led to the decision to mount a larger gun on the BMP-3. The BMP was not developed to serve in guerilla wars, but proved adequate for the job.

The BMP Export

The BMP began to be exported in the early 1970s, first to the Warsaw Pact, and eventually to over a dozen other countries. Figures for Warsaw Pact BMP strength at the time of the Pact's demise in the early 1990s are given on the next page. The BMP has been widely exported in the Middle East in spite of its bad reputation for being uninhabitable in hot weather. Users include Iran (500), Iraq (700+), Kuwait (50+), Libya (450), Syria (1,000+), and Yemen (50+). In Africa, Algeria purchased BMPs in at least two batches in 1975 and 1982 numbering about 500 vehicles; Mozambique had a small number of vehicles in 1984 and the BMP-1 was used in small numbers by Angolan and Cuban troops in the war in

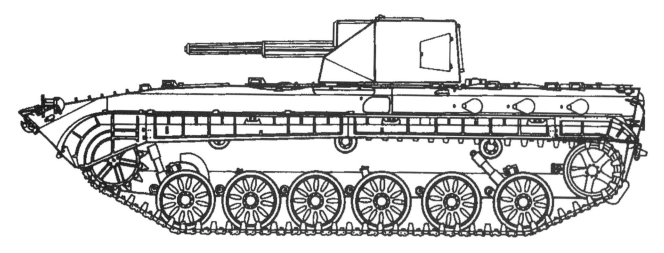

BMP-1 Kliver

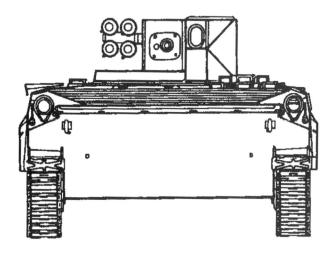

Namibia and Angola with South Africa. In the Americas, Cuba obtained modest numbers of BMP-1s in the 1980s, numbering about sixty. India ordered a number of BMP-1s before deciding on the license production of the BMP-2 as the Sarath. Foreign production of the BMP is covered in the variant section below.

Upgrade Programs for the BMP-1/-2
Kliver and BMD Turrets

A new program for upgrading the BMP-1/-2 is the Kliver turret, developed by the Tula KBP. Although descriptive material on the Kliver has been available since 1996, IDEX-97 was the first time that the actual prototype turret was shown. This is a new one man turret armed with a 30mm gun (the 2A42) and four launch stations for the Kornet anti-tank missile. The design is primarily intended to retrofit old BMP-1s, or to up-arm older BTRs such as the BTR-70 or BTR-80. It is designed around a small diameter turret ring, suitable for small turrets such as BMP-1. The reason for the turret is that the old BMP-1 turret is obsolete, being armed with an inadequate 73mm low-pressure gun and an obsolete anti-tank missile (Malyutka/AT-3).

The Kliver turret has four missiles stowed externally, and has an auto-reload system by which eight additional missiles stowed inside the turret basket can be accessed using a small hatch under the launcher. The Kliver is also being offered to arm small patrol boats, since the Kornet's laser guidance system is useable over water (unlike wire guided anti-tank missiles). The missile has both an anti-armor and a fuel-air explosive warhead, useful for naval or coastal engagements, and entrenched infantry or ATGM crews.

Russia hopes to export this turret upgrade to countries with BMP-1s, including NATO countries such as Greece which, in recent years, received them as aid from Germany.

CFE Counts for BMPs (European Russia): 1990 to 1997

	BMP-1	BMP-2	BMP-3	BRM-1K	Total
1990	8146	5996	33	1,363	15,538
1991	8,208	5,994	35	1,376	15,613
1992	4,271	3,881	29	715	8,896
1993	3,575	3,115	26	582	7,298
1994	3,014	3,169	17	589	6,789
1995	1,928	3,393	25	637	5,983
1996	1,788	3,250	25	626	5,689
1997	1,699	3,364	26	575	5,664

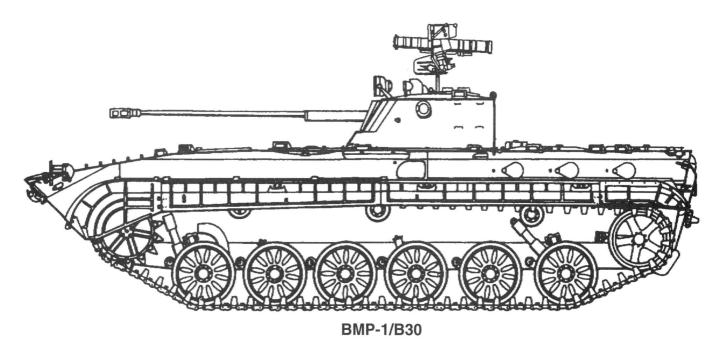

BMP-1/B30

There is apparently a competition in Russia for the BMP-1 upgrade, the alternative being the BMP-1/B-30 which consists of refitting the BMP-1 with the turret from the BMD-2 airborne armored vehicle. This turret is armed with the same 2A42 30mm cannon as the BMP-2, but carries less ammunition. Its anti-armor weapon is an externally mounted firing post for the Fagot/Konkurs (AT-4/AT-5) anti-tank missiles, which exposes the turret gunner to enemy fire.

The BMP-1/-2 Variants

Russian BMP Variants
BRM Boevaya razvedyvatelnaya mashina

The BRM is a scout version of the BMP that replaced the PT-76 amphibious scout tank. It was developed as a competitor to several light tank designs (see light tank section of this study). It first appeared in 1976, hence its NATO code-name, BMP M1976/1. The basic BRM has a large, two-man turret with 2A28 Grom 73mm gun and the associated 1PN22M2 gunner's sight. Unlike the later BRM-1K, it has no radar. The commander, who is seated in the turret, is provided with a day/night sight and a DKRM-1 (1D8) ruby-laser rangefinder. The navigator, who sits behind the driver in the hull, is provided with a TNPK-240A observation device. The basic communications package includes a R-123M, R-130 and dismountable R-148 transceivers, as well as a R-014D teletype. The BRM series uses the TNA-1 Kvadrat 1 navigation device, 1G11N gyro compass and 1T25 land navigation device. Two scouts are seated in the back of the vehicle for dismount operations and to protect the vehicle from rear and side attack during mobile operations. BRMs are issued on a scale of one per each motor rifle or tank regiment, and three per divisional reconnaissance battalion.

BRM-1K Boevaya razvedyvatelnaya mashina (kavalerskiy)

This is an upgraded version of the BRM and was called BMP M1976/2 by NATO. The Soviet developmental designator was izdeliye 676. The most significant change on this vehicle was the addition of a PSNR-5K (1RL133-1) battlefield surveillance radar (NATO: Tall Mike) which operates in the 16.0 to 16.3 Ghz frequency band. It is fitted with a telescoping 50 meter antenna stowed over the hull rear above the exit doors.

BMP-1PG

This is a 1993 upgrade program in which an AGS-17 30mm grenade launcher is added on the rear left side of the turret to provide additional fire support. This version also incorporates many of the BMP-2 powertrain and suspension upgrades.

BMP-1K Boevaya komandnaya mashina

The izedliye 767, called BMP-M1974 in NATO, is a company command vehicle with a UKB radio communications suite including the R-123, R-126 and R-107 transceivers. It was accepted for service in 1972. The company commander's BMP-1K carries the company commander in the usual location, a radio operator and rifleman in the left rear seats and two company staff officers, a medic and a sniper in the right rear seats. Due to this configuration, the firing ports and periscopes on the right side are blanked off, as well as one firing port and peri-

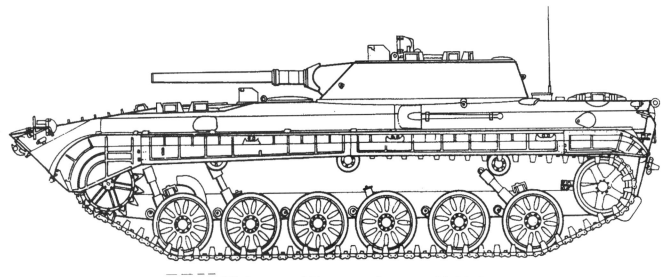
BRM Armored Reconnaissance Vehicle

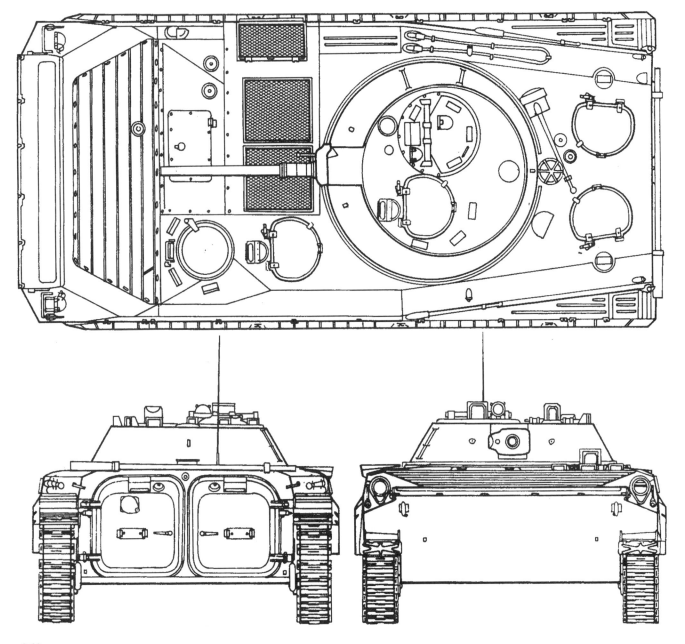

BRM-1K armored reconnaissance vehicle.

BMP-1PG infantry fighting vehicle

scope on the left side. A small telescoping antenna is carried on the right rear hull side side. The more elaborate communication requirements of the battalion and higher staffs led to the development of the BMP-1KSh.

BMP-2K Boevaya komandnaya mashina

The BMP-2K is essentially similar in function to the BMP-1K, but in motor rifle companies using the BMP-2 vehicle. As in the case of the BMP-1K, it has additional radio masts, and has several of the firing ports blanked off.

BMP-1KSh Komandno-shtabnaya mashina

The BMP-1KSh (izdeliye 774) is the standard command and staff vehicle for BMP motor rifle regiments and was called BMP M-1978 by NATO. It entered series production in 1976. This vehicle is fitted with a fixed turret without the usual gun. A large, ten-meter telescoping high-gain antenna (NATO: Hawk Eye) is carried in front of the turret, and can be folded up or down. There are only two rear roof hatches on this vehicle, and an extra generator is fitted on the rear roof to power the radios. The standard radio suite for the vehicle includes a R-130 HF transceiver, a R-111 VHF transceiver, and a R-173 VHF transceiver. These systems are supported by a 1T-219M secure speech coding system and a R-102 automatic calling device. The usual crew for this vehicle is seven men and there is one of these vehicles per motor rifle regiment. This vehicle comes in several configurations for other command roles. The MP-31 (also known

BMP-1 KSh armored command/communications vehicle.

as 1V31) and the 9S743, use different radio configurations, and are fitted with a much larger external generator on the rear hull roof than the standard BMP-1KShM. It is manufactured at Rubtsovsk.

PRP-3 "Bal" Podvizhniy Razvedyvatelniy Punkt

The PRP-3 (izdeliye 773) is an artillery scout vehicle and was called BMP M1975 by NATO. There is one such vehicle in each self-propelled howitzer battalion. It uses a large diameter, two-man turret, similar in appearance (but not identical to) to the BMP-2 turret. The vehicle is armed with a single 6P7 PKT 7.62mm machine gun for self-defense, aimed through the 1P28 periscopic sight on the roof. The basic vehicle sensor is a 1RL126 (NATO: Small Fred) centimetric battlefield surveillance radar, with the antenna located on the rear roof of the vehicle. This radar can detect tanks up to ten kilometers away. The vehicle is also fitted with a 1PN61 night vision sensor in the right sensor package, along with a 1D11 laser rangefinder. The turret is also fited with a TNPO-170A optical periscopic sight. Communications is provided by means of a R-173 VHF transceiver and an 1A3OM command transceiver, supported by a 1T803 secure voice system. Precision navigation is provided by means of a 1V44 (KP-4) course plotter, 1G13 gyro course indicator and IT25 gyrocompass, and a 1V520 ballistic computer is provided for fire control problems. These vehicles usually carry a portable 1D13 laser rangefinder. The driver is provided with a TNPO-350B day sight and a TVNE-1PA image intensification night sight. This vehicle is produced at the Rubtsovsk Machine Building Plant.

PRP-4 "Nard" Podvizhniy Razvedyvatelniy Punkt

The PRP-4 is an improved version of the PRP-3. The most significant change is the addition of a second sensor package on the left side of the turret containing a 1PN59 thermal imaging night vision device and 1D14 laser rangefinder. The right sensor package contains upgraded sensors including the 1PN61 and 1D11M-1 laser rangefinder. The PRP-4M has improved 1PN71 night vision sensors. It is manufactured at Rubtsovsk.

IRM Inzhenernaya razvedyvatlnaya mashina

This is an engineer scout vehicle based on the BMP chassis and was first fielded in 1976. The vehicle bears very little resemblance to the BMP as the entire superstructure is completely new, and the engine has been

PRP-4 artillery reconnaissance vehicle.

IRM combat engineer scout vehicle.

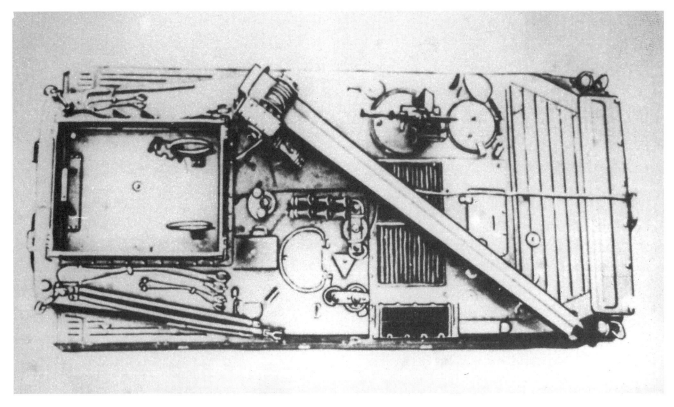

BREM-2 recovery vehicle.

moved to the rear. This vehicle is designed to assist engineer units in scouting rivers to determine their suitability for deep fording operations, pontoon bridge operations and the like. This vehicle is nicknamed Zhuk (Beetle).

BREM-2 Bronirovannaya remontno-evakuatsionnaya mashina

This is a light recovery and repair version of the BMP that entered series production in 1986. A rectangular work platform and container is fitted over the rear troop compartment, and a light crane is fitted immediately forward of this on the left side of the hull roof.

BREM-4 Bronirovannaya remontno-evakuatsionnaya mashina

This is a turretless armored recovery vehicle. A large circular plate covers the turret opening, and a crane is fitted on the rear hull roof. This version produced in Czechoslovakia was designated VPV.

KMT-10 Mine Plows

The BMP is designed to accomodate a set of mine plows on the front hull, and a portion of the BMP-2s in each company are fitted with the appropriate mounting lugs. This is powered off air pressure from the vehicle powerplant. Because of the added weight, an additional trapazoidal flotation panel is added on the bow.

BMP-PPO Podvizhniy Punkt Obucheniya

This is an unusual training vehicle with the turret removed, and eight commanders cupolas added to the roof. This vehicle is used to train BMP squad commanders. It can carry eight squad commanders at once, thereby lowering training costs.

Civilian Variants

As part of the Russian conversion effort, a number of demilitarized versions of the BMP have been produced. One of the better known types is the Berezina, a turretless transporter.

Czechoslovak BMP Variants
BVP-1

The BMP-1 and BMP-2 in Czechoslovak service are designated BVP-1 and BVP-2. Production is undertaken at the Podpolianske Strojarne Detva and at ZTS Dubnica. Total production of BMPs for the Czechoslovak Army amounted to 2,252 as of 1992.

BPzV

This is Czechoslovak reconnaissance version of the BMP-1, used in place of the Soviet BRM-1K. It is fitted with the PSNR-5K (Tall Mike) radar on an external mount on the right rear of the turret. A laser range finder and night sensor is mounted in a very exposed position over the commander's hatch on the left hull side.

BREM-4 recovery vehicle.

PRAM-S

This is a self-propelled 120mm mortar vehicle developed in 1990. The 120mm mortar is fitted in a new fixed superstructure with a traversable mantlet. The mortar system is fed by an autoloader.

DTP-90

This is a turretless maintenance variant of the BVP with various types of equipment stowed in racks on the hull roof. The DP-90 retains the normal BVP-1 turret, without the armament, and is used for light maintenance work.

MU-90

This is a turretless mine-laying version of BMP. The turret has been plated over, and the rear compartment is filled with mine racks.

OT-90

In order to get around the CFE Treaty limitations, the Czechoslovak Army converted about 600 BVP-1 to the OT-90 configuration. This substitutes an OT-64 SKOT-2A turret with the 14.5mm Vladimirov heavy machine gun in place of the normal 73mm gun turret. The troop compartment in this version carries six soldiers. The VT-90 is essentially similar but designed for scout missions with additional communication equipment and diffferent internal stowage.

Boure III

This is a psychological warfare version of the BMP, fitted with a loudspeaker system. There is a new turret-like superstructure in place of the normal turret.

SVO Salvovy Vybusny Odminovac

This is a turretless mine-clearing version of the BVP-1 fitted with an array of twenty-four large rocket-propelled mine clearing charges in an open compartment in the center of the vehicle.

VPV Vyprostovaci Pasove Vozdilo

This is a Czechoslovak version of the Soviet BREM-4 recovery and maintenance vehicle fitted with a rear mounted crane for armored vehicle repair and recovery work.

ZV-90

This is a turretless recovery version, with the turret plated over, and the rear compartment used for tool stowage.

AMB-S

This is an ambulance version, with a large fixed superstructure over the rear of the vehicle instead of the usual turret. It can carry four stretcher cases plus a medic. There is a similar artillery command and reconnaissance vehicle, used by the Czechoslovak Army in place of the Soviet MT-LBu. It has a large fixed superstructure like the AMB-S, and contains extensive radio, navigation and ballistic computer equipment.

Polish BMP Variants
BWP-40

Poland has produced the BMP-1 under license since

the early 1970s at plants in Stalowa Wola under the local designation BWP-1 (Bojowy woz piechoty-1). In 1993, the firms announced a new variant designated BWP-40 undertaken with the Bofors firm of Sweden. Poland and Sweden had traditional arms cooperation agreements in the pre-World War 2 period, and both countries appear to be reviving this tradition.

The BWP-40 substitutes a new, large turret for the normal BMP-1 turret. The turret includes two crewmen, and is armed with a Bofors 40mm L70 gun.

The BWP-40 is regarded by the Poles as a superior alternative to the BMP-2. The 40mm gun is useful in combatting both light armored targets and helicopters; the Poles claim it can engage tanks at ranges up to 2,000 meters using a new Bofors APFSDS round. The fire control system is Swedish designed and consists of an integrated laser-rangefinder, day electro-optical sight and night thermal imaging sight.

The gun has a maximum elevation of +35 degrees to assist in the air defense role. The new turret increases the weight of the vehicle to 15.4 metric tons, about 1.5 tons heavier than the BMP-2. The crew is reduced by one compared to the BMP-2, 3 crewman plus a squad of six in the rear. The reason for the small crew size is that the larger Bofors turret takes up the space occupied by one of the squad machine gunners behind the driver's station in the BMP-2.

German BMP Variants
BMP-1A1 Ost

In 1990, the German government decided to retain a limited number of BMP-1 vehicles in Bundeswehr service from the former East German Army (NVA). These vehicles were rebuilt at the SIVG Neubrandenburg in cooperation with Diehl firm. The changes included incorporation of new communications and electronics, as well as improvements to bring the vehicles up to German safety standards.

Romanian BMP Variants
MLI-84

Romania builds a licensed version of the BMP-1. The most noticable local change has been the addition of a new hatch at the left rear of the vehicle, fitted with a 12.7mm DShK machine gun. This vehicle is powered by a 8V-1240 DT-S engine and the front engine panel has been reconfigured as a result.

Indian BMP Variants
BMP-2 Sarath

In 1983, India reached an agreement with the Soviet Union to begin license production of the BMP-2 in India under the local name of Sarath. The new Shankarpally Ordnance Factory was erected in the Medak district of Andrah Pradesh at a cost of about $350 million. The first vehicles, built from Soviet knock-down kits, was turned over to the Indian Army in 1987, and full-scale local production was completed by 1991. India also manfactures the Konkurs missile used on the BMP-2 at the nearby Bharat Dynamics Ltd. plant in Bhanoor.

Trishul TCV

India is developing a short range air defense missile similar to the Soviet Osa-AKM (SA-8 Gecko), called Trishul. The Army version of Trishul will be launched from a twin rail launcher on a modified BMP-2 Sarath infantry vehicle with its own Flycatcher missile guidance radar, and a second surveillance radar. The basic launcher vehicle is designated the TCV (Trishul Combat Vehicle). The command vehicle for Trishul firing batteries is based on the same hull, and is called the MCP (Mobile command post). Both vehicles use a modified hull with an additional roadwheel station per side. India is also developing a tank destroyer version of the Sarath for use with the new Nag anti-tank missile.

BMP Light Tank

The Indian Army has a requirement to replace the obsolete PT-76 amphibious scout tank. In 1987, The Indian CVRDE (Combat Vehicles R&D Establishment) in Avadi developed a light tank on the BMP-2 Sarath hull using a French GIAT TS-90 turret. This has not yet entered production, and India is also considering the Russian BMP-3 for this role.

Chinese BMP Variants
Type WZ501

The WZ 501 is the basic version of the Chinese BMP. China began manufacturing an unlicensed copy of the BMP in the late 1980s, based on Soviet BMPs received from undisclosed clients. The WZ 501 is a direct copy of the Soviet BMP-1. The subsystems have Chinese designations, for example, the 9M14M Malyutka missile is manufactured in China as the Type 73 Red Arrow, and the 9M32 Strela 2 missile is manufactured in China as the Red Tassel.

Type WZ501A

This version uses a new turret with an elevated 25mm cannon instead of the usual 73mm low-pressure gun.

Type WZ503

This is a low-cost version that substitutes a simple heavy machine cupola for the normal turreted weapon. The hull was raised 100mm higher than the standard

WZ501 to provide additional headroom for the infantry squad, and a large single door is fitted at the rear instead of the usual two split doors. By reconfiguring the rear compartment, the vehicle can carry fifteen troops instead of the usual eleven.

Type WZ504

This is a tank destroyer version, with a four-rail launcher module for the Type 73 Red Arrow (AT-3 Sagger) anti-tank missiles sustituted for the usual turret. This can carry a dismount squad of four troops.

Type WZ505

This is an ambulance version, with a simple box superstructure added to the rear of the vehicle.

Type WZ506

This is a variation of the Type WZ503, but intended for regimental command applications with a more extensive radio communications suite. The standard radio fit includes two Type 889, one Type 892 and one 70-zB or SR119 radio.

NFV-1

This was a joint venture between Norinco and the US firm of FMC, with Norinco providing the basic WZ 501 chassis and FMC providing a one-man turret armed with the M242 25mm Bushmaster autocannon as used on the US Army's M2/M3 Bradley fighting vehicles. The NFV-1 was intended for export, and so far as is known, none were manufactured beyond prototypes.

The MT-LB Armored Transporter

In the 1970's, the Soviet Central Auto and Tractor Directorate embarked on a development program to replace the AT-P, AT-L, AT-S, and AT-T tractors with a new generation of vehicles. The light tractor requirement resulted in the MT-L, as well as an armored derivative, the MT-LB. The MT-LB was intended to replace the AT-P armored artillery tractor and for other utility roles. It was part of family of vehicles manufactured by the Kharkov Tractor Plant which also included the MT-LBu armored command vehicle and the 2S1 Gvozdika 122mm self-propelled howitzer. The AT-P had been developed by Astrov's OKB-40 in Mytishchi and was manufactured from 1954 through 1962 at the Mytishchi Machinery Construction Plant.[25]

One of the main drawbacks of the AT-P is that the gun crew was left exposed in the rear compartment of the vehicle, undermining the operational intent of the vehicle which was to transport to gun and crew into the forward edge of battle area in safety against artillery and small arms. The MT-LB avoided this problem by completely enclosing the troop compartment. The MT-LB was based around many existing components including a commonplace truck engine, and standard suspension elements. It was considerably less expensive to manufacture than the BMP, and so was used as the basis for many utility vehicles. The MT-LB went into production in the early 1970s in the USSR, Poland, and Bulgaria. It is used as the basis for the 9A34/9A35 (SA-13 Gopher) air defense missile vehicle, for the SNAR-10 artillery radar vehicle and in many other roles.

MT-LB armored transporter.

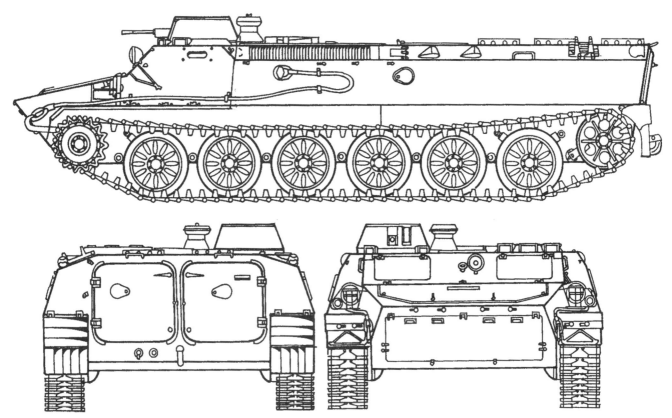

MT-LB Armored Transporter

9A35M3 Strela 10 (MT-LB with SA-13).

9P149 tank destroyer (MT-LB with Shturm-S ATGM).

MT-LBV Arctic armored transporter.

Although not its primary function, the MT-LB has been used as a troop carrier by the Soviet Army. The special arctic MT-LBV version uses a wider track for better floatation on snow. As a result, Soviet motor rifle divisions in the arctic regions north of Leningrad used the MT-LBV in lieu of the BTR-60PB or BMP (including the 54th, 64th, 111th and 131st Motorized Rifle Divisions). At the time of the CFE negotiations, the MT-LB became a bone of contention connected with the counting roles for APCs and IFVs due to its configurational similarity to infantry vehicles. The artillery tractor version is now designated MT-LBT to distinguish it from those used in an infantry transporter role.

The BMP-3

In 1990, a new-generation BMP arrived unexpectedly on the scene, the BMP-3. Although similar to other contemporary armored infantry vehicles in terms of size and armor protection, it is far more heavily armed than any previous infantry vehicle, with a 100mm main gun, supplementary 30mm autocannon and a co-axial 7.62mm machine gun. Indeed, its armament package is more powerful than many tanks from the 1960s.

The engineering design of the BMP-3 can be traced back to the abortive light tank program of the mid-1970s. The Soviet VDV Airborne Assault Force sought a new light tank to replace the ASU-85 assault gun, while at the same time, the Soviet Ground Forces were looking for a scout tank to replace the PT-76.

Two competitive designs were offered for this requirement, the Obiekt 685 from A. Blagonravov's design team at the BMP bureau in Kurgan, and the Obiekt 934 from the BMD design bureau in Volgograd under A. Shabalin. Both light tanks were armed with the same 2A48 100mm

269

Obiekt 688 (BMP-3 prototype).

gun, and were amphibious and air-deployable. In the end, a lower cost solution was reached by adopting the BRM BMP-derivative for the Ground Force requirement and the 2S9 Nona BMD-derivative for the airborne requirement.

Work on the Obiekt 685 was not entirely in vain, as in the late 1970s, the Soviet Ground Forces expressed interest in a next generation infantry fighting vehicle as a response to Western developments such as the US Army Bradley and British Warrior. To meet this requirement, the Obiekt 688 was developed by the A. Blagonravov's design bureau at Kurgan with A. Nikonov as the chief designer. It is not clear if this was a competitive design effort, and other configurations may have been considered. The first prototype of the Obiekt 688 design, completed in 1981, was armed with an elevated, externally-mounted 30mm 2A42 gun, a 30mm grenade launcher, and two anti-tank guided missiles in box launchers resembling the earlier Obiekt 680, (or the French AMX-10P turret and German Marder turret). The Obiekt 688 used a new chassis, derived from the experimental Obiekt 685 light tank but with a new engine. The new UTD-29 diesel engine, developed at the Transmash Diesel Engine Plant in Barnaul, is an extremely flat V-6 configuration designed to sit in the rear of the hull, while allowing the crew to exit over it.

The Obiekt 688 weapons' configuration was eventually rejected, since it offered insufficient firepower advance over the BMP-2. This decision was related to combat experiences with the BMP-2 in Afghanistan. As mentioned above, the BMP-2 had several deficiencies in combat. The 30mm gun was not effective against well entrenched troops. Secondly, when using the 30mm cannon at the high rate of fire, the turret crew became overwhelmed by fumes. Thirdly, the ammunition trays in the BMP-2 were awkwardly located, taking much too long to reload. The only change on the Obiekt 688 armament system was the 30mm grenade launcher, and this was probably viewed as inadequate in range and accuracy to deal with infantry anti-tank teams.

As an alternative, the radically new 2K23 armament system was developed by Shipunov's KB Priborstroyeniya design bureau in Tula. The 2K23 system consists of a 2A70 rifled 100mm gun paired with a 2A72 30mm autocannon, a coaxial 7.62mm 6P7 machine gun, and an associated autoloader and fire control system. The combination gun system had several advantages. The 100mm gun provides the high-explosive firepower needed to deal with entrenched infantry; it is in fact a low velocity howitzer. Secondly, the diameter of this weapon is sufficient to permit the launch of tube-fired anti-tank missiles, derived from the type used with

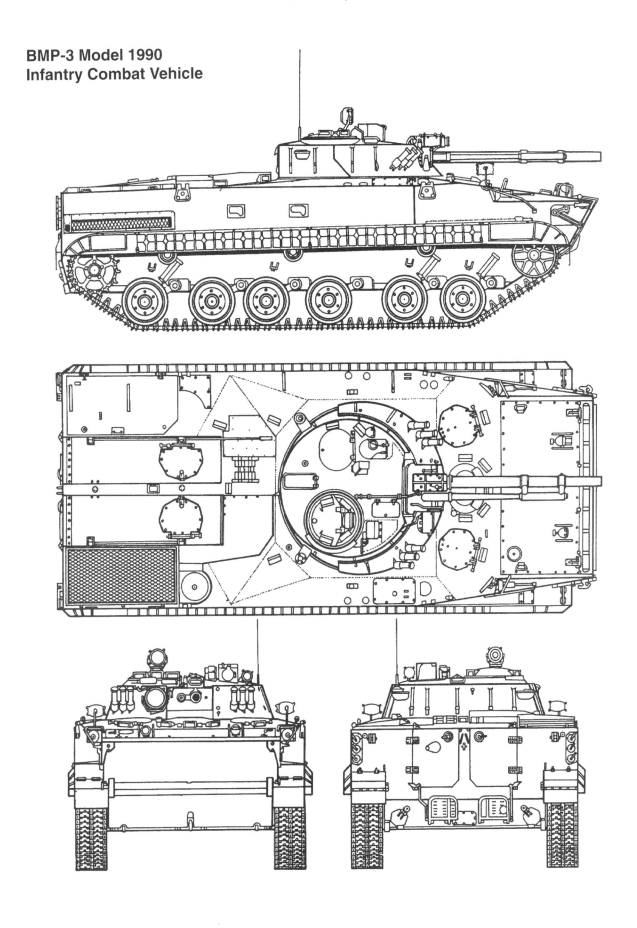

BMP-3 Model 1990
Infantry Combat Vehicle

the D-10T tank gun. This had the advantage of permitting the removal of the turret mounted anti-tank missile standard on previous BMPs, as well as the associated fire control system. Rather surprisingly, this weapon was teamed with a new 30mm gun, the 2A72. This weapon fires the same ammunition as the 2A42 on the BMP-2, but has only one (low) rate of fire in order to avoid the fume problem of the earlier weapon. The reason for the 2A72 is probably the continued Russian concern over the helicopter threat. The 100mm gun can certainly deal with most battlefield targets, but the slow muzzle velocity of its standard unguided round and its ballistic arc would not make it ideal as an anti-aircraft weapon where high speed and high rate of fire are desired.

The 2K23 weapon system was mounted in a conventional turret resembling that of the BMP-2 turret. The new version of Obiekt 688 with the 2K23 armament system was accepted for service in 1986 as the BMP-3. The decision was extremely controversial. The general feeling was that there were simply too many weapons on the vehicle. The Kurgan plant director in a 1994 interview recalled that "It met with more enemies than supporters." The BMP-3 was placed in operational trials since 1986, mainly in Uzbekistan and the Siberian Military District. A small number of BMP-3s became operational with Soviet units in Siberia in 1989. In the European USSR, only thirty-five were deployed by 1992, all of these with higher combined arms command schools or other officer training establishments.

In the European part of Russia in 1994, about sixty-nine were deployed. From the outset, the BMP-3 was intended for both the Ground Forces and the Soviet Naval Infantry. In the Soviet naval infantry, it was probably intended to replace both the aged PT-76 light tank and the BTR-60PB armored transporter.

BMP-3 Description

The most prominent feature of the BMP-3 is its revolutionary new armament system. The primary weapon is the 30mm 2A42 cannon while the secondary weapon is a 100mm low-velocity gun.[26] According to some Russians, the ideal main armament calibre for an IFV is 30 to 35mm cannon supported by the provision of missiles.[27]

As in the case of the 2A42 on the BMP-2, the new armament system has extremely high elevation (-6 to +60 degress) to permit the vehicle to engage aircraft targets, especially helicopters. The 2A70 100mm low-velocity weapon is not related to the standard D-10T tank gun (on the T-54/-55), and uses its own new ammunition. This

100mm round with 9M117 guided projectile.

BMP-3 infantry fighting vehicle.

BMP-3 100mm 3OF32 HE round.

includes the new 3UOF17 round (with the 3OF32 HE projectile), the 3BM25 APFSDS projectile, and the 3UBK10-3 round (with the 9M117 guided missile). The autoloader magazine holds twenty-two 3UOF17 high explosive rounds in a circular tray on the floor of the turret basket, with the projectile nose pointing towards the center. An additional eighteen HE rounds are stored separately in the hull in the rear left corner of the turret ring.[28]

The 100mm weapon has an effective rate of fire (with the high explosive ammunition) of ten rounds per minute. The combined armament is rigidly fixed and is electro-mechanically stabilized in two axis. The 100mm weapon has a very low initial muzzle velocity of only 250 meters per second (820 feet per second) a factor both of the short tube length and small propellant charge. The 2A70 100mm weapons could perhaps better be described as a

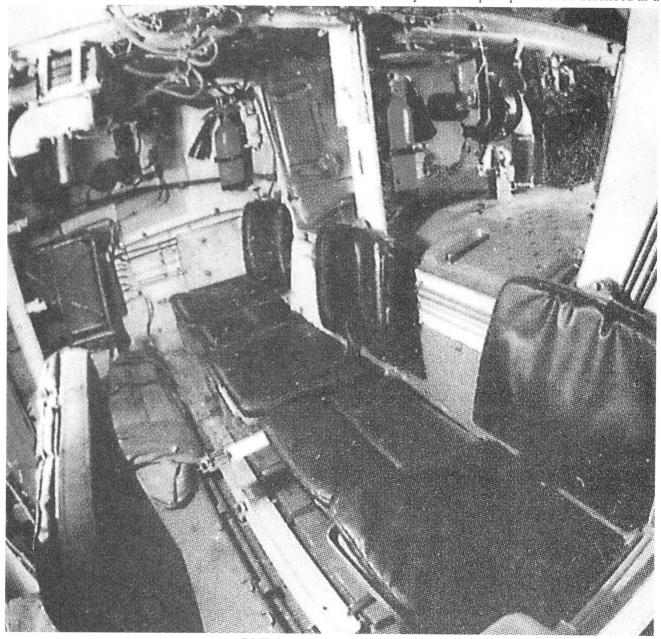

BMP-3 rear crew compartment.

howitzer, and its maximum effective range of four kilometers can probably be attained only with the barrel substantially elevated. The dissimilar ballistic performance of the 2A70 100mm weapon (250 meters per second) and 2A72 30mm autocannon (960 meters per second) means that the both weapons cannot be easily used in conjunction with one another when engaging targets except at point blank range.

The main anti-tank weapon of the 2K23 armament package is the 3UBK10-3 round with the associated 9M117 laser guided missile. The 9M117 is a laser-beam riding missile, essentially similar to the Bastion missile fired from the T-55AM2 but with a different propellant casing. The associated laser emitter for the fire control system is located above the main gun. This missile system is employed in lieu of the external missile launcher found on the BMP-1 and BMP-2. As in the case of other similar laser-guided missiles, the vehicle fire control system automatically elevates the gun tube when firing this round to minimize the amount of dust kicked up in front of the vehicle which can interfere with the laser command link. The 3UBK10-3 missile rounds are stowed in two locations. Three ready rounds are stowed vertically in the left rear of the turret basket behind the gunner; these can be be lifted into the breech by a special rammer separate from the normal autoloader. However, this rammer does not push the round entirely into the breech, and the gunner or commander must push it in the remainder of the way. Another five missiles are stowed in the forward hull compartment bringing the total to eight.

The 2A72 30mm autocannon is mounted coaxially with the 2A70 100mm gun and to its immediate right. The ammunition for the 30mm 2A72 autocannon is contained in two separate trays. The usual load consists of the primary magazine with 305 intermixed rounds of 3UOF8 high-explosive incendiary and 3UOR6 high-explosive-tracer, while the secondary magazine includes 195 rounds of 3UBR6 armor-piercing-incendiary ammunition. The 2A72 30mm gun was developed by the same Tula bureau as the 2A42 on the BMP-2. The guns share nearly identical ballistics. However, the 2A72 has a single fixed rate of fire (330 rounds per minute) to minimize the fume problem (the 2A42 had a selectable rate of 200, 300, or 550 rounds per minute).

The vehicle fire controls include the gunner's PPB-1 periscopic dual-axis stabilized day/night sight with an 8x day channel and 5.5x passive night vision channel. The gunner has a secondary 2.6x monocular periscopic TNPT-1 sight for aerial and ground targets, and the 1K13-2 laser rangefinder/laser guidance system mounted co-axially with the main gun and located over the 100mm 2A70 gun tube. Although not openly described, the system is probably a Nd:YAG type rather than CO_2. The 2K23 weapons system also includes a ballistic computer used when firing the conventional HE ammunition. The vehicle commander, who sits on the right side of the vehicle, can control the weapon system as well, although it is not clear if this control extends to the missile. The commander employs a monocular periscopic day sight for engaging ground and air targets similar to the 1PZ-3 on the BMP-2; he cannot sight the weapons at night. The commander also has a separate day/night sight located in his overhead traversable cupola which incorporates a passive image intensification night channel. The vehicle communications suite, consisting of a R-173 and R-173P radio transceiver, is located in the commander's station.

The basic armament is supplemented by the squad's weapons. There are five ball-mount ports for the AK-74 assault rifles, plus two mounts in the forward corners of the hull for PKT machine guns. These PKT machine guns are operated by the squad's two machine gunners who sit on either side of the driver. When they dismount, they clip the PKT's to a special locking system which enables the vehicle driver to aim and fire the weapons. There is a second set of squad automatic weapons which are taken by the machine gunners when dismounting. The squad is located around the turret. Aside from the two machine gunners already mentioned, there are five seats behind the turret, and two on either side of the turret. Normally, the squad uses three of the back seats and the two sides seats (plus the two forward machine gunners seats). The squad can be increased from seven to nine if necessary, though the vehicle becomes extremely cramped if this is done. Curiously enough, some Russian sources indicate that the complement of the BMP-3 can also be reduced from the normal ten, to nine: five members serving as crew (driver, commander, gunner, two hull machine gunners) and the remaining four serving as a dismountable assault party.[29]

The UTD-29 diesel engine is mounted transversely in the hull rear below the floor. This is a four-stroke, multi-fuel, liquid cooled engine. The engine has ten cylinders located in two banks at an angle of 144 degree, making it a relatively flat design. The transmission is a hydromechanical type with four forward gears (two reverse) and hydrostatic steering. Early production versions of the BMP-3 were apparently powered by the 420 horsepower UTD-29, but current production types use the 500 horsepower UTD-29M. The new engine gives it a higher power-to-weight ratio than the BMP-2, making the BMP-3 a very fast and agile vehicle.

The engine also powers a single stage water-jet system when the vehicle swims. To prepare for swimming, a bow deflector plate is extended forward, and an engine air intake snorkel is elevated on the right rear hull roof. Photos of the BMP-3 in Naval Infantry use show another snorkel mounted on the rear center of the turret roof, presumably for the crew. The hydrojet gives the BMP-3 superior performance in the water compared to the BMP-1/-2 series.

Assessing the BMP-3

Although very heavily armed compared to any other IFV, the layout design of the BMP-3 is poor by Western standards. The Russian rationale for this poor layout is a belief that an IFV must allow infantry to fire from within the vehicle and that there is little reason to dismount in combat.[30] The modern IFV allows the infantry to hold their ground and to protect main battle tanks from within the vehicle. This is one of the reasons why Russian designers of the BMP-3 have moved the infantrymen forward in the vehicle hull. The ideal IFV from the Russian perspective would mimic the infantry men (e.g., in the same way that the infantrymen points his firepower toward the enemy, the IFV should direct it's firepower forward therefore the placement of numerous firing ports at the front of the BMP-3).[31] To further facilitate this approach the vehicle's optics should be such that they impose as little restriction as possible on the vision of it's occupants.

Other apparent layout deficiencies are the relocation of the engine from the front to the rear, the designers have removed a natural source of secondary frontal protection for the crew compartment. If the BMP, Bradley, Warrior, or Marder are struck in the front of the hull by autocannon fire or rocket-propelled grenades, the engine and transmission will absorb much of the damage. While such penetrations will probably disable the vehicle, the crew and infantry squad have a good chance of surviving and fighting. On the BMP-3, such hits will penetrate directly into the fighting compartment, wounding or killing crew and squad members, with a high probability of hitting the large volume of ammunition stowed around the turret basket leading to catastrophic ammunition fires which will gut the vehicle. The designers have attempted to ameliorate this problem by storing much of the vehicle's diesel fuel in the bow in an attempt to offer a measure of passive protection.

The infantry squad seating is remarkably clumsy: the squad light machine gunners on either side of the driver in the hull front exit through two roof hatches, and the hull is a considerable height over the ground. This layout, however, may be due to the intention to have the infantry fight mounted rather than facilitating rapid dismount tactics. The five riflemen in the compartment immediately behind the turret and around the turret exit through the rear. No fewer than four doors have to be opened to exit on the rear roof of the hull and at the hull rear, and the squad exits along a narrow alleyway between the fuel cells and engine radiator. This alleyway wastes a significant amount of internal protected volume at the rear of the vehicle, and if the troops stow their equipment in the exit channel, it will be difficult to get out of the vehicle. In spite of the greater bulk of the BMP-3, the troop compartment is as cramped as the BMP-1/2. The crew will almost certainly be tempted to use the alleyway for stowage with the consequent exit problems. Had the BMP-3 followed the general layout of the BMP-1/-2, these awkward features would have been avoided and would have resulted in a less bulky and more combat-worthy design.

The BMP-3 design has been very controversial in Russian service. The hull is of all aluminum construction, and Russian military depots have very little experience working with this material, which is difficult to weld without specialized equipment. The 2A70/2A72 twin gun is impressive in firepower, but the assymetric recoil has led to cracked trunnions in service use. The BMP-3 originally suffered from significant mechanical teething problems during development, with an average of mechanical failures per 1,000 kilometers of 17.1 (1986); 4.6 (1988); 2.8 (1990); by now it may have reached maturity.

Armor protection on the BMP-3 is essentially similar to the BMP-2, except that the main components are of aluminum construction and an added layer of steel applique has been added to the turret and hull fronts to protect against autocannons such as the 25mm Bushmaster on the M-2 Bradley IFV (it is not clear if this is adequate to stop more advanced APFDS ammunition). Russian sources indicate that the layered front armor offer equivalent protection to fifty millimeters of rolled homogenous armor, compared to 30mm equivalent on the BMP-2.[32]

Night vision features of the BMP-3 are essentially the same as on the BMP-2, with a passive image intensification system. As was shown in the 1991 Gulf War, thermal imaging systems are a preferred choice since they allow combat at long ranges in degraded weather conditions and under typical, smoky battlefield conditions. The Russian electro-optics industry is still behind in mass production of thermal imaging systems, and a French thermal imaging sight was incorporated on the BMP-3s sold to the United Arab Emirates (UAE) in 1993.[33]

Is the BMP-3 a harbinger of the future (like the BMP-1), or a flawed curiosity? The BMP-3 is difficult to compare to contemporary infantry combat vehicles since its Russian designers have not made very clear what they were attempting to create with this unusual design. It resembles a hybrid design, with the heavy firepower of tanks and the light armor and infantry capacity of infantry combat vehicles. In terms of firepower, its 100mm 2A70 provides a high-explosive capability lacking in any other infantry vehicle, but the addition of yet another

BMP-3 in the United Arab Emirates (UAE).

30mm autocannon seems like a bit of overkill. It suffers from the usual Soviet/Russian shortcomings in night fighting capability due to its current lack of thermal imaging sights.

The tube-fired 9M117 missile is a mixed blessing. It gives the BMP-3 long-range anti-armor and anti-helicopter firepower, but at a high cost; the 9M117 missile is twice the cost of the earlier Konkurs. This missile is somewhat faster than wire-guided missiles, but its small diameter warhead and laser guidance link are potential weak points compared to the Konkurs (or TOW 2 on the Bradley).

The BMP-3's armor protection is on par with light armored vehicles of the 1960 and 1970s, and is not as well protected as the new M-2A2 Bradley. However, the BMP-3 weighs nearly 10,000 kilograms less and as a result is twenty-six percent cheaper than the M-2A2 Bradley.[34] Due to the BMP-3's heavy armament, laser-guided munitions, and attractive selling price, two Middle Eastern countries (Kuwait and UAE) have decided to purchase and operate the BMP-3. In the Russian context the BMP-3 is nearly twice as expensive as the BMP-2 in roll-away cost, and probably has higher operational expenses

Close up of thermal imager for UAE BMP-3.

The UAE version of the BMP-3 is different from the standard Russian version, with numerous fittings added to the turret.

Looking into the open rear infantry compartment of a UAE BMP-3.

This Russian BMP-3 was photographed in Omsk Russia during a demontration in 1997.

as well. This comes at a time when the Russian defense budget is shrinking rapidly.

The tactical rationale for the BMP-3 is to allow infantry to fight mounted. However, it is an unusually unbalanced design, with an obsession for firepower at the expense of armor protection and infantry features. Russian armored vehicle design has always been notoriously indifferent to crew conditions, and BMP-3 is certainly no worse than the BMD airborne vehicles.[35] An argument can certainly be made that there is little point to employing armor levels in between the low end (capable of resisting heavy machine gun fire (as on the BMP-3), and tank-level armor, since there are very few intermediate caliber weapons aside from IFV autocannon to defend against.

This has led some observers to wonder if the BMP-3 was intended to supplement, rather than replace, the BMP-2 as an infantry fire-support vehicle, much more like pre-war infantry support tanks. However, this is not entirely plausible since mounting the 2K23 armament package on a BMP-2 hull would give the same firepower advantages without the logistical nightmare posed by having BMP-2s and BMP-3s operating in the same regiment.

Other analysts have suggested the BMP-3 is the outcome of the Soviet Ground Force's acceptance of the bronegruppa concept. Bronegruppa (armored group) tactics are an evolution of BMP tactics, using the vehicles for missions without their infantry dismounts. When a company or battalion of motor rifle troops dismounts and digs in for defensive fighting, the unit commander can take some of his BMPs away to form a central bronegruppa reserve instead of leaving them dug in with their rifle squads. This gives the company or battalion commander a mobile reserve and counterattack force that can be held back until the enemy's objective is clear. Likewise, during offensive operations, the BMP-3 can serve as an independent force when the infantry is fighting dismounted. The BMP-3 would be well suited to this role since its weapon system can deal with a wide range of threats including tanks and attack helicopters. However, there is nothing in the bronegruppa concept that requires a vehicle like the BMP-3, since BMP-2s can perform the same mission. It seems unlikely that such an expensive vehicle was developed for so specialized a tactic.

Another explanation is that the BMP-3 represents a new conception of the modern battlefield where the role of tanks will be diminished due to the advent of so many dedicated anti-armor weapons. With so many anti-armor weapons present, it no longer makes much sense to endure the associated costs of heavy armor since any tank can be penetrated by any of the newer guided munitions. Given the lethality of modern anti-armor weapons, no conceivable protective system can resist them all. The situation could get worse if precision guided munitions like BAT, SADARM, and others finally materialize. Under such a set of circumstances, the alternative might be to step out of the defensive armor technology race. Instead, tank-like firepower can be incorporated into the more numerous IFVs, allowing them to perform both the carrier role of APCs and the firepower role of tanks. This has the secondary advantage of presenting the enemy with double or triple the number of targets, not all of which can be targeted. This is not an entirely novel idea, as Marshal P. Petain in World War 1 had a similar view of early tanks, preferring great masses of cheap, lightly armored FT tanks over small numbers of the more thickly armored (but still vulnerable) Schnieders and St. Chamond tanks, a tactic labeled the "bee swarm." In the 1920s and 1930s, this was echoed by various light AFV proponents, especially in Britain, who saw hordes of cheap tankettes and machine gun carriers as a more cost-effective alternatives to the expensive and cumbersome tanks of the day. One of the problems of this argument is that the Russian tank designers and tank force advocates do not seem to have noticed, and continue to plug away at more advanced protection systems and better-armed tanks.[36]

Some Western observers see the BMP-3 as simply a Russian attempt to skirt around the CFE restrictions by deploying a light tank under the guise of an infantry vehicle. The BMP-3 was jokingly called the TAV (Treaty Avoidance Vehicle) for a time in some NATO circles. The weight of the BMP-3 exceeds the IFV threshold, as does the main weaponry (100mm gun). This argument is not entirely persuasive, since the BMP-3 has been under development much longer than the CFE negotiations. However, this explanation might help account for the switch from the 30mm cannon of the original Obiekt 688 design to the current armament package. Russian planners may have felt that with strict limitations befalling the tank force, it would be best to add as much firepower as possible on the less regulated AFVs. It has noted by some Russian officers that the BMP-3 was a response to Gorbachev's defensive doctrine, first articulated in the late 1980s. With the prospects of curtailed tank forces, they were intent to get as much firepower as possible on whatever vehicle seemed the most suitable, hence the switch from the 30mm gun of the 1981 Obiekt 688 to the 100mm/30mm combination of the series production BMP-3. Infantry vehicles have seen a steady escalation in firepower from the 7.62mm machine gun of the BTR-152 of 1947 through to the 100mm gun of the BMP-3 of 1987.

A final interpretation may be that the BMP-3 represents a case where paper characteristics attractive to the design engineers were foisted on the Russian motor rifle force without adequate consideration of their tactical implications.[37] This has been a frequent complaint among line officers in several of the Russian services in recent years as glasnost has given them a voice. This is supported by a recent Russian account which strongly suggests that the armament package was advocated by the designers, but met considerable resistance from the users.[38]

Several Russian officers have stated that a BMP follow-on is already in development, and that the BMP-3 will not replace the numerous BMP-1/-2s already in Russian service. Russian officers have strongly hinted that they expect a new infantry fighting vehicle superior to the BMP-3 to be in service by the end of the decade. This new IFV will likely be a replacement for the BMP-2/-3 with similar technical characteristics and function.

In addition to a mass-produced BMP-2/-3 follow-on, a second development will likely focus on a heavy BMP optimized for fighting in built-up urban areas. The fighting in the Chechen capital of Grozny has amply illustrated that current BMP designs are not well-suited for fighting in urban and mountainous terrain. And according to some Russian observers, a new heavy urban fighting BMP is needed and should be utilized in simultaneous service with the BMP-3.[39] This new heavy BMP should have the same armor protection as a main battle tank. The Russians have built two prototypes, the BTR-T, based on the T-55 and another on the T-80.

The BMP-3 Variants

The BMP-3 has become the Russian Army's universal light armored chassis and is being used for a wide range of applications. It has also been one of the most successful Russian export items in recent years, having been sold to the United Arab Emirates (UAE), Kuwait, South Korea, and Cyprus. Indeed, there are more BMP-3s in service overseas than in the Russian Army.

BMP-3K Command Vehicle

The BMP-3K command vehicle is basically identical to the standard BMP-3 but is intended to provide improve communication and navigation features for a company or battalion commander. The crew of this vehicle is the usual three, but the vehicle is only intended to carry three officers rather than the usual infantry section. The BMP-3K has several added features. It carries the TNA-4-6 navigation aid which gives the commander current coordinates, bearing grid angle, and destination directional angle superimposed on a topographic map. Once initiated, the system is accurate for about seven hours. The vehicles is fitted with two radios including a R-173 transceiver. A one kilowatt AB1-P28 five-volt APU is fitted to provide electrical power to the vehicle sensors without operating the main engine.

BRM-3 Rys Reconnaissance Vehicle

The BRM-3 Rys (Lynx) is a reconnaissance vehicle, intended to replace earlier types such as the PT-76, as well as specialized reconnaissance vehicles such as the BRM-1 reconnaissance vehicle and PRP-3 and PRP-4 artillery reconnaissance vehicles. It uses the BMP-3 universal chassis. It is manufactured at the Rubtsovsk Machine Building Plant in the Altai region of Russia, one of the two main BMP production facilities.

The BRM-3 is very spacious by Russian standards. The navigator-computer operator sits in the right corner of the hull with a set of map-displays and navigation equipment. The turret is more spacious than in the BMP-3 due to the lack of the usual 100mm gun, and contains a more extensive array of optical and sensor displays as is befitting a vehicle with this role.

The crew is six: reconnaissance group commander, vehicle commander, gunner, driver-mechanic, navigator/computer-operator, and radio operator. The BRM-3 is fitted with a turret similar, but not identical, in size to the normal BMP-3, but armed only with a 2A72 30mm cannon.

Their are four main surveillance devices on the vehicle, housed in the turret including two armored boxes on either side of the turret. The RL133-1 is a battlefield surveillance radar stowed in a bustle at the rear of the turret, that is elevated through a hatch much as on the earlier BRM-1. This has a detection range of 8,000 to 10,000 meters against a tank, and 3,000 to 4,000 meters against personnel. Average range error is twenty-five meters and 00 to 10 mils (thirty-six feet) in azimuth. The 1D14 is a periscopic laser rangefinder integrated with a daylight sight with magnifications of 7.3x and 18x. It has an effective range of 10,000 meters against tanks and up to twenty-five kilometers against major terrain features.

Target coordination range errors are about five meters in range and 00-03 mils (10.8 feet) in azimuth. The 1PN61

BRM-3 (Obiekt 501) armored reconnaissance vehicle.

The BMP-3 Rys was displayed in Abu Dhabi at IDEX '97. The RL-133-1 battlefield surveillance radar can be seen in the raised position

BRM Rys Technical Characteristics

Combat weight (metric tons)	19 metric tons
Crew	6
Engine	UTD-29 muti-fuel diesel
Engine power (kilowatts/horsepower)	367 / 500
Transmission	hydro-mechanical
Maximum road speed (kilometers per hour)	70 forward / 20 reverse
Maximum swimming speed (kilometers per hour)	10
Fuel endurance on highway (kilometers)	600
Ground pressure (kilograms per centimeters squared)	0.6
Main armament	2A72 30mm cannon
Maximum effective range (kilometers)	4 (HE) / 2.5 (AP)
Ammunition (rounds)	600
Rate of fire (rounds per minute)	300
Co-axial weapon	7.62mm PKT (2,000 rounds stowed)

is an image intensification night vision sensor with a laser illuminator. It has seven times optics and can spot tank-sized targets up to 1,500 meters under passive viewing conditions, and 3,000 meters under active illumination. The 1PN71 is a thermal night vision sight operating in the 8 to 14 micron band. The electro-optics have 3.7x and 11.5x power. Maximum range against a tank-sized target is 3,000 meters. Vehicle navigation and orientation equipment includes a 1V520 computer, one T219 coordinator, TNA-4-6 map navigation set, and a 1G50 gyrocompass.

The BRM-3 Rys carries two communication sets and an internal intercom system. The R-163-50U is a UHF transceiver in the 30,000 to 79,999 kHz band with ten pre-set frequencies; it has an effective range of twenty to forty kilometers using a rod or telescopic antenna. The R-162-50K is a HF transceiver in the 2,000 to 29,999 kHz band with sixteen preset frequencies; it has an effective range of 100 to 250 kilometers using a rod/telescoping antenna. These are supported by an internal intercom system and a R-012M call unit.

The BRM-3 Rys is armed with a 30mm 2A72 autocannon and 7.62mm PKT co-axial machine gun stabilized in two planes using the 2E52-1 stabilizer. The 2A72 has a double belt feed for HE-F and AP-T ammunition. There are 600 30mm rounds and 2,000 7.62mm rounds stowed in the vehicle. Fire control is provided by a BPK2-42 periscopic day/night sight; PPB-2 periscopic monocular sight, and a BETs-088 control system. Crew self-defense is provided by six assault rifles, a 9P135M firing post for the Fagot/Konkurs (AT-4/AT-5) ATGM, and fifteen grenades. There are two firing ports in the hull front sides for use of the rifles from under armor.

2S31 Vena 120mm Self-Propelled Gun-Mortar

The 2S31 Vena 120mm self-propelled gun/mortar is based on the BMP-3 chassis. More details of this vehicle will be found in the artillery chapter.

BREM-L Beglianka

The BREM-L Beglianka (Fugitive) armored recovery vehicle was designed at Kurganmashzavod for the evacuation of IFV and other light armored vehicles, while under the hostile fire. It can also be used to provide repair and maintenance under field conditions. BREM-L provides the extraction of bogged down vehicles; self-extraction in the case of bogging down; the towing of vehicles with and without drivers under different terrain conditions and afloat; carrying out repairs (including electric welding/cutting operations with aluminum alloys and steel) in preparing vehicles for evacuation; carrying out hoisting operations to lift a vehicle partially during evacuation and repair; hoisting for the replacement of turrets and power plants, replacing its own power plant; moving with a weight on hook (including turrets) within the repair area; the transportation of vehicles spare parts (including power plants); carrying out the earth-moving

BMP-3 recovery vehicle.

The BREM-L recovery vehicle based on the BMP-3 design shown on display in Abu Dhabi in 1997.

BREM-L Technical Characteristics

Loaded weight	18.7+2 percent
Engine	four-stroke UTD-29 diesel
Engine power (kilowatt/horsepower)	367 / 500
Road clearance (millimeters)	450
Specific ground pressure	59 kPa
Length (millimeters)	7,640
Width (millimeters)	2,710
Height (millimeters)	3,280
Maximum road speed (kilometers per hour)	70
Maximum water speed (kilometers per hour)	9
Average road speed (kilometers per hour)	52
Highway cruising range (kilometers)	600
Crew	3 + 2 reserve
Armament	PKT 7.62mm machine gun
Communication	FM radio (vehicle intercom)

operations in arranging the repair area; and removing soil in preparing a vehicle for extraction. The BREM-L is fitted with a special fully-rotatable five-ton crane; a winch with twenty metric ton capacity (forty metric ton with a pulley block); arc welding equipment; plow/dozer blade; cargo bed with attachments for carrying BMP power blocks of IFV; and rigid coupling and bars for towing.

BMP-3 Multi-Purpose Chassis

The BMP-3 Multi-purpose Chassis is basically a turret-less version of the BMP-3 infantry fighting vehicle intended to serve as the basis for anti-tank missile launchers, air defense systems, self-propelled guns and mortars, command and control vehicles, engineer vehicles and other special applications. It differs from the BMP-3 in the engine area configuration, since the channels for the infantry to exit are no longer required. It weighs 12.1 metric tons in its basic configuration.

BMP-3 9P157 Krizantema and 9P162 Kornet Anti-Tank Destroyers

Russia has built two ATGM carrying anti-tank vehicles based on the BMP-3 chassis. Further details are provided in the ATGM chapter.

BMP-3 Anti-Aircraft Artillery Vehicle

The BMP-3 Multi-Purpose Chassis vehicle was originally brought to IDEX-95 to be mated with a Western European anti-aircraft gun turret but the deal collapsed at the last possible moment. Both French and German companies have offered a similar anti-aircraft gun/missile turret for the BMP-3.[40] These configurations are being offered to satisfy a long-standing United Arab Emirates (UAE), requirement to have an anti-aircraft gun turret mount placed on a BMP-3 hull.

BMP-3 Training Vehicle

A driver's training vehicle for the BMP-3 was revealed at IDEX-95. This is basically a BMP-3 Multi-Purpose Chassis with a training cabin installed in place of the turret. The instructor can sit next to the driver for instruction, or in a seat inside the cabin. The vehicle can hold up to twelve trainees, including four in the cabin. A separate turret trainer and BMP-3 materiel trainer were also mentioned in a brochure.

Future Trends In Infantry Vehicle Development

The situation with infantry combat vehicles is no less complicated than with tanks. In contrast to the Russian tank dilemma, the Soviet Army was in the process of adopting a new infantry combat vehicle, the BMP-3, as the Soviet Union was collapsing. Manufactured in Kurgan, the BMP-3 represents another radical approach to infantry vehicle design. In contrast to the US and Western Europe where the trend has been to improve armor protection on IFVs, the Russian approach has been to accent firepower. The BMP-3 is armed with a combination of a low-pressure 100mm gun and a 30mm autocannon. This configuration was pushed by the industry, not by army requirements, and there is some disenchantment with the results. The combination gun is difficult to maintain, and troop provisions on the vehicle are awkward. To further muddy the issue, the Russian Army was very troubled by the performance of BMPs in the recent Chechen fighting. Although the BMP-3 did not figure prominently in the fighting, it shares nearly the same level of armored protection as the BMP-2. These

BMP-3/BRM-3 Technical Data*

	BMP-3	BRM-3
Crew	3	6
Troop complement	7 (+2)	-
Combat weight	18.7 to 19.0 metric tons	18.7 to 19.0 metric tons
Overall length	7,140 millimeters	7,000 millimeters
Width	3,230 millimeters	3,150 millimeters
Height (turret roof)	2,300 millimeters	2,370 millimeters
Ground clearance	190 to 510 (average 450) millimeters	190 to 510 (avg. 450) millimeters
Ground pressure	0.6 kilograms/cm^2	0.6 kilograms/cm^2
Maximum road range	600 kilometers	600 kilometers
Maximum road speed	70 kilometers per hour	70 kilometers per hour
Maximum swimming speed	10 kilometers per hour	10 kilometers per hour
Power-to-weight ratio	25 horsepower/ton (18.7 kilowatt/ton)	25 horsepower/ton
Gradient	60 percent, 30 percent side slope	60 percent, 30 percent side slope
Obstacle	0.8 meters vertical, 2.2 meters trench	0.8 meters vertical
Engine	UTD-29M 500 hp (368 kilowatt), 770 hp/m^3	UTD-29M
Transmission	hydromechanical 4 forward, 2 reverse gears	hydromechanical
Main armament	100mm 2A70 rifled gun	30mm 2A72
Gun elevation	-6/+60°	-6/+60°
Rate of fire	10 rounds per minute	330 rounds per minute
Fire control	ballistic computer, LRF, environmental sensor	
Initial muzzle velocity	250 meters per second	960 meters per second
Maximum effective range	4,000 meters	
Ammunition stowage	40 (22 rounds HE-Frag in autoloader)	
Anti-tank missile	3UBK10-3 100mm with 9M117 ATGM	none
Effective range	100 to 4,000 meters	n/a
ATGM stowage	8	none
Initial muzzle velocity	960 meters per second	n/a
Secondary armament	2A72 30mm autocannon	n/a
Initial muzzle velocity	960 meters per second	n/a
Rate of fire	330 rounds per minute	n/a
Ammunition stowage	305 HE, 195 AP	
Co-axial machine gun	7.62mm 6P7 PKT	7.62mm 6P7 PKT
Ammunition stowage	6,000	
Unit price	$800,000 (1992 export price)	

n/a=data not applicable

vehicles suffered heavy losses at the hands of the Chechen guerrillas, and there is some question about the armor levels needed in future infantry fighting vehicles.

Regardless of Russian conceptions of an ideal future infantry vehicle, the armored infantry force is faced with the same production and budget dilemmas as the tank force. At the moment, three families of infantry vehicles are in production. The two main BMP plants at Kurgan and Rubtsovsk are still tooled up to manufacture both the BMP-2 and BMP-3 families. The BMP-2 has remained in production mainly due to cost factors. It is only half as expensive as the BMP-3 with an export price of $400,000 versus $800,000. The third vehicle in production, and least expensive of the trio, is the wheeled BTR-80 manufactured at the Arzamas automotive plant. A new version, the BTR-80A with a 30mm gun turret, was unveiled in 1994 and has been in production alongside the normal BTR-80 version with the 14.5mm machine gun turret. A prototype of a larger and heavier wheeled APC, the BTR-90, was displayed in 1995 but has not been accepted for army service.

As is typical in many armies, Russian infantry modernization is likely to receive less priority than tank modernization. There are only twenty-eight BMP-3s in service west of the Urals, and most production in recent years has been for export. The Russian Army is unlikely to be able to meets its light AFV requirements with BMP-3, or even with the less expensive BMP-2 and BTR-80A. As a result, in the near term, the Russian Army will probably continue to buy a mix of the less expensive vehicles along with modest numbers of BMP-3.

Further supply problems of the ubiquitous MT-LB and its derivatives such as the 2S1 Gvozdika 122mm self-propelled howitzer and the 1V12 family of armored com-

BMP-1/-2/-3 Comparative Technical Data

	BMP-1	BMP-2	BMP-3
Crew	2	2	3
Infantry squad	9	7	7
Combat weight (metric tons)	12.6	14	18.7
Power to weight ratio (horsepower/ton)	23.8	21.8	25.0
Ground pressure (kilogram/cm^2)	0.57	0.63	0.6
Length (meters)	6.74	6.74	7.14
Width (meters)	2.94	3.15	3.15
Height (meters)	1.92	2.25	2.3
Ground clearance (meters)	0.39	0.42	0.19 to 0.51
Maximum road speed (kilometers/hour)	80	65	70
Maximum range (kilometers)	600	600	600
Gradient percent)	35	35	60
Vertical obstacle (meters)	0.7	0.7	0.8
Trench (meters)	2.5	2.5	2.2
Engine type	UTD-20	UTD-20	UTD-29
Horsepower	300	300	500
Fuel consumption (liters/kilometers)	0.90	0.92	
Main armament type	2A28 Grom	2A42	2A70
Main gun type	smoothbore	rifle	rifle
Main gun caliber	73	30	100
Gun Stabilization	no	2E36-1	yes
Rate of fire	7-8 rpm	300 or 500 rpm	10 rpm
Gun elevation	-4 to +33°	-5 to +74°	-6 to +60°
Gunner's sight	1PN22M1	BPK-1-42	
Commander's day/night sight	TKN-3B	TKN-3B, 1PZ-3	
Secondary weapon	none	none	2A72 30mm
Co-axial machine gun	PKT 7.62mm	PKT 7.62mm	PKT 7.62mm
Main gun ammunition	40	300	40
Machine gun ammunition	2,000	2,000	6,000
Missile	9M14M Malyutka	9M111 Fagot	9M117
Missile stowage	4	4	8
Missile launcher	9S415	9Sh119M1	2A70
Driver's day sight	TNPO-170A	TNPO-170A	
Driver's night sight	TVNE-1	TVNE-1PA	
Smoke mortar	-	System 902V	System 902V
Frontal turret armor (millimeters)	26 to 33	23 to 33	26*
Side turret armor (millimeters)	19	19	
Hull armor (maximum- millimeters)	19	19	19*
Hull side armor (millimeters)	16-18	16-18	
Hull rear armor (millimeters)	16	16	
Radio	R-123	R-123M	R-173
Unit cost (1992, export)	n/a	$400,000	$800,000

steel armor equivalent

mand vehicles arose due to Ukrainian independence and the corresponding loss of the MT-LB's main production plant in Kharkov. The Russian Army response has been to put out a requirement for a new "universal" chassis to replace the MT-LB in these roles. In the short-term, the Kurgan plant has promoted the BMP-3 chassis. This is being used with several new vehicles including the BRM-3 Rys reconnaissance vehicle to replace the older BRM-1 on the BMP-1 chassis, the 9P157 Krizantema and 9P162 Kornet missile-armed tank destroyers, and the 2S31 Vena 120mm self-propelled weapon, which may replace the 2S1 Gvozdika. However, the BMP-3 chassis is considerably more expensive than the MT-LB, and it remains to be seen whether it is an affordable solution given the shortage of funding.

A potential IFV proposal being considered by Russian designers is a program referred to as the BMP-4, also known as the BTR-T. Some of the aspects of this new vehicle were highlighted in a 1994 interview with the director of the Kurgan plant, Mikhail A. Zakharov when he stated that "We have to seriously think about ergonomics, provide the new machine with modern optics, thermal imaging, and air conditioning." This proposed BMP-3 upgrade or new design would include uparmoring the vehicle with external appliqué armor (similar to the M-2A2 Bradley) and arming the vehicle with only a 30mm (or larger) cannon. This would not be altogether surprising, as the recent Chechen war has led to a call for heavier armor on IFVs. Also, recent Russian accounts of IFV development have characterized the 100mm gun on the BMP-3 as its "secondary" armament, and that the 30mm gun is considered its prime weapon. This may mean that a future BMP will go back to more traditional armament, if for no other reason than cost-cutting, and lean towards heavier protection instead.

In light of experiences in Chechnya and Afghanistan, the future Russian approach to IFV design may likely result in the creation of a hybrid tank/infantry assault vehicle. The Kirov Plant in St. Petersburg has already demonstrated a heavy armored transporter based on the T-80 tank chassis, and this may serve as the basis for a future generation heavy assault vehicle. In addition, other Russian designers have called for the conversion of older MBTs into IFVs (Israel has converted a large number of their captured T-55s into IFVs due to the vulnerability of M-113s to man-portable anti-armor weapons). Such a vehicle, the BTR-T based on the T-55 chassis was unveiled in 1997. What remains to be seen is whether the Russian Army will be able to afford these new design modifications or if these are efforts aimed solely for the export market.

The T-55-based BTR-T

4
ARMORED AIRBORNE

Armored Airborne

The ultimate dream of Russia's earliest airborne visionaries was to create an airlanded combat force no different than conventional army units, meaning a motorized and mechanized airborne force. This dream proved to be impossible for the initial four decades of the Air Assault Force's (VDV) existence due to the limits of Soviet airlift.[1] But by the late 1960s, with the expansion of the VTA airlift force and the development of new transport aircraft, the dream of armored airborne forces came within reach with the development of the BMD armored airborne combat vehicle.

The desire for mechanized airborne forces in the 1960s was based on the Soviet Union's "revolution in military affairs," the post-Stalin renaissance in military thinking in the late 1950s. There was concern about the viability of the airborne force on the nuclear battlefield. This paralleled the tactical developments in the Ground Forces which led to the BTR-60PB and BMP-1. Unprotected paratroopers would soon fall victim to the debilitating effects of residual radiation if landing in areas previously hit by tactical nuclear weapons. Armored vehicles offered a solution to this problem. The steel armor of the vehicle, combined with plasticized lead pellet liners, offered a significant means of protection against this type of radiation. The air within the armored vehicle could be filtered to minimize the risk of inhaling fallout particles. The airborne troops within the vehicles could disembark to carry out traditional dismounted infantry tasks (at some risk of contamination) and return to the relative safety of the vehicle. These factors led to the development of the BMD series of airborne transporters in the mid-1960s.

The main impediment to mechanizing the airborne had been the lack of sufficient airlift capabilities. Until the early 1960s, the VTA airlift force was only capable of lifting conventional paratroop forces. The innovation that made airborne mechanization possible was the Ilyushin Il-76 transport aircraft, the first Soviet jet military transport. Development of the Ilyushin jet began in the mid-1960s as a Soviet counterpart to the US Air Force's Lockheed C-141 Starlifter. The requirements for the new aircraft were simple: double the range, and double the payload over the existing An-12 turbo-prop transport. The first prototype of the new transport began flying slightly behind schedule in 1971, and was introduced into service in 1974. It has formed the backbone of the VTA to this day. From 1974 when the Il-76 was introduced, to 1990, the airlift capacity of the VTA climbed from 16,800 to 28,750 metric tons, giving the force the lift needed to deliver at least a portion of the VDV as a mechanized force.

Early Soviet Armored Airborne Vehicle: ASU-57

During the Second World War, the Red Army's VDV Airborne Force did not have an assault glider with great enough capacity to lift armored vehicles. An attempt was made in 1942 to develop a biplane glider to carry the T-60 light tank, but this proved to be a failure.

After the war, the VDV Airborne Force sponsored the development of assault gliders and high-capacity parachutes which enabled the delivery of heavy loads up to about six tons. This made a lightly armored airborne vehicle practical for the first time. Due to the limitations in airlift at the time, no consideration was given to the development of an airborne infantry vehicle. The highest priority was for an airborne assault gun which could provide anti-tank defense to airborne units as well as a measure of direct fire support. This was the only class of airborne armored vehicle in Soviet service until the expansion in airlift in the early 1970s made widespread airborne mechanization possible.

Around 1947, two design bureaus were given the task of developing light-weight airborne assault guns, the Kravtsev design team in Moscow and the Astrov SKB-40 in Mytishchi (in the Moscow suburbs). The requirement called for a light armored vehicle that could mount either a 57mm anti-tank gun and/or a 76mm divisional gun. This category of vehicles was called ASU (Aviadesantnaya samokhodnaya ustanovka: Airborne self-propelled unit). The program requirement called for the use of inexpensive automotive components where possible since it was suspected that the attrition rate of the vehicles during airdrops would be high.

The Astrov design bureau began work on an air-transportable 76mm assault gun called the ASU-76. This vehicle was armed with a new light-weight D-56T 76mm

ASU-76 airborne assault gun.

K-73 amphibious assault gun.

ASU-57 airborne tank assault gun.

gun, and the vehicle could carry thirty rounds of ammunition (this gun was later used to arm the PT-76 amphibious reconnaissance tank). The suspension components came from Astrov's wartime light tank designs such as the SU-76 assault gun and T-70 light tank, and the propulsion was the common GAZ-51E automobile engine. The armor on most of the vehicle was only three millimeters thick, though the gun shield was thirteen millimeters; there was no overhead armor. The first prototype was completed in 1949. The ASU-76 tipped the scales at 5.8 metric tons which proved to be too heavy for existing airlift means, and the project was shelved.

In the meantime, Kravtsev's design team in Moscow developed a very similar vehicle, the K-73, armed with the new Ch-51 57mm anti-tank gun, developed by a team under E. V. Charnko. This vehicle was even more thinly armored than the ASU-76, and so weighed in at a mere 3.4 tons. It was also designed to be amphibious, and had a small screw propeller than was carried on a folding shaft at the rear of the vehicle. The design was not successful, and in 1949, Astrov's team was instructed to pare down the weight of the ASU-76 and arm it with a 57mm Ch-51 anti-tank gun instead. This emerged in 1951 as the ASU-57. The automotive aspects of the design were undertaken in parallel with work on Obiekt 561, a light artillery tractor which would later emerge as the AT-P. The ASU-57, unlike the K-73, had no amphibious provisions. It was a very small, very lightly armored vehicle with an open rear fighting compartment and a casemate mounted anti-tank gun with limited traverse of only sixteen degrees. The long Ch-51M 57mm gun was selected over the medium-length 76mm gun on the ASU-76 since it offered better anti-tank performance. It could penetrate 100 millimeters of armor at 1,000 meters, while the 76.2mm gun could penetrate only fifty-eight millimeters. On the other hand, the Ch-51 57mm gun fired a much smaller high explosive round; but its role was primarily anti-tank defense, not fire support.

The ASU-57 proved more successful than the Kravtsev K-73 and it was accepted for series production in 1951. It was first publicly shown in 1957. The ASU-57 was very lightly armored, and lacked overhead armor protection. The ASU-57's small size was mandated by the lack of heavy transport aircraft at the time, and it was initially carried in special P-90 aluminum containers, one under each wing of a Tu-4 (B-29 copy) bomber. Later, as transport aircraft like the An-12 were developed, a special parachute pallet was developed which could be dropped out the rear door of the aircraft. There was a considerable amount of small design changes through the ASU-57's production run, which lasted until

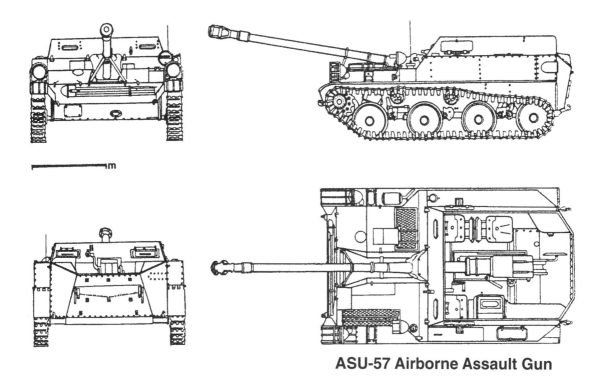

ASU-57 Airborne Assault Gun

1962. The most noticeable was the substitution of the improved Ch-51M gun which had a more conventional muzzle break than the multi-slatted type used on the initial Ch-51 production vehicles. Each airborne regiment had nine ASU-57 in an anti-tank company, while airborne divisions had an anti-tank battery with eighteen further ASU-57s. The ASU-57 was used by the Soviet VDV and was also supplied to the Egyptian Army in small numbers. In 1954, with ASU-57 production well underway, the issue of amphibious capability for the vehicle was revisited. The Astrov design bureau developed an amphibious version, the ASU-57P (Plaviushchiy: swimming). However, to keep the weight down to 3.4 tons, it was necessary to limit the armor even further, and in the end, the type was not accepted for service. It is curious that the Soviets did not develop a light armored vehicle armed with recoilless rifles instead of conventional guns. This was a popular approach at the time in several armies such as the US M-50 Ontos and the Japanese Type 60. The Soviets had actually experimented with such weapons in the late 1930s, but if any attempts were made in the 1950s, they have remained secret.

The light weight of the ASU-57 placed distinct limits on its combat capabilities. The 57mm gun was not adequate to duel with a contemporary main battle tank, except from ambush. New armor piercing ammunition made it capable of penetrating the frontal armor of older NATO tanks like the M-4 Sherman, which were still common in the 1950s. Newer designs like the American M-46 Pershing or the British Centurion could be penetrated from the sides or rear, but the 57mm gun was inadequate to deal with the thicker frontal armor of these new tanks. In spite of these limitations, the ASU-57 was a substantial increase in paratrooper firepower and mobility.

Early Soviet Armored Airborne Vehicle: ASU-85

During the last two years of World War 2, the SU-76 light assault gun had become the primary organic armored support weapon for Soviet rifle divisions. After the war, there were a variety of attempts to replace it with a more modern vehicle including the peculiar SU-100 which placed a fully turreted 100mm gun on the rear of a light armored vehicle chassis. In 1951, the Astrov design bureau, which has developed the original SU-76, was assigned to develop a modern contemporary using components from the new PT-76 light amphibious reconnaissance tank. Designated the SU-85, the new assault gun was armed with the Petrov bureau's D-70 85mm gun. Unlike the SU-76 or ASU-57, the fighting compartment was at the center of the vehicle rather than the rear. The armor was more substantial than on previous light assault guns, forty-five millimeters thick and angled at forty-five degrees. The program was delayed when V.A. Malyshev, the deputy to the Council of Ministers, required the bureau to examine the use of a new horizontal diesel engine in place of the existing V-6 which had been derived from the traditional V-12 used in the T-34 and subsequent Soviet medium tanks. The prototypes of the SU-85 were not completed until 1956, and were powered by the YaMZ-206V six cylinder diesel engine instead of the V-6 which had powered the PT-76. By the time the pro-

ASU-85 airborne assault gun.

totypes were ready, the configuration was already out of date. Like earlier assault guns, the fighting compartment was open-topped. In view of the Soviet Army's new concerns about survivability on the nuclear battlefield, this was completely unacceptable. As a result, the SU-85 underwent a redesign with an armored roof added.[2] This emerged around 1959 and remained in production from 1966 through 1967 in modest numbers. It was accepted for use by the Soviet Army, but by this time, the 85mm gun was outdated when facing newer NATO tanks such as the US Army M-60.

In the end, the VDV became the primary user of this vehicle, renaming it the ASU-85. By the late 1950s, the availability of newer transport aircraft like the AN-12 made it possible to carry larger, better armored, and better armed vehicles than the puny ASU-57 if airlanded rather than air-dropped. The ASU-85 offered significantly better firepower than the ASU-57, so the ASU-85 equipped special assault gun battalions in each of the eight airborne divisions, with thirty-one ASU-85 per battalion. The utility of the ASU-85 was somewhat circumscribed by its size, since it could only be used to support airborne operations if an airstrip could be seized and held to permit the vehicles to be flown in. The ASU-85 was also supplied to the Polish LWP where it was used by the 7th Pomeranian Airborne Division. The ASU-85 chassis design served as the basis for Astrov's later designs including the ZSU-23-4 Shilka (izd. 575) and the 2P25 launcher vehicle for the 2K12 Kub (SA-6 Gainful) air defense missile system (izd. 578).

The BMD Airborne Assault Vehicle

In the wake of the humiliation of the Soviet Union in the Cuban missile crisis in 1963, the Soviet Army was instructed by the Central Committee of the Communist Party to place more emphasis on the means to project power outside of traditional regions of Soviet influence. A major off-shoot of this effort was a rejuvenation of the VDV as a strategic rapid deployment force. Soviet studies of airborne operations during World War 2 concluded that the inability of lightly armed paratroop forces to deal with contemporary mechanized infantry or tank forces severely limited their utility except for peripheral operations. As a result of these analyses, the Soviet Armed Forces decided to begin to gradually mechanize the VDV.

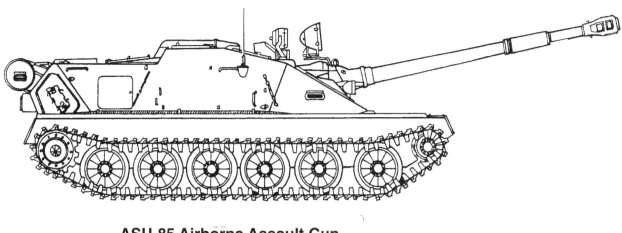

ASU-85 Airborne Assault Gun

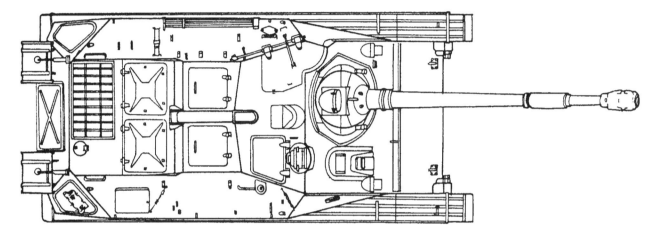

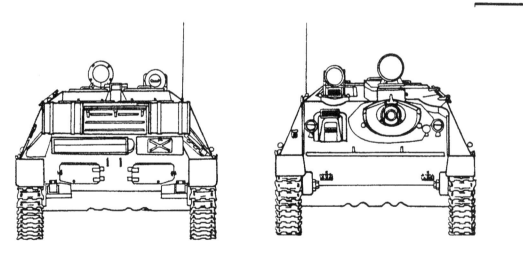

The aim was to offer the new air assault divisions a greater degree of tactical mobility on the ground, better protection against artillery and anti-personnel weapons, and to increase their firepower, especially anti-armor firepower. This required two technical innovations, a new heavy lift transport aircraft (the Ilyushin Il-76 Candid) and an armored airborne vehicle to mechanize the VDV Airborne Forces.

The new vehicle was designated the Boevaya mashina desantnaya (BMD): Airborne combat vehicle. The BMD was designed as an airborne counterpart of the army's best armored infantry vehicle, the BMP. Both vehicles shared the same armament system: a single-man turret with a 73mm Grom low-pressure gun and a launcher for the 9M14M Malyutka anti-tank missile. There the similarity ends. Because the BMD was designed for airborne drop, its size and weight had to be considerably trimmed. The BMP weighed over thirteen tons combat loaded, the BMD requirement called for a vehicle under seven tons, about half the weight.

However, the VDV could not afford to build twice as many BMDs and BMPs, so the troop requirement for the

BMD-1 (Obiekt 915) airborne infantry fighting vehicle.

BMD was not cut in half. The BMP carried eleven troops (including the crew), and the requirement for the BMD insisted on eight.

It is not clear from Russian accounts if alternative designs were considered, such as wheeled vehicles, and nothing has emerged to date on alternative tracked competitors. The design of the BMD was entrusted to the Volgograd Tractor Plant design bureau headed by Ivan V. Gavalov. Gavalov's design team had been an unsuccessful competitor in the Soviet Army's BMP design, losing to the Isakov design bureau from Chelybinsk. But the BMP competition had given Gavalov important experience in designing infantry vehicles. The BMD prototype, codenamed Obiekt 915, was basically a shrunken version of Gavalov's Obiekt 914 BMP design. The one concession to the airlift requirement that was made was to reduce the armor required on the BMD. The BMP was sufficiently armored to protect against heavy machine guns like the American .50 caliber Browning. The BMD requirement was protection against .30 caliber rifle fire. The lower armor requirement, combined with the reduced troop size, allowed Gavalov and his team to substantially reduce the size and weight of the BMD. Development of the BMD-1 first starting in the mid-1960s.

The first production BMDs were completed in 1968. They were formidably armed for so small a vehicle. Indeed, when the BMD was first seen by the West in 1970, it was presumed to be a new light tank rather than an infantry carrier. It's compactness came at a cost: the vehicle was incredibly cramped inside. In the forward portion of the hull was the driver in the center, and the squad leader and squad machine gunner on either side. A single gunner sat in the turret. Behind him were seats for four squad members, shoe-horned into an awkward cavity between the turret and the rear engine compartment. The space in the rear was so cramped that normally only three soldiers would ride in this compartment instead of the theoretical four. There are firing ports for the squad leader, squad machine gunner and riflemen, two facing forward, two to the sides, and one to the rear. As in the case of the BMP, the BMD is fully sealed for operations on a nuclear contaminated battlefield. It is also fully amphibious, and uses a hydro-jet propulsion system.

Limited production of the BMD began in 1968 and it entered operational trials at the VDV's main training center at Ryazan. It was deployed in limited numbers in 1969 with Col. Vladimir Krayev's 7th Guards Airborne Division in Kaunus, Lithuania for further trials. It was the only armored vehicle of its kind ever to enter service anywhere in the world. While other armies had developed airborne tanks — the World War 2 US M-22 Locust and British Tetrarch; the post-war US M-551 Sheridan — no other army ever fielded an airborne infantry vehicle. The US Army was content to keep its one airborne division, the 82nd Airborne Division, a light infantry formation. The 82nd Airborne received one battalion of new M551 Sheridan light tanks (officially dubbed Armored Reconnaissance/ Airborne Assault Vehicle), but armored infantry carriers were never seriously considered. This was due to significant doctrinal differences between the US and Soviet Union. In the early 1970s, Soviet doctrine envisioned the use of the VDV as a significant element in a conventional high intensity war in Europe. Since Soviet doctrine was essentially offensive in nature, the VDV was seen as a shock force suitable for landing deep in NATO's rear. The US Army's orientation in Europe was more defensive, and it was hard to see a paratroop role for the 82nd Airborne in most NATO Central Front scenarios. However, the 82nd Airborne played a critical role in army plans for operations in low-intensity scenarios, especially in the developing world. In this context, armored vehicles seemed a heavy and burdensome hindrance.

BMD Air-Drop Techniques

The BMD was originally air-dropped using the new MKS-350-9 multi-canopy parachute, with the vehicle strapped to a standard cargo pallet. Although it was originally planned to drop the vehicles without their crew, in the early 1970s, there were experiments at Ryazan to drop the BMD with the driver and gunner under Project Kentavr (Centaur). There had always been problems when the crew jumped separately from a heavy load, such as the ASU-57 assault guns. The vehicle often landed miles away from the crew, and this would become a special problem with the BMD where only two of the squad were qualified to operate the vehicle. Project Centaur, directed by Lt. Col. Leonid Zuyev, developed a cushioned seat for the crew, patterned after the Kazbekov seats used in Soviet spacecraft. The first test drop was conducted on 5 January 1973 from an Antonov An-12 transport with Col. Zuyev and the son of the VDV commander riding inside the vehicle. The tests showed that the concept was feasible even though the landing forces involved were severe.

The operational trials of the BMD raised questions about the use of a multi-canopy parachute. It was not very economical in the event it was lost during the exercise. The multi-canopy parachute was heavy, adding about a ton of weight to the aircraft's cargo and there were speed limitations on the aircraft when using it. On landing, it posed a serious hazard to the BMD itself. When the BMD was driven away from the pallet, some of the many tough parachute lines inevitably got caught in the running gear of the vehicle. This could actually stop the

BMD-1 on PRSM-915 parachute pallet.

vehicle, since the suspension of the BMD was relatively delicate compared to other armored vehicles. The solution was to develop some technique to slow the vehicle during the impact so that only a single canopy parachute would be needed.

The Scientific Research Institute for Parachutes in Moscow began exploring new ideas for the delivery of BMDs. The French firm Aerazur had developed a cargo parachute system using air bags or other cushioning devices under parachute pallets. This was examined but eventually rejected. The parachute institute eventually developed a retrorocket system, called the PRSM-915, an acronym for "rocket parachute system for 915-type vehicles."

The PRSM-915 system was based around a conventional pallet. Before the mission, the BMD was driven on to the pallet and strapped in place. The BMD had a hydro-pneumatic suspension system which collapsed and locked the suspension into place for the drop, also lowering the height of the vehicle to make it easier to load into the aircraft. Once firmly on the pallet, the BMD had the PRSM-915 system attached, along with a large, single-canopy parachute. When over the drop zone, the drogue chute from the parachute system was deployed. This small chute streamed out of the aircraft, and eventually exerted enough force to pull the entire BMD, pallet and parachute pack out of the aircraft. Once free of the aircraft, the BMD would free-fall for several seconds as the drogue chute and then main chute deployed. During actual combat jumps, the BMD would be released from as low an altitude as possible, about 1,500 feet. As a result, the time between exiting the aircraft and hitting the ground would be short, less than a minute. After the main chute deployment, lanyards holding a set of four rods under the pallet pull free. This causes the four rods to hinge downward, hanging several feet below the pallet. These rods contain contact sensors on their tips. Because of their location under the pallet, they strike the ground before the rest of the load. When the first one touches the ground, it triggers an electrical signal to the PRSM. The PRSM rocket pack is suspended under the main parachute, and above the BMD and pallet. When the rod strikes the ground, the PRSM retrorocket fires, abruptly slowing the descent of the BMD. This system proved a practical alternative to multi-canopy parachutes for heavy loads, and it entered service in 1975. The system allowed Il-76 jet transports to drop their loads at speeds about twenty-five percent higher than the older multi-canopy parachutes.

With the new airdrop system available, a second set of trials were conducted carrying the two man crew during the parachute descent under the codename Project Reaktavr (Rocketeer). The Rocketeer tests showed that the PRSM-915 retrorocket system was not significantly different from the normal parachute methods, though it too required considerable courage on the part of the crew.

An alternative solution to dropping the crew inside the vehicle was also developed to enable the crew to locate the vehicle quickly. A small radio directional transmitter is mounted in the BMD, and the crew carries a portable receiver preset to a particular frequency. When the BMD lands, it gives off a beeping radio signal that can be picked up by the receiver. Each crew has set their receiver to different frequencies, so that there is no confusion between the different squads.

BMD Deployment

The advent of the BMD changed the organization of the VDV airborne divisions. BMD production was very modest, and insufficient to equip all air assault divisions completely. As a result, through the 1970s, most divisions received only enough BMDs to equip one of their three air assault regiments. It was not until the mid-1980s that enough BMDs had been produced to totally equip all three air assault regiments of the divisions. The key fighting elements of the airborne division were three airborne regiments, each with three airborne battalions. With the introduction of the BMD, the number of troops in the airborne battalion shrunk in size from 600 to 316 men.

This was due in part to the elimination of most of the fire support companies, since the BMDs could provide direct fire support and anti-tank defense by themselves. The new battalion had three airborne companies with seventy-five men, each with ten BMDs. In addition, there was a grenade launcher platoon with AGS-17 grenade launchers on three turretless BTR-Ds, and an air defense platoon with three BTR-Ds armed with a total of nine manportable Strela 2 air defense missiles.

The airborne regiment showed a similar reduction in manpower and increase in firepower. The pre-BMD airborne regiment numbered about 2,060 troops; the mechanized airborne regiment with BMDs saw a decrease to 1,450 troops. At the heart of the new organizational structure were the three BMD airborne battalions which made up the bulk of the regimental strength. Their firepower was enhanced by an artillery battery with six 2S9 Nona-S, an air defense battery with six BTR-ZD/ZU-23 combinations, and an anti-tank battery with nine 9P148 tank destroyers (Konkurs anti-tank missile launchers on a wheeled BRDM-2 armored vehicle). The regiment also contains other necessary support elements such as a reconnaissance company, signal company, engineer company, transport company, parachute supply company, chemical defense platoon, and a medical team.

BMD-1P airborne infantry fighting vehicle.

BMD-1 Variants

BMD-1P

The shortcomings of the 9M14M (AT-3 Sagger) manual command-to-line-of-sight guidance system led to the introduction of the improved 9M111 Fagot and 9M113 Konkurs ATGMs (AT-4/AT-5) in the 1970s. A simple upgrade was developed for the BMD-1, with a small mounting lug added outside the gunner's turret for the 9P135 launcher. The normal rail launcher for the 9M14M Malyutka was deleted at this time. To use the missile, the gunner or one of the squad members mounts the 9P135 launcher on the turret, and then loads one of the cannisterized missiles. This provided the BMD-1 with a more accurate and potent anti-armor weapon, though at the same time, the gunner is exposed when firing the missile. This upgrade is probably undertaken during periodic depot rebuilding, as not all Russian BMD-1s have been brought up to this standard.

BMD-1K

This is the company commander's version of the BMD-1 and carries additional radio equipment and a smaller squad. The comparable vehicle on the BMD-1P hull is designated as the BMD-1PK.

BTR-D

The basic turreted BMD-1 is much too cramped inside to be used for specialized roles such as command vehicles. So in the 1974, the modified Obiekt 916 version was first issued to the troops, called the Bronirovanniy Transporter Desantniy: Armored Transporter-Airborne (BTR-D). This version lacks the normal BMD-1 turret, which frees up a considerable amount of space in the center of the vehicle. In addition, the vehicle hull was lengthened to provide additional internal volume; this is evident from the addition of one more road-wheel on either side. While the BMD-1 can carry seven troops, the BTR-D can carry up to thirteen. The basic BTR-D is used in BMD units to transport heavy weapons squads including AGS-17 automatic grenade launcher teams. The usual crew complement in this case is twelve men, plus the driver, and enough space is still available for the grenade launchers when stowed inside as well as their ammunition. In addition to the basic BTR-D, several specialized versions have also been developed as noted below.

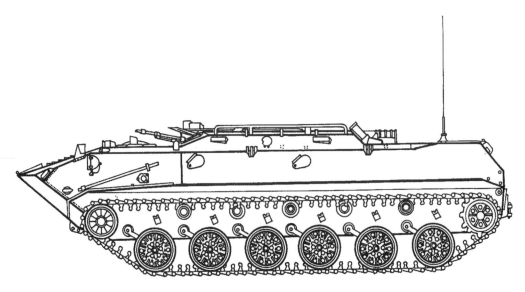

BTR-D Airborne Armored Transporter

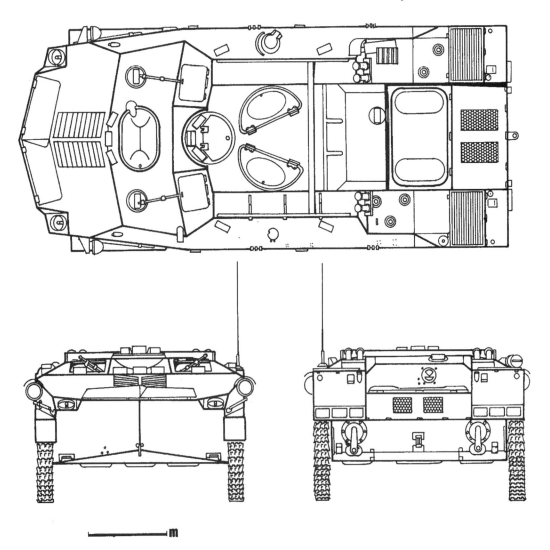

BTR-D armored transporter.

BTR-D Sterkh UAV Carrier

The BTR-D also serves as a launch platform/carrier for the Malakhit Sterkh tactical mini-UAV system. This modified BTR-D is manufactured by the Volgograd Tractor Plant while the UAV is built by Malakhit NPO. The BTR-D serves as the UAVs control station and can control up to two UAVs simultaneously at ranges up to sixty kilometers. The UAV's name is Shmel (Bumblebee) and is powered by a thirty-two horsepower piston engine. The UAV has two modular packages which can perform either reconnaissance or electronic warfare jamming missions. The UAV is recovered by means of a parachute or steerable parafoil and is limited to five to ten landings before major overhaul.

This system was accepted for use in 1990 and small orders have followed. In 1992 Syria placed the first export order.

BTR-ZD

The BTR-ZD air defense vehicle is a BTR-D derivative used to tow regimental ZU-23 23mm twin-barrel anti-aircraft guns. The BTR-D can tow the gun behind, but is also fitted with ramp attachments to allow the gun to be carried portee-style on top the vehicle roof. The BTR-ZD also carries additional air defense weapons including man portable missiles such as the Strela-2M (SA-7 Grail) or newer types such as the Igla-1M (SA-16 Gimlet).

BTR-RD Robot

The BTR-RD is another BTR-D variant used to carry anti-tank missile teams armed with the Fagot/Konkurs (AT-4 Spigot/AT-5 Spandrel) anti-tank missiles. The usual crew complement is five: a driver and two two-man teams, each armed with a 9P135 portable launcher for the 9M111 Fagot and 9M113 Konkurs ATGM. The vehicle has a set of small pintle mounts on the roof which enable the crews to mount and fire the missile launchers from the vehicle.

BMD-1KSh

The BMD-1KSh is a specialized command version of the BTR-D used by battalion and regimental commanders. The larger internal volume of the BTR-D chassis allows the vehicle to carry the necessary R-123 radio, data links and other command and control electronics typical of the modern battlefield. It is distinguishable by additional external radio antennas. The configuration of these antennas has varied, the earlier versions using clothesline" style antennas.

BMD-ODB

One of the most unusual BMD variants is the ODB,

BMD-ODB satellite communications vehicle.

BREM-D airborne recovery vehicle.

a BTR-D fitted with the R-440 satellite communications system. On this vehicle, the whole roof is taken up by a larger circular communications dish that is used by the R-440 system. This vehicle allows the VDV divisional headquarters to communicate with distant headquarters via satellites or other data transfer systems.

BREM-D

The BREM-D (Armored Repair and Towing Vehicle-Airborne) is a specialized armored recovery version of the BTR-D used to support BMD regiments. It has a small crane fitted to the roof to carry out basic repairs.

2S9 Nona

In the early 1970s, there was the desire to replace the outdated ASU-85 assault guns still in VDV service. By this time, these weapons were ineffective against contemporary NATO main battle tanks. With the advent of the BMD, their role had gradually shifted from anti-tank defense to direct fire support. One of the options was to

2S9 Nona-S 120mm airborne vehicle.

replace the ASU-85 with a new light tank being developed by the Soviet Ground Forces to replace the old PT-76 amphibious tank. Two prototypes of this light tank were constructed by two old rivals, Gavalov's BMD design bureau, and Isakov's BMP design bureau. The requirements called for amphibious light tanks armed with a new 100mm smooth bore anti-tank gun. Both light tanks were designed to be air-dropped. In the end, the Soviet Army decided against adopting these vehicles. Instead, the VDV turned to another BMD derivative.

Dedicated tank destroyers were less necessary for VDV units in the 1970s since each BMD-1 had considerable anti-tank firepower, both with its low pressure gun and its anti-tank guided missile launcher. VDV interest shifted towards a self propelled artillery vehicle. Conventional howitzers have substantial recoil which requires a sturdy and heavy chassis when mounted in tracked vehicles. Instead of a conventional howitzer, the VDV selected the revolutionary new 2A60 120mm Nona howitzer/mortar developed by the gun design bureau headed by Yuri Kalachnikov in Perm. This weapon fires 120mm mortar bombs, but unlike a conventional mortar, the ammunition was loaded from the breech end. The weapon could be fired at high elevation, like a mortar, or used in direct fire, like a gun or howitzer. Besides normal mortar ammunition, special projectiles were also developed including a shaped charge anti-tank round that has a stated penetration of 600 millimeters. The new 120mm weapon was mounted on a modified version of the lengthened BTR-D chassis, called the Obiekt 925. It was designated 2S9 Nona-S when accepted for service in 1984. The 2S9 resembled a tank, with a center mounted turret instead of the usual rear mounted turret common to self-propelled artillery. The gun system was fitted with a rammer to help load the heavy 120mm mortar ammunition. One of the clever innovations in the design was a small conveyor belt at the rear of the vehicle that allowed the crew to load ammunition from outside the vehicle when engaged in long fire-support missions.

The 2S9 Nona-S began to enter service in 1984. It is a remarkably powerful weapon for its small size and light weight. It replaced the 120mm towed mortars found previously in VDV regiments. But its direct fire capabilities made the ASU-85 unnecessary and considerably enhanced the fire support available to the paratroopers.

1V118 Reostat/1V119 Spektr

The introduction of the 2S9 Nona-S was accompanied by the deployment of two other specialized artillery vehicles, the 1V118 Reostat and 1V119 Spektr fire control vehicles. These vehicles have a small fixed turret, but lack any armament beyond self-defense machine guns. At the forward portion of the turret is an electro-optical sensor package including a laser range-finder. The 1V118 Reostat serves as a forward artillery observation post, and can also be used by air force forward air controllers when air support is needed by VDV units. It is fitted with a small PSNR-5 (1RL133, NATO: Tall Mike) battlefield surveillance radar which can be mounted above the rear turret roof. The 1V119 Spektr is similar in general ap-

1V118 Reostat airborne artillery reconnaissance vehicle.

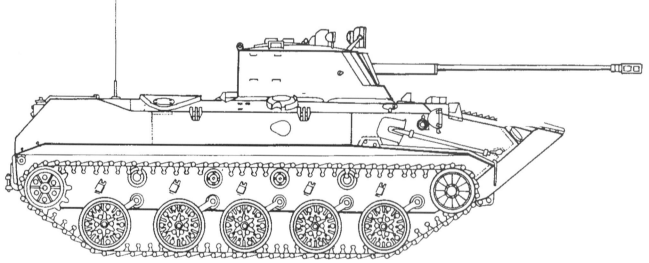

BMD-2 Airborne Infantry Fighting Vehicle

pearance but has the necessary ballistic computers and communications equipment to serve as a fire direction center for the 2S9 Nona-S battery. It lacks the surveillance radar.

BMD-2

Experience with the BMD-1 in Afghanistan revealed serious problems with the design, including its 73mm low-pressure gun armament. As a short term solution, a new turret was developed for the BMD fitted with the more effective 30mm 2A42 auto-cannon found on the infantry's BMP-2 Yozh armored combat vehicle.

Design of the new BMD was directed by Arkadiy Shabalin, the new head of the design bureau at VTZ after Gavalov's retirement. This new version entered development in Volgograd in 1983. The design was essen-

BMD-2 airborne infantry fighting vehicle.

tially the same as the BMD-1, with the exception of the new turret.

Since the hull of the BMD was so much smaller than the BMP, a single-man turret was developed. As in the case of the BMP-2, it was armed with a 2A42 30mm cannon with a co-axial 7.62mm PKT machine. The gun is uses electro-mechanical stabilization in two-axis. The prototypes had a maximum gun elevation of seventy-five degrees, but on the production vehicles this has been reduced to sixty degrees. The smaller turret size neccesitated a smaller ammunition magazine on the airborne vehicle than the BMP, 300 rounds in two trays (usually 180 armor piercing, 120 high explosive).

Like the BMD-1P, the auxiliary anti-tank armament was an externally mounted 9P135 launcher for the Konkurs ATGM. The crew size in the new vehicle was the same as in the BMD-1, two crew plus five dismounts. The new version was accepted for series production in 1985 and was designated the BMD-2. It was deployed to Afghanistan in modest numbers for trials.

The BMD-3 Airborne Combat Vehicle

The combat employment of the BMD-1 in Afghanistan revealed extensive problems with its design. In order to make the vehicle light enough for air transport, the BMD-1 suspension and track had been made very lightweight. This proved to be too flimsy for prolonged operations, and the VDV was obliged to reequip its units in Afghanistan with standard Ground Forces vehicles such as the BMP-2 and BTR-70/BTR-80. The poor showing of the BMD in Afghanistan led to a requirement for a replacement airborne combat vehicle, designated Obiekt 950. With the continued growth of Soviet airlift, a heavier vehicle than the BMD was judged acceptable with the combat weight of the requirement doubling from six to twelve tons. Three can be carried per Il-76M transport aircraft. In many respects, this program aimed at shrinking down the BMP-2 to make it compatible with airborne operations.

The general configuration of the Obiekt 950 followed the BMD-1/2. The Obiekt 950 employed BMP road wheels, but with a hydropneumatic suspension as on the BMD-1 to permit the vehicle chassis to be lowered during the airdrop loading process. The turret and armament selected for the Obiekt 950 was the same as the BMP-2, a 2A42 30mm cannon with co-axial 7.62mm machine gun. One of the few differences in armament was the addition of a ball-mounted AG-17 30mm grenade launcher in the bow with 290 VOG-17 rounds in belts and a further 261 rounds in nearby racks. Instead of the two corner mounted squad automatic weapons, the BMD-3 has a 7.62mm machine gun mounted in a socket mount in the hull front. The new 2V-06 diesel engine offers a high power-to-weight ratio (thirty-six horsepower per ton).

The program was evidently given low priority and

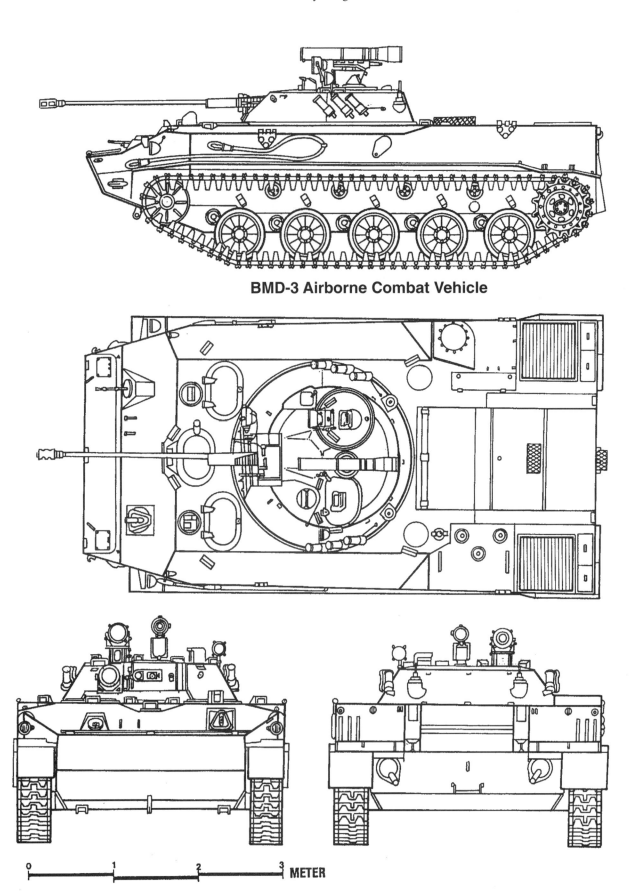

BMD-3 Airborne Combat Vehicle

BMP-3 airborne infantry vehicle.

the the new vehicle did not become available until several years after the planned introduction.[3] The first vehicles were scheduled to enter service in 1990, but production was delayed and the first significant numbers did not begin to appear until 1991. Only sixty-nine BMD-3s have been fielded with the VDV divisions west of the Urals.

The Russian Army plans to base a future generation of VDV armored vehicles around the BMD-3 chassis, much as the BMD-1 served as the basis for the BTR-D, 2S9 and other vehicles. An BMD-3 based armored recovery vehicle and a turretless command/transporter versions have been fielded in very small numbers with other derivatives likely to follow. The only other variant identified so far is the Germes (Hermes) 155mm self-propelled artillery system.

This unusual vehicle is armed with a pair of two 155mm launch tubes with an electro-optical sensor/designator located between them on the turret front. This weapon was developed by the Tula Machine Design bureau which developed the weapon suite on the BMP-1, BMD-1, BMP-2, and BMP-3. The munition fired by this unusual weapon is a sixty kilogram missile with a thirty kilogram shaped charge/fragmentation warhead. The missile is fired at a high angle and dives into the target; the effective range is twelve kilometers, and it takes fifty-four seconds for the missile to reach this distance.

The sensor suite include a battlefield surveillance radar on the upper rear of the turret, a daylight TV, a thermal imaging sight, and a laser rangefinder/designator. The laser can be used to designate the target for the missile, or it can presumably be autonomously designated. The vehicle carries fourteen to sixteen missiles, and the gun system is automated permitting a firing rate of twelve to eighteen rounds per minute when salvo firing. It is not clear if the Germes has actually been built; information to date has been limited to a brochure released in 1994 that provides sketches of the system. Germes may be nothing more than a proposal in search of a customer. The role for the weapon is stated to be coastal defense or repulsing a massed tank attack. It is not at all clear that this weapon is intended for the VDV.

Future Russian Light AFV Development Trends

One possible future outcome may be a shift in Russian armored vehicle development away from heavy main battle tank design towards light AFVs (armored fighting vehicles) especially if the Russian armed forces begin to shift away from a preoccupation with NATO confrontations and towards a greater concern with low intensity regional wars.

The BTR and BMD Family Comparative Technical Data

	BMD-1	BTR-D	BMD-2	BMD-3
Crew+squad	2+5	1+12	2+5	2+5
Weight (tonnes)	6.7	8.0	8.0	13.6
Hull length (millimeters)	5,400	5,885	5,500	6,100
Width (millimeters)	2,630	2,630	2,700	3,134
Height (millimeters) to turret roof	1,970	1,670	2,180	2,250
Clearance (millimeters)	100-450	100-450	100-450	100-530
Engine	5D20	5D20	5D20	2V-06-2
Engine (horsepower)	240	240	240	450
Power/weight (horsepower per ton)	35	30	30	30.5
Max. range (kilometers)	500	500	500	500
Trench crossing (meters)	1.2	1.2	1.2	1.5
Vertical step (meters)	.6	.6	.6	.8
Maximum grade (degrees)	32	32	32	35
Maximum slope (degrees)	18	18	18	25
Ground pressure (kg/cm^2)	.57	.5	.5	.55
Maximum speed (km/hr))	61	61	60	70
Maximum water speed (km/h)	9-10	9-10	9-10	10
Main armament	2A28	n/a	2A42	2A42
Gun caliber (millimeters)	73	n/a	30	30
Maximum range (AP), kilometers	1.6	n/a	2	2
Maximum range (HE), kilometers	1.6	n/a	4	4
Rate of fire (rounds per minute)	7-8	n/a	240/600	240/600
Gun elevation (degrees)	-5+30	n/a	-5+60	-5+75
Ammo stowage (rounds)	40	n/a	300	500+360
Anti-tank missile	9M14M Malyutka	n/a	9M111 Fagot	9M111 Fagot
Missile stowage	3-4	n/a	3	4-6
Co-axial machine gun	7.62mm PKT	n/a	7.62mm PKT	7.62mm PKT
Main gun ammo stowage	2,000	2,000	2,940	2,000
Maximum hull armor (millimeters)	16	16	16	16
Water crossing	amphibious	amphibious	amphibious	amphibious
Radio	R-123	R-123	R-123	R-173

n/a=not applicable

Should this emphasis change, Russian light armored vehicle development is not particularly likely to see any radical changes. This is in part due to the recent development of three new families of light AFVs, the BTR-90, BMP-3, and BMD-3, which are likely to serve as the basis for most light AFVs over the next decade. Kurgan has suggested that the BMP-1/2 chassis serve as the basis for a new family of light AFVs to replace the MT-LB/MT-LBu and 2S1 chassis, which were previously manufactured at the Kharkov Tractor Plant in Ukraine. The Russian defense press has already made it clear that these vehicles will form the families of light AFVs to serve the needs of the Army and the VDV airborne assault force.

One of the most likely outcomes of a light forces orientation will be the adoption of a light assault gun to provide fire support to the BMD-3. The Russian experience in Afghanistan made it clear that the 30mm gun has limitations when used against entrenched troops. The preferred solution was a weapon capable of firing a large high-explosive round, which emerged as the 100mm 2A70 on the BMP-3. It would not be surprising to see a BMD-3 derived assault gun appear over the next few years for this role. This would probably use an existing weapon such as the 2A60 120mm gun/mortar, or the 100mm 2A70 weapon from the BMP-3. Another alternative would be a long-barreled weapon such as the 120mm gun on the experimental 2S31 Vena. The Russians do not appear to be as preoccupied with anti-tank firepower for such vehicles as the US Army with its now-cancelled 105mm armed XM8 AGS project.

The future course of Soviet light wheeled armored vehicle is not obvious at the moment. At the 1994 Nizhni Novgorod show, both the improved BTR-80A and the BTR-90 were shown. Both are armed with a 30mm cannon, one being an incremental improvement in the BTR-80 family, the other an entirely new family of vehicles. It is not clear if these two vehicles are competitors for a

Russian Army requirement, or whether one is the future Russian BTR, and the other a potential export item. Kurgan's plant director has expressed interest in the idea of developing a wheeled BMP, but this appears to be oriented at the foreign export market, especially the Middle East. It is not clear if he was referring to the BTR-90, or yet another vehicle entirely.

The Soviet army was satisfied by a steady incremental evolution from the BTR-60 through the BTR-80. The BTR-90 is part of the natural evolutionary path of wheeled AFV development to supplement tracked IFVs. The most perplexing omission in Russian light AFV design has been the lack of a replacement for the thirty-year old BRDM-2 reconnaissance vehicle. It would not be surprising to see the Russians develop a more heavily armed light armored vehicle for export requirements. This design would probably parallel the French AMX-10RC and South African Rooikat, and serve primarily as the basis for a reconnaissance/cavalry vehicle.

Russian development of light armored vehicles is likely to intensify due to the new demands of peacekeeping along the Russian frontier, especially in Central Asia and the Caucasus, is likely to favor light forces with strategic mobility. In addition, the downsizing of the regular Russian Army has been accompanied by a substantial expansion in paramilitary formations, especially the armed units of the Interior Ministry. These forces have been acquiring light armored vehicles out of Russian Army stocks, and may be the prime candidates for new light armored vehicles. This might include vehicle types developed for the Russian Army, but could eventually include models created primarily for the internal security and border guard roles.

CFE Counts for European Russia BTR-D and BMD-1 to BMD-3 1990 to 1997

	BTR-D	BMD-1	BMD-2	BMD-3	Total
1990	769	1,632	522	0	2,923
1991	769	1,632	522	0	2,923
1992	827	1,525	553	25	2,930
1993	773	1,373	501	30	2,677
1994	530	981	362	40	1,913
1995	440	877	338	82	1,737
1996	424	854	308	82	1,688
1997	465	805	328	89	1,687

5
SELF-PROPELLED ARTILLERY

Introduction

Artillery has held a central place in the military thinking of the Russian army since Tsarist times and so was traditionally called bog voiny (the "god of war"). Although the Russians have often harbored feelings of technological inferiority vis a vis the West in many fields, artillery by contrast has been a source of national pride since the time of Peter the Great.

In discussing the development of Soviet/Russian self-propelled artillery systems since World War 2, this study has adopted the Russians' own definition of what constitutes artillery — a definition which differs somewhat from Western nomenclature. The traditional Soviet/Russian concept of artillery includes many of the same components found in the West: field guns, mortars, and multiple rocket launchers. Additionally, the Russians have also included a class of short-range, tactical ballistic missiles which they refer to as "artillery missiles." These missiles differ from strategic ballistic missiles in several ways: (1) they are relatively short-range, (2) they are intended to act as tactical bombardment weapons (even when delivering nuclear payloads), and (3) are organizationally assigned to the Ground Forces at the division, army, and front levels rather than to the Strategic Missile Forces (RVSN). Because the Russians have long classified such missiles as artillery (and because they are mounted on mobile launchers), we have included a discussion of their evolution along with discussions of more traditional artillery systems like self-propelled field guns, self-propelled mortars, and vehicle-mounted multiple rocket launders.

Mechanized Artillery in World War 2

Soviet artillery in World War 2 was respected by its opponents for the quality of its weapons, even if its tactics were derided for their infatuation with sheer volume of firepower. Mechanization of the Soviet artillery was delayed compared to the process occurring in other European armies of the time. The Soviet Main Artillery Directorate (GAU) showed very little interest in self-propelled artillery during the war. Western observers have frequently mistaken the profusion of assault guns, like the SU-85, SU-100, ISU-122, and ISU-152 for self-propelled artillery. In fact, these assault guns were developed at the bequest of the Main Armor Directorate, manned by tank troops, and employed as surrogate tanks. They were intended primarily for direct fire, and ill-suited for the usual artillery role of indirect fire. They were closest in conception to the German Wehrmacht's Sturmgeschutz assault guns. The only significant design to vaguely resemble American, German, or British self-propelled artillery was the SU-76, a lightly armored vehicle mounting the 76mm divisional field gun. But even this vehicle had been developed as an assault gun, and was used in a role more akin to the pre-war infantry support tanks. The Red Army of World War 2 had no immediate equivalents of the American M-7 105mm howitzer motor carriage, M-40 155mm gun motor carriage; the Canadian Sexton 25 pound self-propelled howitzer; or the German Wespe 105mm or Hummel 150mm self-propelled howitzers.

The Soviet attraction to assault guns was as much due to cost and manufacturing considerations as tactical utility. An assault gun cost less than the turreted tank from which it was derived, but could mount a more potent gun with greater range, and often more protective armor as well. For example, the SU-76, which made up fifty-seven percent of Soviet wartime assault gun production, was based on the obsolete T-70 light tank. Yet the automotive factories which produced the T-70 were not capable of manufacturing medium tanks. The SU-76 was developed initially as an expedient tank destroyer to take advantage of the automotive production facilities, but remained in production since it proved useful in providing infantry units with direct fire support in lieu of unavailable medium tanks. In this sense, it replaced the pre-war infantry tanks such as the T-26. Due to the open configuration of the SU-76M, it could be used in a traditional indirect artillery fire role even though this was seldom its mission.

The SU-85 and the later SU-100 were developed primarily as anti-tank vehicles and were deployed in anti-tank units. They were usually employed in the overwatch role, where their long range firepower was an asset, while their shortcomings such as the lack of a turret, were minimized. In an overwatch role, they would take up static position overlooking enemy positions as tank units advanced in front of them. From their positions, they could counter enemy anti-tank guns or tanks which were firing on the advancing Soviet tanks.

The heavy ISU-122 and ISU-152 were developed for the direct fire support role, with a secondary mission (in the case of the ISU-122) to counter German heavy tanks like the Royal Tiger which could not be dealt with by Soviet medium tanks like the T-34-85. Due to its closed configuration, these vehicles were not suited for use in traditional indirect artillery fire roles.

In 1943, assault gun production constituted only seventeen percent of total Soviet armored vehicle production, in 1944 it peaked at forty-one percent, and declined to twenty-three percent of total Soviet production in 1945.

The only area in which the Red Army showed much interest in mobile indirect-fire artillery was the Guards mortar, better known as Katyushas or multiple rocket launchers. The Red Army pioneered the use of multiple rocket launchers for artillery use, and used them in greater profusion than any other army during World War 2. These artillery rocket launchers were usually mounted on trucks, and by the end of the war, over 10,000 launchers had been manufactured. A total of forty independent battalions, 105 independent regiments, forty independent brigades and seven divisions had been equipped with these weapons, not to mention the numerous divisional guards mortar units. The Katyushas were popular with the Red Army for two reasons. On the logistical side, many Soviet artillery factories had been overrun by the Germans in the opening months of the war. Conventional tubed artillery requires elaborate (and expensive) machining. Katyushas, on the other hand, can be readily manufactured by small machine shops with little specialized equipment. The second attraction of the Katyushas for the Red Army was their applicability to the style of massed firepower favored by Russian artillery men. A single truck mounted launcher could ripple fire a rocket salvo, equivalent to a battery of conventional guns. The Katyushas were mounted on trucks since their very distinctive backblast when fired required frequent movement to avoid hostile counter-battery fire. The limited accuracy of this style of weapons was not a major concern, since they were intended to supplement conventional artillery, not replace it. Conventional artillery could be used in roles, such as counter-battery fire, where accuracy was demanded.

New Directions in the 1950s

In the immediate postwar years, there was still considerable interest in assault guns in the Soviet Army. In the light assault gun category, the SU-76M remained in production until late 1945 and remained in service well into the 1950s. The SU-76 was also used in independent assault gun units, but was mainly deployed in six-vehicle batteries in each rifle regiment. It is still in service with some former Warsaw Pact countries such as Romania,

The SU-100 (Obiekt 416) infantry assault gun.

and in North Korea. By the early 1950s, there was a consensus that its gun was becoming inadequate, and efforts were made to develop replacements. This took two directions.

The Morozov design bureau in Kharkov offered the more radical solution, called the Obiekt 416 or SU-100 which mounted a 100mm gun in a turret mounted at the rear of a new light tracked chassis. This vehicle essentially resembled a light tank, and was somewhat similar to the US Army's T-92 airborne tank of the same period. However, its turreted configuration and weapon raised questions as to why it should be manufactured instead of the existing T-54A tank, and it never was accepted for service.

Four years later in 1956, the Astrov design bureau in Mytishchi developed the Obiekt 573 which consisted of an 85mm gun mounted on a light armored chassis derived from PT-76 light tank components. Although intended for the mechanized units of the Soviet Army, when accepted for service in 1956 it was manufactured only for the VDV Airborne Assault Force, and its designation was changed from SU-85 to ASU-85 as a consequence. (Its development is covered in more detail in the airborne vehicles chapter).

The concept of a light assault gun for direct infantry support gradually faded in the late 1950s for three reasons. First, the growing density of tanks in the Soviet Army reduced the need for a specialized vehicle which had poorer armor and firepower than the standard T-54A tank. Secondly, the anti-tank mission began to be performed by missile-carrying tank destroyers. Thirdly, the Soviet Army began mechanizing the infantry on BTRs which enabled the infantry to carry more firepower including recoilless rifles, rocket grenade launchers, mortars, and other support weapons.

In the medium category, the SU-100 remained in production until 1953 at the Uralmash plant in Sverdlovsk, and production was also initiated in Czechoslovakia, lasting until 1956. The SU-100 was intended primarily for the role of tank destroyer, although it could also be used in the assault gun role. The SU-100 was used mainly by independent assault gun regiments and brigades, although it could be found in the armored regiment of rifle divisions as well. In 1944, the experimental SU-101 was developed as a possible replacement, moving the weapon's compartment to the rear of the hull and employing a modernized chassis derived from the new T-44 medium tank. Two versions were built, the SU-101 with

The SU-100P self-propelled 100mm gun.

a 100mm D-10 gun and the SU-102 with a 122mm D-25 gun. This vehicle had no advantages over the SU-100, (or its experimental 122mm armed version, the SU-122P) and suffered from many of the technical problems of the immature T-44 design, so was never accepted for production. In 1949, work began on a replacement for the SU-100, the Obiekt 600 or SU-122-54 by the Troyanov design bureau in Omsk. This consisted of the new D-49 122mm gun developed by the Petrov design bureau in Perm with the standard T-54A tank chassis. This was the first Soviet tank gun to use a fume extractor, based on examples from US Army M-46 tanks captured in Korea.

The SU-122-54 followed the wartime pattern with a fixed forward casemate and a large socket mount for the gun. The T-54A chassis was modified to better distribute the weight, and wheel spacing was different than on the standard T-54A. The SU-122 had several advantages over the SU-100. The gun was significantly superior in armor penetration, and more suitable for dealing with newer NATO tanks. In addition, a new TKD-09 stereoscopic rangefinder was provided for greater accuracy in long-range engagements. The armor layout was superior to the SU-100 as well as having greater effective thickness. Series production of the SU-122 began in 1954 and lasted until 1956. The SU-122-54 has long been a mysterious design, and it was never identified in NATO vehicle identification handbooks of the 1950s and 1960s. This was probably due both the secrecy attending the design as well as its relatively small numbers. It's short production run was probably brought about by three factors.

With the arrival of missile-armed tank destroyers, the rationale for a large and expensive gun vehicle began to disappear. Nikita Khrushchev was particularly antagonistic to traditional artillery designs, favoring missile weapons. Finally, specialized tracked anti-tank vehicles were disappearing in most armies since the role could be performed by tanks. There have been reports that a 130mm replacement was developed, but there is little evidence from recent Russian accounts. By the 1970s, most of the SU-122-54s were removed from service and converted into improvised recovery vehicles. They were a frequent feature in Revolution Day parades in Moscow in the 1980s were there were always a few lurking in the background to drag off any vehicle that broke down.

Of the three main categories of the assault guns, the most long-lived was the heavy assault gun. The ISU-152 assault gun remained almost continuously in production from 1944 to 1964 at the Kirov Plant in Leningrad and the Chelyabinsk Tractor Plant. It was substantially modernized in 1959 as the ISU-152M. It continued to be used in the same fashion as during World War 2, as a direct fire assault gun. Its stablemate, the essentially similar ISU-122, remained in production until 1955 in Leningrad. There were several attempts to improve upon the wartime assault guns. In 1945, an ISU-152 prototype was built on an IS-3 Stalin heavy tank chassis with a revised armor layout. It offered no significant combat advantages and was not accepted for service. In 1956, another 152mm self-propelled gun was developed as the Obiekt 268 on the basis of the T-10 heavy tank chassis. Although it offered many sophisticated new features including stereoscopic ranging, it was not accepted for production.

Postwar deployment of the heavy assault guns followed much the same lines as wartime use. The ISU-122 and ISU-152 were found in independent heavy assault gun regiments and brigades. One of the few changes from the war years was the formation of heavy armored regiments to support tank, mechanized (and later, motor rifle) divisions. These regiments were a mixed formation of forty-six IS-2 or IS-3 tanks and twenty-one ISU-122 or ISU-152 assault guns. These heavy armored regiments were used to provide long-range fire support to the division's medium tank regiments. This formation gradually disappeared in the late 1950s, being replaced by a homogenous heavy tank regiment as more T-10 heavy tanks became available. Although heavy assault guns remained in use in some units well into the 1970s, by the 1960s most had been relegated to secondary roles, especially by conversion into heavy armored recovery vehicles.

The Soviet Army was one of the last armies to employ assault guns. By the 1960s, the idea of building both tanks and assault guns to perform direct fire missions had been abandoned, with tanks being assigned to carry out both battlefield missions. By the 1960s, tank guns had increased in firepower substantially from World War 2, and were more than adequate for direct fire missions. This had the triple advantage of standardizing production, logistics and tactics.

The Soviet Army was the last major army to deploy self-propelled artillery for the traditional indirect fire role. The first post-war development efforts in the Soviet Army began in 1949 by the Gorlitskiy design bureau at the Uraltransmash plant in Sverdlovsk (now Ekaterinburg). This effort aimed to develop a family of medium armored vehicle sharing a new common chassis. This family included the BTR-112 armored infantry transporter as well as three self-propelled artillery vehicles: the SU-100P (Obiekt 105), SU-152G (Obiekt 108), and SU-152P (Obiekt 116). The new chassis was powered by the 400 horsepower V-54-105 diesel engine, derived from the standard tank engine. All three of these self-propelled artillery vehicles shared a nearly identical configuration. The weapon was mounted in an open compartment at the rear of the chassis to permit easier ammunition reload-

The SU152P self-propelled 152mm gun.

The Kondensator 2P self-propelled 406mm gun.

ing. All three weapons were only partially armored, with a thin armored shield covering the gun crew in all directions except the rear. This protection was not a true turret, and the weapons had a limited traverse of about 143 to 155 degrees. The SU-100P was armed with the D-50/D-10 100mm gun, a derivative of the common D-10T tank gun. The SU-152G used a modified version of the same armament system, the 152mm D-50/D-1 howitzer; only the gun tubes were different. The SU-152P was armed with the new M-53 152mm gun, and carried thirty rounds of ammunition. None of these designs were accepted for production. It is not known why the Soviet artillery branch remained so resistant to mechanization.

It is possible that there was conservative resistance to mechanizing the artillery due to concerns over the mechanical reliability of the tracked chassis. It is also possible that the development program ran afoul of Kremlin politics, either the 1953 anti-Semitic campaign, or Khrushchev's 1956 to 1957 purge of the artillery design bureaus.

There was little if any development of self-propelled artillery in the late 1950s with one exception. The Soviet artillery force began experimenting with the possibility of nuclear artillery, probably based on similar US Army programs such as the 280mm "Atomic Cannon." At least two separate efforts were undertaken: by the Kotin design bureau at the Kirov Plant in Leningrad and by the Balzhiy design bureau at Chelyabinsk. Both vehicles were based on Stalin/T-10 heavy tank components. The Chelyabinsk design, designated Kondensator 2P, mated the Grabin bureau's SM-54 406mm gun on the new Obiekt 271 tracked chassis. This weapon had an effective range of twenty-eight kilometers. The Leningrad design bureau's Oka combined the Shavyrin mortar design bureau's 420mm breech-loaded mortar with another Stalin tank/T-10 derived heavy chassis. The Oka's enormous mortar had a maximum range of forty-five kilometers according to recent Russian accounts. Both were built in very small numbers. Their service with the special artillery regiments of the High Command Reserve (RVGK) was very short-lived. Nikita Khrushchev was particularly unhappy with the concept, singling it out in his memoirs as an example of the reactionary tendencies of the artillery branch in the face of modern missile weapons. The Grabin artillery design bureau was closed due in part to this episode; the Shavyrin bureau survived as by this time it was working on guided anti-tank missiles. Both nuclear artillery weapons were retired in the early 1960s as effective tactical ballistic missiles such as the R-11 (Scud) and Luna (FROG-3) became available.

With the "revolution in military affairs" taking place in the late 1950s, the Soviet Army reexamined the issue of artillery mechanization. A fully turreted, self-propelled artillery piece was the obvious solution to the problems posed by tactical nuclear weapons. The Uraltransmash design bureau in Sverdlovsk dusted off their decade old self-propelled gun designs. In the meantime, the "izdeliye 100" series chassis had entered production as the carrier for the new Krug (SA-4 Ganef) tactical air defense missile system. This chassis was used for the new SU-152 self-propelled gun in 1965. The new vehicle mated the long M-69 152mm gun in a fully enclosed turret on the "izdeliye 100" hull. The SU-152 could carry twenty-two rounds of ammunition. This design was not accepted for production. Instead, the Soviet Army opted for a similar design, also developed at Sverdlovsk, using a 152mm howitzer instead of the M-69 152mm gun. This emerged in 1969 as the 2S3 Akatsiya.

Unlike armored infantry vehicle development, where the lack of wartime development was followed by a blossoming of interest after the war, there was no sudden spurt of interest in mechanizing Soviet artillery after the war. Soviet artillery development after the war focused almost exclusively on evolutionary development of its wartime weapons and tactics. The administration of the Soviet artillery branch, under Marshal N.N. Voronov, was preoccupied with other matters, especially the absorption of missile technology into the Soviet armed forces. All early Soviet missile technology was managed by the GAU (Main Artillery Directorate), later renamed the GRAU

The SU-152 self-propelled 152mm gun.

by Khrushchev (Main Missile and Artillery Directorate) to reflect its new mandate. These programs included both the tactical ballistic missile and air defense missile development programs. The combination of the heavy drain of technical resources into these programs, as well as Khrushchev's antipathy to traditional artillery weapons, probably accounts for the Soviet Army's sluggish performance in self-propelled artillery development.

Early Post-War Missile Development

In 1946, the Soviet Army formed the "Special Purpose Brigade" under the command of Maj. Gen. A. Tveretskiy, armed with the first Soviet R-1 missiles. These were copies of German A-4 (V-2) ballistic missiles and the early units fired them from special trains. As quantity production of the R-1 (SS-1a Scunner) and the improved R-2 (SS-2 Sibling) took place, additional Special Missile Brigades were formed and deployed at army and front level to supplement the Katyushas.

These early ballistic missiles based on wartime German technology were very time consuming to fuel and prepare for launch and they were not reliable enough to entrust to nuclear weapons delivery. In the early 1950s, a requirement was issued to OKB-1 in Kaliningrad, the main ballistic missile development center headed by Sergei Korolev, to develop a family of tactical battlefield missiles capable of carrying small nuclear warheads. At the same time, Kotin's armored vehicle design bureau in Leningrad began work on a tracked launcher vehicle for the new missiles derived from the IS-2 tank. This resulted in the 8K-11 missile system (SS-1b Scud A) which were deployed in army or front-level missile brigades. In parallel, a program was undertaken to field more compact missile systems for divisional-level missile artillery units which resulted in the Luna/FROG series. This development program is covered in detail later in this study.

Early Post-War Multiple Rocket Launcher Development

In the immediate post-war years, the GAU directed a major modernization program for its multiple rocket launchers. The Barmin design bureau, which had developed the wartime rockets, became preoccupied with the development of strategic missile launcher systems. As a result, a new design institute, the NII Splav in Tula under A.N. Ganichev, took over these development efforts. The BM-13-16 on the ZiS-151 truck replaced the BM-13N (on the Lend-Lease Studebaker US 6x6 truck) beginning in 1949. The long-range BMD-20 was introduced, based on wartime experiments with the BM-13SN. In 1955, the BM-14 appeared, to replace the BM-13; the BM-24 heavy guards mortar was introduced the same year and became the standard weapon in the heavy MRL class. The multiple rocket launchers were employed both in independent artillery units, and as organic artillery units in tank, rifle and mechanized divisions. Each division usually had a battalion of guards mortars to supplement conventional towed artillery.

Mechanized Field Artillery Requirements After 1965

The coup against Khrushchev in 1964 removed one of the barriers to conventional artillery modernization. Around 1965, the GRAU began to lay down requirements for a new generation of artillery vehicles. After two decades of distraction with the monumental task of building up the tactical nuclear forces of the Army, the artillery branch finally turned to the more mundane task of modernizing its conventional tubed artillery. The need for mechanized artillery had become more manifest by this time. NATO had been significantly improving its artillery force, notably including the US M-109 155mm and M-110 203mm self-propelled howitzers and the British Abbot self-propelled 105mm howitzer. More importantly, NATO's ability to conduct counter-battery fire was improving. With the advent of more and improved artillery location radars, Soviet artillery sites could be quickly identified and targeted. New ammunition developments, especially chemical weapons, and improved conventional munitions (ICM) cargo rounds made artillery crews especially vulnerable to counter-battery fire.

The Soviet Ground Forces had received several excellent towed artillery pieces in the 1960s, notably the D-30 (2A18) which had appeared in 1963. But towed guns took precious time to emplace, move and re-site when faced by counter-battery fire. Towed guns were not survivable on a nuclear battlefield. What was required was mobility to avoid counter-battery fire, and armored protection to resist counter-battery fire should it come. The obvious direction was self-propelled howitzers using proven howitzer designs like the Petrov bureau's fine D-30 122mm howitzer and D-20 152mm howitzer.

The requirement for this new generation of weapons was coordinated between the GRAU (gun development) and the Main Auto-Tractor Directorate (GAVTU) which had traditionally been responsible for artillery tractor development. The configuration selected for both vehicles was the conventional self-propelled howitzer configuration pioneered by the US Army with the gun mounted in a fully rotating turret on the hull rear. The GRAU decided for logistical reasons to develop the new vehicles on the basis of two common chassis. The 2S1

Gvozdika (Carnation) 122mm howitzer vehicle, using a derivative of the 122mm D-30 howitzer, was to be based on a derivative of GAVTU's new MT-LB artillery tractor design. The 2S3 Akatsiya (Acacia) medium howitzer vehicle, using a derivative of the D-20 152mm gun, was to be based around the same chassis izdeliye 300 series developed by the GABTU for the GMZ minelaying vehicle, and the 2K11 Krug/SA-4 Ganef air defense missile system.

The 2S1 Gvozdika Self-Propelled Gun

The 122mm self-propelled gun was developed under the codename "izdeliye 26" with the GAVTU Main Auto-Transport Directorate responsible for the chassis design. The chassis was based on the MT-LB armored transporter, which had been developed on the basis of a GAVTU requirement to replace the obsolete AT-P light artillery tractor. Production and engineering design of this program was undertaken by the Kharkov Tractor Plant. It was accepted for production in 1971 and was designated 2S1 Gvozdika (Carnation).

The 2A31 122mm howitzer system was derived from the D-30 (2A18) 122mm howitzer designed by F.F.Petrov's bureau.[1] Although mounting a smaller gun than the 2S3 and lighter, the 2S1 is nearly as large. The size is deceptive; the bulky hull was incorporated to provide enough buoyancy to allow the 2S1 to float. The 2S1 does not have a specialized amphibious propulsion system, but uses its tracks. When prepared for swimming, a small track cover (normally stowed on the rear of the turret) is placed over the front hull side to better direct the water flow over the tracks, and a set of swim vanes is attached behind the tracks. Covers are also placed around the engine air intakes to prevent water ingestion into the engine compartment.

As in the case of the 2S3, the four-man 2S1 crew is usually supplemented by two additional loaders outside the vehicle during prolonged firing to assist in ammunition handling. The 2S1 fires the full range of Soviet 122mm ammunition. It can fire a typical high explosive round like the OF-462 (total weight of 21.7 kilograms with 3.5 kilograms of high explosive) to a maximum range of 15.2 kilometers. The 2S1 is used in motor rifle and tank divisions to replace towed 122mm howitzers. The tank divisions were the first to receive the 2S1s and initially had a single battalion in their artillery regiments, with three batteries of six vehicles each (eighteen 2S1s per battalion).

As production has continued, Category 1 motor rifle divisions received up to six battalions of 2S1, two in their artillery regiment, and one in their tank regiment and each motor rifle regiment. Category 1 tank divisions were re-equipped with up to six battalions, two in the artillery regiment, and one each in the three tank regiments and the motor rifle regiment. Total production of Soviet self-propelled guns since 1972 has been over 10,000. The vast majority of these have been the 2S1 and 2S3, with the 2S1 being the more numerous.

Production of the 2S1 was also undertaken in Poland and Bulgaria. The Poles have developed a modernized version using polyurethane appliqués for improved buoyancy and small propellers for improved water

2S1 Gvozdika 122mm self-propelled howitzer.

Soviet/Russian Armor and Artillery Design Practices: 1945 to Present

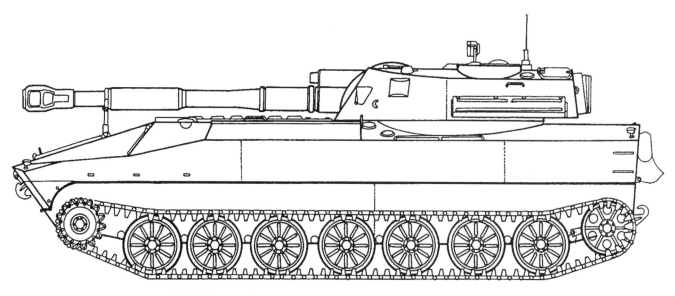

2S1 Gvozdika 122mm Self-propelled Howitzer

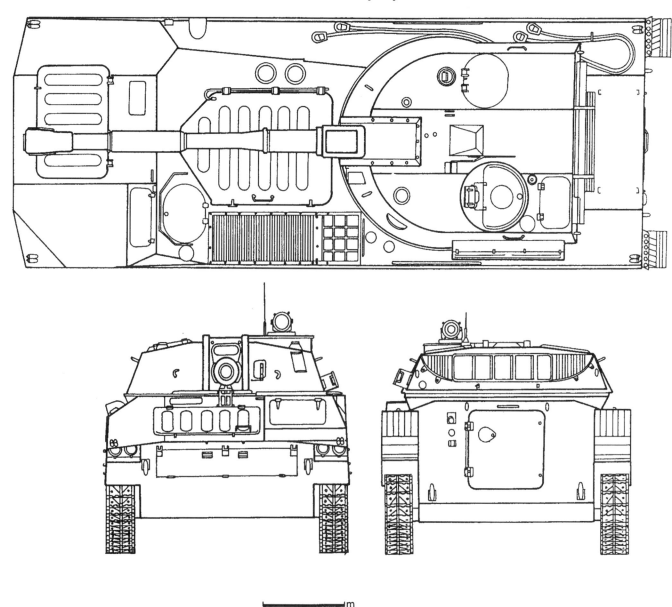

2S1 Technical Data

Crew	4 men
Length (millimeters)	7,260
Width (millimeters)	2,850
Height (millimeters)	2,725
Combat weight (metric tons)	15.7
Caliber	2A31 122 millimeter gun
Tube length without muzzle brake (millimeters)	4,270
Muzzle velocity (meters per second)	686 (HE) to 723 (HEAT)
Gun traverse (degrees)	360
Gun elevation (degrees)	-3 to +70
Ammunition stowage (rounds)	40
Effective firing range (kilometers)	15.2
Maximum firing (rounds per minute)	1 to 2 on-board to 4 to 5 from ground
Powerplant	YaMZ 238N V-8, four stroke, diesel (220kW)
Maximum road speed/cross (kilometers per hour)	61.5/30
Maximum cross-country (kilometers per hour)	30
Maximum swimming speed (kilometers per hour)	4.5
Fuel capacity (liters)	550
Maximum road range/cross-country (kilometers)	500/450
Ground clearance (millimeters)	400
Fording (meters)	1.0
Maximum inclination (degrees)	35
Maximum slope (degrees)	25
Trench (meters)	3.0
Vertical obstacle (meters)	0.7

performance. A significantly modified version using a different chassis was manufactured in Romania as the Model 89 122mm self-propelled howitzer. In the early 1980s, a new 122mm howitzer vehicle entered development, but details to date are lacking.

The 2S3 Akatsiya Self-Propelled Gun

Development of a 152mm self-propelled howitzer can be traced back to the abortive programs of the Gorlitskiy design bureau at the Uraltransmash plant in Sverdlovsk in 1949. (Uraltransmash is the former Metallist Zavod Number 40 in Sverdlovsk: now Ekaterinberg). This program was revived in 1965. The first new design, the Obiekt 120 or SU-152, was based on a version of the same chassis developed in 1949, and subsequently revived in the early 1960s as the Obiekt 123 for the 2K11 Krug/SA-4 Ganef air defense missile system. The Obiekt 120 mounted a new M-69 152mm gun in a fully enclosed turret with an autoloading system. A small number of these vehicles were produced for trials purposes, but the type was not accepted for Soviet service. No details are available as to why the program was canceled. It was followed in the late 1960s by a closely related program, the Obiekt 303, which used a 2A33 152mm howitzer derived from the D-20 instead of a 152mm gun. In 1971, this new design was accepted for Soviet service as the 2S1 Akatsiya (Acacia).[2]

The "Akatsiya" 2S3 152mm self-propelled howitzer was manufactured for nearly eighteen years. A special tracked chassis (Izdeliye 303) was developed for the howitzer. Later the 2S5 152mm long-range self-propelled gun was mounted on that same chassis. The welded armored hull provides bullet and shrapnel protection for the crew and internal equipment. The 2S3 is air-transportable and two of the self-propelled guns can be carried in an AN-22.

The gun has been twice modernized in the process of production. In 1975, the design of the mechanized ammunition loader was improved which permitted them to increase firing rate and was renamed 2S3M. The 2S3M was adapted to allow the loading of its ammunition supply from the ground. In this case, the crew is increased by two men. In 1987, the weapon's model number was changed once again to 2S3Ml. This signified equipping the self-propelled gun with a device to receive/display command information and with a new gunsight. Besides conventional munitions, the 2S3 can use precision-guided rounds like the Krasnapol laser projectile.

An entrenching blade is built into the nose front portion of the vehicle. The power plant is a multi-fueled engine with water cooling, fuel injection, and reheating. The engine's output is 520 horsepower. The 2S3 uses a four-man crew (internally), but generally an additional two crewmen would be carried onboard the supporting

2S3M Akatsiya Self-propelled Howitzer

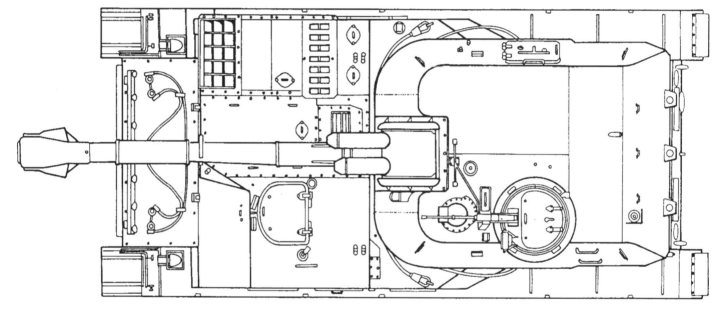

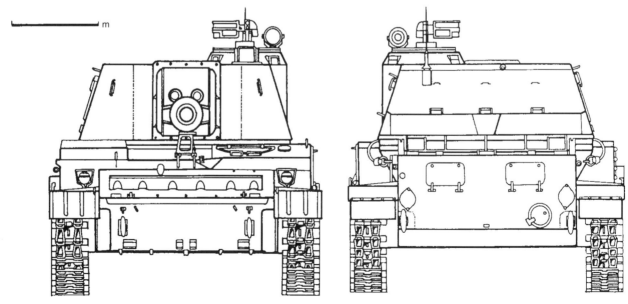

ammunition trucks. These two loaders stand at the rear of the vehicle, and feed ammunition into the vehicle through small rear ports. These ports feed into a conveyor system which passes the ammunition into the fighting compartment.

In 1975, the 2S3 was improved as the 2S3M with power-assisted carousel to assist ammunition loading. The 2S3 has a rate of fire of about three to four rounds a minute, and a sustained rate of about sixty rounds per hour; the 2S3M has a rate of fire of six rounds per minute. It can fire the normal OF-540 high-explosive-fragmentation projectile (43.5 kilograms, 6.4 kilograms of high explosive) to a range of 17.3 kilometers. Its maximum range is about thirty kilometers when firing rocket-assisted projectiles. It is capable of firing a variety of chemical rounds and also can fire nuclear projectiles.

The 2S3 is similar in performance and characteristics to the US M-109 self-propelled howitzer. The 2S3 Akatsiya was used to replace the towed D-1 152mm howitzer battalions in motor rifle division artillery regiments. The reconfigured regiments have a battalion of eighteen 2S3s in three six gun batteries. In the tank divisions, the 2S3 Akatsiya is used to replace a battalion of D-30 122mm howitzers. The 2S3 Akatsiya is also replacing some towed guns in artillery divisions and in army-level artillery regiments and brigades. The 2S3 was the second most common self-propelled artillery piece in the former Soviet Army, with 2,012 in service west of Urals prior to the 1991 breakup (compared to 2,292 2S1s). The "Msta-S" 2S19 152mm self-propelled howitzer, is replacing the 2S3.[3]

The Czechoslovak Army decided against adopting the 2S3, and opted instead for a wheeled 152mm howitzer vehicle, developed cooperatively between Tatra (chassis) and Skoda (gun system). This vehicle, called the vzor 77 152mm samohybna houfnice DANA, is based on a version of the Tatra 813 heavy truck. The gun is mounted in an armored turret with an automatic loading system. It is one of the most unconventional self-propelled artillery vehicles since 1945. It is used by the CSLA, and has been exported to the Soviet Union, Poland, and Libya.

2S3M Technical Data

Crew	4 men
Length (millimeters)	7,765
Width (millimeters)	3,250
Height (millimeters)	3,050
Combat weight (metric tons)	27.5
Caliber	2A33 152.4 millimeter gun
Anti-aircraft machine gun	7.62 PKT
Tube length with muzzle brake (millimeters)	5,195
Number of rifling grooves	48
Muzzle velocity (meters per second)	680
Gun traverse (degrees)	360
Gun elevation (degrees)	-4 to +60
Ammunition stowage (rounds)	46
Anti-aircraft machine gun rounds stored	1,500
Effective firing range (kilometers)	17.0
Maximum firing rate (rounds per minute)	3
Emplacement time (minutes)	3
Powerplant	V-59-V-12, 520 horsepower direct injection, inertial supercharge, multi-fuel diesel
Engine power (kilowatt)	382
Maximum road speed (kilometers per hour)	60
Maximum cross-country (kilometers per hour)	25
Fuel capacity (liters)	1,830
Maximum road range (kilometers)	500
Maximum cross-country range (kilometers)	270
Ground clearance (millimeters)	450
Fording (meters)	1.0
Maximum inclination (degrees)	25
Maximum slope (degrees)	30
Trench (meters)	3.0
Vertical obstacle (meters)	0.7

2S3 Akatsiya 152mm self-propelled howizter.

The 2S5 Giatsint Self-Propelled Gun

In the early 1970s the former Soviet Union developed two 152 mm artillery systems. One was self-propelled and had the industrial designation 2S5 with the more common name of Giatsint (Hyacinth) and the other was towed and had the designation 2A36. Both of these systems were subsequently placed in production in the mid-1970s. The 2A36 was seen as early as 1976 by Western sources and hence given the designation of the M1976. However, it was not seen in public until May 1985 when it was observed being towed through Red Square behind a KrAZ-260 (6 x 6) truck. Since its debut, significant numbers of 2A36s have been built.[4]

One of the final members of the first generation of Soviet mechanized artillery vehicles to appear was the 2S5 Giatsint (Hyacinth) 152mm self-propelled gun. Development was begun in the early 1970s as the Obiekt 307. The design consists of hull derived from the same Uraltranmash chassis as used with the 2S3, mated to a version of the 2A36 towed 152mm gun, the 2A37. Unlike the 2S3, the 2S5 has the gun mounted externally with little crew protection. The 2S5 Giatsint was accepted for service in 1974 and was first identified by NATO in 1981. The 2S5 was adopted to replace a portion of towed guns in the heavy artillery brigades at army level.

The chassis of the 2S5 is all-welded, steel armor construction that is believed to have a maximum thickness of fifteen millimeters. This provides the crew with pro-

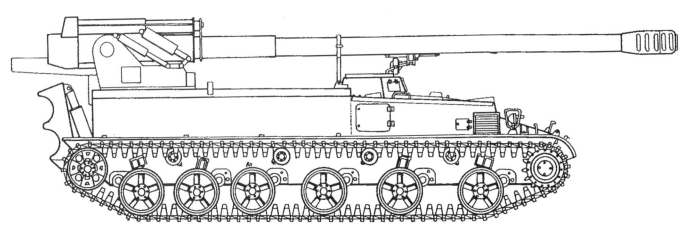

2S5 Giatsint 152mm Self-propelled Gun

tection from small arms fire and shell splinters. The chassis of the 2S5 is also used for a number of other applications including the GMZ armored minelaying system. As with a many former Soviet Army vehicles, including the 152 mm 2S3 self-propelled artillery system and the T-72 MBT, a dozer blade is mounted under the nose of the 2S5. This is used for clearing obstacles and preparing firing positions without specialized engineer support.

The driver is seated at the front of the vehicle on the left and has a single hatch cover that opens to the rear and can be locked in the vertical position. In front of the hatch cover are day periscopes, one of which can be replaced by an infra-red periscope for night driving. The vehicle commander is seated in a raised superstructure to the rear of the driver and he has a cupola that can be traversed through 360 degrees. Mounted externally on the forward part of the cupola is a 7.62 mm machine gun that can be operated by remote control.

The engine compartment is to the right of the driver with the air-inlet and air-outlet louvers in the roof. The engine, which is coupled to a manual transmission, will run on diesel or jet fuel. The other three crew members are seated in the crew compartment at the rear of the hull and enter and leave via a rear ramp that is ribbed.

The rear crew compartment is provided with roof hatches and roof-mounted periscopes to give observation to the sides of the vehicle. The long barreled 152 mm ordnance, which is fitted with a five-part multi-baffle muzzle brake, is mounted on the roof at the rear. When traveling, it is held in position by a lock. The gun layer is seated to the left of the gun and is provided with a shield to his immediate front only. Elevation and traverse is electric with manual controls for emergency use.

When deployed into the firing position, a barge spade is lowered to the ground to provide a more stable firing platform. The 2S5 takes three minutes to come into action and a similar time to come out of action. Projectiles and charges are loaded separately and crew fatigue is reduced by means of a semi-automatic loading system. This consists of a electrically driven chain rammer located to the left of the breech. The projectile and charge are loaded into the trays where the rammer loads both. The 2S5 can be supplied with either on-board ammunition via the rear of the vehicle or ammunition stockpiled on the ground. The loading system is operated by remote control with the operator normally standing to the left rear of the system.

The hull contains a special ammunition conveyor system which is used in combination with an externally mounted autoloader due to the weight of the ammunition. The mechanical ammunition handling system enables a maximum rate of fire of five to six rounds a minute to be achieved and according to former Soviet sources, a battery of 2S5s can have forty projectiles in the air be-

2S5 Technical Data

Crew	5 to 7 men
Length (millimeters)	8,330
Width (millimeters)	3,250
Height (millimeters)	2,760
Weight (metric tons)	28.5
Caliber	2A37 152mm gun
Gun traverse (degrees)	15 right and left
Gun elevation (degrees)	-2 to +57
Ammunition stowage (rounds)	30
Unit of fire (rounds)	60
Effective firing range (kilometers)	28.5
Maximum firing rate (rounds per minute)	5 to 6
Emplacement time (minutes)	3
Powerplant	V-59-V-12 520 horsepower, direct injection, inertial supercharge, multi-fuel diesel
Power to weight ratio (horsepower per ton)	19
Engine power (kilowatts)	382
Maximum road speed (kilometers per hour)	63
Maximum range (kilometers)	500
Fording (meters)	1.05
Gradient (percent)	58
Side slope (percent)	47
Trench (meters)	2.5
Vertical obstacle (meters)	0.7
Unit cost	$1,500,000 (1992 export price)

2S5 Giatsint 152mm self-propelled gun.

fore the first projectile lands on the enemy position. A unit of fire is sixty rounds.

A total of thirty projectiles and charges are carried. The projectiles are stowed vertically in a carousel device in the left side of the rear compartment while the three rows of charges, each of ten, are also stowed vertically yet have a horizontal conveyor belt that returns under the floor of the vehicle. The charge consists of a conventional cartridge case containing the actual charge. Once the gun is fired, the breech automatically opens and the spent cartridge case is ejected to the rear.

The effective range of the gun is twenty-seven kilometers using conventional ammunition and thirty-seven kilometers using rocket-assisted projectiles. It is capable of firing nuclear projectiles as well as conventional and chemical rounds. In addition to the normal high explosive fragmentation (HE-FRAG) projectile, other types include concrete piercing and improved conventional munitions. The former Soviet Union has also developed the Krasnapol, a 152mm laser designated guided projectile to engage armor targets at extended ranges. The Krasnapol round can be fired by both the 2S3 and 2S5.

While a welcome improvement over the earlier towed artillery systems, the 2S5 does have a number of disadvantages. These include a lack of protection for some of the gun crew when in the firing position, and sustained fire missions in a NBC environment would be difficult. In addition, the limited traverse of the main armament means that targets outside of the thirty degree arc (fifteen degree left and right) cannot be quickly engaged.

The 2A36 has also been exported to Finland and Iraq with large numbers from Iraq being captured or destroyed during Desert Storm. In contrast to the public appearances of the 2A36, the 2S5 was never seen during annual Moscow military parades. However, it is believed that the 2S5 entered service with the Soviet Army around 1980 and total production is thought to have exceeded 2,000. Now, as part of its efforts to obtain foreign currency and keep at least some of its production lines open, the former Soviet Union is offering the 2S5 for export at around 1.5 to 2.0 million dollars per system. The "Msta-S" 2S19 152 mm self-propelled howitzer, will continue to replace the 2S3 and 2S5 in frontline units.

The 2S7 Pion Self-Propelled Gun

The most potent of the Soviet self-propelled guns to emerge was the 2S7 Pion (Peony) which was first identified by NATO in 1975. The 2S7 was developed under the designation Obiekt 216 by the design bureau at the Kirovskiy Plant in Leningrad (St. Petersburg) under

2S7 Pion 203mm self-propelled gun.

Nikolai Sergeyevich Popov. An improved model, the 2S7M, is also known to have been built and deployed in Russia. In addition, the 2S7, with an estimated range of well over thirty kilometers and new ammunition, gave Soviet artillery the capability to fire nuclear rounds. This was not the case with guns of 122mm or with the old 2B4M. With its armored tracked chassis, the 2S7 continued the trend in the Soviet army away from towed field guns, begun in 1973 with the introduction of the 152mm 2S3 self-propelled gun; the trend continued with the deployment of the 122mm 2S1 self-propelled gun shortly thereafter. In the Soviet Army, it was deployed at front level in special high command reserve (RVGK) heavy artillery brigades with 48 2S7 Pions per brigade. Prior to the Soviet breakup in 1991, there were 305 2S7s in Soviet Army service west of the Urals.[5]

The 2S7 Pion vehicle has the driver, commander, and gunner in the forward compartment and the four other troops in the vehicle's center compartment. Before firing, the chassis is lowered using the idler for greater stability. An armored cab projects over the front of the chassis. This adds to the overall height of three meters and also acts as a counterweight to the gun which is mounted far to the rear in a movable cradle. When traveling, the cab accommodates the driver, gunner, and two other men of the seven-man gun crew. They enter the cab via two roof hatches. The V-12, liquid-cooled, four-stroke diesel engine is situated at the rear of the cab. The long exhaust ducts to the left and right of the cab limit the fording capability of the 2S7 to 1,200 millimeters. Maximum road speed of the forty-six-ton 2S7 is fifty kilometers per hour although the normal cruising speed is around twenty kilometers per hour. Maximum road range is 650 kilometers although this figure is not achievable in combat since the 2S7 does not have an auxiliary power unit and the vehicle's engine has to be used as a power supply during firing.

Behind the engine, in the middle of the vehicle, are two additional hatches on the left and right. These lead to the second crew compartment for the other three members when the vehicle is on the move. This compartment also contains the antenna base of the radio. The R-173 is fitted in the 2S7M. The 1V116 command vehicle is used when traveling. Ammunition is stowed in a compartment at the rear of the vehicle. A small cart is carried on the vehicle to assist in handling the ammunition. Loading is assisted by a power rammer system. The chassis can be locked down in a lowered position with the idler at the

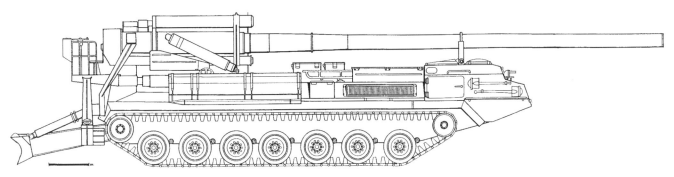

2S7 Pion Self-propelled 203mm Gun

2S7M Technical Data

Crew	7 men
Length (millimeters)	13,000
Width (millimeters)	3,380
Height (millimeters)	3,000
Weapon	2A44 203mm gun
Range (kilometers with a 110 kilogram HE round)	37.5/55.0 with rocket-assisted
Muzzle velocity (meters per second)	960
Rate of fire (rounds per minute)	2
Elevation (degrees)	0 to +60
Azimuth (degrees)	+30
Preparation time (minutes)	6
Ammunition (on-board rounds)	8
Engine	V-12 cylinder four-stoke diesel (575 kW)
Maximum vehicle speed (kilometers per hour)	50
Range (kilometers)	500
Fording (millimeters)	1,200
Ground clearance (millimeters)	400
Gradient (degrees)	25
Side slope (degrees)	15
Combat weight (metric tons)	46

rear touching the ground.

The 2S7 Pion is capable of firing conventional projectiles to a range of 37.5 kilometers and rocket-assisted rounds to a range of forty-seven kilometers, making it one of the longest ranged artillery systems in use today. The 2A44 203mm gun is mounted on the rear, and has power elevation and traverse, controlled from a small crow's nest on the left rear corner of the hull. Loading is power-assisted using a power rammer fitted to the right rear of the hull.

The 2A44 gun, weighing 14.6 tons, is fitted to the rear. The rifled gun, is nearly twelve meters long and is secured during traveling by a manually operated lock on the cab roof. The gun does not have a muzzle brake and thus no fume extractor. The gun fires separate ammunition and has a screw-type breech block opening to the right. With a muzzle velocity of 960 meters per second on maximum charge, the 2S7 can fire conventional HE rounds 37.5 kilometers. The average barrel life for the 2A44 is 450 rounds. A spade running the width of the vehicle is located at the rear and is hydraulically lowered before firing commences to ensure stability.

The typical combat load is forty rounds which are carried primarily on a truck. The one meter long 110 kilogram ZOF-43 projectiles are loaded from the ground with a movable loading system; four rounds are carried as reserve on board. Each round is mounted via the right rear and remotely controlled by the loader. An integrated system rams the round directly. Completely combustible bag charges are used. The range is determined by the charge which weighs from forty-five kilograms to the smaller charge of 24.9 kilograms.

The 2S7 does not have self-defense systems such as machine guns, mountings for the crew's personal weapons or smoke dischargers. The rate of fire of up to two rounds per minute, although fairly high for such a caliber, is still relatively low compared with those of modern 152 and 155 mm guns.

The 2S7 Pion exists in both a 2S7 Pion and 2S7M Pion-M variant. The 2S7M was developed in 1983 and increased the rate of fire from two rounds per minute to three rounds per minute. The 2S7M has a sustainable rate of about thirty rounds per hour. This system has an upgraded chassis and can carry eight more rounds on board. It also includes a data link system for receiving and employing firing data from command centers. The modification also increased the vehicle and system durability. The future viability of this vehicle is in question due to the improving capabilities and reduced cost of employing precision guided standoff missiles and MLRS.[6]

The 2S19 Msta-S Self-Propelled Gun

The 2S19 is a new 152mm self-propelled gun, combining a version of the 2A65 towed 152mm gun with a new chassis based on T-72 and T-80 tank components. The 2S19 was developed to combine the roles of the 2S3 and 2S5 in a single vehicle. A major aspect of the program was to develop an automated ammunition handling system and improved fire controls to provide the capability to avoid counter-battery fire by rapid relocation.[7]

In the 1985, work began on a new self-propelled 152mm howitzer that would replace the well-known 2S3

Akatsiya. Even though it had undergone several modernization and update programs, the 2S3 had fallen behind other contemporary artillery weapons. In part, this was due to the fact that its 2A33 howitzer was a modified version of the D-20 towed 152mm howitzer developed in the 1955, and the chassis was part of the GM-300 series which had first seen service as the chassis for the 2K11 Krug (SA-4) surface-to-air missile complex in 1964.

Therefore, the decision was made to develop a completely new self-propelled artillery weapon (SAU). The weapon began development under the codename Ferma (Dairy Farm) at the Uraltransmash Works, under the direction of chief designer Yuriy Tomashov (deputy chief designer Mikhail Tretyakov). The resulting work produced a completely new vehicle, which was accepted for service in 1989 as the Msta-S (where the "S" stands for self-propelled; this is to differentiate it from the 2A65 Msta-B, which is the Buksiruemyi or towed version). For that reason, the Uraltransmash Works also created the 2Kh5l special purpose training system to educate the crews that will use the 2S19. The first series produced 2S19s were made at the Uraltransmash Works plant in Yekaterinburg, but production has now been shifted to the Bashkiri Machine Plant (STEMA) in Sterlitamak, a new plant geared towards self-propelled artillery manufacture. Msta is named after a river in the Ilmen district, a break in the previous Soviet practice of naming self-propelled guns after flowers or plants.[8]

The armor on the 2S19 provides reliable protection for the crew, ammunition supply, and all equipment from small arms fire or shell fragments. The 2S19 consists of an armored hull and turret. The diesel engine, transmission, control systems, and suspension are located within the hull. The running gear of the chassis is formed from standardized parts from the T-72 and T-80 tanks.

The 2S19 is powered by a twelve-cylinder V-type V-84A diesel engine producing 840 horsepower; this engine can use any of six different types of fuel. It is a high-speed four-cycle liquid-cooled diesel motor. The V-84A is also used in the latest models of the T-72 tank. It can also be replaced by the V-46 family of engines (which produce 780 vice the 840 horsepower of the new engine). The gearbox has seven forward speeds and one in reverse. The electricity is provided via four accumulator batteries producing twenty-seven volts direct current power.

The running gear itself (per side) consists of six road wheels, an idler wheel with track tensioning device, a drive wheel with bolted-on toothed drive rings, and five return rollers. All of the road wheels, idler wheels, and drive wheels are identical to those used on the T-80 tank. The suspension uses long torsion bars, which cause the corresponding road wheels on the right and left sides to be offset. The vehicle uses the suspension components of the T-80. The first, second, and sixth road wheels are equipped with regulated telescopic shock absorbers, which are controlled when firing the weapon by blank-

2S19 152mm self-propelled gun

2S19 Technical Characteristics

Developer	Uraltransmash Design Bureau
Year accepted for service	1989
Producers	Uraltransmash Works and the plant in Sterlitamak
Year entered production	1989
Weight (tons)	42
Crew	Five men
Crew when using ammunition from ground supplies	Seven men
Gun dimensions:	
—Length over gun tube (millimeters)	11,917
—Width over tracks (millimeters)	3,380
—Width over side skirts (millimeters)	3,584
—Height to roof of turret (millimeters)	2,985
—Firing height (millimeters)	2,270
—Distance between track centers (millimeters)	2,800
—Length of track on ground (millimeters)	4,704
—Track width (millimeters)	580
—Ground clearance (millimeters)	435
—Road wheel diameter (millimeters)	670
Armament:	
—Main gun	2A64 152mm Howitzer
—Anti-aircraft machine gun	12.7mm NSVT Machine Gun in PZU-5 AA Mount
—Vehicle close-in protection	Five stowed AK-74 Automatic Rifles
Ammunition storage:	
—Main gun rounds	50 rounds
—Anti-aircraft rounds	300 rounds
—Gun loading	Automatic for projectiles, semi- for propellant
Engagement angles for main weapon:	
—Elevation (degrees)	40 to +680
—Traverse (degrees)	360
Range of fire:	
—With HE-Frag ammunition (meters)	24,700
—With rocket assisted ammunition (meters)	28,900
—Rate of fire (rounds per minute)	7 to 8
Regulated rates of fire:	
—First hour (rounds)	up to 100
—Each additional hour (rounds)	up to 60
—Time to go from travel to firing mode (minutes)	1-2
Engine	V-84A
Power output (horsepower diesel)	840
Auxiliary power unit (kilowatt)	16
Power to weight ratio (horsepower per metric ton)	18.5 to 20.0
Highway speed (kilometers per hour)	60
Range without refueling (kilometers)	500
Fording depth (unprepared) (meters)	1.5
Fording depth (width of 1,000 meters) (meters)	5.0
Gradient (degrees)	25
Maximum gradient (percent)	47
Maximum incline (percent side slope)	36
Maximum step (meters)	0.5
Maximum trench crossing width (meters)	2.8
Service guarantee life (kilometers)	5,000
Engine guarantee life (hours)	350
Unit cost:	$1,600,000 (1992 export price; 2A65 towed howitzer $600,000)

ing off their oscillations. This means that the weapon does not need to use auxiliary firing jacks or spades for stabilization.

The 2S19 uses the 580 millimeter wide track with rubber inside facing, which is identical with that used on the T-80 tank and 2S7. The suspension can handle firing all types of projectiles without the necessity of having to previously prepare a firing position. The enclosed welded turret contains the 2A64 howitzer with its aiming and pointing mechanisms, an automated shell selection and handling mechanism, the conveyor which can be used to bring up shells from the ground, the ammunition supply with its programmable selector, an execution device which brings the selected shell from the ammunition supply into line with the 2A64 howitzer, the onboard power supply with automatic power regulation, the filtering and ventilation system, the communications suite (internal intercom, external wire and radio), the PZU-5 antiaircraft machine gun mounting, a mechanism to hermetically seal the breech of the howitzer, and the device for flushing gases from the fighting compartment.

The armament of the vehicle consists of the 2A64 152mm howitzer mounted in the turret, and the 12.7mm NSVT anti-aircraft machine gun mounted in the PZU-5 antiaircraft mount on the commander's cupola. The 2A64 is a modified version of the 2A65 towed howitzer.

Fire can be conducted with this weapon using the 3OF-45 high-explosive fragmentation round (up to 24,700 meters) and the 3OF-61 (to a range of 28,900 meters), the 3OF-23 cassette round (carrying forty-two anti-armor sub-munitions and reaching out to 26,000 meters), active radio jamming projectiles from the 3NS30 family (to 22,300 meters), the 3vDTs8 nuclear projectile, and others. It can also use all standard ammunition for the D-20 and 2S3 howitzers, as well as guided projectiles such as the 30F39 Krasnopol laser guided projectile, which includes the 3V0F64. Target illumination for these rounds can be carried out by an artillery observer with the 1D15 (PP-3) or 1D22 laser devices. Thanks to the semiautomatic shell handling system, the crew of the weapon can achieve a rate of fire of seven to eight rounds per minute with the internal ammunition supply and six to seven rounds per minute when conveying shells up from the ground.[9] This translates into an eight-tube battery of these howitzers being able to place nearly three tons of steel on target in one minute. At maximum range, that is nearly seventy shells in the air at one time before the first one hits the target.

At a firing range 200 kilometers from Abu-Dhabi, a demonstration was held of the firepower of the 2S19 (Msta-S) at the 1993 International Defense Exhibition (IDEX-93). Using forty Krasnopol laser-guided projectiles, the 2S19 hit thirty-eight targets at a range of fifteen kilometers.

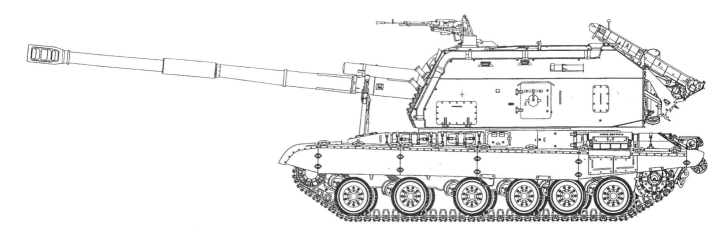

2S19 Msta-S
Characteristics of Basic Ammunition for the 2S19 SP Howitzer

Designator	Range (meters)	Weight (kilograms)	Weight of Filler (kilograms)	Length (mm)	Muzzle Velocity (mps)	Type of Round
3OF-45	24,700	43.56	7.65	864	810	HE-Frag
3OF-61	28,900	42.86	7.80	864	828	Rocket assisted
3OF-23	26,000	42.80				42 AT sub-munitions
3NS-30	22,300	43.56	8.20			Active jamming round
3OF-39	20,000	50.00	6.60			Laser-guided

Since the main ammunition supply is located in the turret, it can be selected and loaded no matter which direction the turret is pointing in relation to the hull without a problem. The loading mechanism also supports the conduct of fire at any angle of traverse or elevation at the maximum rate of fire. To assist the loader — who would have to handle a 43.56 kilogram load (the OF-45 projectile) the projectiles are loaded automatically. The shell racks are designed to handle various types of ammunition, to include special rounds. To search for the required projectile on the ammunition conveyer, and to control the entire process of loading the weapon, there is a loading system control mechanism. Besides that, it can count the number of each type of projectile carried. There are two separate conveyors, one each for projectiles and charges. Each of is served by its respective loader, and that increases the rate of fire. The ejection port for the

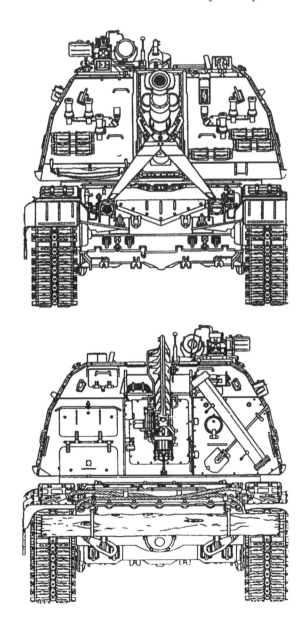
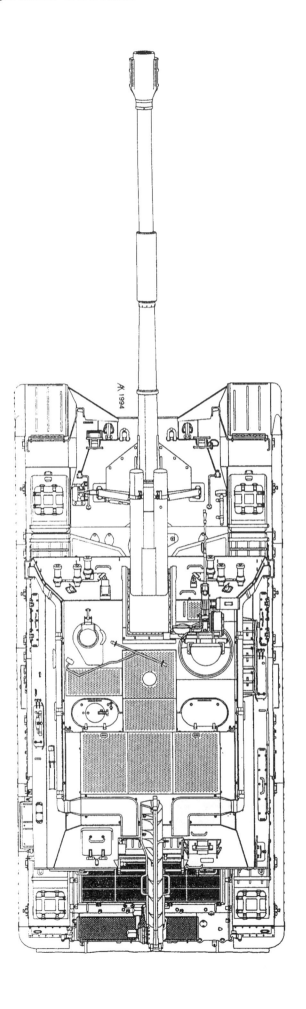

A 2S19 in Abu Dhabi in 1995. The overall size can be seen here.

expended casings is located in the turret above the howitzer. Having the ground conveyor provides for the conduct of fire using the selection of on-board ammunition and not wasting the on-board supply. When preparing for movement, the ground projectile conveyor is stowed on the back of the turret, whereas the ground charge conveyor is stored inside the turret.

The 2S19 has two sights: the 1P22, which is located in a rotating armored cover in the roof of the turret, and the 1P23 direct fire sight, the observation window for which is located in the glacis plate of the turret. The 1P22 is a 3.7 power optic and has an automatic stabilization feature in the horizontal plane, which works when the vehicle is not canted more than five degrees in either direction. In travel mode, to the right of the sight is a protective curtain which can be wrapped around the sight to protect the view piece. The 1P23 has a power optic and can track through angles of elevation of -4 to +55 degrees. The 2Eh46 howitzer drive mechanism is electric and automatically controlled in the vertical plane; horizontal movement is directed from a control panel. The automatic return to the correct angle of elevation after each round is fired simplifies the work of the gunner. When firing, the gunner only has to perform a single operation — he has to keep the panoramic sight leveled on his aiming point. The commander also has the mechanisms for aiming and firing the weapon. When the electrical system is turned off, there are backup manual controls for the weapon for loading and aiming.

The 2S19 is equipped with a data transmission and receiving system. The data can be sent via landline or by radio via the 1V122 vehicle over distances of up to 500 meters. This includes a system to control gunlaying, which is coordinated with the fire direction center vehicle.

The PZU-5 is provided for defense against lightly armored vehicles, helicopters and aircraft, and is analogous to the same system provided to the T-64 tanks and several models of the T-80 tank. It can be fired from inside the turret via remote controls. The 12.7mm NSVT Utes (6P11) machine gun has an sighted range of 2,000 meters and a rate of fire of 700 to 800 rounds per minute. Its range of elevation is -3 to +70 degrees. It is provided with five belts of ammunition, with sixty rounds in each belt. To provide for operations when the main engine is shut off, the vehicle is provided with an AP-18D autonomous power supply unit — this is a gas turbine generator producing sixteen kilowatts of power. It can operate without interruption for up to eight hours.

The 2S19 is part of an artillery firing battery complex (OBAK). The complete OBAK consists of a commander's vehicle (based on a BTR-80 chassis), a senior officer of the battery's vehicle (based on a Ural-43201 truck), and up to eight 2S19 weapons. Any size artillery combined unit may be formed from these OBAK organizations.

The 2S19 is offered for export. In this condition, the vehicle is priced at $1.6 million. Special export versions of the 2S19 in 155mm caliber are also being offered, which will then permit the weapon to handle any NATO ammunition type of that caliber.

By February 1991, the Soviet Army had accepted some 371 2S19s and 2A65s weapons into service and by 1 January 1995 there were 563. Russian artillery designers have began working on an improved version of the

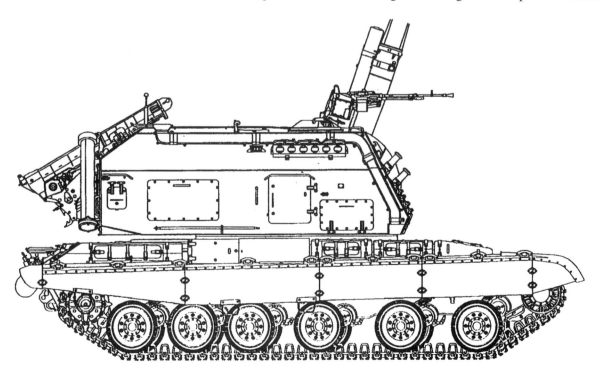

weapon, the 2S30 Iset. In addition, enhancements to the 2S19's ability to fire guided projectiles such as Krasnopol, utilizing a remotely piloted vehicle (RPVs) as a source of laser target illumination is also being development for current 2S19s and its future follow-on system.

The Bereg 130mm Coastal Defense Artillery System

Currently many of the fielded self-propelled artillery systems are not capable of dealing effectively with coastal artillery defense because of problems associated with targeting sea-based targets. The problem is that in order to hit a series of small amphibious targets, like landing craft and amphibious-capable AFVs and tanks, a coastal-based artillery system requires a constant stream of data on the position and the speed of targets and many conventional artillery system constructed today are not capable of handling this demanding fire control task. In order to deal with this deficiency, special fire control systems and coastal radar stations are required for coastal defense artillery systems to be effective.

One novel solution designed to deal with the unique requirements of coastal artillery is the new Bereg 130mm self-propelled coastal defense artillery system constructed by the Barricade Production Association. The Bereg system is based on the same 130mm gun and turret mount as that found on the Russian Sovremennyy-class destroyer (anti-surface warfare ship). The chief difference between the naval and land gun versions is that the naval gun mount is water-cooled while the Bereg is air-cooled.

The Bereg coastal defense artillery gun mount is designated A-222 and utilizes six crew members; a commander, gun layer, and four loaders. The turret can engaged enemy forces over a wide sector of ±120 degrees and the gun can be laid for elevation from -5 to +50 degrees. The gun has a maximum range of twenty kilometers. Two stowage racks are located at the rear loaders section of the turret mount and minimal basic load is forty based-fused fragmentation and high-explosive rounds. In addition, the system can also utilizes three other types of rounds: a nose-fused anti-aircraft round, a practice round, and a training round.

The Bereg coastal defense artillery complex is composed of a central radar, target detection, and fire control vehicle, called the central station; a combat support vehicle, and six self-propelled artillery mounts equipped with a 130mm gun mount. The total crew strength of this complex is six gun crews of six men totaling thirty-six men, a central station utilizing seven men, and a combat support vehicle of four men, for a complex grand total of forty-seven men.

All vehicles of the Bereg system are mounted on a similar cross-country chassis, the MAZ-543M — an 8 x 8 wheeled truck. This vehicle is capable of a maximum road speed of sixty kilometers per hour and has a fuel distance of 650 kilometers. The high-cross country capacity of the chassis, the use of night vision equipment, and automated navigation equipment enable the vehicle to move rapidly from one firing position to the next. These actions can also be accomplished twenty-four hours a day, and during periods of active or passive jamming or in any climatic conditions. The entire Bereg complex can change its relative firing positions in as little as five

Bereg 130mm self-propelled gun complex.

and as long as thirty minutes depending upon the distance traveled.

The central stations radar's can be utilized to search and track targets on ocean or land. The fire control data can be sent to the self-propelled gun mounts utilizing four methods: (1) the central station's fire control system, (2) one to two observation posts, (3) a fire control helicopter, or (4) the firing vehicle's built in systems. The 130mm gun mount has the ability to deliver fire independently and to gather its own fire control solution utilizing its own set of optical sights, the commander's vision and designation devices, ballistic computers, and a laser rangefinder.

The central station consists of the fire control system BR-136, with radar and target tracking channels and the required communications equipment to provide this data to the firing units. The BR-136 weighs 5,000 kilograms. The central station can be positioned more than a 1,000 meters in distance and 300 meters in height from the six-gun systems and utilizes a UHF data link to transmit fire control information. The central station utilizes five separate compartments: engine, antenna station, high-frequency power unit, radio-operator, and fire control operations station. The engine compartment contains a diesel-electric generator which provides standby power when it is not able to receive it from the combat support vehicle or centralized power lines. The diesel-electric generator is an AC power of 30 kilowatt, a voltage of 220 to 280 volts and a reducer converter rated at a power of fifteen kilowatt and a secondary voltage of 220 volts.

To power the radar and the fire control system the central station utilizes power from diesel generators located in the combat support vehicle. In addition, each vehicle is also equipped with its own auxiliary power unit (APU) which reduces the dependency on external vehicles and allows the Bereg complex to maintain high-states of

Bereg 130mm self-propelled gun.

Bereg Gun Systems

Effective firing range (kilometers)	Up to 20
Effective detection and location of targets (kilometers)	Up to 35
Time required to change over to fire position (minutes)	5 to 30
Maximal distance between central station and gun mounts	
—In range (meters)	1,000
—In height (meters)	300
Maximum speed of target engagement (knots)	200
Radar coverage (degrees)	360
Targets tracked by radar and targets engaged simultaneously	4/2
TV/Laser system coverage (degrees)	± 135
Moving sea target destruction time with 0.8 probability (minutes)	1 to 2
Total crew size (with 6 guns and 2 support vehicles)	47 men
Maximal road running speed (kilometers per hour)	60
Fuel distance (kilometers)	650

Central Station

Target acquisition minimum range (kilometers)	35
Mass (kilograms)	43,400
Length (millimeters)	15,200
Width (millimeters)	3,240
Height (millimeters)	4,415
Crew size	7 men

Self-Propelled Gun Mount

Caliber	130mm
Firing rate (shots minimum)	10
Organic load of ammunition (shots minimum)	40
Ammunition types	HE/AA/Practice/Training
Firing angle (degrees)	± 120
Elevation angle (degrees)	-5 to 50
Mass (kilograms)	43,700
Length (millimeters)	12,950
Width (millimeters)	3,100
Height (millimeters)	3,925
Crew size	8 men

Combat Support Vehicle

Power (kilowatt)	2 by 30
Men protected by survival systems	10 men
Fuel, food, and water reserve (days)	7
Mass (kilograms)	43,500
Length (millimeters)	15,936
Width (millimeters)	3,230
Height (millimeters)	4,415
Crew size	4 men

operational readiness over the course of several days.

The antenna section houses the receiver and transmitter: a circular-scan radar antenna with its rotator and an optical-electronic target detection/tracking channel range-finding and sighting device, mounted on an extensive base. The radar's antenna can rotate at speeds of between fifteen to twenty-five rotation per minute. The expected service life of the radar is twenty years. The antenna is raised through the central-station's van-roof and weighs 470 kilograms. The operator's compartment house's the commander, central station commander, fire-control operator, artillery electrician, and radio operator.

The gun is leveled through the use of either manual or internally controlled mechanical jacks, which enable the crew to ensure a flat plane for the rotation of the radar antennae.

The BR-136 fire control systems generates angles of elevation and traverse for each of the six-guns in the system while also providing point corrections. Through the use of a central laying mode laying angles are generated for all six guns with allowances made for ballistic, meteorological corrections, and gun separation. The central station in a fully automated process, computes fire-control updates and relays this information in near-real time to each gun. The total laying angle error, utilizing this mode, results in a maximum error of three mrads. Adjustments are sent-back after each firing with new data adjusting each gun's parameters based on the radar-observed shell splashes.

The BR-136's radar is capable of utilizing two modes of operation either a circular or sector-scan and can track up to four targets while placing fire on two of those four targets. The circular scan radar's information is displayed on a rectangular azimuth scan indicator which provides two electronic cursor for the two automated tracking channels. Each of these two automated channels is three operated-assisted modes: (1) manual laying, (2) aided tracking, and (3) automated. In the automated mode there are two settings; programmed target lock-on and automatic tracking.

The Bereg complex's fire control system is capable of instantly transferring fire to the next set of targets in a track. The radar provides a continuous 360 degree coverage over water and land. The radar and fire control system are able to detect targets out to thirty-five kilometers and the gun is capable of hitting targets out to twenty kilometers and is able to engage targets traveling at speeds of up to 200 knots. In addition to the radar, the complex also employs a television-based optical scanning system, weighing 235 kilograms, and is designed to search a sector of ±135 degrees of the sea surface day or night. The first firing and tracking channel has provisions to assist the radar utilizing a TV/Laser combination to track any radar-detected targets.

The fire control channels of the Bereg complex also allow the gun system to operate against sea or coastal targets. When all the guns of the system are netted together the BR-136 enables the gun systems to fire in succession and at rates starting at four and every even number up to twelve rounds per minute. The firing process can be monitored by the gun commander utilizing various displays in the gun turret mount.

The combat support vehicle houses a removable power supply unit with two diesel-electric sets. The storage capacity of the fuel tanks, water, and food allows for continuous operation for up to seven days. Also located in the combat support vehicle is equipment which monitors radioactive/ chemical contamination and the arrival of each of the six self-propelled artillery gun mounts to their designated areas. In addition, the van houses two, four-berth compartments, a washer and dryer, and a kitchen which is capable of feeding up to four men. The combat support vehicle is manned by a crew of four. Located on top of the van is a mount for a 7.62mm PKT

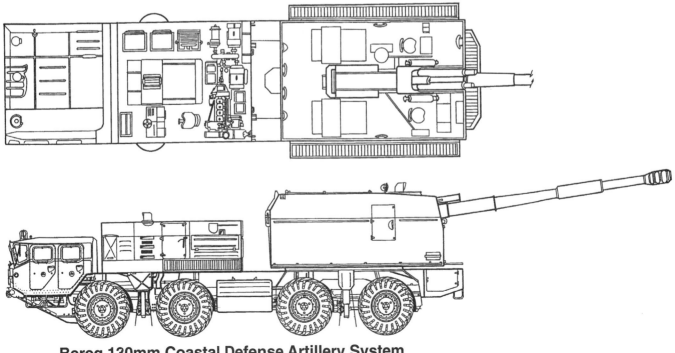

Bereg 130mm Coastal Defense Artillery System

machine gun and is intended to provide a point-defense capability for of this vehicle.

The MAZ-543M chassis is built in Minsk, Belarus at the Minsk Automotive Zavod (MAZ). The electronics fire control system is built in the Ukraine by an enterprise which privatized the experimental model, to facilitate on-going work. This system has successfully complete its state acceptance tests and is scheduled to be procured by Russia's coastal defense forces.[10] However, production and hence, deliveries of these systems have stalled, not because the major sub-components are being built and supplied by the Ukraine and Belarus but because a Russian enterprise, the joint stock company "Shemerlinsk Specialized Motor Vehicle Plant," refused to build the truck bed for the MAZ-543M 8 x 8 truck.[11]

This complex is currently being offered for export to several Middle Eastern and littoral Asian states. However, pricing data for this complex was not available from company representatives or from the advertising placard when it was debuted at the International Defense Exhibi-

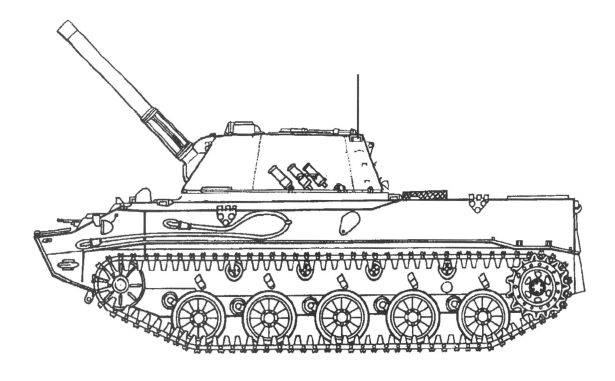

Germes 155mm Self-Propelled Gun System

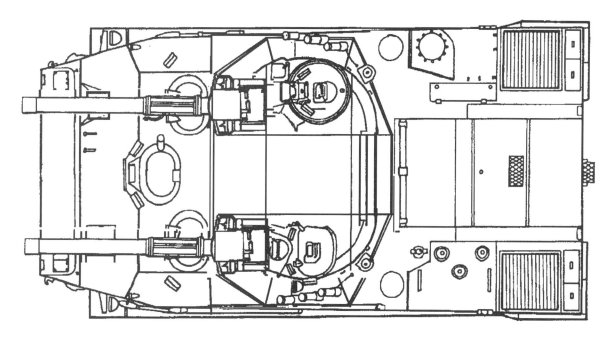

Germes Gun System

Firing range (kilometers)	12
Hit probability (percent)	0.8 to 0.9
Firing method	Salvo of Burst
Missiles per salvo (rounds)	2 to 4
Operational conditions	day or night from -50°C to +50°C
Reaction time (seconds)	10
Combat performance (targets per minute)	12 to 18
Ammunition load (rounds)	14 to 16
Vehicle weight (metric tons)	20

Guided Missile for the Germes Gun System

Missile weight (kilogram)	60
Caliber	155mm
Warhead type	Shaped-charge/ Blast-fragmentation
Warhead weight (kilograms)	30
Explosive weight (kilograms)	12
Guidance pattern (2-channel laser designator)	Semi-active homing
Terminal trajectory	Dive
Flight time to a range of 12 kilometers	54 seconds

tion (IDEX-93) in Abu Dubai, United Arab Emirates in March 1993. There are currently no reports regarding export sales of this system.

The Germes 155mm Self-Propelled Gun System

The Tula Design Bureau has begun to look for a foreign partner to develop a highly mobile multipurpose 155mm self-propelled gun system called "Germes."[12] Unlike the Bereg coastal artillery system the Germes does not have an operational prototype built to date. It is interesting to note that the gun caliber is 155mm rather than the standard 152mm utilized by most Russian artillery today. This indicates that this system is being offered to meet the requirements of foreign customers who already operate tube and self-propelled 155mm artillery systems. The Germes system is designed to counter massive armor attacks, provide coastal defense against marine assault, and provide platoon-level support in destroying point-targets. In addition, promotion literature claims that three Germes equipped fighting vehicles possess the same combat capability as two to three batteries of 155mm tubed artillery systems.

The Germes system utilizes two low-pressure 155mm howitzers with an on-board surveillance radar and laser target designator. The twin-mounted 155mm tubes are mounted on a modified BMD-3 chassis which is relatively light at twenty tons. The ability to mount such a gun system on a relative light and small BMD-3 chassis is done through the use of laser-guided and rocket-assisted missile rounds. The sixty kilogram projectile is advertised to have much greater range and power than the Krasnopol laser-guided 152mm artillery round utilized by the 2S3M, 2S5, and 2S19. The Germes system is capable of launching its rounds in either an indirect or direct mode, but is optimized for targeting point rather than area targets.

Recent reports suggest that India and South Korea have expressed tacit interest in the Germes development program, but history has shown that such arrangements seldom have produced results. This program's ability to garner foreign participation will determine its eventual ability to produce a prototype since it is clear that the Russian Ministry of Defense is unwilling or unable to finance such a project.

Post-War Self-Propelled Mortars Developments

Russian historians credit the invention of the modern mortar to a Czarist Army officer, Lenoid Gobiato. Gobiato's mortar was used for the first time in September and October 1904 against Japanese forces entrenched around Port Arthur during the Russo-Japanese War of 1904 to 1905.

Russia has historically been one of the world's leading producer of mortars and is currently the world's leader in self-propelled mortar systems. During World War 2, Soviet defense industry produced over 348,000 mortars of different caliber, more than all other belligerent countries combined. During World War 2 the Red Army employed the 160mm Model 1942, the largest mortar in general use (the exception being a handful of German super-heavy mortars such as the Karl Gerat). Several even larger mortars were under development including the ZIS-27 and S-16 240mm by the Grabin design bureau and the OB-29 240mm mortar by the OKB-172 bureau. In 1944, the Shavyrin design team at NII-13 in Leningrad began work on the M-240 mortar. Work was held up by the war, and the project was resumed in 1947 when Shavyrin reestablished his mortar design bureau in Kolomna. Development of the M-240 was completed in 1949 and it was accepted for Soviet Army service in 1950.

The first Soviet self-propelled mortar was the super-heavy "Oka" 420 mm mortar developed in the mid-1950s to deliver an early nuclear projectile. This system had a twenty meter tube, fabricated from one piece of steel and delivered a round out to a world-record distance of some forty-five kilometers. The chassis was derived from IS Stalin heavy tank components. A small number of these self-propelled weapons were manufactured and they appeared on parade in the early 1960s. The mortar was not adopted by the Soviet army due to Khrushchev's disfavor with such weapons and preference for missiles.

The towed M-240 mortar was used in artillery regiments of rifle divisions in the 1950s, but was withdrawn from divisional use in the late 1950s when the rifle divisions were mechanized as motor rifle divisions. The M-240 mortar was then transferred to army or front level heavy mortar battalions. The M-240 was not a very practical weapon. Its 3OF-864 high explosive round weighed 130 kilograms and so was very difficult for the crew to handle lacking a small crane.

The rejuvenation of breech-loading mortars was probably due to a revival in Soviet interest in nuclear projectiles. Although the Soviet Ground Forces were amply supplied with nuclear armed rockets like the Scud and Frog systems, nuclear projectiles are not as cumbersome as the rockets, and have much greater accuracy. In addition, a 240mm mortar round from the M-240 mortar system has a destructive force which exceeds conventional artillery rounds of similar caliber. Therefore, they are more useful for strikes against protected targets like entrenched command centers, enemy forces on reverse slopes of hills or enemy defensive fortifications.

The 2S4 Tyulpan Mechanized Mortar

The development of the first practical self-propelled heavy mortars began in the early 1960s by Yuriy V. Tomashov's Design Bureau at the Uraltranmash plant under the designation of Obiekt 305. This combined the chassis used on the 2K11 Krug/SA-4 Ganef chassis with a modified version of Shavyrin's M-240 mortar, the 2B8, developed by Yuriy N. Kalachnikov at the Perm Machine Building Plant.[13] The resulting vehicle was designated

2S4 Tyulpan 240mm self-propelled mortar.

2S4 Technical Data

Maximum firing range	
—High-explosive fragmentation round (meters)	9,650
—Base-Bleed high-explosive fragmentation round (meters)	18,000
Rate of fire	
—With a medium position of the barrel at 60 degrees	62 seconds
—With an extreme position of the barrel at 80 degrees	77 seconds
Angle in elevation	
—Along the vertical (degrees)	+50 to +80
—Along the horizontal	+/- 10
Combat load	
—Mortar rounds	40
—Cartridges for the PKT machine-gun	1,500
Engine data	
—Engine type	V-59 diesel
—Maximum horsepower	520
—Power-to-weight ratio (horsepower per ton)	15.71
Maximum road speed (kilometers per hour)	Up to 60
Range based upon fuel (kilometers)	500
Fuel capacity (liters)	850
Crew	
—Men in self-propelled vehicle	4 men
—Men in support vehicle	5 men
Weight	
—Combat weight (kilograms)	27,500
—High-explosive mortar round (kilograms)	130
—Base-bleed mortar round (kilograms)	225
Vehicle clearances	
—Fording (meter)	1.00
—Vertical obstacle (meter)	0.70
—Trench (meter)	2.79
—Slope (percent)	32
—Gradient (percent)	65
—Step (meters)	1.10

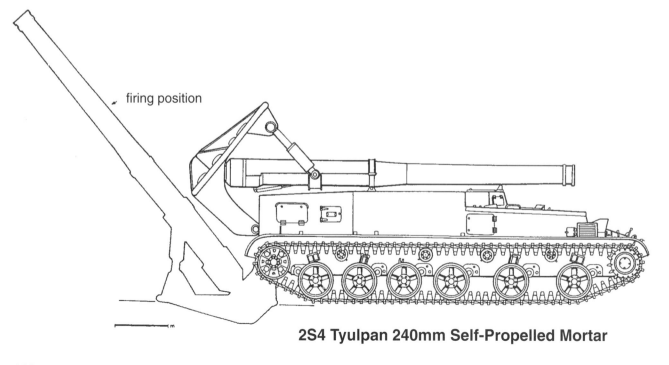

2S4 Tyulpan 240mm Self-Propelled Mortar

the 2S4 Tyulpan (Tulip) and was accepted for service in the early 1970s. The 2S4 was first identified by the West around 1975, hence the STANAG designation M1975 240mm self-propelled mortar.

The 2B8 240mm mortar system has a minimum effective range of 800 meters and a maximum effective range of 9.65 kilometers when firing a 130 kilogram HE-fragmentation round, and eighteen kilometers when firing a rocket-assisted 228 kilogram HE-fragmentation round. The system takes sixty-two seconds to load when at medium elevation (sixty degrees) and seventy-seven seconds when at high elevation (eighty degrees). The 2S4 mortar has an electro-hydraulic elevation drive which permits it to be fired from +50 to +80 degrees and in azimuth in either direction of ten degrees. The 2S4 contains stowage for forty standard mortar rounds (or twenty long-range rocket rounds) and 1,500 rounds of PKT 7.62mm ammunition. The 2S4 weighs 27.5 metric tons, has a maximum road-speed of sixty kilometers per hour, and a range of 500 kilometers.

The 2S4 Tyulpan's design is unconventional. In the travel mode the barrel is laid down parallel to the roof and the mortar is then able to be transported in even the roughest of terrain. The 2B8 is mounted on a hydraulic cradle on the rear of the vehicle, which is lowered to the ground, base-plate-first for firing. The mortar fires toward the rear of the vehicle. Like the majority of its smoothbore counterparts, the 2S4 is a breech-loading mortar. Inside the hull of the vehicle, a mechanized system located around two drums of mortar rounds (up to twenty rounds in each drum) by the turn of a drum raise a round through a hatch in the vehicles roof. The mortar breaks at the breech, and pivots on the trunnion using a hydraulic system, so that the open breech end is facing the rear of the hull and the autoloader.

The projectile and requisite bag charges are then pushed into the barrel by a special telescoping rammer. The round is fired utilizing either a mechanical or electrical trigger. There are two basic types of rounds utilized in the combat load of the 2S4: a conventional high-explosive fragmentation round and a base-bleed round. The latter round is equipped with a rocket engines located on the bottom of the round which are set off after firing to increase the round's flight range. In addition, the system can employ other types of projectiles to include, chemical, concrete-piercing, and "spetszariyad" tactical nuclear rounds.

Details of the deployment of the 2S4 are lacking, but it is probably used in heavy mortar battalions in special heavy artillery regiments at army or front level. The 2S4 mortar system has no equal in caliber to it in any foreign inventory. The 2S4 was one of the rarest types of Soviet self-propelled artillery. Only fifty-five were in service with the Soviet Army west of the Urals prior to the breakup of the USSR in 1991. In addition, the system has been exported to the former Czechoslovakia (now Czech Republic and Slovakia), Iraq, and reportedly to Lebanon. The Lebanese 2S4s were supplied to the Christian Army via Iraq.

The 2B9 Vasilyek 82mm Mortar Mounted on a MT-LBu

The 2B9 Vasilyek (Cornflower) automatic mortar was introduced in the early 1970s. The 2B9 is a breech-loaded, recoil-operated mortar which utilizes a rapid four-round clip of 82mm high-explosive fragmentation (HE-Frag)

2B9 Technical Data

Designation	2B9
Crew	5 to 6
Gun caliber	82mm
Traveling weight (kilograms)	645
Firing position weight (kilograms)	645
Dimensions in traveling mode:	
—Length (millimeters)	4,115
—Width (millimeters)	1,576
—Height (millimeters)	1,180
Mortar bomb weight (kilograms)	3.2
Maximum muzzle velocity (meters per second)	270
Maximum range (kilometers)	4.27
Minimum range (kilometers)	0.80
Maximum rate of fire (rounds per minute)	100
Operational arc	
—Traverse (without moving) (degrees)	+/-30
—Traverse (with moving) (degrees)	360
—Traverse (with moving) (degrees)	-1 to 85

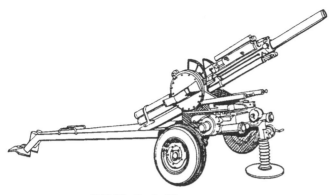

2B9 Vasilyeh 82mm mortar.

rounds. The rate of fire for a four round clip is around two seconds. The Soviet Army had fielded improvised mortar vehicles during the Afghanistan war. Units equipped with the 2B9 Vasilyek automatic mortar removed the wheels and mounted the weapon on top of the rear deck of MT-LBu artillery prime movers. By using improvised attachments, the 2B9 could be fired from the roof. This concept has obviously proven popular, as during the 1994 fighting in Chechnya, such mortar-MT-LBu combinations were still in use.

The 2B9 is still in service with Russian airborne units who have maintained it in their assault units. The 2B9 is gradually being replaced in frontline service by the 120mm 2B11 mortar system. In 1990 Hungary debuted development of a self-propelled version based of the MT-LBu Artillery Command and Communications Vehicle (ACRV). Hungary is offering their modified 2B9/MT-LBu self-propelled mortar vehicle 'for sale' overseas. In addition, the Hungarians are offering this system with a HEAT round to attack armor targets.

The 2S9 Nona-S

The Soviet VDV Airborne Force had a long standing requirement for an airmobile fire support vehicle to re-

2S9 Nona-K 120mm self-propelled weapon.

place the outdated ASU-85 tank destroyer. Development of this vehicle was undertaken by Shabalin's design bureau in Volgograd. The associated 2A60 gun/mortar combination weapon was developed by Avenir G. Novozhilov of the Central Scientific Research Institute for Precision Machine Construction (TsNII TochMash) and Yuriy N. Kalachnikov from the Perm Motovilikha Machine Construction Plant. The chassis for this airborne assault vehicle is based on the lengthened BTR-D transporter of the BMD-1 airborne vehicle family chassis. The chassis was modified to permit it to be locked down while firing. When accepted for service in 1981, the new vehicle was designated the 2S9 Nona-S (the towed 2B11 version being designated Nona-K: s=samokhodnaya, self-propelled; k=kolesnaya, wheeled).

The 120mm 2A60 is a unique weapon, firing mortar bombs rather than conventional ammunition, but unlike most mortars is rifled rather than smooth bore. It is breech loaded, and can fire horizontally like a conventional howitzer, or vertically, like a mortar. This unusual option was selected since the designers felt that a thin-walled mortar bomb offered more high-explosive on target for a given caliber than comparable howitzer ammunition, making it well suited to airborne operations where weight is at a premium. Russian sources boast that it offers the explosive effect of 152mm howitzers. Although lacking the range of conventional artillery, the weapon is more compact and lighter, and the range is adequate for typical fire support missions. Although turret mounted, the weapon cannot be traversed through the entire 360 degrees, but has a limited forward firing arc.

The 2A60 120mm combination weapon system on the Nona series was designed to accept a wide range of Western mortar munitions. Indeed, the designers pro-

2S9 Technical Data

Designation	2S9
Russian Name	Nona-S
Crew	4
Combat weight (kilograms)	8,700
Length (millimeters)	6,020
Width (millimeters)	2,630
Height	
—Maximum ground clearance (millimeters)	2,300
—In firing configuration (millimeters)	1,900
—Ground clearance (millimeters)	400 to 500
Maximum speed	
—Road (kilometers per hour)	60
—Water (kilometers per hour)	9
Maximum range	
—Road (kilometers)	500
—Water (kilometers)	75 to 90
Terrain obstacles	
—Gradient (percent)	60
—Slope (percent)	33
Gun elevation/depression (degrees)	+80/-4
Gun traverse (degrees)	35 left and right
Ammunition storage (rounds)	60
Engine	
—Diesel	5D20 engine
—Horsepower	240

120mm Ammunition

Type	HE	HE-RAP	HEAT
Length (millimeters)	828	835	960
Weight (kilograms)	19.8	19.8	13.17
Minimum range (kilometers)	1	-	0.5
Initial muzzle velocity (meters per second)	367	367	560
Rate of fire (rounds per minute)	7	7	7
Armor penetration (millimeters)	n/a	n/a	600

vide a data plate on the vehicle listing the ballistic characteristics of French, and other European 120mm mortar bombs on the presumption that airborne forces are likely to be able to exploit captured enemy arms dumps.

The 2A60 combination weapon consists of a rifled barrel with a breechblock, a cradle with a safety guard made of aluminum-magnesium alloy, a counter-recoil system, and an elevation mechanism. The breechblock is of a combined, semiautomatic type with wedge locking mechanism and powder gases obturator. A loading assistance device eases the loader's efforts in ramming the projectile into the chamber. The mechanical arm of the autoloader is actuated by the loader with the press of a button. Such a device is essential when the gun is firing at extreme angles when the barrel would be nearly vertical.

The ammunition for the Nona system was developed by the design team headed by Anatoli Obukhov at the Bazalt GNPO (state research and production enterprise). The maximum range of the weapon system with a high-explosive projectile is 8.7 kilometers; the minimum range is 400 meters. The initial muzzle velocity for a round is between 367 to 560 meters per second. This weapon can fire between 8 to 10 rounds a minute.

One of the main drawbacks of the 2S9 vehicle is its relatively small ammunition stowage capacity of only twenty-five rounds. This is ameliorated by the use of a simple conveyor that can be deployed at the rear of the vehicle to facilitate rapid reloading during fire missions.

The 2S9 was first deployed in combat in Afghanistan where it was widely viewed as a success. As a result, the 2A60 120mm combination weapon was adapted to a Ground Forces vehicle, the 2S23 Nona-SVK to arm Soviet/Russian motor rifle units.[14]

The 2S12 Truck-Mobile Mortar

Several self-propelled mortars (such as the Vasilyek 2B9) were fielded in the early 1980s in the former Soviet Union and several Warsaw Pact countries (i.e., Hungary, Bulgaria, and Romania). During this era, Russia introduced a new mortar that was carried on the bed of a GAZ-66 truck. The mortar carried on the truck was the towed 120mm 2B11 Sani wheeled mortar. When the 2B11 is mounted on the GAZ-66 truck it is redesignated the 2S12.[15] The 120mm 2B11 is a mortar system designed to replace the World War 2-based M-1943 120 mm mortar.

The 120mm 2B11 mortar has an effective range of 7,100 meters and a weight, including carriage, of 297 kilograms. A new long-range charge greatly increases the mortar's range, propelling the new sixteen kilogram OF-843B mortar bomb at a muzzle velocity of 325 meters per second. The long-range charge cannot be used with older mortar rounds designed for the M-1943. With a charge 6, the 2B11 has a range of 5,840 meters with a muzzle velocity of 275 meters per second, matching the maximum performance of the M-1943.

The mortar's main components are the barrel and breech, bipod, baseplate, double loading safety device, MPM-44 sight, and carriage. The round welded-steel plate weighing seventy-eight kilograms is the heaviest individual part of the 2B11. The four carrying handles on the outside of the baseplate, but angled inwards, eas-

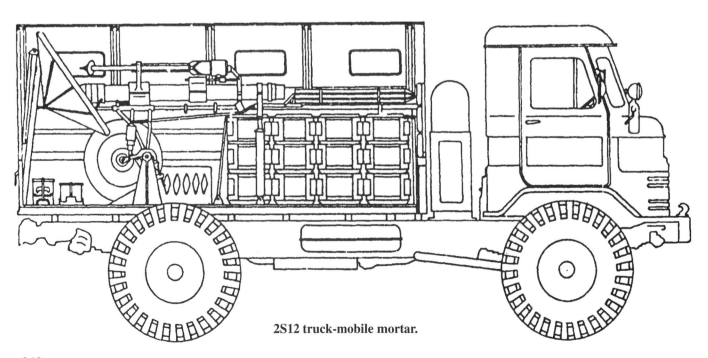

2S12 truck-mobile mortar.

2B11/2S12 Technical Data

Designation	2B11/2S12
Crew	5
Gun caliber	120mm
Traveling weight (kilograms)	297
Firing position weight (kilograms)	210
Dimensions in traveling mode:	
—Length (millimeters)	2,230
—Width (millimeters)	1,070
—Height (millimeters)	1,360
Mortar bomb weight (kilograms)	16
Maximum muzzle velocity (meters per second)	325
Maximum range (kilometers)	7.1
Minimum range (kilometers)	0.48
Maximum rate of fire (rounds per minute)	15
Operational arc	
—Traverse (without moving) (degrees)	+/-5
—Traverse (with moving) (degrees)	+/-26
—Traverse (with moving) (degrees)	45 to 80
Gun elevation/depression (degrees)	+80/-4
Gun traverse (degrees)	35 left and right
Ammunition storage (rounds)	60

ily distinguish it from the M-1943's baseplate.

The carriage of the 2B11 is made of thin steel-welded tubes and weighs only eighty-seven kilograms. It is much lighter than that of the M-1943 which is almost as heavy as the mortar itself. The new carriage is not as robust as its heavier predecessors, however, this lighter design facilitates its portage on the GAZ-66 truck. The vehicle also carries the mortar's ammunition and the loading/unloading rails. The 2B11 can in extreme cases be towed for short distances and then only at low speeds. Under combat conditions with rapid and frequent displacements, the mortar can be towed by a lower barrel clamp. The truck also transports the 2B11's five-man crew, consisting of the driver, gunner, layer, loader, and ammunition handler.

The production and deployment of the 2B11 ended the brief appearance of the 2B9 Vasilyek 82mm automatic mortar which seemed positioned to become the new fire support weapon for Russian motorized rifle troops. However, it appears that the Russians have only produced a small number of these systems and do not appear to be deploying or exporting the 2B11 or 2S12 in significant numbers. In addition to Russia, both the Bulgarians and Hungarians are marketing the 2B11 and are offering the mortar mounted on a MT-LB rather than the Russian GAZ-66 truck.

The 2S23 Nona-SVK

The self-propelled 2S9 Nona-S 120mm combination weapon began its service with the Russian VDV airborne forces in 1981. This system was utilized in Afghanistan where Russian combat reports stated that it saved airborne units on numerous occasions. The ability of the weapon system to fire from an almost vertical position

2S23 Technical Data

Crew	4
Combat weight (metric tons)	14.5
Armament	2A60 120mm gun/mortar
Ammunition/weight (rounds/kilograms)	30/17.3
Rate of fire (rounds per minute)	10
Maximum range (kilometers)	8.70
Elevation (degrees)	- 4 to +80
Traverse (degrees)	+/- 35
Maximum road speed (kilometers per hour)	80 on land; 10 in water
Unit price	$450,000 (1992 export price)

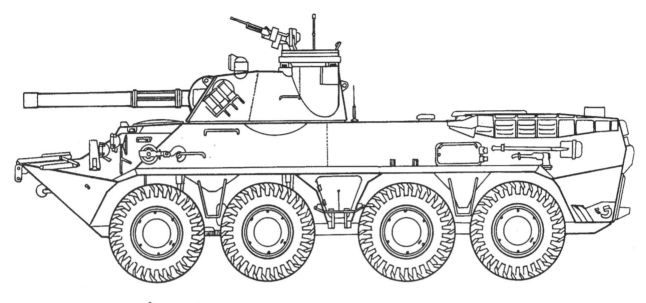

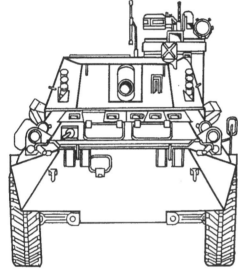

2S23 Nona-SVK

allowed it to engage enemy forces in mountainous terrain, including units on reverse slopes that were inaccessible to other more conventional weapons.

Drawing on their combat experiences in Afghanistan the Motovilikha Plant in Perm also produced a towed 120mm combination weapon called the 2B16 Nona-K which was accepted into Soviet service in 1986. These weapon systems were developed by TsNII Tochmash in Klimovsk by a design team headed by Yuri Kalachnikov in cooperation with the production facility at the Motovilikha Plant. The success of the 2S9 Nona-S in Afghanistan attracted the attention of the Soviet Ground Forces, and in the late 1980s, an effort was begun to adapt this weapon to a common Ground Forces chassis. The chassis finally selected was the BTR-80 wheeled armored transporter, though other vehicles may also have been examined. Designated 2S23 Nona-SVK, the new self-propelled combination weapon was the first time a wheeled armored vehicle chassis was used by the Soviets for a self-propelled artillery system. The 2S23 was intended for use both by the Ground Forces, and the Naval Infantry. The 2S23 Nona-SVK was accepted for service in 1990.

The 2S23 can carry up to thirty rounds of various high-explosive, illuminating, smoke, and incendiary rounds, compared to twenty-five rounds on the 2S9 Nona-S. The rounds are stowed in the middle of the fighting compartment. The 2S23 is based on an BTR-80 wheeled armored transporter and carries a crew of four (a commander, gunner, loader, and driver). Chassis integration of the 2S9 and 2S23 was undertaken by the I.V. Gavalov Design Bureau located in Volgograd Tractor Plant. Although the turret of the 2S9 and 2S23 are similar in appearance, they are in fact different designs. The 2S23 is distinctive due to a prominent cupola for the vehicle commander which has optical ports for observation while the commander remains protected inside.

Secondary armament is a 7.62mm PKT machine gun mounted on the commander's cupola ring. There are 5,000 rounds of 7.62mm ammunition for this machine gun. The 2S23 also carry carries four Kalsahnikov assault rifles or carbines (for self-defense), two Igla (SA-16) air defense missiles, fifteen hand grenades, and twenty 30mm flares. The 902V smoke-screen system utilizes six 3D6 smoke grenade launchers on either side of the turret.

The 2S23 is designed to perform a number of combat tasks such as destroying entrenched infantry, providing counterbattery artillery support, attacking field fortifications, destroying missile launchers, and command

2S23 Nona-SVK 120mm self-propelled weapon.

posts. The 2S23 gun can fire in a variety of climatic conditions in temperatures ranging from +50 to -50 centigrade. The BTR-80 chassis is powered by a 260 horsepower engine (192 kilowatt) liquid-cooled diesel engine. The four-axle, full drive vehicle is capable of negotiating trenches over two meters wide and vertical walls, half a meter high. The boat shaped all-welded hull allows the vehicle to stay afloat for twelve hours in the water under normal engine operations utilizing powerful hydrojets at water speeds up to nine kilometers per hour. The 2S23 has a maximum range on a single load of fuel of over 600 kilometers.

In recent discussion in the Russian defense press, Russian artillery officers have indicated that procurement of the 2S23 is one of their top four priorities (the others being the 2S19 Msta, the Smerch multiple rocket launcher, and the Zoopark artillery location radar).[16]

2S31 Vena 120mm Self-Propelled Gun-Mortar

The 2S31 Vena 120mm self-propelled gun/mortar was first disclosed in 1993, and details of the program have dribbled out ever since. IDEX-97 was the first time that the actual vehicle was displayed. This is a lightweight self-propelled artillery weapon using the latest version of the unusual 120mm gun/mortars that have characterized Russian mechanized artillery for rapid reaction forces. The three earlier examples of these weapons were the 2A51 120mm weapon on the 2S9 Nona-S (BMD chassis), the towed 2B16 Nona-K, and the 2A60 120mm weapon on the 2S23 Nona-SVK (BTR-80 chassis). These weapons are intended to offer the firepower of a 152mm howitzer, but in a lighter weapon that can be mounted into a light armored vehicle.

In contrast to the earlier weapons, the new 2A80 120mm weapon of the 2S31 has a longer rifled barrel. This type of weapon is capable of firing both conventional 120mm mortar bombs, as well as a special family of thin-walled 120mm artillery projectiles. All of these weapons were developed by the TsNII TochMash (Central Scientific Research Institute for the Precision Machinery Industry) which is the main Russian small arms development agency, located in Klimovsk. The weapons are manufactured by the Motovilikha artillery plant in Perm. The ammunition is developed by GNPO Bazalt.

When first revealed in 1993, the 2S31 utilized a redundant Obiekt 934 light tank chassis from the aborted PT-100 light tank competition of 1975-76. This chassis was developed by the I. V. Gavalov design bureau at the Volgograd Tractor Plant which also developed the BMD chassis used for the 2S9 Nona-S. This light tank never went into series production, and it would appear that the use of this hull on the 2S31 prototype was simply a convenient way to mount the turret as a proof-of-concept

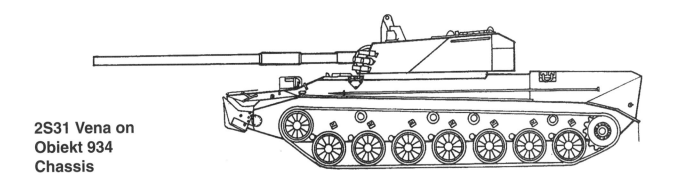

2S31 Vena on Obiekt 934 Chassis

demonstration and was not intended to serve as the basis for the production system. Dr. Aleksandr V. Khinikadze, the General Director for the TsNII TochMash in Klimovsk, has stated that the developers of the new 120mm weapon system have had very little to do with the development of the vehicle's chassis. Although Dr. Khinikadze stated that the eventual chassis for the 2S31 would be on a tracked vehicle that was intended to replace the MT-LB as the standard universal light armored tracked vehicle.

A new Russian light armored tracked chassis was required due to the loss of the Kharkov Tractor Plant, located in the Ukraine, which was the only producer in the former Soviet Union, of the MT-LB, 2S1 chassis, and MT-LBu ACRV. This loss left Russia with only two plants for this class of vehicle (Kurgan and Rubtsovsk). Designers of the 2S31 from TsNII-TochMash in Klimovsk, at a 1994 arms exhibition stated that the series production configuration of the 2S31 would be mounted on a new universal tracked chassis, which was subsequently revealed to be the BMP-3 derivative.

The 2S31 Vena gun system is highly automated. The vehicle is fitted with an onboard computer, linked to an automated survey and navigation system. Some of this equipment was absent in the vehicle present at IDEX-97, having been intentionally left off due to secrecy concerns.

2S31 Vena 120mm self-propelled gun on an Obiekt 934 light tank chassis.

Self-Propelled Artillery

2S31 Vena on BMP-3 chassis.

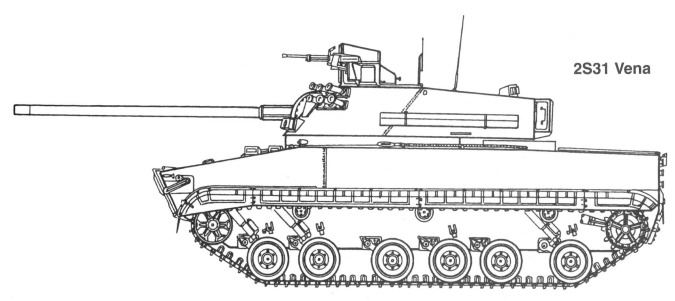

2S31 Vena

According to company representatives, the computerized fire control system allows the weapon to be automatically traversed and elevated based on target data inputs entered earlier into the computer.

There is a standard ammunition stowage of seventy rounds. Of these ten rounds are the Kitolov semi-active laser guided projectiles, and the remainder a mixture of ammunition types. The ten Kitolov rounds are stowed vertically on the left rear corner of the turret basket. A total of thirty-six ready rounds are available in two eighteen-round revolver racks at the rear of the turret in a bustle. On the vehicle at IDEX-97, the right ammunition tray contained mortar bombs while the left tray contained the special 120mm gun ammunition. The two revolver trays in the turret bustle are part of an ammunition loading assist system. The trays feed the round on to a rack to the left of the loader, who can then push them up into the breech of the weapon.

The remaining ammunition is stowed in covered tubes at the rear of the fighting compartment, forward of the engine bulkhead. There are a total of about forty-five rounds stowed in this area. There is also a small ammunition ready rack on the floor of the turret which may hold ammunition or bag charges. All types of ammunition used by the weapon can be automatically loaded except for the Kitolov PGM which is too long for the loading trays. The bag charges are kept in a set of stowage containers in the lower hull, and can be manually loaded. The loading assist feature was not demonstrated at the show, but presumably requires the gun to return to a near horizontal position, similar to the method used in Russian tank gun autoloading systems. As is to be expected in a system using mortar projectiles, the 2A80 gun/mortar can be elevated up to eighty degrees.

The turret has a crew of three: commander to the right front, loader to the right center, and gunner to the left. There is a fourth seat behind the 7 o'clock position of the turret basket in the left rear corner of the fighting com-

2S31 Technical Details

System designation	2S31 Vena
Design bureau	TsNII TochMash, Klimovsk
Vehicle chassis	BMP-3
Vehicle weight (kilograms)	19,100
Crew	4
Weapon system	120mm 2A80 rifled gun/mortar
Gun elevation (degrees)	-4 to +80
Gun azimuth (degrees)	360
Rate of fire (rounds per minute)	8 to 10
Ammunition	70+ rounds including 10 Kitolov PGM
Ready ammunition	Two eighteen-round magazines
Maximum range (projectiles) (kilometers)	13
Maximum range (mortar bombs) (kilometers)	7.2
Vehicle self-defense	7.62mm PKT machine gun on commander's cupola
Vehicle passive defense	Laser warning receiver, 12 smoke dischargers

This 2S31 was photographed at IDEX'97 in Abu Dhabi and was among several versions of the BMP-3 that were on display.

partment which doubles as a toilet. This is the first time such a feature has been seen on a Russian vehicle. It is evidence of a serious effort made by the designers to make the vehicle fightable in chemical warfare conditions. This is probably due more to export considerations than Russian Army interest. The primary market for this weapon is the Middle East, and chemical defense is a serious concern in the area considering the use of chemical weapons in the 1980-88 Iran-Iraq war.

There are several types of projectiles specially designed for use from these 120mm gun/mortars. These were all developed by the Bazalt GNPO in Moscow. There are two new rounds developed specifically for the 2S31. The new HE-Frag round weighs twenty-six kilograms, has an effective range of two to eighteen kilometers, and has a lethal area against personnel in open terrain of 2,200 meters squared. A new ICM round also weighs twenty-six kilograms, has an effective range of 1.5 to 11 kilometers, and has a lethal area against personnel in open terrain of 10,000 meters squared. There are thirty-five submunitions in this round and they are dual-purpose with an armor penetration of over 100mm. There are several existing rounds for the older 120mm gun/mortars which presumably can be fired by the 2S31 including a 120mm HEAT anti-tank round, and the 3VO54 and 3VOF55 120mm HE-Frag projectiles. The HEAT round weighs 13.2 kilograms and has an effective range of about 1,000 meters in the direct fire mode. The 3VOF54 round weights 19.8 kilograms and has an effective range of about 8.8 kilograms. The 3VOF55 round also weighs 19.8 kilograms, but has a small rocket-assist charge in its base, which boosts its effective range to thirteen kilometers. The 2A80 gun/mortar can also fire a full range of Russian and NATO 120mm mortar ammunition. The mortar rounds have a range typically of six to seven kilometers depending on the type.

The gun fire controls include a day/night 1P51 sighting system with a 1D22S laser range finder/designator. These are located in a sighting complex attached to the commander's cupola mounted on the right front corner of turret. The gunner sits on the left side of the vehicle with his panoramic sight, direct fire telescopic sight and gun controls. There is a small hatch on the right side of the turret which can be used for resupplying the vehicle with ammunition without exposing the crew.

The 2S31 is one of the first Russian vehicles of its type fitted with a self-defense system. This consists of an array of laser detectors similar to those used on the tank-based Shtora system for determining if a vehicle has been illuminated by a laser designator or range finder. It is fitted with twelve smoke dischargers on the forward portion of the turret to create a smoke-screen to defeat laser guided PGMs. It is not fitted with the electro-optical dazzler used with the tank mounted Shtora system, presumably because direct-fire wire-guided anti-tank missiles are not a major concern. Company representatives have in the past stated that this system significantly increases the survivability of the vehicle.

The 2S31 is also fitted with unusual NBC wash-down system. There are small attachments on the turret exterior which can spray water or other chemicals to decontaminate the vehicle. As in the case of the vehicle toilet, this is reflection of an effort to give the vehicle a serious chemical defensive capability. The vehicle is fitted with a typical Soviet filtration and overpressure system as well for this purpose.

Future Trends in Russian Self-Propelled Artillery

The Russian Army, whether Tsarist or Soviet, has called artillery the "God of War." The respect given this branch is evident in the diverse and versatile range of self-propelled artillery fielded by the Soviet Army in the 1980s. At the time of the Soviet Union's collapse, the army had hoped to standardize the artillery force around a handful of new pieces. The new weapon for the motor rifle and tank divisions was to be the new 2S19 Msta-S 152mm self-propelled, manufactured by Uraltransmash in Ekaterinburg in cooperation with the new STEMA plant in Sterlitamak. The 2S19 Msta-S is a sophisticated new weapon system with highly automated fire controls and based on a new chassis derived from the T-72 and T-80 tank chassis.

At least two other 152mm self-propelled guns are being considered, the improved 2S19 Msta S1, and the 2S30 Iset. The latter weapon has not been displayed and details about it are lacking. As in the case of the tank plants, the artillery plants have suffered in the budget turmoil. Recently, the Msta production plants had 218 self-propelled and towed Msta 152mm guns waiting for delivery, but without payment. This includes eighty-four 2S19 Msta-S self-propelled howitzers contracted by the Russian Army in 1994 for R75 billion and 117 additional weapons without orders or payment.

The 2S19 Msta-S remains in production for Russian requirements, but only 282 are in service west of the Urals, and there seems little hope that enough can be produced to enable standardization of the artillery force any time in the near future. In 1994, Uraltransmash proposed updating the existing 2S3 Akatsiya force as the 2S3M1 with a new, longer gun tube offering range and ballistic performance more similar to the 2S19.

As a lighter alternative for the rapid reaction divisions, Russian defense industry is attempting to interest the army in the new 2S31 Vena 120mm self-propelled weapon. The 2S31 is the latest in a line of these unique gun/mortar systems and is also the first Russian self-propelled gun with an organic electronic defense suite to minimize its vulnerability to counter-battery fire by precision guided munitions.

The Soviets also developed a new 122mm self-propelled howitzer in the late 1980s, the 2S17, but this program appears to have been shelved. Like the 2S1 and 2S3 self-propelled artillery systems which were developed into or from towed artillery systems, the recently developed towed 2A61 152mm howitzer (based on the D-30 base unit) could be placed on a light self-propelled chassis. It would appear that many of the future developments of Russian self-propelled artillery will be confined to upgrades of existing systems and confined to the development of only one or two new self-propelled artillery pieces through the end of this century.

Artillery Support Equipment

The Russians utilize a number of armored support vehicles which provide fire data to self-propelled and towed artillery units. Below are the major unit types which provide this support.

The Armored Command and Reconnaissance Vehicle

The growing sophistication of artillery fire control made possible by ballistic computers, artillery location radars and land navigation systems led to a requirement for armored command vehicles to accompany the new generation of self-propelled artillery in the mid-1960s. The original technical-tactical requirement for the 2S1 and 2S3 incorporated a requirement for an associated command and control vehicle. This led to the basic 1V12 Mashina command complex of armored command vehicles, called the ACRV M1974 (Armored Command and Reconnaissance Vehicle) by NATO.

The 1V12 series is based on the MT-LBu chassis, developed on the basis of the MT-LB series at the Kharkov Tractor Plant. The lower chassis of the MT-LBu is virtually identical to that of the 2S1, that is to say a lengthened hull with seven road-wheels rather than the six found on the MT-LB. The MT-LBu hull is large and boxy, befitting a command vehicle where the staff has to stand to inspect maps. There are several versions of the MT-LBu for use in artillery units, each tailored to particular artillery battalion staff requirements. The 1V13 (ACRV M1974-1), is a battery fire direction center and there is one per 2S1 or 2S3 battery (three per battalion). During operations, this vehicle is co-located with the guns, and

IV13 (ACRV M1974-1) artillery command reconnaissance vehicle.

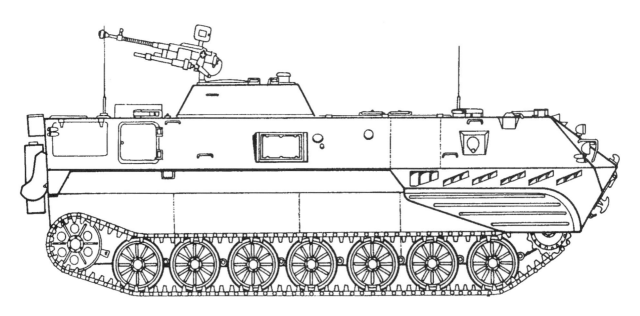

1V13 Artillery Command and Reconnaissance Vehicle

is crewed by the battery officer with his associated fire control and communications troops. This vehicle serves as the link between the battery and higher command elements.

The battery commander employs the 1V14 vehicle (ACRV M1974-2). This vehicle is similar to the 1V13, but has a sensor package on the right side of the turret containing a 1PN-44 night sight, and a turret roof optical sensor which incorporates a 1D11M-1 laser rangefinder.

There are three in each artillery battalion, one with each battery commander. It is used as a command and observation post. The 1V14 is fitted with a 1D11M-1 laser rangefinder in the turret and often carries a dismountable laser rangefinder like the DAK-1 (NATO Codename: Sage Gloss). The 1V14 is normally located aside the command vehicle of the tank or motor unit that the battery is supporting. For example, if a battery is supporting a tank regiment, the battery 1V14 would be aside the regimental command tank or other command vehicle. A close relative of the 1V14 is the 1V15 (ACRV M1974-2) which is the battalion commander's vehicle. The most significant difference between these vehicles is that the 1V15 has additional radio equipment to communicate with higher command, and there is an associated telescopic antenna.

The 1V16 (ACRV M1974-3) is the battalion fire direction center. It is commanded by the battalion chief of staff, and contains the battalion's electronic 9V59 field artillery computer. There is only one of these per battalion, and it would be located at the battalion headquarters.

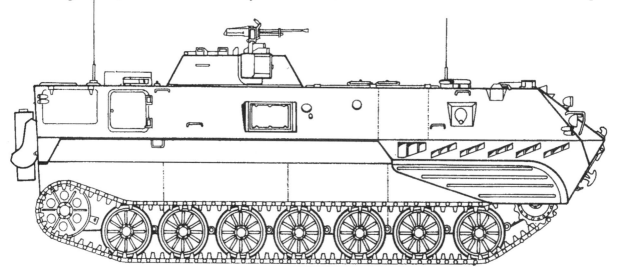

1V14/1V15 Artillery Command and Reconnaissance Vehicle

IV14/IV15 (ACRV M1974-1) aretillery command reconnaissance vehicle.

The initial 1V16 version resembled the 1V13 except for the absence of the right side high frequency antenna box and the PV-1 periscope sight on the roof. The improved 1V16M does not have a turret at all.

The 1V12 Mashnia family of command vehicles has undergone continual modernization. In 1983, a basic upgrade program was introduced, resulting in the 1V12M Mashina-M series (1V13M, etc.). The changes included the incorporation of a new externally mounted generator, and a new automated command transmitter/receiver on several of the vehicles in the family.

Although the 1V12 Mashina command vehicles are

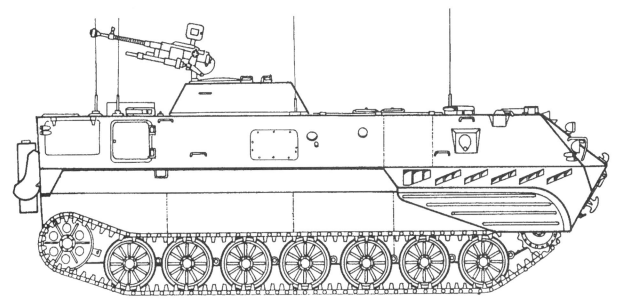

IV16 Artillery Command and Reconnaissance Vehicle

1V19 artillery command reconnaissance vehicle.

the best known of the vehicles based on the MT-LBu chassis, there are a large number of other command vehicles and special purpose vehicles based on the MT-LBu, including the MP-21 Rangir air defense command vehicle family.

Towed artillery units have been provided with a similar family of armored command vehicles, the 1V18 battery command observation post and the 1V19 battalion command observation post. Both of these are based on modified BTR-60PB armored transporter chassis. Likewise, VDV airborne artillery units are provided with vehicles based on the BMD chassis, the 1V118 Reostat forward command observation post and 1V119 Spektr fire direction center.

1V118 Reostat/1V119 Spektr

The introduction of the 2S9 Nona-S was accompanied by the deployment of two other specialized artillery vehicles, the 1V118 Reostat and 1V119 Spektr fire control vehicles. These vehicles have a small fixed turret, but lack any armament beyond self-defense machine guns. At the forward portion of the turret is an electro-optical sensor package including a laser rangefinder. The 1V118 Reostat serves as a forward artillery observation post, and can also be used by air force forward air controllers when air support is needed by VDV units. It is fitted with a small PSNR-5 (1RL133, NATO: Tall Mike) battlefield surveillance radar which can be mounted above the rear turret roof. The 1V119 Spektr is similar in general appearance but has the necessary ballistic computers and communications equipment to serve as a fire direction center for the 2S9 Nona-S battery. It lacks the surveillance radar.

PKNP Kushetka-B

The Kushetka-B PNKP (Podvizhniy komandno-nablyudatelniy punkt: Mobile command-observation station) is a BTR-80 variant that serves as part of the OBAK Battery Automated Fire-control System for the new 2S19 152mm self-propelled gun. This system was developed by the Radiopribor NPO in Zaporozhe, Ukraine in cooperation with VNII Signal in Kovrov. The Kushetka-B is based on the GAZ-59032 chassis, which uses the basic BTR-80 suspension but has an enlarged hull superstructure.

The Kushetka-B is forward deployed to conduct fire reconnaissance and observation. It is fitted with a laser designation/ranging system for use with precision guided munitions such as the Krasnopol semi-active laser guided projectile. Communications equipment includes two R-171M UHF radios, one R-163-50U UHF radio, one R-163-10V UHF radio, one R-163UP receiver, one R-134M HF radio, and data and telephone enciphering equipment. The crew of the vehicle is five. The Kushetka-B is used in conjunction with a battery fire control station based on the Ural-43203 truck. The fire control system was developed by the VNII-Signal in Kovrov.

PRP-3 "Bal" Podvizhniy Razvedyvatelniy Punkt

The PRP-3 (izdeliye 773) is an artillery scout vehicle and was called BMP M1975 by NATO. There is one such vehicle in each self-propelled howitzer battalion. It uses a large diameter, two-man turret, similar in appearance (but not identical to) to the BMP-2 turret. The vehicle is armed with a single 6P7 PKT 7.62mm machine gun for self-defense, aimed through the 1P28 periscope sight on the roof. The basic vehicle sensor is a 1RL126 (NATO: Small Fred) centimetric battlefield surveillance radar, with the antenna located on the rear roof of the vehicle. This radar can detect tanks up to ten kilometers away. The vehicle is also fitted with a 1PN61 night vision sensor in the right sensor package, along with a 1D11 laser rangefinder. The turret is also fitted with a TNPO-170A optical periscope sight. Communications is provided by means of a R-173 VHF transceiver and an 1A3OM command transceiver, supported by a 1T803 secure voice system. Precision navigation is provided by means of a 1V44 (KP-4) course plotter, 1G13 gyro course indicator and IT25 gyrocompass, and a 1V520 ballistic computer is provided for fire control problems. These vehicles usually carry a portable 1D13 laser rangefinder. The driver is provided with a TNPO-350B day sight and a TVNE-1PA image intensification night sight. This vehicle is produced at the Rubtsovsk Machine Building Plant.

PRP-4 "Nard" Podvizhniy Razvedyvatelniy Punkt

The PRP-4 is an improved version of the PRP-3. The most significant change is the addition of a second sensor package on the left side of the turret containing a 1PN71 thermal imaging night vision device and 1D14 laser rangefinder. The right sensor package contains up-

Self-Propelled Artillery

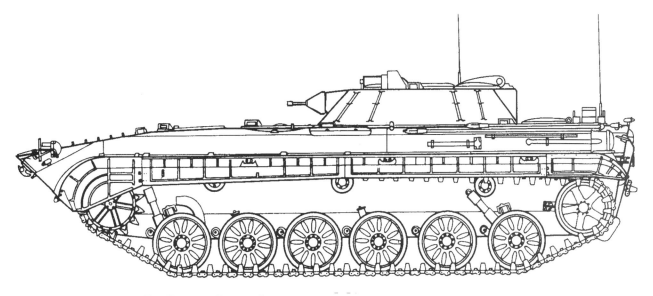

PRP-3 Artillery Command and Reconnaissance Vehicle

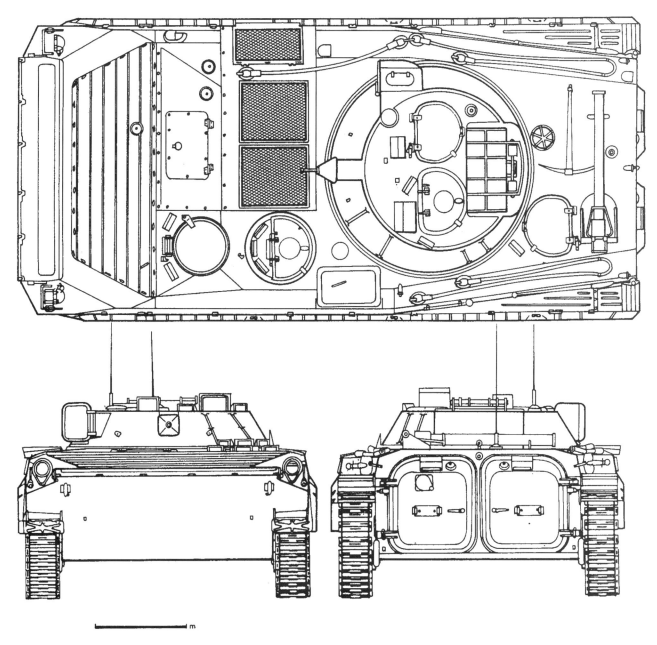

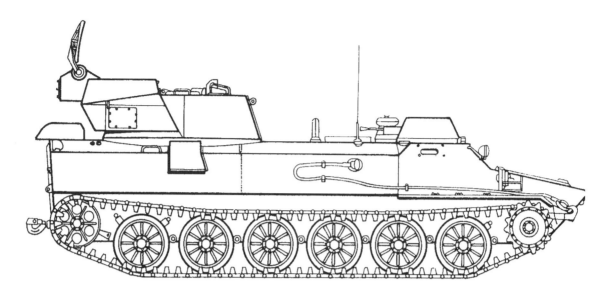

SNAR-10 Artillery Location Radar Vehicle

graded sensors including the 1PN61 and 1D11M-1 laser rangefinder. The PRP-4M has improved 1PN59 night vision sensors. It is manufactured at Rubtsovsk.

SNAR 10 Artillery Location Radar Vehicle

One of the most widely deployed battlefield surveillance radars of the former Soviet Union was the self-propelled SNAR-10 (stankiya nazemnoy artilleriysko razvedki) or "Big Fred" by NATO. The system is a battlefield surveillance radar with a limited artillery locating function. The SNAR-10 is unable to calculate the trajectories of projectiles or locate enemy artillery firing positions from their impact areas. For such tasks, Russia now utilizes the IL-219 Zoopark-1.

However, the SNAR-10, can provide location data on stationary and moving battlefield targets. It is normally deployed close to infantry or artillery observation posts. The SNAR-10 can provide data on friendly artillery mission impact patterns and can record and registration fire. The SNAR-10 can also detect targets at sea as well as slow, low-flying helicopters and remotely piloted vehicles (RPVs).

The SNAR-10 was introduced into service by the former Soviet Army in 1975. The system is currently being withdrawn from frontline service due to the fielding of the IL-219 Zoopark-1. By the early 1980s, the SNAR-10 entered service with several members of the former Warsaw Pact as well as the former Yugoslavia Army. Today, this system is unlikely to be in operational service with any these countries.

The SNAR-10 has a four-man crew. The SNAR-10 is based on the amphibious MT-LBu tracked vehicle. Because of the heavy turret, the SNAR-10 cannot swim. The SNAR-10's cruising range is up to 475 kilometers with 475 liters of diesel fuel stowed in two tanks along each side-wall in the hull. On the move, the power supply is provided by the engine's generator. Once in position, power is provided by a special four kilowatt generator.

The SNAR-10's communication equipment consists of two HF/VHF R-123M radios (20 to 51.5 MHz), two field telephones and the R-124 internal radio. To pinpoint position and direction, the SNAR-10 has a built-in navigation system and additional topographic equipment similar to the BRM and the BMP-3 Rys surveillance and reconnaissance vehicles.

The radar of the SNAR-10 works in the 35 GHz frequency range (this range is not utilized by other battlefield sensors and is unlikely to be detect or jammed). The dipping antenna is designation 1RL-127-1 while the entire system is referred to as 1RL-232-1. The radar is a two-dimensional pulse doppler radar, with the complete system placed inside the turret (except for the power supply).

The radar works either as a pulse radar or as a doppler radar, the operator making the selection. In the pulse mode, it is impossible to distinguish between moving and stationary targets, and it is, therefore, not ideal for the surveillance of a saturated battlefield. To detect moving objects, the doppler mode is more suitable.

The SNAR-10 was at one time a very capable system due to several improvements but is not as capable as the new Russian Zoopark-1 artillery counter-battery system. The SNAR-10 is being retired in Russian service.

Zoopark-1 IL219 Counterbattery Radar

The Zoopark-1 (IL219) counter-battery radar system was designed by the Tula Design Bureau and was developed to find the coordinates of firing positions of mortars, artillery systems, and tactical missiles, as well as multiple-launch rocket systems (MLRS). The Zoopark system utilizes the MT-LBu all-purpose light-armored prime mover chassis, the same basic chassis as that found on the 1V12 Mashina (ACRV M1974). The Zoopark chassis supports a large three-dimensional multi-function phased array radar and a crew of five. The Zoopark system has been in full scale production since 1991 and is intended to work with the 152mm self-propelled 2S19 Msta-S Howitzer, and will be the standard counterbattery radar interface for all self-propelled and tubed artillery in service with the Russian Army.[17] In addition, the complex comprises the 1L30 maintenance van on the Ural-43203 truck and a trailer-mounted ED30-T230P-RPM power station.[18]

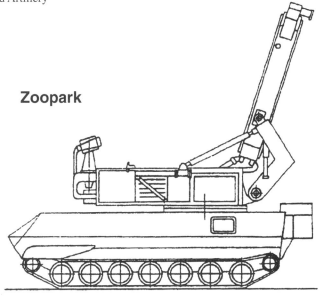

Zoopark

Zoopark artillery location radar vehicle.

IL219 Technical Data

Designation	IL219 Zoopark-1
Instrument detection range (kilometers)	45
First shot detection range	
(80 percent probability at X number of kilometers)	
—81mm mortar	12
—120mm mortar	15
—105mm field-artillery	8
—155mm and 203mm artillery systems	10
—122mm multiple launch rocket system	12
—240mm multiple launch rocket system	20
—Lance-type tactical missiles	35
Radar	
—Phase shifters	3,320
—Modules	104
Simultaneously tracked target trajectories	18 per minute
Number of targets tracked	4
Coverage	
—Azimuth (degrees)	60
—Elevation (degrees)	40
Complex service life (years)	8
Transport base	MT-LBu

The Zoopark is being marketed as being able to carry out nearly instantaneous detection of enemy tube and rocket artillery and to be able to determine the firing positions. The system can currently track up to eighteen targets (trajectories) per minute, with simultaneous tracking of up to four targets. The surveillance range for mortars is fifteen kilometers, for guns is ten kilometers, and for MLRS is twenty kilometers. The average error in the determination of coordinates is thirty meters and the average deployment time is five minutes. The two ACRV models will process the data from Zoopark utilizing onboard computers and use this information to compute a fire control solution which is then directly transmitted to the 2S19 and other Russian artillery pieces. This technique was utilized by the US Army in Desert Storm through the use of the Firefinder radars (AN/TPQ-36 and AN/TPQ-37) and their communications link to counterbattery artillery sites. Such a connection allowed for rapid targeting by the US Army of any Iraqi artillery locations within minutes of their firing. It is believe that this system will offer many of the similar advantageous offered by the US Firefinder system.

Multiple Rocket Launch Systems (MLRS)

The Russian's have been one of the leading countries in the world to develop, adopt, and employ multiple rocket launch systems (MLRS). Dating back to world famous "Katyushas" rocket launcher utilized by the Russians against the Germans on the Eastern Front during World War 2. Furthermore, the Russians were the first country in the world to create and manufacture MLRS. The patent for the Katyushas system was granted on 9 April 1939 (patent number 3603) to three Russian designers: Ivan Gvay, Andrey Kostikov, and Vasiliy Aborenkov.[19]

The first system was designated BM-13 by the developers, this system was capable for firing rocket from its 16 launch rails in salvos within seven to ten seconds, with some accuracy. During World War 2 a number of MLRS designs were developed in the Soviet Union. In addition to the BM-13 (Katyushas), there were the BM-8-36, BM-8-24, BM-13N, BM-13-12, and BM-13SN. After the war, developers of MLRS continued their work. In the 1950s the BM-14 (140mm system with a range of 9.8 kilometers) and the BM-24 (240mm system with a range of 16.8 kilometers) were developed. The accuracy of these system were greatly increase due to inflight rocket rotation of the missiles.

Development of multiple rocket launchers during the war was undertaken by the Barmin design bureau at the Kompressor Plant in Moscow. After the war, Barmin's team became increasingly preoccupied with the design of strategic missile launch systems. As a result, in 1957 the work was transferred to the Reaktivniy NII (Scientific Research Institute for Rockets) headed by A. N. Ganichev. This organization evolved into today's Splav GNNP (Main Scientific Production Enterprise) in Tula. The chief designer after Ganichev's retirement in 1983

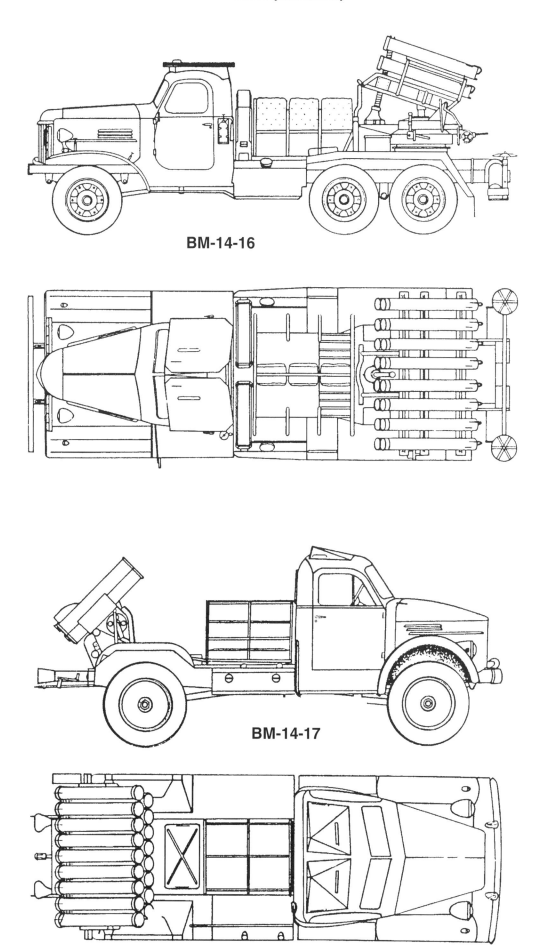

BM-14-16

BM-14-17

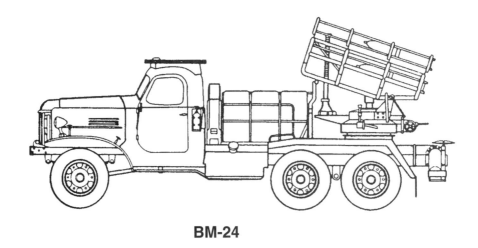

BM-24

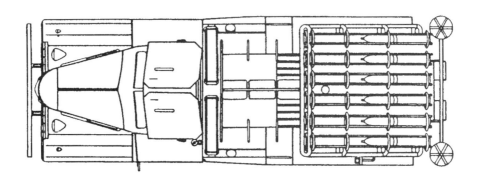

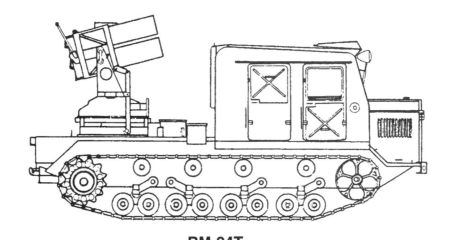

BM-24T

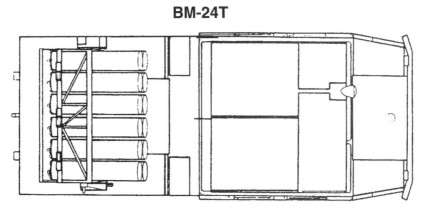

BM-21 Grad 122mm multiple rocket launcher.

was Gennadiy Denezhkin. In 1957, the institute was given the assignment for developing a new MLRS that would be two to three time more effective as the older BM-13 and BM-14 series. That system was the BM-21 Grad (Hail).

The BM-21 Grad

In the area of multiple rocket launchers, an effort was made to cut back on the bewildering variety of calibers and types with the introduction of the new Grad (Hail) system in 1957. Although some older systems, like the BM-13 were modernized, such as by mounting them on newer trucks like the ZiL-131, the emphasis was on adopting a new common type for standardization. The Grad system was a new 122mm rocket system mounted on the Ural-375D truck as the BM-21 and was intended to replace the multiple rocket systems at divisional level. GNPP Splav was the prime developer, with the Miass Automotive Plant developing the launcher system and other components associated with the truck launcher vehicle. It was accepted for service in 1963 and first saw combat during fighting with the Chinese on Damanskiy Island in the Ussuri river area in the late 1960s.

The BM-21 uses the 9P132 40 tube launching system. The basic rocket used by the Grad system is the 9M22, also known as the M-21-OF and DB-1B high explosive rocket. Other rocket types with different warheads are available, as is a shorter (1.9 verses 3.3 meters) round with reduced range, the 9M22K (M-14-OF or DKZ-B). This system has become the standard Soviet divisional multiple rocket launcher since 1964. A modernized version of the system, the BM-21-1 Grad 9P137 was fielded in the late 1980s which employs the improved Ural-4320 truck.

The BM-21 equips a multiple rocket launcher battalion in each division's artillery regiment. This battalion has three batteries, each with six BM-21, for a total of eighteen BM-21 per division. The salvo firepower of one of these batteries is massive. A single battery of six launchers can fire 240 rockets totaling eleven tons of metal and high explosive into an area roughly 950 by 600 meters in a twenty seconds ripple fire. It takes about ten minutes to reload the BM-21 between salvoes. In comparison, a battery of its 122mm towed howitzer counterpart, the D-30 (2A18) would take seventy minutes to deliver the same volume of fire. The main disadvantage of the BM-21 compared to conventional tubed artillery is that the rockets suffer from much more dispersion at range. Taking the oval dispersion footprint of a single BM-21 and D-30 as an example, the linear dispersion of a BM-21 is about seventy-four meters deep compared to only twenty-six meters for a D-30, while the lateral dispersion is 118 meters across for the BM-21 and only about 9.7 meters for a D-30 at fifteen kilometers range. For this reason, the multiple rocket launchers are really only useful for area bombardment, not for precision artillery attacks.

In the early 1970s, a new lightweight thirty-six tube

BM-21 Grad 122mm multiple rocket launcher.

launcher was developed, which generally uses the 9M22K short missile. Known as the Grad-1, it was first on the 9P139 tracked launcher using the MT-LB chassis. This version was never very common and instead the standard Grad-1 system was the 9P138 mounted on the ZIL-131 truck. This lightweight version was designed to be more suitable for use in mountainous areas, and is fitted with thirty-six tubes instead of the usual forty tubes on the BM-21, missing the two central tubes in the two bottom rows. Another special lightweight twelve-tube version was also developed for the VDV Airborne Force, and mounted on a GAZ-66B light truck, the BM-21-V. The new Damba coastal defense version of the basic BM-21 Grad system, using rockets tailored for naval missions,

BM-21 Family Technical Data

System	BM-21	BM-21
Designation	9K56	9A51
Year in Service	1963	1987
Codename	Grad	Prima
System weight (kilograms)	13.7	13.9
Crew	6	3
Vehicle range (kilometers)	1,000	990
Vehicle speed (kilometers per hour)	80	85
Caliber	122mm	122mm
Rocket weight (kilograms)	66.5	70
Warhead weight (kilograms)	18.4	26
Cluster sub-munitions	no	no
Quantity of rounds	40	50
Area covered by a salvo (hectares)	3.5	19
Maximum range (kilometers)	20.5	20.5
Minimum range (kilometers)	5	5
Vehicle mount	Ural-375D	Ural-4320
Transloader weight (kilograms)	13.7	13.9
Rockets on transloader	40	50

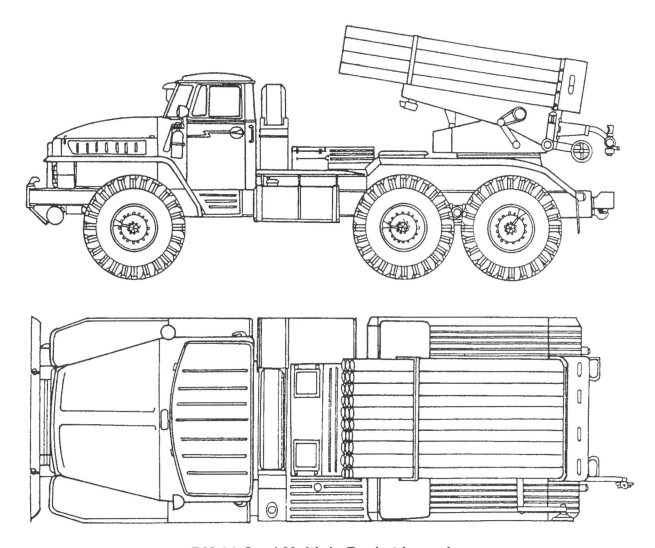

BM-21 Grad Multiple Rocket Launcher

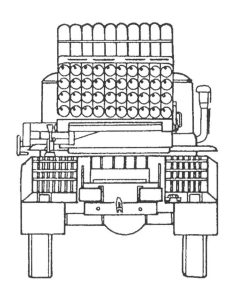
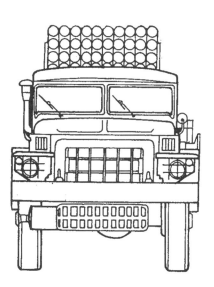

has been promoted by GNPP Splav since 1992.

In 1994, at the Nizhni-Novgorod arms exhibition, Splav GNPP unveiled a new modernization program for the BM-21 Grad. The improved rocket system will increase the effective range to thirty-five kilometers. New warheads would include an anti-personnel fragmentation sub-munitions, an anti-armor warhead with HEAT/Frag warhead, an anti-armor warhead with a guided sub-munitions, an anti-armor mine dispensing rocket, and an anti-personnel mine dispensing rocket.

In the 1980s, the GNPP Splav developed a new generation BM-21 rocket system called Prima. The associated 9M28 rocket can be fired from existing launcher vehicles but it is primarily intended for the new 9A51 launcher vehicle consisting of a twelve-tube launcher mounted on the Ural-4320 truck. The rocket differs from the earlier 9M22 series in that during the terminal descent, the warhead separates from the rocket body and descends on a small parachute. The warhead airbursts over the target, providing a more lethal attack than the conventional 9M22 series.

The Czechoslovak People's Army had developed their own multiple rocket launcher, the M-51 mounted on the V3S truck. The CSLA later decided to adopt the Soviet BM-21 system, but was opposed to using an unarmored truck. They instead developed a version of the Tatra 813 truck with an armored cab. This version entered service in 1972 as the RM-70. The RM-70 is considerably larger than the BM-21, and besides the advantages offered by its armored cab, it is also fitted with a semiautomatic loading system with an additional forty rounds of rockets for rapid reload capability. The RM-70 is the standard multiple rocket launcher of the CSLA and the former East German NVA. The RM 70/85 is a modified version of the standard Czechoslovak RM-70 122mm multiple-rocket launcher, but with a soft cab rather than the usual armored cab. The Romanian Army used an adaptation of the Czechoslovak M-51 on a Zil-151/157, but in the 1960s adopted a locally produced version of the BM-21, mounted on the Bucesi SR-114 truck about the same time.

The 9K57 (BM-22) Uragan

Besides the long range the self-propelled artillery 2S5 and 2S7, Soviet army and front level artillery units began to receive the new 9P140 Uragan (Hurricane; sometimes misidentified as the BM-27) in 1975. The BM-22 is intended as a replacement for the obsolete BMD-20 and BM-25 long range multiple rocket launchers.[20]

Like all Russian multiple rocket launcher (MRL) sys-

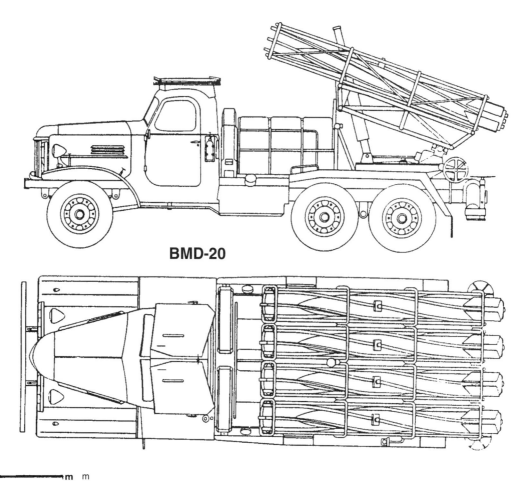

BMD-20

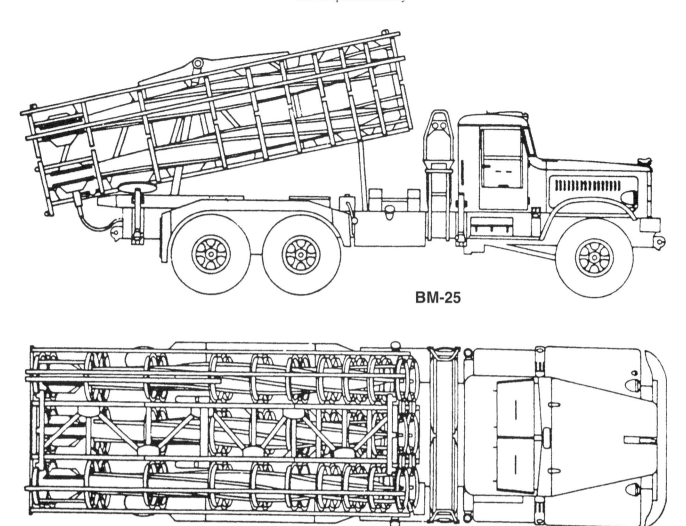

BM-25

tems, the 220 mm BM 9P140 (sixteen-round) Uragan (Hurricane) was developed at the Splav Government Scientific Production Concern at Tula which is to the south of Moscow. The system was designed in the early 1970s with development being completed in 1975. It entered service with the Soviet Army shortly afterwards and was allocated the STANAG designation of the M1977 by NATO, this being the year it was first identified.

The complete system has the industrial designation of the 9K57 which comprises the launcher, rockets, and the essential rocket resupply vehicle. Until the Splav 300 mm BM 9A52 (twelve-round) Smerch rocket system entered service with the Soviet Army in the late 1980s, the 220 mm BM-22 (9P140) was the largest system of its type in service.

The main advantage of the 220 mm BM-22 (9P140) is that it is mounted on an 8 x 8 chassis with improved tactical and strategic mobility. Its rockets have a much longer range and have a variety of warheads enabling a wider range of battlefield missions to be undertaken using the same launch platform.

The only comparable system in the West is the US Multiple Launch Rocket System (MLRS) used by the US Army and a number of other countries and which is also made under license in Europe for the French, German, Italian, and UK armies. MLRS is based on a full tracked armored chassis and has a turntable mounted at the rear with two pods of six 227 mm rockets which in the current Phase I configuration have a maximum range of thirty-five kilometers. Each rocket contains 644 M77 Dual-Purpose Improved Conventional Munitions (DPICM).

The 9K57 Uragan is based around the 9P140 launcher vehicle which consists of a BAZ-135 truck chassis and a 16 tube launcher. Uragan has been designed to operate in temperatures ranging from -40 to +50 degrees centigrade. The system is based on a modified ZiL-135LM (8 x 8) truck chassis which has excellent cross-country mobility and is used for a wide range of other battlefield missions including TEL (Transporter, Erector, Launcher) for the FROG-7 surface-to-surface rocket system (and its associated reload vehicle), Sepal cruise missile carrier, cargo truck, and tractor truck to name but a few. The chassis was developed at the Likhachev Motor Vehicle

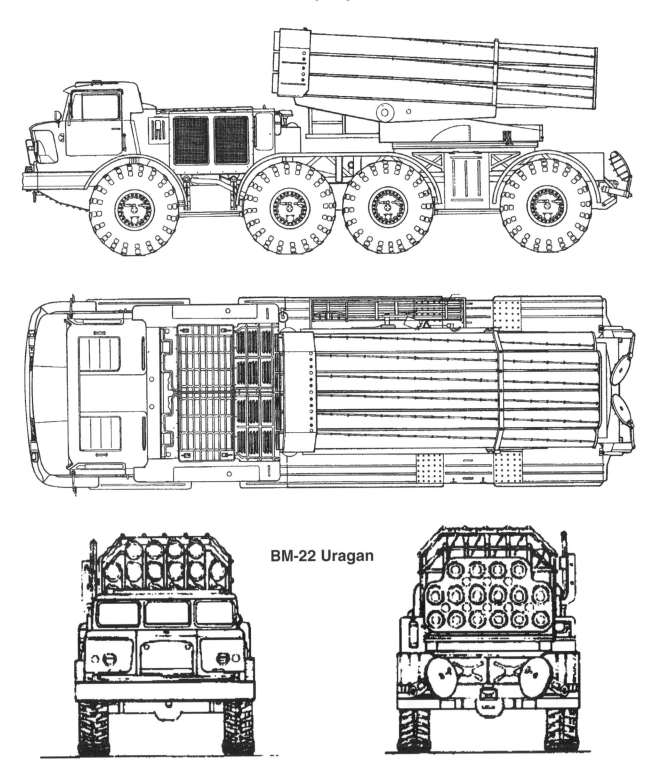

BM-22 Uragan

Plant near Moscow but production was undertaken at the Bryansk Automobile Works for which reason the family is sometimes referred to as the BAZ-135L4. The unarmored crew compartment is at the front with the engines to the rear, the turntable mounted rocket launcher is mounted at the very rear of the chassis and is traversed to the front for traveling.

The vehicle is powered by two engines each of which drives the four wheels on one side of the vehicle only. Steering is power assisted on the front and rear axles and a central tire pressure system is fitted as standard. This allows the driver to adjust the tire pressure to suit the terrain being crossed without leaving his seat.

The rocket launcher pod consists of an upper layer of four tubes and two lower layers of six tubes so giving a total of sixteen launch tubes. The sighting system is mounted on the left side of the launcher and access to this is via a folding ladder. The crew compartment accommodates the launch preparation and firing equipment and the operator can select single rockets or salvo firing.

9P140 Uragan 220mm multiple rocket launcher.

9T452 Uragan transloader vehicle.

Uragan (9K-57) Technical Data

System	BM-22 (9K-57)
Designation	9P140
Year in Service	1975
Codename	Uragan
System fully-loaded weight (metric tons)	20
System empty weight (metric tons)	15.1
Crew	4
Length traveling (millimeters)	9,630
Length firing (millimeters)	10,830
Width traveling (millimeters)	2,800
Width firing (millimeters)	5,340
Height traveling (millimeters)	3,225
Height firing (millimeters)	5,240
Vehicle range (kilometers)	570
Ground clearance (millimeters)	580
Track (millimeters)	2,300
Engine	2-8 cylinder, four-stroke petrol developing 177 hp at 3,200 rpm
Vehicle speed (kilometers per hour)	65
Maximum road speed (kilometers per hour)	65
Maximum range (kilometers)	570
Fuel capacity (liters)	768
Fording (millimeters)	1,200
Gradient (percent)	57
Vertical obstacle (millimeters)	600
Trench (millimeters)	2,000
Vehicle mount	BAZ-135LM
Caliber	220mm
Maximum range (kilometers)	35
Minimum range (kilometers)	10
Vehicle mount	BAZ-135LM
Rocket weight (kilograms)	280
Warhead weight (kilograms)	100
Cluster sub-munitions	312
Quantity of rounds/number of tubes	16
Area covered by a salvo (hectares)	46.2
Transloader designation	9T452
Transloader weight (kilograms)	20
Transloader empty weight (kilograms)	15.22
Transloader crew size	3 men
Rockets on transloader	16

Unguided Artillery Rocket Launched by the Uragan MLRS

Rocket	Warhead	Type	Payload
9M27F	9N218F	unitary HE	100 kilograms-warhead contains 51 kilograms of HE
9M27K1	9N218K1	ICM	30 N9N210 bomblets*
9M27K2	9N2I8K2	carrier	24 PGMDM anti-tank mines
9M27K3	9N218K3	carrier	312 PFM-l anti-personnel minelets

*These are of the anti-personnel/anti-material type

When deployed in the firing position, steel shutters are normally raised over the front windscreens and two stabilizers are lowered at the very rear to provide a more stable firing platform. The launcher is mounted on a turntable at the rear with powered elevation from 0 to +55 degrees and powered traverse of thirty degrees left and right.

The new 9M27 rockets have a maximum range of about forty kilometers, about double that of the divisional level BM-21. The warhead types available probably include both conventional and chemical munitions. The conventional rounds include a 9M27F unitary high-explosive-fragmentation warhead (fifty-one kilograms of high explosive); and several different 9M27K cassette sub-munitions types including a cluster munition with thirty HE-Frag sub-munitions; a cluster munition with twenty-four anti-tank mines, and a cluster munition with 312 anti-personnel mines.

The DPICM scatterable anti-tank mine is also used in a number of other applications and consists of a liquid explosive charge contained in a thin flexible cover. It uses a pressure activated fuze which is also used in the PFM-1 anti-personnel mine. Unconfirmed reports indicate that there may well be a chemical rocket for this system and recent Russian literature also indicates that a fuel air explosive (obemno-detoniruyushchaya smes) has been developed.

Each rocket is 220 mm in diameter, 5,178 millimeters long and weighs 280 kilograms at launch with the warhead weighing 100 kilograms. Minimum range is ten kilometers while maximum range is thirty-five kilometers. The rockets are fin and spin stabilized and have wrap round fins at the rear that unfold after the rockets are launched.

A full salvo from the Uragan takes less than twenty seconds and can saturate an area of 426,000 square meters or about forty-two hectares. Maximum time quoted for the system to be prepared for action from the traveling configuration to ready to fire is three minutes with a similar time to come out of action.

The BM-22 is also the first Soviet multiple rocket launcher designed from the outset to include a reload assistance vehicle (a transloader for the BM-21 was developed long after initial fielding of the basic launcher vehicle). The 9T452 transloader vehicle is also mounted on a modified BAZ-135 chassis. It feeds reload rockets into the empty tubes of the 9P140 in much the same manner as is employed on the Czechoslovak RM-70 system. This system was adopted not only for speed of reloading, but because the BM-22 rockets are too large for practical reloading by an unaided crew. However, it is estimated that the reload time is twenty to thirty minutes, a long time when compared to the US MLRS system.

Uragan was used in action during the Soviet occupation of Afghanistan and is known to have been exported to Syria. Russia has been offering the 9P140 launcher for around $650,000 with each rocket costing $5,600.

The 9K58 Smerch (9A52-2 Smerch-M)

The most recent heavy MLRS system developed by the GNPP Splav is the 9K58 Smerch (Tornado) which entered service in 1987. The 9K58 Smerch system is

9A52 Smerch 300mm multiple rocket launcher.

9A52 Smerch 300 multiple rocket launcher.

based around the 9A52 launcher vehicle, based on a MAZ-543 heavy 8 x 8 truck fitted with a twelve-tube 300mm rocket launcher system; other major components are the transloader and fire control vehicles. The launcher assembly for the 9A52 launcher vehicle was developed and manufactured by the Motovilikha Plant in Perm with the chassis coming from the Minsk Automobile Plant (MAZ) in Belarus. The improved 9K58-2 Smerch-M system uses the 9A52-2 launcher vehicle. About twenty plants are involved in Smerch manufacture.

The 300mm 9A52 Smerch has a characteristic four-axle, unarmored wheeled chassis with the twelve-tube launcher system mounted on the cargo bed. Before firing commences, the rear of the vehicle is stabilized using two hydraulic legs which descend between the third and fourth axle. A combat-ready 9A52 Smerch weighs in at 43.7 tons and is based on the MAZ-543 vehicle. This four-axle off-road truck has been in production since 1967 and provides a cargo capacity of twenty tons. Steering is controlled by the two front axles.[21]

The MAZ-543 has good off-road characteristics due to its all-wheel and central tire pressure control system. The vehicle is powered by a V-12 four-stroke diesel engine. The maximum speed for the MAZ-543 is sixty kilometers per hour and range is set at 850 kilometers. Two crew of the four-man complement of the 9A52 Smerch are accommodated in a tandem layout in the command cab, while the other two sit side-by-side in the fire control cab located at the forward end of the cargo bed.

The 9A52 Smerch launcher can fire in an arc of thirty degrees to either side of its in-line axis; correspondingly, the vehicle does not have to be parked pointing directly in the line of fire. Unlike the FROG and SS-21 short-range missile systems, this capability allows re-targeting even when the launcher is ready to fire or after it has already fired from its existing position. The rockets can be fired individually or in salvos. Discharging the entire complement of rockets takes thirty-eight seconds.

The 300mm missile for the Smerch system is designated the 9M55, with the basic high-explosive-fragmentation type being the 9M55F and the sub-munitions type being the 9M55K. The weigh of the rocket is 800 kilograms and includes seventy-two sub-munitions warhead weighing 300 kilograms. The 9M55K, is thought to also have a fuel-air explosive (FAE) warhead. Although rockets with anti-personnel and anti-tank mine dispensers are available for the BM-22, it would appear that none such exist for use with Smerch. It seems logical to assume that work is also in progress on developing a warhead with smart sub-munitions. However, it is not known how far this work has progressed; however, no such sub-munitions has yet been shown at any defense exhibition. The rocket engine for the 9M55 was developed by the design team under Vitaliy Kolesnikov at the Perm Asso-

Smerch (9K-58) Technical Data

System	9K-58
Designation	9A-52/9A-52-2
Year in Service	1987
Codename	Smerch/Smerch-M
System fully-loaded weight (metric tons)	43.7
Crew	4
Length traveling (millimeters)	12,000
Width traveling (millimeters)	3,050
Height traveling (millimeters)	3,050
Vehicle range (kilometers)	900
Ground clearance (millimeters)	440
Engine	V12 cylinder, four-stroke diesel developing 386 kW at 2,200 rpm
Vehicle speed (kilometers per hour)	60
Fording (millimeters)	1,100
Gradient (percent)	57
Vertical obstacle (millimeters)	600
Trench (millimeters)	2,000
Vehicle mount	MAZ-543
Caliber	300mm
Maximum range (kilometers)	70
Minimum range (kilometers)	20
High-explosive round	9M55F
Cluster warhead round	9M55K
Rocket weight (kilograms)	800
Warhead weight (kilograms)	280
Time from travel to firing position (minutes)	3
Time required to fire a full salvo (seconds)	38
Cluster sub-munitions	72
Quantity of rounds/number of tubes	12
Duration to reload 12 tubes (minutes)	20
Area covered by a salvo (hectares)	67.2
Transloader designation	9T234/9T-234-2
Transloader weight (kilograms)	20
Rockets on transloader	12

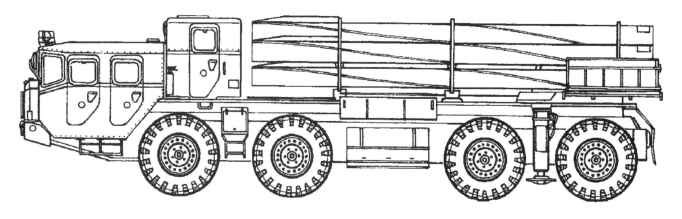

9A52 Smerch

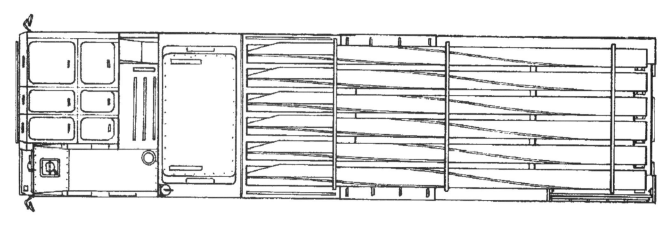

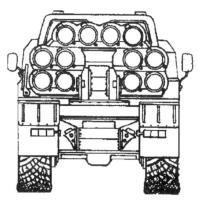

CFE Counts for MLRS (European Russia): 1990-1997

Year	BM-21 Grad	9P138 Grad-1	9P140 Uragan	9A52 Smerch	Total
1990	1,571	50	531	154	2,306
1991	1,544	73	602	160	2,379
1992	827	58	333	20	1,238
1993	658	4	333	56	1,051
1994	495	27	314	71	907
1995	354	1	385	105	845
1996	486	1	381	105	973
1997	356	13	415	105	889

9T452 Smerch transloader vehicle.

ciation imeni Kirov. The onboard ballistic computer was developed by the design team headed by Yuriy Savykin at the Signal VNII Scientific Research Institute in Kirov.

Depending on the range of the target, a salvo of twelve 9M55K rockets can cover an area of between 400,000 meters square (e.g., approximately 650 by 650 meters) and up to 700,000 meters square (e.g., 800 by 800 meters). A total of 864 high-explosive-fragmentation bomblets are distributed over this area, giving a theoretical bomblet density of one per 460 meters squared (e.g., 21.5 by 21.5 meters). Since this is a relatively low density normal practice is to employ an entire battery firing in a coordinated fashion.

There are two necessary conditions to effectively utilize a long-range MLRS: (1) sufficient up-to-date targeting information must be available, and (2) such information must remain correct until the launcher is ready to fire. The second condition is made much easier by the quick reaction capability of the 9A52 Smerch. This system needs only three minutes after arrival at a fully prepared launch site until it is ready to fire. Thanks to both its high cross-country and road speeds, 9A52 Smerch can reach its launching sites quickly.

Each 9A52 Smerch launcher is accompanied by one transport and loading vehicle (TLV) 9T234, which is also based on the MAZ-543. A unified command vehicle has also been developed for 9A52 Smerch, the 1K123 Fire Detection System. This is used by all command echelons in the MLRS formation.

9T234-2 Oplot Transloader for 9A52 Smerch

Accompanying each 9A52-2 Smerch-M launcher is a 9T234-2 transloader. The transloader is based on the same MAZ-543M chassis. The transloader's drop-sided body can carry twelve reload rounds and utilizes a hydraulic crane on it rear decking to load and unload the launcher. The rockets have to be reloaded individually using the loading crane mounted behind the launch vehicle's cab. This rotating crane has a lifting capability of 850 kilograms. It is not possible to replace the tube assembly either in part or in full. The entire loading procedure for a rocket launcher therefore takes at least twenty minutes.

The vehicle weights 40,500 kilograms fully load or 30,000 kilograms empty. The maximum highway speed is sixty kilogram per hour and the maximum range on its internal fuel tanks is 850 kilometers. The MAZ-543M utilizes a diesel engine and can ford bodies of water up to 1 meter in depth.

1K123 Fire Detection System for 9A52 Smerch

The 1K123 is an automated system which is intended for command and control of the 9K58 Smerch 300mm long-range multiple-launch rocket system. The 1K123 system utilizes similar capabilities to that of the US Tacfire system, but the Russians claim that it has numerous advantageous over the US system. For example, the 1K123 does not require the extensive preparation time for combat operations or to rely commands that is required by the US Tacfire system.²²

The 1K123 system includes a command and control artillery post vehicle for the artillery unit. This includes: two command-staff vehicles for a commander and his chief of staff, up to three command and battalion staff vehicles for the battalion commander and up to eighteen vehicles for the battery commanders. The basic vehicles are based on box-bodied three-axled unarmored KamAZ-4310 trucks with the K4310 van body. Each truck accommodates the fire control computers, data transfer equipment, data processing equipment and radio gear. The 1K123 tows a single-axle generator trailer, so the system is independent of external power sources. The communications range supported by the radio sets in the command-staff vehicles is fifty kilometers when stationary and twenty-five kilometers when moving and using the ultra-shortwave radio band. The distance is 350 kilometers when stationary and up to fifty kilometers in motion in the shortwave radio band.

"Artillery" Missiles

Evolution of Short-Range Army, Front Level, and Divisional Artillery Missile Asset.

With the advent of the "revolution in military affairs" brought about by the deployment of tactical nuclear weapons by NATO, the Soviet Army began to develop weapon systems to deliver their own tactical nuclear warheads. In the mid-1950s, there were several competitors in this field. The most unusual were the super-heavy gun and mortar programs mentioned earlier in this study such as the 420mm superheavy mortar developed by the Shavyrin design bureau in Kolomna. However, these met the severe disapproval of Khrushchev and never entered service in any significant quantity. There is little information on when the Soviet army received its first tube-fired tactical nuclear artillery rounds.

Due to the size of early tactical nuclear weapons, the most practical solution for delivery of weapons from ground-based systems was the missile. Early Soviet bombardment missiles, the R-1 and R-2 were primitive and unwieldy derivatives of the wartime German A-4 (V-2). These were not seriously considered for nuclear weapons delivery. The first nuclear armed missile was the R-5 (SS-3), which was deployed with strategic missile units, and not the Ground Forces. The Ground Forces began development of three categories of tactical nuclear missile delivery systems: the FKR-1 and FKR-2 cruise missiles (Samlet/Sepal); the Mars/Filin/Luna (Frog) divisional artillery rockets and the R-11 (Scud) army/front-level ballistic missile. The cruise missiles are not discussed here as their service was short-lived. We have also omitted the longer ranged missiles (such as the SS-12 and SS-23) removed under the INF Treaty as their

Filin (3R-1) rocket (FROG-1).

long range role, dubbed "operational" by the Soviets, is outside the scope of this study. Instead, we have focused on the most common of these delivery systems, the FROG, Scud, and Scarab. Although these weapons were initially developed for nuclear delivery, they have always had a secondary role in the delivery of other warheads including conventional explosive and chemical warheads. In recent years, with the advent of improved guidance, the conventional mission of these systems has increased.

Evolution of the Free Rocket Over Ground Family

Filin (3R-1) Rocket (FROG-1)

The Filin (Eagle-Owl) first entered service in 1955 and was displayed in public in November 1957. The 3R-1 Filin was a two-stage, solid-fueled rocket with a launch weight of 4.93 metric tons, a warhead of 1.2 metric tons, and a range of twenty-six kilometers. It appears that the Filin was solely intended to delivery tactical nuclear weapons and its large size probably reflects the heavy weight of early Soviet nuclear devices. It does not appear to have been introduced in any significant numbers and eventually disappeared from service with the advent of the later Mars and Luna systems. The Filin was easily distinguishable from later free rocket systems by its 2P2 launcher vehicle which was based on the IS-2 heavy tank chassis.

Mars (3R-2) Rocket (FROG-2)

The smaller Mars systems was developed in parallel with the Filin. The Mars was lighter in weight and has less range than the contemporary Filin. It also differed from the Filin in that the 2P4 launcher vehicle was based on a modified PT-76 light tank chassis. The main advantages of the Mars over the Filin were its light weight and greater mobility in the field. The Mars, like the Filin, also appears to have been produced in modest numbers, possibly due to the inherent immaturity of the system or of its fission warhead.

The exact relationship between the Filin and the Mars remains unclear. It is possible that they were competitive efforts which explored alternative configurations for meeting the requirements for a division artillery missile. It is also possible that they were developed to carry different models of tactical nuclear warheads then being developed by the KB-11 nuclear weapons design bureau at Arzamas-16. The Mars, example, carried a warhead of only 500 kilogram whereas the Filin employed a 1,200 kilogram warhead.

In any event the Mars configuration was the design approach adopted for later Luna series of free rockets. This may have been in recognition that the Mars configuration adequately met divisional artillery requirements while the R-11 (SS-1b Scud A) addressed longer-range needs.

Early Luna Models (FROG-3/5)[23]

The first divisional missile artillery system to appear in significant numbers was the improved 2K6 Luna system which was first seen in 1960, and was accepted into

FROG Family Transporter-Erector-Launcher (TEL)

Crew	5
Chassis	Modified PT-76 amphibious tank
Weight (metric tons)	16.4
Length (millimeters)	10,550
Width (millimeters)	3,150
Height (millimeters)	3,000
Maximum road speed (kilometers per hour)	40
Road range (kilometers)	250
Propulsion (horsepower)	240 V-6 diesel
Armor maximum (millimeters)	14

FROG-3/-5 Rockets

NATO designation:	FROG-3	FROG-5
Soviet designation:	3R-8	3R-10 (R-30)
Weight (metric tons)	2.8	1.9
Length (millimeters)	11,000	9,100
Diameter (millimeters)	535	400
Fin span (millimeters)	1,050	1,050
Maximum range (kilometers)	40	55
Minimum range (kilometers)	15	15

the Soviet army in 1961. The improved Luna system consisted of an updated 2P16 transporter-erector-launcher (TEL) and a new family of R-30 rockets which used a more energetic solid fuel engine. The new Luna system had an effective range of forty to fifty-five kilometers, compared to only eighteen kilometers for the Mars/FROG-2. The 2P16 Luna TEL resembled the earlier Mars TEL, both being derived from the PT-76 amphibious tank. The 2P4 Mars TEL shared an identical suspension with the PT-76; the 2P16 Luna TEL had a modified chassis with return rollers and a modified rear stabilization and lockout system. Like the 3R-2 Mars, the 3R-8 (FROG-3) Luna rocket had the distinctive swelled warhead so characteristic of these early nuclear bombardment rockets. However, the shape of these warheads differed, the Luna having a more simple tubular cross-section.

Appearing at the same time as the Luna in 1960 was the 3R-9 Luna-1 (FROG-4) rocket. The 3R-9 Luna-1 was easily distinguished by its smaller warhead section which was identical in diameter to the rocket fuselage. This rocket was accepted for service in 1964, but soon disappeared from view.

The final member of the improved Luna series to appear was the 3R-10 Luna-2 (FROG-5) which was first publicly displayed in 1963. This rocket resembled the FROG-4 but lacked its prominent nose probe. The 3R-10 and 3R-9 rockets were essentially identical except for the warhead sections. The initial 3R-8 Luna differed from these later two types in having a larger bulged warhead. The added weight of the original 3R-8 Luna/FROG-3 meant that it had a shorter maximum range than the 3R-10 Luna-2/FROG-5. The more bulbous 3R-8 Luna/FROG-3 nose carried an earlier, larger one-ton tactical nuclear warhead, while the 3R-10 Luna-2/FROG-5 was armed with a smaller, more advanced tactical nuclear device weighing 450 kilograms.

At least four warhead options for the early Luna missiles were available: high explosive, chemical, nuclear and training. The nuclear warheads have been described in Western sources as a twenty kiloton type (Luna), and a selectable yield type with a range of twenty to one hundred kilotons (Luna 2).

When originally deployed in the late 1950s, the FROG rocket system was organized as a composite rocket artillery battalion in selected tank divisions. This battalion had a firing battery with two FROG launchers and another battery with two BMD-20 multiple rocket launchers with about 190 personnel. The improved Luna system did not enter widespread service in the Soviet Ground Forces until 1962. By this time, the organization of the rocket units was changed to homogenous battalions. The new Luna battalions were based around three TEL launcher vehicles each. The battalion had a variety of distinctive support equipment, including a ZIL-157V transloader for rocket reapply, a K-51 crane, and a modified GAZ-69 vehicle for site surveying. The meteoro-

9P113M Luna-M (FROG-7) artillery rocket system.

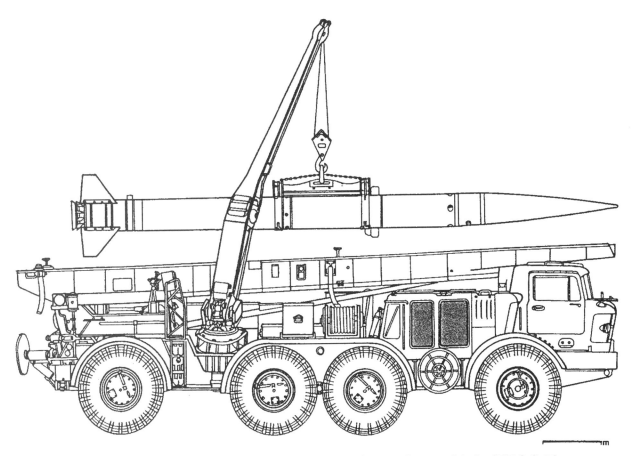

9P113 Luna-M Transporter-Erector-Launcher vehicle (FROG-7).

logical radar used in the Luna battalions was the RVS-1 Malakhit (Bread Bin). The Bread Bin was mounted on a one-axle trailer and was towed by a GAZ-63. It operated in the 1,700 MHz frequency and had a range of about twenty-five kilometers.

The TEL launcher was loaded in a rear area while a surveying section prepared a launch site. The Luna system took about fifteen to thirty minutes to lockout, erect, test and launch after arrival at the pre-surveyed site. Reloading time was largely dependent on the configuration of the resupply formation, but averaged about one hour.

The improved Luna series was the first of these divisional artillery rockets to be exported in any significant numbers. The early recipients of this missile included all six Warsaw Pact members. Soviet client states began requesting the Luna, but early attempts were rebuffed. Egyptian requests in 1958 were turned down. However, by the late 1960s, at which time the newer Luna-M was entering Soviet service, the Kremlin finally relented. Reportedly, FROG-3/-5 were provided to Algeria, Cuba, and North Korea.

The Soviet Ground Forces received about 200 improved Luna systems, and these remained in service well into the late 1980s, being mostly removed by 1989. Including Warsaw Pact and foreign exports, total launcher production was probably about 300.

The 9K52 Luna-M System (FROG-7)

The Luna-M (FROG-7) system was developed in the early 1960s as a divisional artillery rocket system, probably by Ganichev's Reaktivniy NII in Tula. The Luna-M was primarily intended for the delivery of tactical nuclear weapons, though the gradual evolution of Soviet doctrine in the post-Khrushchev years meant that some attention was paid to the use of the Luna-M system in the conventional delivery role.

In 1965, a new rocket system, designated the Luna-M was first deployed. This differed significantly from the earlier Luna-1/-2 system. The most obvious change was the launcher vehicle. The new TEL, designated 9P113, was based on the eight-wheeled ZIL-135LM vehicle rather than a tracked chassis. This entered series production at the Bryansk Automobile Plant, and hence is now designated BAZ-135. The launcher assembly was designed and manufactured at the Barrikady plant in Volgograd. The new launcher had an organic reloading crane onboard, permitting faster turnaround between launches. The new rocket, designated 3R-11, had superior range and accuracy to the earlier 3R-2/3R-9 series.

FROG-7 Rocket

Rocket/warhead designation
    9M21B (AA22 nuclear warhead)
    9M21F (9N18F HE warhead)
    9M21E (9N18E sub-munitions warhead)
    9M21Kh (9N18Kh chemical warhead)

Launch weight (metric tons)	2.5
Warhead weight (kilograms)	420 to 457
Length (meters)	8.95 to 9.4
Diameter (millimeters)	544
Span (meters)	1.7
Maximum range (kilometers)	65
Minimum range (kilometers)	15
Operating range	-40 to +50 degrees C

FROG-7 Transporter-Erector-Launcher (TEL)

TEL designation	9P113
TEL chassis	ZIL-135 LM
Weight (metric tons)	19
Length (millimeters)	10,690
Width (millimeters)	2,800
Height (millimeters)	With missile on-board 3,350
	Without missile on-board 2,860
Cross-country speed	15
Road speed	40
Vehicle range (kilometers)	650
Ground clearance (millimeters)	475
Wading (meters)	1.2
Radios	R-123, R-124

Rocket Transporter Vehicle

Designation	9T29
Vehicle chassis	ZIL-135 LM
Weight (metric tons)	20
Length (millimeters)	10,200
Width (millimeters)	2,850
Height (millimeters)	3,230
Range (kilometers)	650
Radio	R-123

The 9M21 Rocket

The basic weapon of the 9K52 system is the 9M21 series of artillery rockets, originally called the 3R-11 and 9R11. The rocket is solid-fuel, and spin-stabilized. On ignition of the system, the main solid rocket engine segments are ignited. Immediately behind the warhead section is a spin-stabilizer compartment that has four off-angle venire chambers. When ignited on launch, the venires give the rocket a spin to improve its stability and accuracy. The rocket has an accuracy (CEP) on the order of 400 meters at range.

The initial version of the Luna-M used the 3R-11 rocket, also known by its military designator as the R-65 rocket (NATO: FROG-7A). In 1968, an improved version of the rocket was developed, designated R-70 (FROG-7B). The R-70 was lengthened from 8.9 to 9.4 meters. The most significant improvement in the design of the rocket was the addition of a pair of air brakes at the rear of the rocket which can be locked open. The purpose of this system is to lower the minimum range of the rocket. This version also has the capability of switching warhead sections. Warhead handling and training is undertaken with a 9N218EhTS (Ehkvivalentniy takticheskiy sniard: Dummy tactical warhead).

A training rocket is also available, the PV-65 (prakticheskaya vystrel-65). For actual launch training, a sub-scale M21-OF rocket can be fired, which is the type of rocket used with the BM-21 Grad multiple rocket launcher.

The Luna-M rocket was eventually redesignated under the current system, from 3R-11 to 9R11 and finally to 9M21. There are four 9M21 variants currently in service. The standard rocket is the 9M21F. This is fitted with a 9N18F high-explosive/fragmentation warhead. The 9M21E is fitted with a 9N18E dispenser warhead, carrying shaped-charge dual-purpose sub-munitions. Another warhead is designated 9N32M, but its application is not known. The 9M21B can be armed with one of three tactical nuclear warheads. The AA-22 was the initial warhead type and has selectable yields at three, ten, and twenty kilotons. The AA-38 warhead was an improved type, with the same three selectable yields. The AA-52 warhead had four selectable yields at five, ten, twenty, and 200 kilotons. There is a chemical warhead version (designation not known but possibly 9M21Kh). This warhead is 436 kilograms with 216 kilograms of VX agent. The warhead uses a VT fuze with burster charge to airburst the round, creating a cloud of dense aerosol and droplets.

Luna-M Unit Organization

With the arrival of the new Luna-M system, Soviet divisional rocket artillery battalions began to shift from a three launcher to a four launcher configuration. Each battalion is organized into a headquarters battery and two firing batteries. Each battery numbers about 170 personnel (twenty officers/160 enlisted men).

The headquarters battery, numbering about eighty personnel, provides logistical support for the firing batteries. It includes the battalion's supply, medical and repair needs. This battery includes four 9T29 transporter vehicles, sometimes called RTM (raketno-tekhnicheskiy vzvod: missile technical unit). These are based on the ZIL-135RTM chassis and carry three 9M21 rockets. In combat, the battalion carries (on average) seven rockets for every launcher vehicle. Although the 9P113 TELs can directly transload rockets from the RTM to the launch rail due to their organic crane, the HQ battery also has a 9T31M1 crane vehicle for general support, based on the Ural-375D truck. Maintenance is handled at battalion level by the RM-1 maintenance complex based on three ZiL-157 vehicles. Other general service vehicles for the battalion include the RVD-1 optical maintenance vehicle (on Ural-375D) and the PKPP maintenance/check vehicle (on ZiL-131).

The two firing batteries each have a headquarters, meteorological section, survey section and two launcher sections. The battery level command vehicle is the 9S445M. This consists of a GAZ-66 truck with a shelter containing a 9V57M-1 (or VM-3M1) fire control computer; R-123, R-104UM and R-108M radios and two TA-57 telephones. In the 1970s, the meteorological section operated an RVS-1 Malakhit and an RMS-1 meteorological radar. But these days, the more common configuration is a single RMS-1 (End Tray) low altitude wind meteorological radar towed behind a GAZ-66 and supported by an APU on a two-axle trailer behind another vehicle. The RVZ-1 operates at the 1,700 MHz frequency. The survey section uses the GAZ-69TM/TMG/TMG-2, GAZ-66T or UAZ-452T survey vehicle for site preparation.

Each launcher section is based around a single 9P113, most often called an LTM (lafetno-takticheskaya mashina: launcher vehicle). The basic crew for the LTM is four men. Once the survey section prepares the launch site, the LTM section drives up and fixes the 9P113 launcher. It takes about fifteen to thirty minutes to prepare the launcher prior to firing. Due to the range of the Luna-M system, Soviet doctrine called for deploying the Luna-M battalion in the division rear, usually in the same echelon as the other divisional artillery about twenty to twenty-five kilometers behind the forward edge of battle.

The Luna-M is the longest-ranged artillery system available to the divisional commander and is usually retained for special missions. It's accuracy is not sufficient to encourage its use with a simple high-explosive warhead, unless bombarding a fairly large and vital target. It is a far more valuable system for area bombardment of targets with specialized warheads including chemical, nuclear or sub-munition.

According to a recent Russian article, the Russian Army now has 370 9P113 Luna-M launchers and 1,450 tactical nuclear warheads.[24]

Evolution of Short Range Ballistic Missiles (SRBMs)

The SS-1b/c (Scud-b/c)

The Scud missile stemmed from a Soviet Army requirement in the early 1950s for an operational-tactical missile for the delivery of conventional and nuclear warheads. The Russian term "operational-tactical" refers to system which is suitable for use at army level, bridging the gap between short-range tactical rockets such as the FROG family, and operational missiles like the SS-12 Scaleboard used by front-level brigades, and capable of striking targets over 500 kilometers away. The new missile was intended to replace the R-1 and R-2 missiles, which were Soviet derivatives of the World War 2 German V-2 ballistic missiles. The V-2 and its evolutionary descendants were difficult to employ on the battlefield due to the awkwardness of their cryogenic fuel system

SS-1b (Scud A) on a modified IS-2 TEL.

which required complicated support equipment. A V-2 launch battery, which included 152 trucks, seventy trailers and over 500 troops, could only launch nine V-2 missile per day.

The development of the Scud missile was undertaken by OKB-1 in Kaliningrad, in the suburbs of Moscow. This design bureau was the Soviet Union's premier missile design center and was headed by Sergei Korolev the "father" of the Soviet space program. The new system, called 8K11 in Soviet terminology (NATO: SS-1b Scud A), used the new R-11 missile. Although the Scud is often described in the popular press as a derivative of the German V-2 missile, it is in fact a descendent of the lesser known German Wasserfal anti-aircraft missile. The engine of the R-11 was based on the Wasserfal engine and used nitric acid and kerosene propellants instead of the cryogenic fuel of the V-2. This new propulsion system made it much simpler to employ in the field. The first test flight of the R-11 was conducted at Kapustin Yar on 28 April 1953, and the system was accepted for Soviet Army service in 1955. The 8U218 launcher selected for the 8K11 system was developed by the Kotin Design Bureau in Leningrad, and based on a derivative of the IS-2 heavy tank, the Obiekt 218.

A new plant for R-11 missile production was set up in the town of Miass, near Chelyabinsk. A young thirty-year old engineer, Viktor Makayev, was assigned as the chief designer at the Miass plant, and would later head all attempts to modernize the Scud missile system. Production began in 1955, and the cost of a basic conventional R-11 missile was 800,000 rubles. The nuclear armed versions cost four to eight million rubles depending on the type of nuclear warhead, which were available in the twenty to one hundred kiloton range. The improved R-11A version was developed in the late 1950s, featuring a revised fuel tank layout.

The R-11 missile was deployed in operational-tactical missile brigades (OTBR) which were deployed at army level (army is the Soviet term for a grouping of several divisions; a front is several armies). They had nine launcher vehicles each, supported by about 200 trucks and 1,200 troops. A brigade typically had twenty-seven R-11 missiles on hand and was easily capable of firing all these missiles in the course of one day's operation. Only about 100 of these 8K11 launchers were deployed, due to shortcomings in the system. A naval version of the R-11 was also developed for submarine launch, the R-11FM. This submarine version missile was later exported to China in small numbers.

In the mid-1950s, the Makayev Design Bureau in Miass began development of an improved Scud system, the 8K14 Elbrus. The US/NATO designation for this sys-

SS-1c (Scud B) on a MAZ-7310LTM 8 x 8 heavy truck.

tem is SS-1c Scud B and this has become numerically the most common version of the Scud missile system. The Elbrus system used the new R-17 missile. The aim of this program was to extend the range of the missile from the R-11's 180 to 300 kilometers, while at the same time increasing accuracy from the R-11's dismal four kilometers CEP (Circular Error Probability) to one kilometers CEP at full range. The range extension was made possible by adopting a turbopump for the engine, instead of the air pressure fuel injection system of the R-11 which necessitated heavy tankage reinforcement. The engine controls were improved to provide a more even fuel burn. A new strap-down inertial guidance package was developed. Its major elements consist of a 1SB9 vertical gyroscopic stabilizer package, 1SB10 vertical gyroscopic package with lateral adjustment integrator, a 1SB13 calculation package and a set of 1SB14 rudder control servos. These new features offered a very significant accuracy improvement compared to the R-11. The R-17 is rated by the Russians as having a range CEP of under 700 meters when used against targets 200 kilometers away, although Western intelligence places this closer to 900 meters. Russian sources state that range CEP varies from 180 to 610 meters depending on launch range, and 100 to 350 meters of lateral CEP. The R-17 was first test launched in March-April 1957 at Kapustin Yar. Production of the R-17 was first undertaken at Miass, but in the late 1960s shifted to Votkinsk, as Miass began to specialize exclusively in submarine-launched ballistic missiles. Guidance production was undertaken at Izhevsk.

The 8K14 Elbrus missile system was first based on a modified version of the IS-2 heavy tank chassis used with the earlier 8K11 system called the 2P19, and deployment began in 1961. However, in 1960, Khrushchev ordered Soviet heavy tank production to be terminated, so a new launcher vehicle was needed. In 1965, the R-17 began appearing on a new wheeled launcher, the 9P117, based on the MAZ-543LTM 8 x 8 heavy truck, and developed by the Titan Central Design Bureau at the Barrikady Plant in Volgograd. The wheeled chassis had many advantages over a tracked chassis. It caused less vibration to the missile and its delicate testing equipment aboard the launcher vehicle, as well as offering better reliability, and lower operating costs with only a modest decrease in cross-country performance. The 9P117 launch vehicle is officially named Uragan (Hurricane) but is popularly called Kashalot (Sperm Whale) by its Soviet crews due to its enormous size. The R-17 missile with the 9P117 launcher was initially designated 8K14, the same as the original version with the tracked launcher. During later modernization of the Elbrus system, it was redesignated as 9K72 Elbrus M, and the improved R-17 missile was

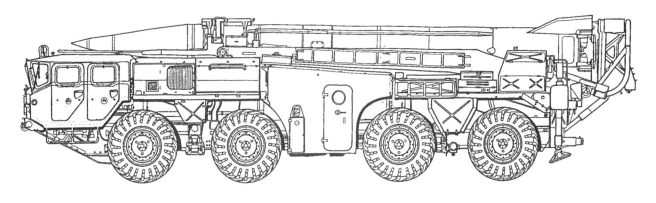

9P117
MAZ-7310LTM SS-1c Scud B Tactical Ballistic Missile TEL

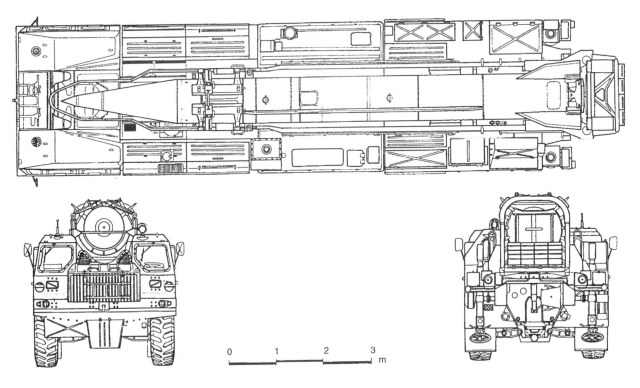

designated R-300.

There are several warheads available for the R-17 and R-300 missile, including conventional HE, chemical and nuclear. The 8F44 conventional high-explosive warhead weighs 989 kilogram, and was developed primarily for export clients on the modified R-17E/R-300E export missile. Although the R-17 is significantly more accurate than the R-11, it is still incapable of hitting precise military targets using a conventional warhead. The damage effect of the warhead is expected to include a crater 1.5 to four meters deep and twelve meters wide. The izdeliye 269A was the standard nuclear warhead for the R-300 and the yield of this and other optional types has been reported to be from five to eighty kilotons. It was developed by the All-Union Scientific Research Institute for Physics Technology in Kasli, near Chelyabinsk. The

R-17's chemical warhead contains a payload of 555 kilograms of thickened VX agent, and the total warhead weight is 985 kilograms. It uses a proximity fuze and a burster charge to disperse the agent before impact with the ground. There may be other versions available with different fillings. Due to the high terminal dive speeds (on an order of Mach 3), the aerosol dispersion system must be precisely timed, which is a considerable engineering challenge.

Besides the basic R-17, the Makeyev OKB developed an extended range version of the R-17 capable of reaching 500 to 600 kilometers, and first tested from the Kapustin Yar test range in 1965. It had much poorer accuracy than the basic R-17. Its performance overlapped that of the 9M76 Temp (SS-12 Scaleboard) missile, so it was never produced in large numbers. It was first given the temporary US designator of KY-3, but it was later called the SS-1d Scud C.

The accuracy of the basic 9K72 missile system is too poor to hit precision targets with conventional warheads. So in 1967, the TsNIIAG (Central Scientific Research Institute for Automation and Hydraulics) began development work on a precision guided version of the R-17. In this article, this version is called the R-17 VTO, but the actual Soviet designation is still undisclosed. The project was headed by Zinoviy M. Persits, who had earlier been involved on the Shmel (AT-1 Snapper) anti-tank missile. The first approach was to use optical comparison. A photograph of the target would be provided from an air reconnaissance mission, and inserted in a special 4 by 4 millimeter holder. A test version was built, but proved to be heavy and ineffective. The Soviet artillery branch was unhappy with the concept, recognizing the practical difficulties of such a system. Under field conditions, it would be difficult to provide photos of targets in specific size formats, and in addition, such a warhead was effective only under clear weather conditions. The army suggested a radio command system, but this was not favored by the developers.

In 1974, with miniaturized computers becoming available, it was decided to try to digitize the image. In this fashion, targets could be easily changed in the warhead from a computer library. A prototype was completed in 1975 and tested under an Su-17 strike aircraft. The first live test of a missile took place on 29 September 1979 with the R-17 VTO hitting within a few meters of the designated target. In the following years, the system was modified so that the warhead compartment separated from the missile fuselage. New control surfaces were added to that the warhead could make terminal corrections. This version of the R-17 VTO was first test fired on 24 September 1984, and both this launch and a subsequent one on 31 October were unsuccessful. The Main Artillery and Missile Directorate supported continuing the program in spite of severe criticism. In 1985, after extensive redesign efforts, it was realized that the problem stemmed from the buildup of a thin layer of dust in the inner surface of the optical lens at the nose of the missile. The 1985 tests were successful.

Testing and development concluded in 1989 and the R-300 VTO was accepted for Soviet Army service. The R-300 VTO parallels the general Soviet doctrinal drift away from dependence on tactical nuclear warheads. The new system was developed mainly to permit the use conventional warheads, especially sub-munition buses to deliver bomblets. Such a weapon could be used for high value targets such as command centers, air defense missile batteries, runways and other targets. This version is called the SS-1e Scud D in NATO. In 1992, this warhead section was offered for export to some Scud clients, but current Russian government commitments to the Missile Technology Control Regime (MTCR) are likely to restrict such exports.

Export of the 8K11 and 8K14 system to the members of the Warsaw Pact began in 1961. By the 1980s, some 140 launcher vehicles were exported, equipping thirteen brigades: Poland (4 brigades), Czechoslovakia (3); Germany (1); Romania (3); Hungary (1); and Bulgaria (3). Several of these countries removed the Scud brigades in the early 1990s, including Germany and Czechoslovakia.

The first exports of the 8K14E export system (SS-1c Scud B) system began in 1973. These used a modified 9P117 launcher vehicle with a less accurate manual command and control interface, and a related R-17E missile. The first customer for the 8K14E was Egypt which received them in time to use them in small numbers in the 1973 Mid East war. Later shipments to the Mid East included Syria (one brigade- 18 9P117); Libya (72 9P117), North Korea (one regiment, 24 9P117), Iraq (2 brigades, 36 9P117), Yemen, and Afghanistan. Iran later obtained Scud systems from other sources (reportedly Libya) and the Croatian Army is rumored to have obtained a small number from Hungary. Unlicensed copies of the R-17 missile are produced in North Korea, and modified versions of the R-17 have been assembled by Iraq and North

SS-1b/c (Scud-b/c) and SS-23 (Spider) Missile Systems

Russian system designation	8K11	8K14 (9K72)	9K714
Russian missile designation	R-11 (R-150)	R-17 (R-300)	9M714
US designation	SS-1b	SS-1c	SS-23
NATO codename	Scud A	Scud B	Spider
Length	10.7 meters	11.2 meters	7.31 meters
Diameter	0.88 meters	0.88 meters	0.97 meters
Span	1.8 meters	1.8 meters	n/a
Launch weight	5,400 kilograms	5,860 kilograms	4,630 kilograms
Maximum range	180 kilometers	300 kilometers	400 kilometers
Circular error of probability	4,000 meters	1,000 meters	700 meters

Korea. Export prices of the R-17E varied, but are generally believed to have been about $1 million in the 1980s.

Operational Use of the Scud Missile System

The 9K72 missile system is deployed in army-level operational-tactical missile brigades (OTRB). In the 1980s, a typical brigade had twelve to eighteen 9P117 launcher vehicles organized into three launch battalions. The difference in size depends on the number of firing batteries in each battalion; some have two firing batteries with two 9P117, others have three firing batteries with two 9P117s each. Each 9P117 launcher vehicle is usually assigned a crew of seven, though an additional ten-man OSNAZ security detachment is assigned when nuclear warheads are being used.

The Scud missile brigade is supported by a variety of specialized vehicles. The 9S436 is the battalion command vehicle, and consists of a ZIL-131K truck with shelter body, command radios and the 9V51B firing data computer. Missile fueling is undertaken by the 8G1 fuel truck (TS-1 or TT-1 aviation kerosene) and 8G17 oxidizer truck (inhibited red fuming nitric acid), both based on the Ural-375 truck. The 9V41 is a ZIL-131KO truck with a shelter body containing check-out instrumentation for the missile system, including the control surface servo system. The 9F21 is a ZIL-157KE-1 truck with a special isolated shelter body fitted with air-conditioning for the transport and handling of nuclear warheads. The 2T3M is a ZIL-157 tractor used to tow the semi-trailer for carrying additional R-17 and R-300 missiles. The 2ShCh1 is a ZIL-131 or ZIL-157 truck with shelter body which contains the maintenance and test equipment for servicing the 9P117 launcher vehicle. Besides such vehicles peculiar to the Scud brigade, the unit also uses other vehicles found in other types of missile units including the 9T31M1 seven-ton crane vehicle, 8T210 crane vehicles, UKS-400V compressor truck and VAZ-452 land-navigation vehicle.

The preparation of the R-300 missile for launch is broken down into six readiness levels, the first three being called arsenal readiness, and the last three being called field readiness levels:

Readiness Level 6: At this stage, the missile is in storage, the warhead is in its transit container and the 9P117 is parked and unmanned. Periodic maintenance and testing is conducted every two years.

Readiness Level 5: The missile and its components are removed from storage and component checks are started. The warhead is transported to the missile assembly point in the 9F21 or similar truck.

Readiness Level 4: The warhead is mated to the missile fuselage and the missile is fueled with propellant and oxidizer. In the meantime, the command and control network is established between the missile battalion and headquarters.

Readiness Level 3: The R-300 missile is loaded from the 2T3M semi-trailer onto the 9P117 launcher vehicle by the 9T31M crane. The VAZ-452 conducts a geodesic survey of the launch site for proper positioning of the 9P117 launcher vehicle.

Readiness Level 2: The 9P117 launcher vehicle moves to the launch site already surveyed by the VAZ-452. The brigade's meteorological section prepares and launches RKZ-1 (NATO: Ball Point) radiosonde balloons. These are tracked by the section's RMS-1 (End Tray) or ARMS-3 meteorological radars in order to determine wind velocity and direction, temperature, humidity, and air pressure. This data is passed to the 9S436 command vehicle for computation of necessary guidance corrections to the missile's inertial navigation system. In 1989, the Ulybka radar system was introduced into some Soviet brigades. It does not require the active illumination of the radiosonde, but uses a modified balloon which actively emits. This was introduced since the use of the earlier RMS-1 radar signals tends to alert an enemy's signals intelligence units that preparations are being made for a missile strike and may assist in pinpointing the launch site prior to launch.

When the 9P117 arrives at the pre-surveyed launch site, the section commander establishes a radio link with the battalion 9S436 command vehicle for further instructions as well as needed meteorological data, and the aiming officer begins to compute the missile launch data. The remainder of the crew begin to erect the missile which takes about three minutes. Four crewmen are positioned at each fin of the missile to make certain that the missile is erected in a perfectly vertical orientation. The missile alignment is checked using a 8Sh18 equipment set which includes a theodolite collimator. The crew attaches the firing cables from the rear of the 9P117 to the rear of the missile and completes their checks of the missile. Fuel for the engine turbopump is fed into the system. The 8V117 ignition controller is placed a safe distance from the launcher, and a cable is spooled out to it. The missile crew now awaits the launch instructions from the battalion command vehicle.

Readiness Level 1: The missile's inertial navigation system is turned on and allowed to warm-up, and the electrical system for the warhead is turned on. In the event a nuclear warhead is being used, the safety and arming device is actuated by inserting the proper codes. The crew removes the 2Sh2 thermal blanket from the warhead. The missile is now ready to launch. From time of arrival at the pre-surveyed site during Readiness Level 2 to missile launch takes about ninety minutes.

Scud Replacement

The Soviet Army began to develop a replacement for the Scud system in the early 1970s. The two main problems with the Scud were its very slow reaction time, some ninety minutes to prepare and fire, and its poor accuracy when employing conventional warheads. The new system, codename 9K714 Oka, entered development at the KB Mashinostroeniya (Machine Industry Design Bureau) in Kolmona, under the direction of S. P. Nepobidimy. A new design bureau was selected, as by this time, the Makayev bureau in Miass was specializing in naval submarine launched strategic missiles. The new 9K714 has reaction time of under thirty minutes, and uses a solid-fuel 9M714 missile that requires little maintenance. The four-ton missile has a more sophisticated guidance system which allows launches with surface wind conditions up to seventy kilometers per hour which would have prevented a Scud launch. The Oka has terminal accuracy of under 700 meters at a full 400 kilometers range. The range of the system was increased from the 300 kilometers range of the R-300 (Scud) to 400 kilometers on the 9M714. Two versions of the missile were developed, the 9M714B with a 772 kilogram nuclear warhead, and the 9M714K, with a 716 kilogram 9M74K cluster sub-munition warhead. Missile production was undertaken by the same Votkinsk plant as the R-17/R-300 Scud missile. The 9M714 missile is launched from a fully enclosed 9P71 launcher, built on the BAZ-6944 8 x 8 truck chassis at the Lenin TMS Plant in Petropavlovsk. A total of 450 Oka missiles and about 130 9P71 launchers were manufactured at a total program cost of four billion rubles.

The 9K714 Oka began entering Soviet Army service in 1980 and replaced seven Scud brigades by 1987. Seventy Oka missiles were exported to Warsaw Pact countries beginning in 1985 including Germany (twenty-four missiles, four 9P71 launchers), Bulgaria (eight systems), and Czechoslovakia (ten systems). The Oka is known as the SS-23 Spider under the US/NATO designation system. The service life of the 9K714 Oka system was short-lived due to the Intermediate Nuclear Forces Treaty between the US and USSR in 1987. The concession by the Soviet Ministry of Foreign Affairs to include the Oka system under the terms of the INF Treaty created a furor in the Soviet Army and amongst Russian military industrial leaders. The INF Treaty was supposed to effect systems with ranges of 500 kilometers and greater, while the Oka's maximum range was about 400 kilometers. In spite of military pressure, the Gorbachev administration decided to abide by its promises, whether mistaken or not. This led to the destruction of all 106 Soviet 9P71 launcher vehicles, eighty-six 9T230 missile transporter vehicles and the remaining 239 9M714 combat missiles by 1989.

Work on a successor to the 9K714 Oka began shortly afterwards and was called Iskander. The system is essentially similar to the Oka, although its range may be somewhat less to prevent problems with the INF Treaty. By 1993, the Scud system has proven to be obsolete, and many of the 9P117 launcher vehicles are being retired

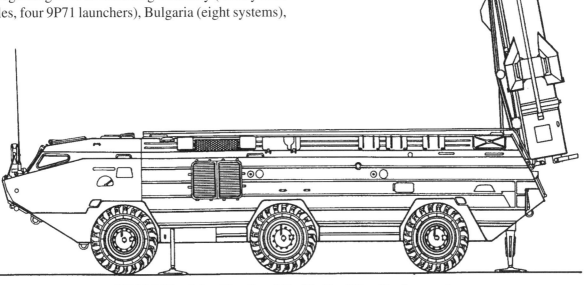

9P129 Tochka Tactical Ballistic Missile Launcher

due to age. In addition, large numbers of Scud missiles were shipped to Afghanistan from 1988 to 1989, depleting the inventory of these missiles. A successor to the Scud could either be an extended range version of the Tochka (SS-21) or the Iskander.

FROG Follow-On: The Tochka (SS-21)

Development of the 9K79 Tochka (Point) tactical ballistic missile system began in the late 1960s by the Machine Industry Design Bureau in Kolomna (KBM: KB Mashinostroyeniya) under S. P. Nepobidimy to replace the earlier 9K52 Luna-M (FROG-7) ballistic rocket system. The Luna-M suffered from two significant shortcomings. It lacked a guidance system, relying on spin-stabilization, and so had poor accuracy. Western sources usually put the a CEP (Circular error probability) at 500 to 700 meters at its maximum range of sixty-five kilometers, and recent Russian sources indicate that the likely impact area would measure about 2.8 kilometers in depth from range error and 1.8 kilometers in width in azimuth error. (By way of comparison, the US Army's MGM-52 Lance has a CEP of 225 meters at a range of seventy-five kilometers.) The Luna-M's poor accuracy made it nearly useless for the delivery of conventional warheads. Secondly, the Luna-M's missile was completely exposed on its launch rail, making it vulnerable to artillery damage and also creating climatic control problems when employing temperature-sensitive nuclear warheads. An attempt to correct the latter problem was made in the early 1960s with an experimental version of the Luna-M's normal 9P113 wheeled launch vehicle which was modified by the addition of a fully enclosed superstructure with an opening launch roof. This did not entirely cure the problem, leading to the requirement for an entirely new system.

The KBM Machine Industry Design Bureau in Kolomna was the descendent of the Shavyrin mortar design bureau. In the late 1950s, it was shifted to missile

SS-21 (9P129) Tochka-U tactical ballistic missile launcher.

development after its super-heavy 420mm Oka self-propelled mortar was vehemently criticized and canceled by Khrushchev. Although originally involved in the development of anti-tank missiles, its past associations led it to the development of artillery missile systems such as the Tochka and the later 9K714 Oka (SS-23 Spider). It is possible that the Tochka was developed competitively with another system from the NPP Splav design bureau, which had probably designed the earlier Luna systems.

The requirements for the new system stressed the need for accuracy, quick reaction time, a high degree of automation, and durability. The system had to be capable of operating in temperature extremes of ± 50 degrees centigrade, and be able to be launched under adverse weather conditions including surface winds up to seventy kilometers per hour. Open Russian accounts of the program indicated a requirement for a CEP accuracy of 220 meters; in fact the requirement was more demanding — 160 meters. Test launches of the new missile began in the early 1970s, and the system was accepted for Soviet Army service in 1975. When the Tochka was first spotted in service by Western intelligence in 1976, it was dubbed SS-21 Scarab under the US/NATO codename system.

Tochka Organization

The Tochka was intended as a divisional missile system, with a tactical missile battalion (Takticheskо-raketniy divizion) per tank or motor rifle division. This battalion consists of a battalion headquarters, a headquarters battery, and two launch batteries, and possesses a total of four 9P129 transporter-erector-launcher (TEL) vehicles. The headquarters battery provides technical and logistical support for the launch batteries and employs two different types of missile transport vehicles. This includes four 9T238 missile transporter semi-trailers, towed by a ZIL-131-4401 tractor which are used to transport missiles and warhead sections from the rear areas to the battalion.

To transport the assembled missiles from the battalion to the forward-deployed launch batteries, the headquarters battery has four 9T218 transloaders. These are similar in appearance to the 9P129 launch vehicle, except using a canvas covered rear bed, and it is based on the related BAZ-5922 chassis. The 9T218 transports two missiles and is fitted with a 9T315/9T316 crane for transferring the missile to the 9P129 launcher vehicle. This transloader may have some capability to launch the missile under extreme conditions. The other vehicle peculiar to the Tochka battalion is the 9V844 maintenance system, based on two ZIL-131 trucks, which is used to conduct routine maintenance on the missile and launcher systems.

The two Tochka launch batteries are identically equipped, and each control two of the battalion's four 9P129 launcher vehicles. The launch batteries are organized into a battery headquarters, a meteorological section, a survey section and a launcher section. The headquarters section includes the battery command vehicle (on a GAZ-66 van-bodied vehicle), and an 9V818 AKIM vehicle (Avtomaticheskaya kontrolno-ispytatelnaya mashina: Automated testing vehicle, based on the ZIL-131 truck) which is used to conduct testing on the missile and missile guidance system.

The meteorological section is used to conduct atmospheric surveys which provide the data needed by the missile's inertial navigation system to adjust for crosswind and other factors. When first deployed in the mid-1970s, these sections were equipped with the RMS-1 (End Tray) or ARMS-3 meteorological radars. Prior to the missile launch, the meteorological section launches RKZ-1 (NATO: Ball Point) radiosonde balloons. The radar is used to track the balloon to determine wind velocity and direction, and instrumentation on the balloon provides data on temperature, humidity and air pressure. The met section transfers this data to the battery command vehicle where the launch officer enters it into a ballistic computer to prepare the proper launch correction data for the missile's inertial navigation system. The accuracy of this data is vital to overall missile accuracy.

In 1989, the new Ulybka radar system was introduced into some Tochka battalions in place of the older RMS-1 End Tray. It does not require the active illumination of the radiosonde, but uses a modified balloon which actively emits radio signals which can be tracked by the Ulybka. This was introduced since the use of the earlier RMS-1 radar signals can alert an enemy's signals intelligence units that preparations are being made for a missile strike and may assist in pinpointing the launch site prior to launch. The other critical ingredient in missile accuracy is accurate site survey. This is conducted by the battery's survey section, generally using the VAZ-452 survey vehicle. This section identifies likely launch sites, and then marks them prior to the arrival of the launcher vehicles. The newer 9P129-1 launcher vehicle for the improved Tochka-U system is fitted with an integral 1T25 topogeodetic survey system, obviating the need for pre-survey.

Each of the two launch batteries deploys two 9P129 transporter-erector-launcher vehicles, based on the BAZ-5921 chassis. The basic automotive chassis is produced by the Briansk Automotive Plant (BAZ) and system assembly is completed at the Lenin Heavy Machine Industry Plant in Pavlograd. The BAZ-5921 chassis is related to the 9A33 (BAZ-5937) launcher vehicle used with the 9K33 Romb (SA-8 Gecko) air defense missile system. The 9P129 launcher vehicle contains a crew compartment in the forward section of the vehicle, and a missile compartment behind this. During transport, the missile is entirely enclosed, with the warhead resting inside a temperature-controlled casing. The overhead doors above the missile are probably lightly armored to protect the missile against overhead artillery airburst. The 9P129 vehicle, unlike the 9P113 used with the Luna-M/FROG-7, is fully amphibious, using a rear-mounted water-jet system. The four man crew is protected by a FVU-100N-24 NBC filtration system. Vehicle systems can be powered off the vehicle's 5D20B engine or an onboard fifty kW D-144-81 auxiliary power gas-turbine with VG-7500N generator.

The missile is mounted on an elevating launch rail that is hydraulically actuated. When ready for launch, the missile is elevated to an angle of eighty degrees. The 9P129 is highly automated, containing integral missile control and checkout equipment, and an electronic computer for transferring data from the battalion met section into the missile prior to launch. As mentioned earlier, the improved 9P129-1 being used with the Tochka-U system has its own onboard topogeodetic survey system, which allows the vehicle to be accurately and rapidly positioned without the need for site survey by the battery survey section. According to Russian sources, the time from the march to first missile launch is seventeen minutes, and refire time for the Tochka-U is forty minutes.

Although the Tochka was originally intended as a divisional artillery missile system, the Intermediate Nuclear Forces Treaty disrupted Soviet deployment plans. Beginning in 1980, the Soviet Army was replacing the obsolete 9K72 (SS-1c Scud B) operational-tactical missile system with the 9K714 Oka (SS-23 Spider) in army-level missile brigades. However, the 1987 INF Treaty led to the elimination of the Oka, creating an equipment shortfall in army-level missile brigades. As a result, since 1989 some Tochka missile battalions have been "booted upstairs" into new army-level missile brigades instead of divisional deployment. These missile brigades have three battalions with twelve to eighteen launcher vehicles.

The Tochka missile system was first deployed in the western military districts of the USSR in 1976 and with Soviet units in Germany in 1981. Production of the Tochka system was scheduled to be completed in 1987, but was extended, probably due to the INF Treaty-induced changes mentioned above. To date, about 300 launcher vehicles have been deployed with the army units of the former Soviet Union, enabling them to equip seventy-five tank or motor rifle divisions with a standard Tochka battalion.

The 9M79 Missile

The 9M79 Tochka missile is manufactured by the Petropavlovsk Machinery Plant, in Petropavlovsk, Kazakhstan. The 9M79 missile is propelled by a conventional solid-propellant rocket engine with composite solid propellant fuel, and a steel combustion chamber. Stabilization is provided by both conventional mid-body fins, and novel grid fins at the rear of the fuselage. This type of grid fin was first seen on the 9M79, but has subsequently been seen on many other Russian missiles including both ballistic missiles (9M714 Oka, RSD-10 Pioner) and even air-to-air missiles (R-77/AA-12).

Typical of many Soviet designs, the 9M79 Tochka is larger than comparable American solid-fuel tactical missiles (such as the MGM-140 Army TACMS), showing less attention to volume restrictions than many Western designs. Aside from the low-risk rationale for such a design approach, the larger volume of the design also allowed for increasing performance at a later date using improved composite propellants and engine design as took

SS-21's 9M79 Tochka-U missile.

SS-21 (Tochka) Transporter-erector-launcher (TEL)

Launcher	9P129
Launcher chassis	BAZ-5921
Crew	4
Length (millimeters)	9,485
Width (millimeters)	2,782
Height (millimeters)	2,373
Weight with missile (kilograms)	18,085 k
Maximum road speed (kilometers per hour)	60
Maximum swimming speed (kilometers per hour)	10
Maximum range (kilometers)	650
Radio	R-123M
Internal communications	R-124

SS-21 (Tochka) Missile Systems

Missile system codename	**Tochka**	**Tochka-U**
Missile designation	9M79	9M79-1
Length (millimeters)	6,400	6,416
Diameter (millimeters)	650	650
Span (millimeters)	1,448	1,448
Launch weight (kilograms)	2,000	2,010
Warhead weight kilograms)	482	482
Minimum range (kilometers)	15	20
Maximum range (kilometers)	70	120

place on the 9M79-1 Tochka-U.

The 9M79 missile uses a strap-down pre-programmed inertial guidance package, mounted in the nose behind a laser altimeter. The system manual indicates that this provides it with a range CEP of 525 feet (160 meters) although some Western intelligence assessments suggest better accuracy. Flight course corrections are provided by a balanced combination of thrust and aerodynamic steering. There are four grid fins at the rear of the missile which are driven by the thrust vector system control to carry out course correction in flight, mainly influencing range. In addition, the rocket engine is fitted with simple jet vanes which can be employed by the thrust vector flight control system to influence direction.

The 9M79 missile can be fitted with several warhead types. The basic 9M79F version of the missile employs a 9N123F high explosive warhead weighing 1,060 pound. (482 kilograms) with a 265 pound (120 kilograms) high-explosive fill. The improved 9M79K version of the missile employs a 9N123K sub-munition warhead with sub-munition bomblets. The 9M79B "special" version of the missile carries an AA60 tactical nuclear warhead. Russian press accounts in 1992 suggested that there were about 310 of these tactical nuclear warheads for the Tochka.

No chemical version of the 9M79 missile has been

CFE Counts for Self-Propelled Artillery and Mortars (European Russia): 1990 to 1997

	2S1	2S3	2S4	2S5	2S7	2S9	2S12	2S19	2S23	Total
1990	2,309	2,010	66	499	311	452	1,188	14	0	6,849
1991	2,292	2,012	55	496	305	509	1,401	15	0	7085
1992	1,163	1,154	10	401	155	536	587	6	0	4,012
1993	1,068	1,131	11	424	107	407	569	96	0	3,813
1994	1,037	1,004	9	399	106	332	445	173	1	3,516
1995	742	1,001	21	365	81	358	190	373	2	3,133
1996	695	1,026	21	449	80	346	181	372	2	3,172
1997	643	1,087	19	449	32	329	171	411	2	3,143

officially acknowledged, but it is believed that such a warhead variant exists. The traditional Soviet pattern has been to reserve about four percent of tactical missiles for chemical warheads. US sources have also indicated that the Tochka can be fitted with a scatterable mine warhead, but this has not been acknowledged in Russian sources. There is also a 9N123K-UT (uchebno-trenirovanaya) inert training practice warhead. There have been repeated reports of a terminally guided warhead using radar scene matching, but there has been no authoritative confirmation of this system.

It seems likely that advanced warheads for the Tochka are being developed. One of the most likely candidates would be a sensor-fuzed smart anti-armor munitions, comparable to the SADARM being developed for the US Army's M270 MLRS and MGM-140 TACMS. In 1992, the NPO Bazalt design bureau first displayed its SPBE-D sensor fuzed munitions, in a cluster bomb configuration using the RBK-500 tactical munition dispenser. This munition weighs 14.5 kilograms and uses a dual-frequency infrared sensor for target detection, with and explosively-formed penetrator as a kill mechanism. The basic configuration of the SPBE-D is closer to the US Air Force SFW (Sensor Fuzed Weapon) than the SADARM, however. Such a munition could presumably be adapted to the Tochka missile for use in attacks against armor formations.

At least three generations of the 9M79 missile have been developed to date, the original 9M79, the improved 9M79M, and the current 9M79-1 Tochka-U. The 9M79-1 Tochka-U uses an improved composite propellant which increases the nominal range of the system from the original seventy kilometers to 120 kilometers without a significant change in basic weight or payload. The 9M79-1 is probably capable of ranges of 150 kilometers depending on warheads and trajectories. Like the earlier versions of the system, the 9M79-1 can be fitted with the various warhead types mentioned above. In these cases, it is identified by the suitable suffix (for example, 9M79-1F for a missile fitted with the 9N123F conventional HE warhead).

The Tochka-R Anti-Radar Missile

One of the secondary missions for the Tochka system is to assist in the suppression of enemy air defenses by attacking high value air defense systems. In the NATO context, this would have meant targeting US Army and Bundeswehr Patriot batteries. The threat of this system was the primary catalyst for the US Army to develop the PAC-1 version of the Patriot missile, in order to protect Patriot sites. In the mid-1980s, the KBM design bureau in Kolomna began working on a new variant of the Tochka system specially tailored for anti-radar role, designated Tochka-R. This missile uses a new warhead section. *Jane's Intelligence Review* has reported that the warhead is an electromagnetic type that explosively generates an energy pulse to destroy a nearby radar's electronics.[25]

6

TOWED ARTILLERY AND ANTI-TANK SYSTEMS

Towed Artillery Systems

Soviet towed artillery in World War 2 was respected by its opponents for the quality of its weapons, even if its tactics were derided for their infatuation with sheer volume of firepower. The Soviet Main Artillery Directorate (GAU) showed very little interest in self-propelled artillery during the war. This was because self-propelled artillery in World War 2 was primarily used as mobile anti-tank guns. Hence World War 2 self-propelled artillery development fell under the control of the Main Armor Directorate instead.

Many Soviet artillery factories had been overrun by the Germans in the opening months of the war delaying the production of new or existing types of tubed artillery systems. In addition, conventional tubed artillery required elaborate (and expensive) machining. This resulted in long-lead times and expense associated with the production of tubed artillery. Due in part to this reason, Soviet artillery relied on Multiple Launch Rocket systems (MRLs), such as Katyusha rockets, to deliver the Red Army style of massed firepower favored by Russian artillery men. A single truck-mounted launcher could ripple fire a rocket salvo, equivalent to a battery of conventional guns. The limited accuracy of this style of weapons was not a major concern, since they were intended to supplement conventional artillery, not replace it. Conventional artillery could be used in roles, such as counter-battery fire, where accuracy was demanded.

With the "revolution in military affairs" taking place in the late 1950s, the Soviet Army reexamined the role of towed artillery. Soviet artillery development after the war focused almost exclusively on evolutionary development of its wartime weapons and tactics. The administration of the Soviet artillery branch, under Marshal N.N. Voronov, was preoccupied with other matters, especially the absorption of missile technology into the Soviet armed forces. The combination of the heavy drain of technical resources into these programs, as well as Khrushchev's antipathy to traditional artillery weapons, probably accounts for the Soviet Army's sluggish performance in most artillery developments during this period.

The coup against Khrushchev in 1964 removed one of the barriers to conventional artillery modernization. Around 1965, the Main Rocket Artillery Directorate (GRAU) began to lay down requirements for a new generation of towed artillery systems. After two decades of distraction with the monumental task of building up the tactical nuclear forces of the Army, the artillery branch finally turned to the more mundane task of modernizing its conventional tubed artillery.

The need for improved artillery systems had become more manifest by this time. NATO had been significantly improving its artillery force, notably including the US M-109 155mm and M-110 203mm self-propelled howitzers and the British Abbot self-propelled 105mm howitzer. More importantly, NATO's ability to conduct counter-battery fire was improving. With the advent of more and improved artillery location radars, Soviet towed artillery sites could be quickly identified and targeted. New ammunition developments, especially chemical weapons, and improved conventional munitions (ICM) cargo rounds made artillery crews especially vulnerable to counter-battery fire.

The Soviet Ground Forces received several excellent towed artillery pieces in the 1960s, notably the D-30 (2A18) which appeared in 1963. But towed guns took precious time to site, move, and re-site when faced by counter-battery fire. Towed guns were not survivable on a nuclear battlefield. What was required was mobility to avoid counter-battery fire, and armored protection to resist counter-battery fire should it come. The obvious direction was self-propelled howitzers using proven howitzer designs like the Petrov design bureau's fine D-30 122mm howitzer and D-20 152mm howitzer. However, the Russians never completely abandoned towed artillery systems for cost reasons. In part, towed artillery systems were cost effective when dealing with any potential invasion or war in the Far East. Due to cost and effectiveness reasons, the Soviets and Russians continued development in towed artillery systems through the fielding of the D-20/D-30 in the 1950s/1960s, the 2A36 in the 1970s, the 2A65 in the late-1980s, and finally the lightweight 2A61/Model 389 in the late-1990s.

In the 1990s the Russians are once-again looking to reduce cost and see towed artillery systems as a cost-effective means of delivering fire to support operations on the Russian periphery. With the threat axis to Russia changing from Western Europe to its border republics, low-technology weapons are perfectly suited to deal with such threats. In the recent Chechen War, towed artillery systems were extensively utilized to bombard major buildings in downtown Grozny. Because Russia continues to see such conflicts as the most likely threat, Russia will likely continue to look to accurate, lightweight, and cost-effective towed artillery systems to fill the low-technology/low-cost niche for Russia's mobile forces for the next twenty years.

The M-46 130mm Field-Gun[2]

Since 1954, the M-46 130mm field gun has replaced the A-19 field gun in Soviet artillery divisions, and it has subsequently been made available to other countries. The 130mm field gun is the only one of its caliber deployed in significant numbers in any army worldwide. In Russia, the M-46 has in turn been largely replaced by the 2A36 152 mm gun and the 2A65 152mm howitzer. On 1 January 1996 CFE treaty data indicates only one M-46, but includes 524 2A36s and 330 2A65s. However, the M-46 is being stored in large numbers east of the Urals in depot facilities.

The towing vehicles for the M-46 are either trucks or MT-LBs, both vehicles carry the crew, equipment, and part of the ammunition. Because ground clearance is limited, cross-country capability is reduced and firing positions are usually close to roads or well-cut paths.

The M-46's main gun is a rifled monobloc barrel. Its muzzle brake, which absorbs about thirty-five percent of the recoil energy is screwed onto the barrel which is mounted on rails and pivots in the cradle. The barrel is closed at the rear by the breech ring. The firing pin is released only if the breech ring is completely closed. The breech mechanism can only be opened if the recoil mechanism has been correctly locked into its holder in the breech ring. The arc of traverse for the M-46 is + or -25 degrees.

The maximum range of the M-46 with a 33.4 kilogram high-explosive (HE) projectile at full charge is 27.1 kilometers and has a muzzle velocity of 930 meters per second. The propellant case, weighing 25.7 kilogram has a two-part charge. The 33.4 kilogram armor piercing-tungsten (AP-T) projectiles are used for direct firing against a 2.7 meter high armored targets and can penetrate up to 250 millimeters of homogeneous steel up to a distance of 1.3 kilometers.

The gun crew consists of the commander, six gunners, and the vehicle driver. The gun crew is protected behind an approximately seven millimeter thick V-shaped shield mounted on the carriage. The slits for the optics are only opened as needed and normally have metal covers.

The gunner stands to the left of the gun. He operates both hand wheels to control the gun, and also has the PG-IM standard panoramic sight and the OP4M-35 telescope. It is designed for armor piercing projectiles up to 4,000 meters and for high-explosive projectiles up to 8,000 meters.

The chassis and the two welded boxed section split arms are built on the lower carriage. The chassis consists of two individually mounted wheels with torsion bar suspension. The carriage also helps to compensate for gun tilt. The M-46 remains on its wheels even during firing since it is fitted with neither supports or a rotary plate.

The M-46 can achieve a maximum rate of fire of eight rounds per minute. However, this is possible only for short periods; not at high angles of elevation; and only with previously prepared rounds and cartridges. The M-46 is intended to engage targets deep behind enemy lines and thus, primarily is a counter-battery artillery weapon.

The M-46 has been extensively modified and the former Soviet Union fielded a type with a longer barrel and improved cradle/recoil system. In addition, several countries are currently offering larger gun barrel upgrade packages to convert the M-46 130mm to a 155mm standard. Two countries who are aggressive in this area are China's NORINCO and Soltam of Israel. In addition, several countries offer new ammunition types to include

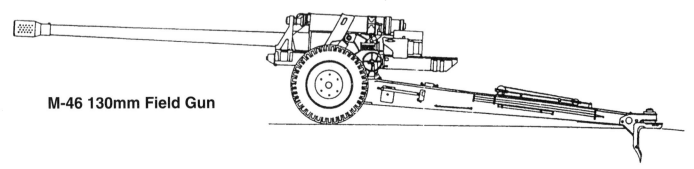

M-46 130mm Field Gun

NORINCO of China (who offer a family of ten different kinds of 130mm projectiles), Giat Industries of France, and Denel of South Africa. For an indepth look at Russian artillery ammunition, see the Chapter on Anti-Armor Developments.

The M-46 is currently deployed throughout the world. It is also produced under license as the Type 59 and Type 59-1 130mm gun in China while Egypt has produced a reversed engineered version of the latter as the 130mm M59-1M. The M-46 and its variants are still sold to countries in South America, Africa, the Middle East, and Asia.

The D-30 122mm Howitzer[3]

The D-30 (2A18) 122mm howitzer system was designed by Petrov's Design Bureau at the Artillery Plant Number 9 and entered service in the late-1960s. The system was designed to replace the 122mm howitzer M-30 which was designed just prior to the start of World War 2. The major areas of improvement to the D-30 over the earlier M-30 model was its ability to traverse the gun a full 360 degrees and its greater firing range. The D-30 has undergone numerous changes in its design life and later models are referred to as D-30A or D-30M. According to the 1 January 1996 CFE data, there are 837 D-30s in European Russia.

Both the D-30 and the 2S1 utilize the gun designated 2A18. The D-30 is towed muzzle first and can be sited and fired quickly. A firing jack is mounted under the center of the gun which raises the wheels of the carriage off the ground. The three outer arms are spread out and

M-46 and D-30 cartridge cases.

M-46 Technical Data

Crew	8 men
Gun shield	yes
Barrel length with muzzle brake(millimeters)	7,600
Length (millimeters)	11,730
Width (millimeters)	2,450
Height (millimeters)	2,550
Track (millimeters)	2,060
Ground clearance (millimeters)	400
Weight (kilograms)	8,450
Towing vehicle	AT-S, ATS-59, and M1972 Medium Tracked Artillery Tractor
Maximum towing speed (kilometers per hour)	50 road and 10 to 20 cross-country
Caliber	130mm gun
Maximum range (meters)	27,150
Maximum penetration of AP-T (millimeters)	250 (500 meters RHA at 90 degrees)
	150 (4,000 meters RHA at 90 degrees)
Muzzle velocity (meters per second)	930
Gun traverse (degrees)	25 right and left
Gun elevation (degrees)	-2.5/+45
Effective firing range (kilometers)	27.0 (standard) to 40.0 (rocket assisted projectile)
Maximum firing rate (rounds per minute)	5 to 8

D-30 122mm towed howitzer.

the firing jack is raised until all three arms are off the ground. Once off the ground, each arm is staked into the ground. The fire-control instruments for the D-30 include a type PG-1M panoramic sight, a type OP4M-45 telescopic sight, and a type K-1 collimator. The D-30 has its recoil system mounted over the barrel of the gun (an unusual feature for guns of this type).

The D-30 fires the full range of Soviet/Russian 122mm artillery ammunition. In addition, the D-30 can fire some of the projectiles used by the older M-30, however, unlike the M-30, the D-30 can fire a fin-stabilized HEAT projectile. It can fire a typical high explosive round like the 3OF-462 (total weight of 21.7 kilograms with 3.5 kilograms of high explosive) to a maximum range of 15.2 kilometers. Additional projectiles which the D-30 can fire include chemical, illuminating (S-463), smoke (D-462), leaflet, flechette, and incendiary projectiles. The D-30 is also capable of firing rocket-assisted projectiles out to a maximum range of twenty-two kilometers. For an indepth look at Russian artillery ammunition, see the Chapter on Anti-Armor Developments.

The D-30 is produced in China, Egypt, Iraq (called Saddam), and the former Yugoslavia (called D-30J). In addition, NORINCO of China also produced an extensive line of ammunition for the D-30 which can be fired by the Chinese or Russian versions of the D-30. Russia is also offering an improved version of the D-30 which is designated D-30A or D-30M with the gun being designated 2A18M.

The new D-30M has a number of improvements to the older D-30 which include a new double-baffle muzzle brake and a new central control baseplate which is square rather than round shaped. In addition, the D-30M has changed its gun cradle, carriage, and has a modified recoil system. The new cradle and carriage allows the D-30A or D-30M to travel at road speeds up to eighty

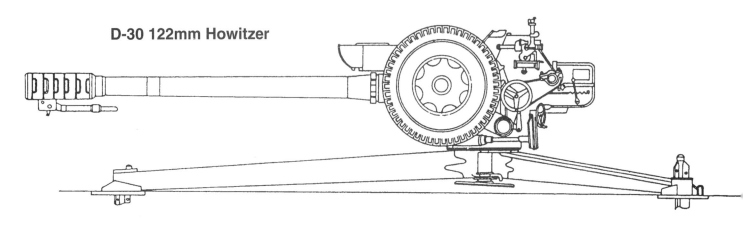

D-30 122mm Howitzer

Air-mobile D-30 122mm towed howitzer.

D-30A Technical Data

Designation	D-30A	D-30
Gun designation	2A18M	2A-18
Crew	7 men	7 men
Gun shield	yes	yes
Barrel length (millimeters)	4,875	4,875
Length (millimeters)	5,400	5,400
Width (millimeters)	2,200	1,950
Height (millimeters)	1,800	1,660
Track (millimeters)	1,850	1,850
Ground clearance (millimeters)	325 to 345	325
Weight (kilograms)	3,210	3,210
Towing vehicle	Ural-375 (6 by 6) and MT-LBu	Ural-375 (6 by 6) and MT-LBu
Maximum towing speed (kilometers per hour)	60 road	80 road
Caliber	2A18M 122mm gun	2A18 122mm gun
Gun traverse (degrees)	360	360
Gun elevation (degrees)	-7/+70	-7/+70
Effective firing range (kilometers)	15.4 (standard) to 21.9 (RAP)	15.3 (standard)
Sustained rate of fire (rounds)	75	75
Maximum firing rate (rounds per minute)	7 to 8	6 to 8

Iraqi D-30 122mm towed howitzer.

D-30 cartridge cases.

kilometers per hour. Russia is also offering upgrade packages to existing D-30s to the D-30A or D-30M standard.

The D-20 152mm Gun-Howitzer

The 152mm Gun-Howitzer D-20 was developed shortly after World War 2 by the Petrov Artillery Design Bureau at Artillery Plant Number 9, located in the town of Yekaterinburg and was first seen in public during the May Day parade in Moscow in 1955.[4]

The D-20 uses the same carriage and recoil system as the D-74 122mm field gun. The D-20 has an irregular gun shield and the top of the shield can be lower to reduce the overall height of the gun. The gun is mounted on a circular firing pedestal and allows for the gun to be traversed a full 360 degrees. In addition, the gun is secured with two arms which can be moved rapidly due to wheels on each arm. This system can be towed by the AT-S medium tracked artillery tractor or by the Ural-375 (6 x 6) truck. The D-20's maximum road speed is sixty kilometers per hour and fifteen kilometers per hour cross-country.

The ordnance of the 152mm towed gun-howitzer is the same gun as that used on the 2S3. It fires a case-type charge with separate loading. It can fire the normal 3OF-540 high-explosive-fragmentation projectile (43.5 kilograms total weight with 6.4 kilograms of high explosive) to a range of 17.3 kilometers. Its maximum range is about thirty kilometers when firing rocket-assisted projectiles (RAP). Other types of ammunition include:

D-20 152mm gun-howitzer.

chemical, high-explosive rocket-assisted projectile (with a range of twenty-four kilometers), illuminating, smoke, and tactical nuclear rounds. In addition, the D-20 also fires a spin-stabilized HEAT round, flechette, scatterable mines (anti-tank and anti-personal), and semi-active laser rounds called Krasnapol. For an indepth look at Russian artillery ammunition, see the Chapter on Anti-Armor Developments.

According to the 1 January 1996 CFE data, there are 213 D-20s in European Russia. A long-barreled version of this system was produced by the former Yugoslavia and is designated M-84 in service. The D-20 system is also manufactured in China by NORINCO and is designated the Type 66.

The 2A36 152mm Gun[5]

In the 1970s, the Soviet Union developed a new 152mm towed gun which was allocated the industrial number of 2A36. The 2A36 was placed in production in the mid-1970s. The 152mm 2A36 replaced the 130mm M-46 in Soviet Army service. The 152mm ordnance of the 2A36 is identical to that used in the 2S5 self-propelled artillery gun which entered service with the Soviet Army in 1980.

The 2A36 was seen as early as 1976 by Western sources, and hence, was given the designation of the M1976. Within the former Soviet Army the 2A36 is commonly referred to as the Giatsint (Hyacinth). However, it was not seen in public until May 1985 when it was observed

D-20 Technical Data

Crew	10 men
Gun shield	yes
Barrel length (millimeters)	5,195
Length (millimeters)	8,690
Width (millimeters)	2,400
Height (millimeters)	1,925
Track (millimeters)	2,030
Ground clearance (millimeters)	380
Weight (kilograms)	5,700
Towing vehicle	Ural-375 (6 by 6) truck and AT-S Medium Tracked Artillery Tractor
Maximum towing speed (kilometers per hour)	60 road or 15 cross-country
Caliber	152mm gun
Gun traverse (degrees)	15 right and left
Gun elevation (degrees)	-5/+63
Effective firing range (kilometers)	17.4 (standard) to 24.0 (rocket assisted projectile)
Unit of fire (rounds)	65
Maximum firing rate (rounds per minute)	5 to 6

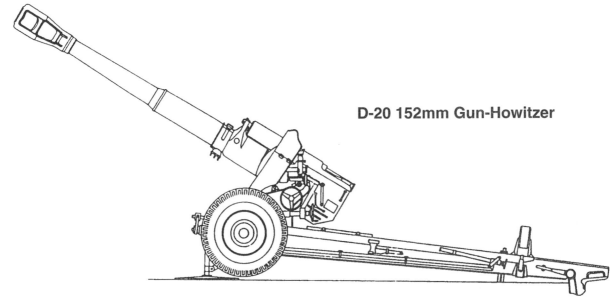

D-20 152mm Gun-Howitzer

2A36 152mm gun.

(6 x 6) truck. Since then, well over 1,000 2A36s have been built.[6] According to the 1 January 1996 CFE data there are 524 2A36s in European Russia.

The 49-caliber barrel of the 2A36 is fitted with a multi-slotted muzzle brake weighing 141 kilograms with the recoil system consisting of a buffer and recuperator which are mounted above the ordnance towards the rear. The horizontal sliding breech mechanism opens to the right automatically. The ordnance pivots on the cradle which is made of cast and welded steel construction. Elevation and traverse is manual, with the former being from -2 degrees to +57 degrees, with traverse being twenty-five degrees left and right. The direct and indirect sights are mounted on the left side of the carriage, as are the elevation and traverse mechanisms. Some of the eight man crew of the 2A36 are provided with protection from shell splinters by the armored shield that slopes to the rear and extends over the wheels. Mounted below and to the rear of the breech is the load assist system which is referred to as a quick firing loading system by the Russians.

A fused projectile is placed on the loading tray which is mounted to the right of the breech. When the breech is opened this slides to the left until it lines up with the

2A36 152mm gun.

breech. The projectile is then rammed into the gun, the loading tray then slides back to the right. The cartridge cased charge is then located in a similar manner and when the breech is closed the rammer automatically returns to the side and the weapon is then ready to fire. If the hydraulic rammer fails then loading takes place manually, although the rate of fire is reduced. The charge consists of a conventional cartridge case containing the actual charge. Once the gun is fired, the breech automatically opens and the spent cartridge case is ejected to the rear.

The 2A36's rate of fire is five to six rounds a minute. The effective range of the gun is twenty-seven kilometers using conventional ammunition and thirty-seven kilometers using rocket-assisted projectiles. It is capable of firing nuclear projectiles, as well as, conventional and chemical rounds. In addition to the normal forty-six kilogram high explosive fragmentation (HE-Frag) projectile with a 800 meters per second muzzle velocity other types include chemical, concrete piercing, improved conventional munitions, and tactical nuclear projectiles. For an indepth look at Russian artillery ammunition, see the Chapter on Anti-Armor Developments. The former Soviet Union has also developed a 152mm laser designated guided projectile to engage armor targets at extended ranges. The round, known as Krasnapol, can be fired by the 2A36, 2A65, 2S3, 2S5, and 2S19. The ammunition system used by the 152mm 2A36 is of a new design and is not interoperable with earlier Soviet artillery systems (i.e., 152mm self-propelled 2S3).

The 152mm 2A36 is towed by the KrAZ-255B or KrAZ-260 (6 x 6) trucks. In addition, it can also use fully tracked vehicles such as the AT-T, AT-S or ATS-59 which carry the crew plus a quantity of ready use projectiles, charges, and fuses. Using the KrAZ-260 as the prime

2A36 Technical Data

Crew	8 to 10 men
Gun shield	yes
Barrel length (millimeters)	8,197
Length (millimeters)	12,930
Width (millimeters)	2,788
Height (millimeters)	2,760
Track (millimeters)	2,340
Ground clearance (millimeters)	475
Weight (kilograms)	8,800
Towing vehicle	KrAZ-255B and KrAZ-260 (6 by 6) truck, AT-T, AT-S, and ATS-59 Medium Tracked Artillery Tractor
Maximum towing speed (kilometers per hour)	80 road or 35 to 45 cross-country
Caliber	2A36 152mm gun
Muzzle velocity (meters per second)	800
Gun traverse (degrees)	15 right and left
Gun elevation (degrees)	-2/+57
Effective firing range (kilometers)	27.0 (standard) to 40.0 (rocket assisted projectile)
Unit of fire (rounds)	60
Maximum firing rate (rounds per minute)	5 to 6
Sustained rate of fire (rounds per minute)	1

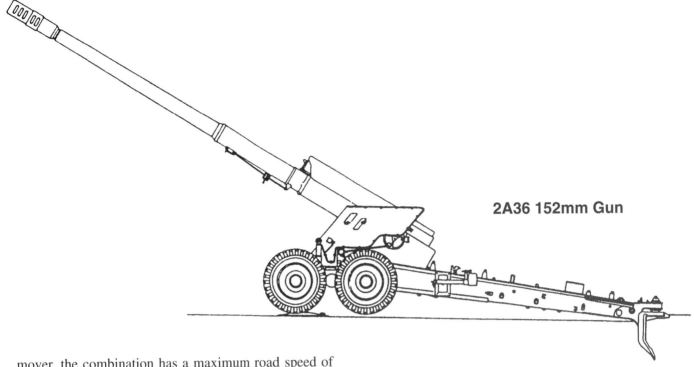

2A36 152mm Gun

mover, the combination has a maximum road speed of eighty kilometers per hour with a cross-country speed of between thirty-five to forty-five kilometers per hour.

The 2A36 has also been exported to Finland and Iraq with large numbers from Iraq being captured or destroyed during Desert Storm. Now, as part of its efforts to obtain foreign currency and keep at least some of its production lines open, Russia is offering the 2A36 for export at around $500,000 to $600,000 per system. The "Msta-B" 2A65 152mm towed howitzer, will continue to replace the 2A36 in front-line Russian Army service.

The 2A61 152mm/Model 389 155mm, Towed Howitzer[7]

The Russians, in early 1993, debuted information on a new lightweight 152mm towed artillery system designated 2A61. This towed gun system was being developed by the Petrov Artillery Design Bureau at Artillery Plant Number 9. The 152mm 2A61 lightweight howitzer is being developed to meet a Russian mobile forces requirement to have a large-bore gun that is lightweight and can rapidly be brought to a firing state.

The 2A61 is based on the same carriage as the 122mm D-30 and has been modified to accept a new larger 152mm gun. The 2A61 is fitted with a large muzzle brake with a double "T" configuration with a towing eye located just below it. One of the new features of this system is the use of a ramming system which enables the 2A61 to achieve a high rate of fire of between six to eight rounds per minute.

When deployed, the gun's wheels are raised and the three arms are placed on the ground for support. The gun has an elevation of +70 degrees, a depression of -5 degrees, and has a traverse of 360 degrees. A small gun shield is provided to protect the seven member gun crew from small-arms and shell fragments. The towing ve-

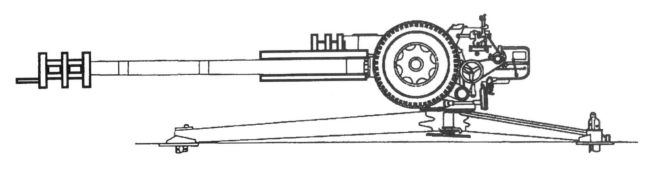

2A61 152mm/Model 389 155mm Towed Howitzer

2A61 Technical Data

Crew (men)	7
Gun shield	yes
Length (millimeters)	6,360
Width (millimeters)	2,200
Height (millimeters)	1,970
Track width (millimeters)	1,840
Ground clearance (millimeters)	350
Combat weight (kilograms)	4,350
Towing vehicle	ZIL-131 or MT-LBu
Maximum towing road speed (kilometers per hour)	80
Caliber	2A61 152mm gun
Gun traverse (degrees)	360
Gun elevation (degrees)	-5/+70
Maximum rate of fire (rounds per minute)	6 to 8
Gun's maximum range (meters)	15,000
Gun's minimum range (meters)	4,000

hicle for this system can be either the ZIL-131 (6 x 6) truck or the multi-purpose MT-LBu.

The 2A61 152mm howitzer was displayed publicly for the first time at a 1997 regional arms show in Omsk. In fact, the actual weapon for display was a slightly modified version, the "Model 389," which is the same as the 2A61 152mm howitzer except chambered for NATO-standard 155mm ammunition. By improving the recoil system and strenghtening the carriage and trunnion, the weapon can now be fitted with a 152mm or 155mm tube, offering a substantial increase in firepower for a 3.7 percent increase in overall weight (+600 kilograms).

2A65 152mm howitzer.

The Model 389 is now being examined for self-propelled applications. A prototype of the 2A61 suitable for self-propelled artillery has already been developed and tested at Artillery Plant Number 9's test range, though not yet fitted on a specific chassis. It appears that the objective of the program is to use the basic 2A61 on an upgrade of the 2S1 122mm self-propelled howitzer, and to use a planned version, with a barrel lengthened to six meters to upgrade the 2S3 152mm self-propelled gun.

2A65 152mm Howitzer[8]

A decision was made in the late-1980s to develop a completely new towed artillery weapon. The resulting work produced a 152mm towed artillery howitzer called the 2A65 Msta-B, (the "B" stands for Buksiruemyi or towed version). Msta is named after a river in the Ilmen district, a break in the previous Soviet practice of naming self-propelled guns after flowers or plants.[9]

The 2A65 is mounted on a split-arm carriage and when deployed rests on three points: a circular mounted plate on the carriage in front and two rear spades. Each of the rear arms has a wheel to allow the arms to be moved rapidly. When movement is finished, the wheels are tilted to sit on top of the arms.

The main gun is fitted with a muzzle brake and has a semi-automatic breech mechanism and ramming system. The elevation and traverse of the gun are done manually and the system has a direct and indirect sighting system. The 2A65 fires the same ammunition types as those used with the 2S19. Fire can be conducted with this weapon using the 3OF-45 high-explosive fragmentation round (up to 24.7 kilometers) and the 3OF-61 (to a range of 28.9 kilometers), the 3OF-23 cluster round (carrying forty-two anti-armor sub-munitions, reaching out to twenty-six kilometers, and penetrating up to 100 millimeters of armor), active radio jamming projectiles from the 3NS-30 family (to a range of 22.3 kilometers), and the 3vDTs8 tactical nuclear projectile. It can also use all standard ammunition for the D-20 and 2S3 howitzers, as well as, guided projectiles such as the 30F-39 Krasnapol laser-guided projectile. Target illumination for these laser-guided rounds can be carried out by an artillery observer with the 1D15 (PP-3) or 1D22 laser devices.

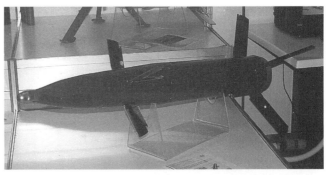
Krasnapol laser-guided artillery projectile.

2A65 high-explosive round.

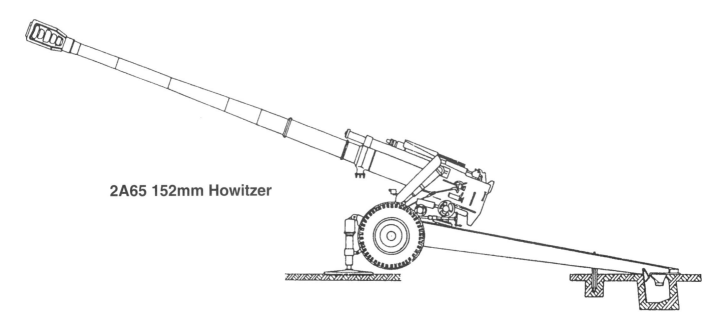
2A65 152mm Howitzer

2A65 Technical Characteristics

Crew	Eight men
Crew when using ammunition from ground supplies	Ten men
Gun shield	yes
Weight (kilograms)	7,000
Gun dimensions:	
—Length over gun tube (millimeters)	11,917
Muzzle velocity (meters per second)	810 to 828
Armament:	2A64 152mm Howitzer
Engagement angles for main weapon:	
—Elevation (degrees)	+70/-3.5
—Traverse (degrees)	27 left or right
Range of fire:	
—With HE-Frag ammunition/RAP (meters)	24,700/28,900
—Rate of fire (rounds per minute)	7 to 8
Regulated rates of fire:	
—First hour/ each additional hour (rounds)	up to 100/up to 60
—Time to go from travel to firing mode (minutes)	1 to 2
Towing highway/cross-country speed (kilometers per hour)	80/20
Unit cost:	$600,000 (1992 export price)

Characteristics of Ammunition for the 2A65 Towed Artillery System

Designator	Range (meters)	Weight (kilograms)	Weight of Filler (kilograms)	Length (millimeters)	Muzzle Velocity (mps)	Type of Round
3OF-45	24,700	43.56	7.65	864	810	HE-Frag
3OF-61	28,900	42.86	7.80	864	828	Rocket assisted
3OF-23	26,000	42.80				42 AT sub-munitions
3NS-30	22,300	43.56	8.20*			Active jamming round
3OF-39	20,000	50.00	6.60			Laser-guided

The 2A65 is part of an artillery firing battery complex (OBAK). The complete OBAK consists of a commander's vehicle (based on a BTR-80 chassis), a senior officer of the battery's vehicle (based on a Ural-43201 truck), and up to eight 2A65 weapons. Any size artillery combined unit may be formed from these OBAK organizations.

By February 1991, the Soviet Army had accepted some 371 2S19s and 2A65s weapons into service and by 1 January 1995 there were 281 2A65s in service with the Russian Army. In addition, enhancements to the 2A65's ability to fire guided projectiles such as Krasnapol are being developed which utilize a remotely piloted vehicle (RPVs) as the source for laser target illumination.

Russia artillery designers have also began work on a new 152mm towed gun version of the 2A65. Similar in development to the 2A65 gun being adapted to the 2S19 it appears that a new long-barrelled gun to replace the 2S5 is being developed and is called the 2S30 Iset. This gun may also be utilized in a towed version as was the case with 2A65.

The 2A65 is offered for export and costs roughly $600,000 a piece. Also being offered are special export versions of the 2A65 and 2S19, which feature a NATO-standard 155mm caliber gun tube, which facilitates the use of NATO-standard ammunition.

Anti-Tank Gun Systems

At the outset of World War 2, Soviet anti-tank artillery was deficient in quantity and stayed that way for much of the war.[10] Technically, the Red Army was not as well equipped for anti-tank defense as their German opponents. Divisional anti-tank weapons were inadequate in the later years of the war. While the 57mm and later 100mm guns could have solved this problem, they were not produced in sufficient quantities to equip many rifle divisions.

The Soviet Army began a broad effort to improve its anti-armor capabilities in the wake of World War 2. These qualitative improvements in anti-tank weapons extended upwards in the division. Rifle regiments replaced their 76mm ZIS-3 divisional guns with six 85mm divisional guns which had significantly better anti-armor capabilities. These anti-armor improvements were evolutionary rather than revolutionary. Tank armor continued to increase, so that by the mid-1950s, NATO main battle tanks had an effective frontal protection equivalent to about 300 millimeters of steel armor; the penetration capability of the heaviest towed anti-tank guns of the period such as the Soviet 100mm gun firing armor piercing (AP) rounds was about 250 millimeters of RHA.[11]

In 1956, a dedicated program began in the Soviet Union to deploy ATGMs as an alternative to towed anti-tank guns. The 1956 program can be traced to several factors. From the tactical side, the "revolution in military affairs" brought about by tactical nuclear weapons sounded the death knell for conventional towed artillery in the view of many tacticians. Towed anti-tank guns were especially vulnerable to radiation, since the crew was completely exposed. There was no easy way to enclose the crew short of mounting the weapon in a heavy armored vehicle. Of equal importance, Nikita Khrushchev was becoming infatuated with missiles as a revolutionary alternative to conventional tubed-gun weapons.

Khrushchev saw anti-tank missile launchers as an alternative to heavy tanks, and tactical ballistic missiles as an alternative to heavy artillery. Khrushchev intended not only to revolutionize the army, but to modernize the defense industrial base at the same time. Some bureaus, closely associated with conventional gun-tubed weapons, were simply closed. The artillery design bureaus were heavily hit, most notably Grabin's bureau which had developed many of the most successful World War 2 anti-armor weapons including the ZIS-2 57mm anti-tank gun, the ZIS-3 76mm divisional gun, and F-34 76mm tank gun.

The development of anti-tank guided missiles did not completely end development of anti-tank gun-tube weapons. The towed anti-tank 85mm gun was replaced by the smoothbore 100mm T-12 and MT-12.[12] However, in the late 1970s and 1980s the Soviets began to look at producing towed anti-tank guns which would match the caliber of their latest main battle tank's guns. The 2A45 was produced in part to allow for longer range, improve lethality, and reduce the logistical strain. This was also made possible by the development of long-range through-tube laser guided munitions like the 100mm Kastet and 125mm Kobra, Svir, and Refleck gun-tube missiles.[13]

The T-12 100mm and the 2A45 125mm anti-tank guns are being offered for sale on the world market and are capable of firing out to greater ranges with these newly developed laser-guided rounds. But it is highly unlikely that many countries see the need for such weapons in the era of even cheaper, lighter, and as capable anti-tank guided missiles.

The T-12/T-12A 100mm Anti-Tank Gun[14]

The Soviet Union developed and fielded anti-tank guns despite the fact that there are no such weapons utilized by the West. In addition, to keep pace with Western armor improvements, the gun caliber of anti-tank guns had to increase. This, in turn, has made the handling of such guns more difficult.

One such gun system placed into Soviet service was the T-12 anti-tank gun which entered service in 1955. Its carriage and other features were improved in 1972 and the modified gun became the T-12A; this version is the standard anti-tank gun in service. It is used by anti-tank units of armored and motor rifle regiments to protect open flanks against counterattack during rapid advances. Today, the T-12A is deployed in Russia, Hungary, and the former Yugoslavia.

The gun normally requires a six-man crew: the commander, the driver of the towing vehicle, the gun layer and loader, and two ammunition crewmen. Unlike anti-tank guided missiles (ATGM) units which are able to engage targets at long range, no anti-tank gun crews would risk an exchange of fire with a main battle tank. The element of surprise is the main advantage of an anti-tank gun against a main battle tank's superior mobility, pro-

T-12/T-12A 100mm Anti-Tank Gun

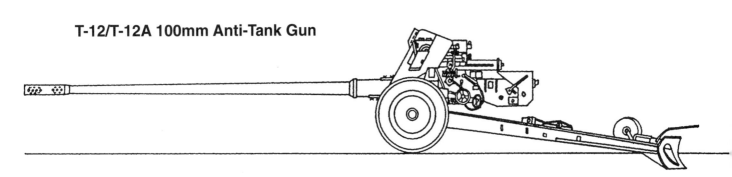

T-12/T-12A 100mm anti-tank gun.

T-12/T-12A Technical Data

Designation	T-12A	T-12
Crew (men)	6	6
Gun shield	yes	yes
Length (millimeters)	9,650	9,480
Width (millimeters)	2,310	1,795
Height (millimeters)	1,600	1,565
Track width (millimeters)	1,910	1,465
Ground clearance (millimeters)	380	380
Combat weight (kilograms)	3,050	2,750
Towing vehicle	MT-LBu	Ural-375D (6 x6)
Maximum towing road speed (kilometers per hour)	70	60
Maximum cross-country towing (kilometers per hour)	25	15
Caliber	2A29 100mm gun	2A19 100mm gun
Gun traverse (degrees)	+/- 27	+/- 27
Gun elevation (degrees)	-6/+20	-6/+20
Barrel recoil (millimeters)	780	780
Maximum rate of fire (rounds per minute)	14	14
Muzzle velocity (meters per second):		
—APFSDS	1,575	1,575
—HEAT	975	975
—HE	700	700
Gun range (meters):		
—Direct APFSDS	3,000	3,000
—Direct HEAT	5,955	5,955
—Indirect HE	8,200	8,200
—Laser-guided round	5,000	5,000
Round penetration (millimeters of RHA at 90 obliquity):		
—APFSDS (500 to 3,000 meters)	230 to 140	230 to 140
—HEAT	350	350
—Laser-guided round (4,000 meters)	550 to 600	550 to 600

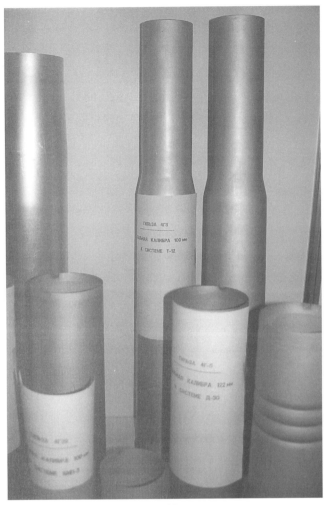

T-12 cartridge case.

tection, and firepower. This is one of the many reasons why such systems are not deployed in the West.

The T-12A system has had to keep pace with Western improvements in armor protection and sensor technologies (e.g., thermal imagers). The T-12 and the T-12A can provide direct as well as indirect fire. Indirect engagements are clearly a secondary role due to the maximum angle of elevation being only twenty degrees. The weapons are, therefore, not subject to CFE treaty limitations as the caliber might otherwise suggest. Moreover, as both guns are already deployed, but are not primarily indirect firing weapons, they do not count as artillery systems.

The primary differences between the T-12 and T-12A can be found in the carriage suspension and equilibrator which have no influence on weapon performance. In construction, the T-12 series comprises four main components: a barrel with breech ring, a semi-automatic breech, a top carriage with sight, and a lower carriage.

The 100mm caliber smoothbore monobloc barrel is 6,300 millimeters long. The chamber, 915 millimeters in length, comprises several cones fitted together. The front strengthened part of the barrel, has eighty holes and serves as a muzzle brake. Employing the gun's recoil and loader assist together with cartridged ammunition results in the T-12A's high rate of fire of up to fourteen rounds per minute.

The aiming system consists of the S71-40 mount with the panoramic sight; the PG-1M for indirect aiming; the OP4M-40U direct sight and the APN-6-40 (or older APN-5- 40) for indirect aiming at night. The main sight for aiming the T-12A is the direct sight. This has a magnification of 5.5 and angle of view of eleven degrees. For night firing, the night vision system has a 6.8 magnification and angle of view of approximately seven degrees.

The improved carriage is the main difference between the T-12 and the T-12A. This was required by the switch to a new towing vehicle, the MT-LBu multi-purpose tracked vehicle. The T-12A is a highly accurate, rugged gun. Its semi-automatic loading system, and cased ammunition permit high rates of fire with low manning levels. The weapon's small size allows individual firing emplacements to be easily camouflaged. The V-shaped gun shield attached to the carriage protects the crew against fragments and small arms fire.

The APFSDS (sub-caliber), HEAT and HE cartridged ammunition are available for both the T-12 and T-12A. A retrofit kit, which extends the combat range of the T-12A to at least 4,000 meters, is currently being marketed under the name Kastet. This system allows a through tube-laser guided round to be fired. The round is similar to the smoothbore T-55 laser-guided round called Bastion (9M117) and is given the system designation of 9K116. The laser guidance unit for the Kastet system is mounted on a tripod rather than the carriage and is placed next to the gun.

The ability to fire both laser-guided gun-tube weapons and conventional ammunition, and the inclusion of thermal imaging technology, make the T-12A a potential threat to armored vehicles. However, many of today's modern tanks with frontal armor protection of between 800 to 900 millimeters of rolled homogeneous armor (RHA) are not threatened from the front by the T-12A with its current ammunition. This caliber ammunition has reached its physical limits.

An anti-tank gun makes sense only if a main battle tank's frontal armor can be penetrated at ranges of more than 1,000 meters. Consequently, the Russians recently increased the caliber of its anti-tank guns. Accordingly, the 125mm smoothbore gun of the T-72 and T-80 has been mounted on a field gun carriage. This anti-tank gun is called the Sprut-B. Highly effective laser-guided gun-tube weapons are also available for this gun. Only the addition of guidance systems and a radio are required for their deployment. With the advent of the 125mm 2A45M anti-tank gun, the future of the T-12/T-12A is uncertain.

2A45M Sprut-B 125mm anti-tank gun.

2A45M Sprut-B 125mm anti-tank self-propelled gun.

The 2A45M Sprut-B 125mm Anti-Tank Gun[15]

Soviet and Russian military designers focused much attention on the development of tube artillery and anti-tank gun/missile systems. These developers believe that this type of weapon system is the most effective type to defeat tanks in different combat settings. Therefore, they continue to improve the armor penetration of anti-tank guns through the use of larger caliber guns which allow for higher muzzle velocities and round weight.

In the late 1980s the Petrov Design Bureau at Artillery Plant Number 9 created the 125mm smoothbore anti-tank gun. This bureau also designed the famous 122mm D-30 artillery systems, as well as, many other main battle tank gun systems. The 2A45M was given the nicknamed of Sprut-B (Octopus). The 2A45M is powered by an auxiliary power unit (APU) which allows the gun to be moved under its own power on the battlefield.

The 2A45M is effective in destroying both mobile/stationary ground and armor targets in either direct or indirect modes. The high muzzle velocity of its 125mm projectiles allows for better armor penetration capability and is the salient design feature of the 2A45M. The high rate of fire, from six to eight rounds per minute, is reliably ensured by the availability of automatic systems and convenient elevating and traversing handwheels. Owing to the availability of special mechanical-assist mechanisms these handwheels are easy to operate. This saves the gunner's time and labor over extended operating periods, especially when firing at moving targets. The carriage has three split arms which provides for rapid and broad fire coverage, owing to large angles of elevation, ranging from -6 to +25 degrees and a traverse of 360 degrees.

The 2A45M's gun has a vertically arranged wedge breechblock with a mechanical semiautomatic control system. The safety mechanism is housed in the lower part of the breechblock. This device eliminates the firing of the gun with a partially closed breechblock. The breechblock is opened manually before the first shot, and only then becomes recoil-operated. After the firing pin is cocked-back and fired, a spent cartridge-case is ejected from the gun. There is a special mechanism for blowing through the tube after firing to prevent a backfire. A special thermal-protection sleeve is attached to the barrel to lessen the harmful atmospheric heating/cooling effects on fire accuracy.

The 2A45M uses separate loading rounds intended for the tank gun D-81 utilized by the T-64, T-72, T-80, and T-90. Special equipment, type 9S53, is used to fire laser-guided round 3UBK-14 from the 9K120 Reflecks or Svir missile complexes which allow accurate destruction of targets out to a distance of some five kilometers with up to 700 millimeters of penetration.

The 2A45M is rigged with various aiming sights to ensure accurate fire. The optical direct fire sight OP4M-48A is used for direct fire during the day. The night vision sight 1PN53-1 is used for night firing. The iron sight 2Ts33 is used together with the panoramic sight PG-1M for firing from indirect laying positions and is also used, if the optical sight is rendered useless.

The gun's recoil mechanism ensures reliable breaking and return of the gun to its initial position, thereby considerably reducing the effect of the fire on the cradle. This recoil mechanism utilizes a hydraulic recoil brake, spin-type recuperator, and pneumatic recuperator. These mechanisms are all located in the cradle's case above the gun barrel. The gun has a mechanical assist system to

2A45M Technical Data

Designation	Sprut-B
Crew (men)	7
Gun shield	yes
Length (millimeters)	7,120
Width (millimeters)	2,660
Height (millimeters)	2,090
Track width (millimeters)	2,200
Fording (millimeters)	900
Ground clearance (millimeters)	360
Combat weight (kilograms)	6,500
Towing vehicle	Ural-4320 or MT-LBu
Maximum towing road speed (kilometers per hour)	80
Auxiliary power unit mode (kilometers per hour)	14
Range for APU mode (kilometers)	50
Caliber	2A45M 125mm gun
Gun traverse (degrees)	360
Gun elevation (degrees)	-6/+25
Maximum rate of fire (rounds per minute)	6 to 8
Muzzle velocity (meters per second):	
—APFSDS	1,700
—HEAT	905
—HE	850
Gun range (meters):	
—Direct APFSDS	2,100
—Indirect HE	12,200
—Laser-guided round	5,000

2A45M Sprut-B 125mm Anti-Tank Gun

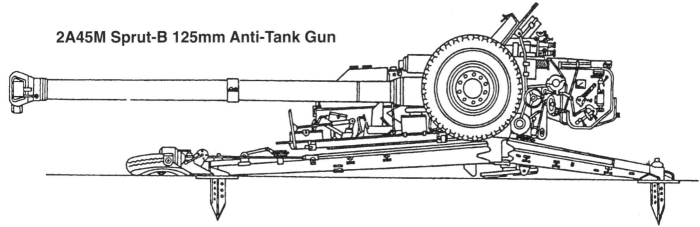

transfer it from combat to a traveling position. This system comprises a hydraulic jack and hydraulic cylinders.

The 2A45M's jacks can lift-up the cradle's arms and place them in a traveling or firing position. The hydraulic cylinders lift the gun to the its maximum ground clearance and raises/lowers the wheels. The 2A45M is brought from combat to its traveling position and back in two ways. It can either utilize a hydraulic motor driven by the auxiliary power unit (APU) or a manual pump. The gun can be placed into combat in as little as 1.5 minutes and it takes nearly two minutes to be placed into a traveling position.

The gun may be towed by Ural-4320 or by the multipurpose full-tracked MT-LBu. In the towed position, the road clearance equals 360 millimeters. This system is also capable of fording rivers up to 900 millimeters deep. The maximum towing speed for the 2A45M on highways is eighty kilometers per hour.

The gun features an auxiliary power unit, which ensures rapid self-movement on the battlefield. This significantly increases the 2A45M's maneuverability in combat. The power unit is developed on the basis of the MeMZ-957A engine. The power unit is located on the right side of the gun and is housed in a special casing. The driver's seat is arranged on the left side together with the gun's driving controls for self-movement. The maximum self-propelled speed on roads is fourteen kilometers per hour and the fuel endurance is around fifty kilometers. In this mode the gun can carry six rounds as its basic ammunition load.

The APU is wheeled and utilizes tires designed for various terrain types. Each wheel is fitted with a brake which insures greater maneuverability and cross-country capacity for the gun. The double-action hydraulic telescopic shock absorbers are mounted on two power driven wheels, which considerably increase the smoothness of the gun's movement across various types of terrain.

The electrical needs of the gun are provided by the on-board APU. This unit provides the numerous electrical needs of the gun such as providing lights, firing the round, and engine ignition during self-propulsion of the gun. Storage batteries are placed under the driver's seat.

The 2A45 is fitted with an armored shield to protect the crew and gun's mechanisms from bullets, shell fragments, and other types of projectiles. There is a slit on its left side for the panoramic sight. The gun is manned by a crew of seven men. The 125mm anti-tank gun 2A45 is unique in its size and versatility, it is simple in design, reliable in operation, and easy to maintain. However, the future viability of anti-tank guns are in doubt given the proliferation of highly effective, lightweight, and inexpensive ATGMs and RPGs.

CFE Count of Russian Towed Artillery in European Russia

	M-46	D-30	D-20	2A36	2A65	2B16	Total
1990	13	1,904	937	488	297	38	3,677
1991	13	2,001	807	626	356	38	3,841
1992	14	1,321	576	394	423	39	2,767
1993	15	1,078	406	251	104	39	1,893
1994	15	771	281	245	150	37	1,499
1995	1	598	333	447	255	2	1,636
1996	1	830	261	507	285	2	1,886
1997	1	839	213	524	330	2	1,909

7

FINAL OBSERVATIONS

Preceding chapters examined the evolution of Russian armor and artillery since World War 2 on a system-by-system basis. This one, by contrast, shifts to a broader focus in order to uncover basic, long-standing design principles which transcend the development of particular systems. The result of this effort is a set of pithily-written "laws" which describe the nature of the design process itself, identify overarching technical priorities, and discuss apparent design trade-off criteria. Finally, this discussion explores how changing times may affect the continuing validity of these traditional principles for shaping future generations of Russian armor and artillery systems.

Observations About the Design Process

1. The process as a whole (regardless of type of system) was often more of a bottoms-up process than a top-down process. As a practical matter, this means that the process is better characterized as one driven by technology push rather than requirements pull. Indeed, the customer seems to have exercised virtually no control and minimal influence upon the design process.

2. Equipment designers enjoyed an unexpectedly high degree of freedom to initiate new designs (and some times even to build prototypes) without prior customer approval. This is doubly surprising because the Soviet Union was a society of scarcity where resources were carefully rationed by central planners and because the military was long thought to exercise iron control over the research and development (R&D) process through its requirements generation and validation process.

3. Designers generally favored a gradual evolutionary approach to system development, even if this meant building a large number of interim prototypes along the way. Periodically, however, designers would depart from this pattern and make daring technological leaps (e.g., the T-64 tank and the BMP-1 infantry fighting vehicle) where one platform would serve as a technology demonstrator for multiple new, and radically different, technologies.

4. Political intervention by regional and central government officials often played an important role in shaping the overall nature and subsequent emphasis of R&D programs. (Khrushchev's decisions to cancel the heavy tank program and to emphasize missile technology effected tank and armored fighting vehicle designs long after he was out of power.)

5. Threat assessments often prompted the initiation of entirely new systems and components.

6. The process employed multiple design bureaus (four to five tank design bureaus, three ATGM design bureaus) which were constantly generating prototypes, many more than were probably necessary to meet military requirements.

7. The building of competing prototypes to fulfill the same requirements seems to have been more a result of a policy of keeping design bureaus busy rather than an attempt to use competition as a tool for getting the best vehicle. Indeed, the military often bought some number of all of the competitors. This, in turn, lead to terrible operational inefficiencies and lower operational effectiveness.

8. The "drive-away" price was an important consideration in the selection of several armor and artillery systems.

9. Little attention, however, was paid to the total life cycle costs of acquisition decisions. Thus, the initial savings achieved by purchasing the cheapest model were often offset by higher operational and maintenance costs later.

10. Overall, the armor and artillery acquisition process appears to have been far less efficient than previously thought by Western analysts.

Overarching Technical Priorities

1. Quantifiable measures of merit like armor thickness, gun size, and vehicle weight were the favorite ways of evaluating designs.

2. Less objective qualities like ergonomics, safety, product reliability, and maintainability were seldom important factors in judging new models. These factors only came into play after the fact if a system displayed very serious and obvious flaws in one of these areas.

3. Designs were optimized to meet Russian military needs and the climatic conditions likely to be encountered by Russian forces. Consequently, little attention was paid in the past to requirements of the export market.

4. Overall vehicle weight has long been an important constraint on the design of tanks, infantry fighting vehicles, and airborne assault guns. This constraint has sometimes forced Russian designers to adopt novel technologies to stay within acceptable weight limits. For example, Russian tank designers seem to have put the DROZD active tank defense system on the Naval Infantry versions of T-55 tanks instead of better armor in order to keep the vehicles' weight within the tolerances of Russian landing craft.

5. Limiting the internal volume of armored vehicles was another major principle of Russian design practice. This was such an important consideration that the Russians were willing to accept serious design penalties such as limited on-board storage of ammunition and placing extra fuel tanks outside the armor shell of vehicles.

6. Maintaining a firepower advantage over the enemy was a major design driver for tanks and armed infantry fighting vehicles.

7. Where possible, Russian designers preferred to improve earlier models through retrofits and factory rebuilding programs. Russian designers also liked to employ add-on kits which could be applied by forces in the field. (Early versions of Russian reactive armor which were bolted onto tanks in the field is an especially good example of this practice.)

8. Russian designers were often willing to pioneer the deployment of novel active and passive protection schemes for armored vehicles.

9. Russian designs have also been willing to accommodate rather radical approaches for satisfying armament requirements of particular vehicles. Examples of this willingness include deployment of the first smoothbore tank gun with APFSDS, the first use of a low-pressure gun on the BMP-1, and the use of gun/mortar combinations on light armored fighting vehicles.

10. Designs relied overwhelmingly upon native Russian design resources (both human and institutional), Russian-developed technologies, or Russian-versions of foreign concepts.

Continuing Applicability of These Principles

The post-Cold War world poses a radically new situation for Russian armor and artillery designers. Some of the major challenges they face are: (1) declining military budgets, (2) shrinking force sizes, (3) breakup of the former Soviet Union into many different independent states, (4) decentralization of industrial decision-making, and (5) loss of a principle enemy against which designs can be optimized.

This new situation is so radically different that it is tempting to believe that all of the old principles will be abandoned. However, such will probably not be the case. Instead, one should expect a more mixed response with some approaches being abandoned, others modified, and (as many as possible) retained. It should also be noted that the new situation offers opportunities for Russian designers as well as imposes limitations and creates problems.

The recent Chechen conflict has amply illustrated the problems faced by Russian designers. The apparent failure of the BMP-2, BTR-80, and T-80 to perform well in built-up urban areas is in part due to the fact these vehicles were not optimized for urban combat but for blitzkrieg-style warfare. These design deficiencies point out the problem of a radically changing threat environment for Russia. An environment which will require Russian armor designers to rethink their design trade-offs on current and future tank models. These design considerations will clearly require technical solutions, in what form or what specific improvements will be made will be at the heart of the trade-off debate held amongst Russian tank designers and their military masters through the end of this year and decade.

Perhaps the greatest changes will come in the focus of design activity. That is, declining domestic demand for military products will force designers to consider the requirements of the export market to an unprecedented degree. There are already signs that this is beginning to take place. For example, the T-80 tanks exhibited at Abu Dhabi displayed dust recuperators for desert operations to make it more appealing to prospective Arab customers. Similarly, a thermal imager was added to the BMP-3 to win an order from the United Arab Emirates (UAE) which insisted that prospective armored fighting vehicles have a night-fighting capability.

Elimination of export restrictions and the new emphasis on foreign sales will also seriously erode traditional Russian emphasis on self-reliance. Indeed, this is already beginning to happen. In the case of the thermal imager mounted on the BMP-3s sold to the United Arab Emirates, for instance, the device came from SAT, a French company. The financial need of Russian weapons developers is also forcing them to search for foreign partners to finish existing projects and/or to commence new ones. The original developers of the active tank defense system, for example, are aggressively searching for foreign partners to share development costs.

Emphasis on foreign military sales has implications for other long-standing Russian design principles. The two most notable changes will probably be a new emphasis on reliability and maintainability since the Russians will now be competing with foreign firms with much better reputations in these areas. Declining worldwide military budgets in general, and the Russian military budget in particular, will probably force Russian designers to pay greater attention to the life-cycle cost implications of their efforts. Greater attention to ergonomic considerations will also probably take place, again in response to international perceptions that better crew performance was one of the keys to the Allied victory in the recent war with Iraq.

It is also tempting to think that decreasing Russian force size, coupled with severely budget constraints, will result in the appearance of fewer models within that force. The economic logic of the situation suggests that the time is really right for the appearance of the "universal" tank long discussed by some Russian designers. Political realities like the desire of local government officials to continue tank production in their region, lack of meaningful centralized control, and the historical freedom of action for designers suggest that it will be hard to achieve that goal. This will be doubly true if designers can secure foreign capital to underwrite at least part of the development costs. This judgment is consistent with the present situation where tank designers in different locations are working simultaneously on upgraded designs for the T-72, the perfection of T-80 designs, and the development of T-90 — far more configurations than make economic sense in light of the domestic Russian military requirements or world market demands.

Competition for foreign sales will also probably ensure continued (and possibly even increased) Russian willingness to investigate and field novel armor vehicle protection systems and guns. Additionally, short-term worldwide interest in upgrading existing systems will probably encourage continuing Russian interest in developing add-ons kits and upgrades which can be added to vehicles during depot level maintenance.

Clearly, change is the order of the day for some aspects of the Russian development process and design principles. Whenever and wherever possible, however, Russian designers will strive to retain past approaches and design priorities. This is because Russians are, by nature, very conservative traditionalists. Additionally, many of the senior people in the process are relatively old and set in their ways. Thus, many of the aforementioned principles will continue to guide future Russian armor and artillery developments.

APPENDIX A
SOVIET/RUSSIAN MORTAR, ANTI-TANK, AND ARTILLERY AMMUNITION

Towed or Self-Propelled Mortars (120 to 240mm)

Caliber	Warhead Type	Projectile Model Name	Round	Using Weapon	Projectile Length	Projectile Fuzed Weight	Round Weight	Fuze	Filler Type	Filler Weight	Initial Velocity	Index Effect	Duration	Firing Velocity
120 mm	HE	3F-843	?	2B11/2S12/PM38	749 mm	16.45 kg	?	GVMZ PD	TNT	3.90 kg	?	—	—	?
120 mm	HE	3OF-843	?	2B11/2S12/PM38	656 mm	16.02 kg	?	GVMZ-7	TNT	?	—	—	?
120 mm	Frag-HE	3OF-843A	?	2B11/2S12/PM38	624 mm	15.98 kg	?	GMVZ-7 PD	Amatol 80/20	1.58 kg	?	—	—	?
120 mm	Frag-HE	3OF-843B	?	2B11/2S12/PM38	640 mm	16.00 kg	?	GVMZ-7 PD	TNT 50/50	1.40 kg	325 m/s	—	—	480 to 5840 m
120 mm	Illuminating	S-843	?	2B11/2S12/PM38	724 mm	16.86 kg	?	T-1 TSQ	Illum	2.00 kg	?	430 m	45 s	450 to 5800 m
120 mm	Smoke	?	?	2B11/2S12	640 mm	16.10 kg	?	?	Smoke	3.00 kg	?	200 m	150 s	450 to 7000 m
120 mm	Frag-HE	?	?	2B11/2S12	640 mm	16.10 kg	?	B-35	Amatol 80/20	3.20 kg	?	0 m	—	450 to 7000 m
120 mm	Incendiary	?	?	2B11/2S12	720 mm	16.30 kg	?	?	Incend Comp	2.70 kg	?	6 islands	60 s	450 to 5800 m
120 mm	Frag-HE	3VOF49	3OF-54	2S9/2S23	828 mm	19.80 kg	?	B-35	RDX	4.90 kg	367 m/s	614 to 2226 m²	—	1000 to 8855m
120 mm	HE-RAP	?	?	2S9/2S23	835 mm	19.80 kg	?	B-35	RDX	3.00 kg	367 m/s	546 to 1860 m²	—	1000 to 13000 m
120 mm	HEAT	?	?	2S9/2S23	960 mm	13.17 kg	?	?	?	?	560 m/s	Pen 600 mm	—	Max Effect 500 m
160 mm	HE	3F-852	?	M160	1194 mm	39.59 kg	?	GVMZ-7 PD	TNT	7.39 kg	345 m/s	—	—	800 to 8500 m
160 mm	HE	3F-853A	?	M160	1119 mm	41.18 kg	?	GVMZ-7 PD	Amatol	7.73 kg	343 m/s	—	—	750 to 8050 m
160 mm	HE	3F-853U	?	M160	1120 mm	41.18 kg	?	GVMZ-7 PD	TNT	8.99 kg	343 m/s	—	—	750 to 8040 m
240 mm	HE	3F-864	53-F-864	2S4/M240	1565 mm	130.84 kg	?	GVMZ-7 PD	TNT	31.93 kg	362 m/s	—	—	800 to 9700 m
240 mm	HE-RA	?	ARM-03F2	2S4/M240	1565 mm	130.84 kg	228.00 kg	GVMZ-7 PD	TNT	31.93 kg	362 m/s	—	—	800 to 18000 m
240 mm	HE	3F5/3VF4	Smelchak 1K11	2S4/M240	1565 mm	134.20 kg	?	GVMZ-7 PD	TNT	31.93 kg	362 m/s	—	0.1 to 3 s	3600 to 9200 m

Towed Anti-Tank Guns (100 to 125mm)

Caliber	Warhead Type	Projectile Model	Round	Using Weapon	Projectile Length	Projectile Fuzed Weight	Round Weight	Fuze	Filler Type	Filler Weight	Initial Velocity	Index Effect @ 1000m	Duration	Firing Range
100mm	AFPFSDS-T	3BM-2	?	2A19 (T-12)/2A29 (T-12A)	?	5.65 kg	19.50 kg	None	Tungsten	?	1575 m/s	225 mm armor	—	100 to 3000 m
100mm	APFSDS	3BM-25	3UBM-11	2A19 (T-12)/2A29 (T-12A)	?	5.02 kg	20.70 kg	?	?	?	1430 m/s	—	—	?
100mm	HVAPDS-T	3BM-8	?	2A19 (T-12)/2A29 (T-12A)	?	9.50 kg	23.50 kg	GLB-2 PD	RDX	1.04 kg	975 m/s	400 mm armor	—	100 to 5955 m
100mm	HEAT-FS	3BK-5M	?	2A19 (T-12)/2A29 (T-12A)	?	12.36 kg	?	VP-9	RDX	1.00 kg	380 m/s	380 mm armor	—	?
100mm	HEAT	3BK-17M	3UBK-9	2A19 (T-12)/2A29 (T-12A)	?	10.00 kg	21.90 kg	O or F-412 PD	A-IX-1	2.16 kg	1075 m/s	—	—	?
100mm	Frag-HE	3OF-412	?	2A19 (T-12)/2A29 (T-12A)	?	15.84 kg	28.60 kg	?	TNT	1.70 kg	900 m/s	—	—	Indirect to 8200 m
100mm	Frag-HE	3OF-32	3UOF-10	2A19 (T-12)/2A29 (T-12A)	?	15.60 kg	30.00 kg	Integral Fuze	A-IX-2	3.60 kg	900 m/s	50-660 mm armor	26 to 41 s	Out to 13400 m
100mm	Missile	3UBK-12	Kastet 9K117	2A19 (T-12)/2A29 (T-12A)	518 mm	18.40 kg	24.50 kg	None	Hard Steel	3.60 kg	1825 m/s	—	—	100 to 4000 m
125mm	APFSDS-S	3BM-9	3VBM-3	2A45M (SPRUT-B)	?	5.62 kg	19.50 kg	None	Penetrator	3.80 kg	1800 m/s	150 mm armor	—	?
125mm	APFSDS	3BM-12	3VBM-6	2A45M (SPRUT-B)	547 mm	5.93 kg	20.90 kg	None	Tungsten Carbide	7.10 kg	1800 m/s	150 mm armor	—	?
125mm	APFSDS-TC	3BM-15	3VBM-7	2A45M (SPRUT-B)	?	?	19.70 kg	None	Depleted Uranium	7.10 kg	1800 m/s	250 mm armor	—	?
125mm	APFSDS	3BM-17	3VBM-8	2A45M (SPRUT-B)	585 mm	10.95 kg	20.90 kg	None	Tungsten	1.85 kg	1700 m/s	—	—	100 to 2000 m
125mm	APFSDS-DU	3BM-32	3BM-38/3VBM-13	2A45M (SPRUT-B)	621 mm	10.80 kg	20.70 kg	VG-15 PD	RDX	4.20 kg	1700 m/s	770 mm armor	—	100 to 4000 m
125mm	APFSDS-T	3BM-42	3BM-44/3VBM-17	2A45M (SPRUT-B)	675 mm	19.08 kg	29.58 kg	?	RDX-Copper	1.85 kg	905 m/s	220 mm armor	—	?
125mm	HEAT-FS	3BK-12M	3VBK-7	2A45M (SPRUT-B)	695 mm	17.20 kg	24.30 kg	VG-15 PD	HMX-Copper	1.76 kg	905 m/s	260 mm armor	—	?
125mm	HEAT-FS	3UBK-14	Svir 9K119	2A45M (SPRUT-B)	677 mm	19.02 kg	29.58 kg	B-15	Depleted Uranium	1.85 kg	905 m/s	ERA+300 mm armor	—	?
125mm	HEAT-FS	3BK-14M	3VBK-10	2A45M (SPRUT-B)	680 mm	19.00 kg	28.50 kg	V-15 PD	?-Copper	1.64 kg	915 m/s	—	—	?
125mm	HEAT-FS	3BK-18M	3VBK-16	2A45M (SPRUT-B)	680 mm	19.00 kg	29.00 kg	V-15 PD	TNT-Copper	3.15 kg	850 m/s	Laser 200 to 5000 m	—	Out to 9400 m
125mm	HEAT-DU	3BK-21B	3VBK-17	2A45M (SPRUT-B)	680 mm	19.00 kg	28.40 kg	V-15 PD	RDX	3.34 kg	850 m/s	—	—	Out to 12200 m
125mm	HEAT-FS	3BK-29	?	2A45M (SPRUT-B)	670 mm	23.00 kg	33.00 kg	V-429E PD						
125mm	Frag-HE (FS)	3OF-19	3VOF-22	2A45M (SPRUT-B)	675 mm	23.20 kg	32.50 kg	V-429E PD						
125mm	Frag-HE (FS)	3OF-26	3VOF-36	2A45M (SPRUT-B)										

Appendix A: Soviet/Russian Mortar, Anti-Tank, and Artillery Ammunition

Towed Artillery (122 to 152 mm)

Caliber	Warhead Type	Projectile Model Name	Round	Using Weapon	Projectile Length	Projectile Fuzed Weight	Round Weight	Fuze	Filler Type	Filler Weight	Initial Velocity	Index Effect @ 1000m	Duration	Firing Range
122 mm	HEAT	BP-463	?	M-30	497 mm	14.80 kg	40.8 kg	GKV PD	RDX	2.18 kg	?	—	—	?
122 mm	HEAT	3BK-463UM	?	M-30	657 mm	21.26 kg	40.8 kg	GPV PD	RDX	2.15 kg	?	—	—	?
122 mm	Frag-HE	3OF-462	53-VOF-471G	M-30/D-30	559 mm	21.76 kg	8.5–29 k	RGM-2 PD	TNT/Amatol	3.46 kg	5 to 690 m	—	—	5350 to 17360 m
122 mm	HE-RAP	3OF-462	?	M-30/D-30	?	?	?	?	?	?	?	—	—	21900 m
122 mm	HEAT-T	3PB-463	?	M-30/D-30	497 mm	14.80 kg	40.8 kg	GKV PD	RDX	2.18 kg	515 m/s	200 mm armor	—	Max Effect 1000m
122 mm	Smoke	3D-462	?	M-30/D-30	545 mm	22.55 kg	40.8 kg	KTM-2 PD	TNT-0.16kg/WP	3.60 kg	?	—	—	5350 to 17360 m
122 mm	Smoke	3D-4	?	M-30/D-30	499 mm	21.70 kg	40.8 kg	RGM-2 PD	TNT-0.16kg/WP	3.60 kg	?	—	—	5350 to 17360 m
122 mm	Illum	S-463	?	M-30/D-30	551 mm	21.96 kg	40.8 kg	T-7 Time	Illum Compound	1.00 kg	5 to 690 m	25 s of 450000 candle	25 s	5350 to 17360 m
122 mm	VHF Jammer	?	?	M-30/D-30	415 mm	3.80 kg	40.8 kg	?	R-046L Jammer	—	?	20-100 MHz to 700 m	1 hr	5350 to 17360 m
122 mm	HE	3OF-56	3V0F81/3V0F82	D-30	502 mm	21.76 kg	40.8 kg	RGM-2	RDX	4.31 kg	?	—	—	5350 to 17360 m
122 mm	HE	3OF-56-1	3V0F81/3V0F82	D-30	502 mm	21.76 kg	40.8 kg	RGM-2 PD	RDX	4.31 kg	?	—	—	5350 to 17360 m
122 mm	HEAT-FS-T	3BK-13	?	D-30	637 mm	18.20 kg	40.8 kg	V-15 PD	RDX	1.80 kg	?	—	—	?
122 mm	HEAT-FS	3BK-6M	?	D-30	706 mm	21.58 kg	40.8 kg	GPV-2 PD	RDX	2.15 kg	740 m/s	460 to 580 mm armor	—	Max Effect 800 m
122 mm	Chem	3D-462	?	D-30	545 mm	22.20 kg	40.8 kg	T-7 TSQ	Sarin	1.30 kg	5 to 690 m	—	—	5350 to 17360 m
122 mm	Chem	3D-462	?	D-30	545 mm	23.10 kg	40.8 kg	T-7 TSQ	Lewisite	3.30 kg	5 to 690 m	—	—	5350 to 17360 m
130 mm	SAP-HE	3PB-42	?	M-58	605 mm	33.40 kg	?	V-350 BD	RDX	2.76 kg	?	? mm of naval armor	—	?
130 mm	SAP-HE	3OF-3S-42	?	M-58	620 mm	33.46 kg	59.10 kg	VGU-1 PD	TNT	3.60 kg	930 m/s	? mm of naval armor	—	6000 to 27490 m
130 mm	Frag-HE	3OF-482M	?	M-46	670 mm	33.40 kg	59.10 kg	RGM-2 PD	TNT	3.60 kg	930 m/s	—	—	6000 to 27490 m
130 mm	HE	3OF-33	3V0F43/3V0F44	M-46	670 mm	33.40 kg	59.10 kg	V-429	RDX	4.17 kg	930 m/s	—	—	6000 to 27490 m
130 mm	Smoke-Target	3DTs-1	?	M-46	672 mm	32.80 kg	59.10 kg	V-429 PD	WP	3.23 kg	930 m/s	—	—	6000 to 27490 m
130 mm	Chem	3DT-1	?	M-46	672 mm	33.40 kg	59.10 kg	V-429 PD	Sarin	1.60 kg	930 m/s	—	—	6000 to 27490 m
130 mm	Chem	3DT-1	?	M-46	672 mm	32.80 kg	59.10 kg	V-429 PD	VX	1.40 kg	930 m/s	—	—	6000 to 27490 m
130 mm	Illuminating	3SP-46	?	M-46	573 mm	25.80 kg	59.10 kg	TM-16L MT	Illum Compoun	2.38 kg	930 m/s	40 s of 500000 candle	40 s	6000 to 27490 m
152 mm	Frag-HE	3OF-540	?	D-20	706 mm	43.56 kg	59.50 kg	RGM-2 PD	TNT	6.24 kg	655 m/s	—	—	6710 to 17400 m
152 mm	HE	3OF-25	3V0F32/3V0F33	D-20	?	43.56 kg	60.00 kg	V-90	RDX	6.88 kg	655 m/s	—	—	4600 to 17410 m
152 mm	HE-BB	3OV-64	3V0F43/3V0F44	D-20	?	43.56 kg	60.00 kg	V-90	RDX	6.88 kg	655 m/s	—	—	4600 to 18500 m
152 mm	HEAT	3BP-540	3VBP2	D-20	?	27.00 kg	41.14 kg	GPV-3	?	4.00 kg	?	—	—	?
152 mm	Smoke	3D-540	?	D-20	690 mm	43.51 kg	59.50 kg	RGM-2 PD	WP	6.62 kg	655 m/s	—	—	6710 to 17400 m
152 mm	Chem	3D-540	?	D-20	690 mm	40.00 kg	59.50 kg	RGM-2 PD	Sarin	2.80 kg	655 m/s	—	—	6710 to 17400 m
152 mm	Chem	3D-540	?	D-20	690 mm	42.50 kg	59.50 kg	RGM-2 PD	Lewisite	5.40 kg	655 m/s	—	—	6710 to 17400 m
152 mm	Nuclear	3OF-45	3V0F72/3V0F58	D-20	864 mm	43.56 kg	?	?	Nuclear-0.2 kT	?	?	0.2 KT	—	6710 to 24700 m
152 mm	ICM	3O-23	3V0F28/3V0F29	D-20/2A36	820 mm	42.80 kg	?	MT	Bmb-350g/45g Exp	42 Bombs	?	30 deg-100mm armor	—	26500 m
152 mm	HE	3OF-29	3V0F39/3V0F40	2A36	?	43.51 kg	80.80 kg	V-429	RDX	6.73 kg	945 m/s	—	—	28500 m
152 mm	HE-BB	3OF-64	3V0F96/97/98	2A36	?	43.56 kg	60.00 kg	V-90	RDX	6.88 kg	655 m/s	—	—	4600 to 18500 m

Towed Artillery (122 to 152 mm) (continued)

Caliber	Warhead Type	Projectile Name	Round	Using Weapon	Projectile Length	Projectile Fuzed Weight	Round Weight	Fuze	Filler Type	Filler Weight	Initial Velocity	Index Effect @ 1000m	Duration	Firing Range
152 mm	HE-BB	3OF-59	3VOF86/3VOF87	2A36	?	43.51 kg	80.80 kg	V-429	RDX	6.73 kg	945 m/s	—	—	30000 m
152 mm	VHF Jammer	3NS-30	3VNS36/37/38	D-20/2A65	818 mm	43.56 kg	80.80 kg	MT PD	R-045L-Jammer	8.20 kg	?	1.5-120 MHz to 700 m	—	22000 m
152 mm	SADARM	?	?	D-20/2A65	?	40.70 kg	?	?	IR or MMW Seeker	?	?	small cluster munitions	—	20000 m
152 mm	Frag-HE CLGP	3OF-38	Santimetr 2K24	D-20/2A65	?	49.50 kg	?	?	RDX	8.50 kg	?	—	.05 to 3	2000 to 12000 m
152 mm	Frag-HE CLGP	3OF-39	Krasnapol 9K25	D-20/2A65	1305 mm	19.70 kg	50.80 kg	?	RDX	6.30 kg	?	Laser 200 to 5000 m	6 to 15 s	3000 to 20000 m
152 mm	Frag-HE	3OF-540	?	2A65	710 mm	43.51 kg	59.50 kg	RGM-2 PD	TNT	6.24 kg	655 m/s	—	—	6710 to 24700 m
152 mm	Frag-HE	3OF-45	3VOF58/7273	2A65	864 mm	43.56 kg	80.80 kg	RGM-2 PD	RDX	7.65 kg	810 m/s	1.5 to 2 x > OF-540	—	6710 to 24700 m
152 mm	Frag-HE BB	3OF-61	3VOF91	2A65	864 mm	42.86 kg	80.80 kg	KZ-88 PD	RDX	7.80 kg	828 m/s	1.5 to 2 x > OF-540	—	6710 to 29000 m

Appendix A: Soviet/Russian Mortar, Anti-Tank, and Artillery Ammunition

Self-Propelled Artillery (122 to 203mm)

Caliber	Warhead Type	Projectile Name	Round	Using Weapon	Projectile Length	Projectile Fuzed Weight	Round Weight	Fuze	Filler Type	Filler Weight	Initial Velocity	Index Effect@1000m	Duration	Firing Range
122mm	Frag-HE	3OF-462	53-VOF-471G	2S1	559 mm	21.76 kg	8.5-29 k	RGM-2 PD	TNT/Amatol	3.46 kg	5 to 690 m	—	—	5350 to 17360 m
122mm	HE-RAP	3OF-462	?	2S1	?	?	?	?	?	?	?	—	—	21900 m
122mm	HEAT-T	3PB-463	?	2S1	497 mm	14.80 kg	40.80 kg	GKV PD	RDX	2.18 kg	515 m/s	200 mm armor	—	Max Effect 1000 m
122mm	Smoke	3D-462	?	2S1	545 mm	22.55 kg	40.80 kg	KTM-2 PD	TNT-0.16kg/WP	3.60 kg	?	—	—	5350 to 17360 m
122mm	Smoke	3D-4	?	2S1	499 mm	21.70 kg	40.80 kg	RGM-2 PD	TNT-0.16kg/WP	3.60 kg	?	—	—	5350 to 17360 m
122mm	Illum	3S-463	?	2S1	551 mm	21.96 kg	40.80 kg	T-7 Time	Illum Compoun	1.00 kg	5 to 690 m	25 s of 450000 candle	25 s	5350 to 17360 m
122mm	VHF Jammer			2S1	415 mm	3.80 kg	40.80 kg		R-046L Jammer		?	20-100 MHz to 700 m	1 hr	5350 to 17360 m
122mm	HE	3OF-56	?	2S1	502 mm	21.76 kg	40.80 kg	RGM-2	RDX	4.31 kg	?	—	—	5350 to 17360 m
122mm	HE	3OF-56-1	?	2S1	502 mm	21.76 kg	40.80 kg	RGM-2 PD	RDX	4.31 kg	?	—	—	5350 to 17360 m
122mm	HEAT-FS-T	3BK-13	?	2S1	637 mm	18.20 kg	40.80 kg	V-15 PD	RDX	1.80 kg	?	—	—	?
122mm	HEAT-FS	3BK-6M	?	2S1	706 mm	21.58 kg	40.80 kg	GPV-2 PD	RDX	2.15 kg	740 m/s	460 to 580 mm armor	—	Max Effect 800 m
122mm	Chem	3D-462	?	2S1	545 mm	22.20 kg	40.80 kg	T-7 TSQ	Sarin	1.30 kg	5 to 690 m	—	—	5350 to 17360 m
122mm	Chem	3D-462	?	2S1	545 mm	23.10 kg	40.80 kg	T-7 TSQ	Lewisite	3.30 kg	5 to 690 m	—	—	5350 to 17360 m
122mm	Chem	3D-462	?	2S1	545 mm	23.10 kg	40.80 kg	T-7 TSQ	VX	2.90 kg	5 to 690 m	—	—	5350 to 17360 m
122mm	Cargo	3D-462	?	2S1	545 mm	22.20 kg	40.80 kg	T-7 TSQ	Cargo Leaflets	3.60 kg	5 to 690 m	—	—	5350 to 17360 m
122mm	Frag-HE CLGP	?	Kitolov-2	2S1	1225 mm	12.00 kg	25.00 kg	?	CLGP	5.50 kg	?	—	—	12000 m
130mm	AP/Anti-Ship	A3-F-44	A3-UF-44	Bereg	1367 mm	33.40 kg	52.80 kg	VG-130 ?	?	3.55 kg	850 m/s	Mechanical base	—	23000 m
130mm	AP/Anti-Ship	A3-ZC-44	A3-UZC-44	Bereg	1369 mm	33.40 kg	52.80 kg	VG-130 ?	?	3.56 kg	850 m/s	Mechanical time	—	23000 m
130mm	AA/Anti-Aircraft	A3-ZC-44	A3-UZC-44R	Bereg	1364mm	33.40 kg	52.80 kg	AP-51L	?	3.56 kg	850 m/s	CM 8 m/Air 15 m	—	23000 m
152mm	Frag-HE	3OF-540	?	2S3	706 mm	43.56 kg	?	RGM PD	TNT	6.24 kg	655 m/s	—	—	6710 to 18500 m
152mm	Frag-HE	3OF-540	?	2S3	706 mm	43.56 kg	59.50 kg	RGM-2 PD	TNT	6.24 kg	655 m/s	—	—	6710 to 17400 m
152mm	HEAT	3BP-540	3VBP2	2S3	?	27.00 kg	41.14 kg	GPV-3	RDX	4.00 kg	655 m/s	—	—	?
152mm	HE	3OF-25	3VOF32/3VOF33	2S3M	?	43.56 kg	60.00 kg	V-90	RDX	6.88 kg	655 m/s	—	—	4600 to 17410 m
152mm	HE-BB	3OV-64	3VOF43/3VOF44	2S3M	?	43.56 kg	60.00 kg	V-90	RDX	6.88 kg	655 m/s	—	—	4600 to 18500 m
152mm	Smoke	3D-540		2S3/2S19	690 mm	43.51 kg	59.50 kg	RGM-2 PD	WP	6.62 kg	655 m/s	—	—	6710 to 17400 m
152mm	Chem	3D-540	?	2S3/2S19	690 mm	40.00 kg	59.50 kg	RGM-2 PD	Sarin	2.80 kg	655 m/s	—	—	6710 to 17400 m
152mm	Chem	3D-540	?	2S3/2S19	690 mm	42.50 kg	59.50 kg	RGM-2 PD	Lewisite	5.40 kg	655 m/s	—	—	6710 to 17400 m
152mm	VHF Jammer	3NS-30	3VNS36/37/38	2S3M/2S19	818 mm	43.56 kg	80.80 kg	MT PD	R-045L Jammer	8.20 kg	?	1.5-120 MHz to 700 m	1 hr	22000 m
152mm	SADARM			2S3M/2S19	?	40.70 kg	?	?	IR or MMW Seeker	?	?	all cluster munitions	—	20000 m
152mm	Nuclear			2S3M/2S19	864 mm	43.56 kg	?	?	Nuclear- 0.2 kT	?	?	0.2 kT	—	6710 to 24700 m
152mm	HE-BB	3OF-45	3VOF72/3VOF58	2S5	?	43.56 kg	60.00 kg	V-90	RDX	6.88 kg	655 m/s	—	—	4600 to 18500 m
152mm	HE	3OF-64	3VOF96/97/98	2S5	?	43.51 kg	80.80 kg	V-429	RDX	6.73 kg	945 m/s	—	—	28500 m
152mm	HE-BB	3OF-29	3VOF39/3VOF40	2S5	?	43.51 kg	80.80 kg	V-429	RDX	6.73 kg	945 m/s	—	—	30000 m
152mm	ICM	3O-23	3VOF86/3VOF87	2S5	?	42.80 kg	?	MT	Bmb-350g/45g	42 Bombs	?	—	—	26500 m
152mm	AP-T	3BR-540	3VOF28/3VOF29	2S5	?	48.78 kg	?	MD-7	None	1.20 kg	600 m/s	30 deg-100mm armor 124 mm armor	—	?

Self-Propelled Artillery (122 to 203mm)(Continued)

Caliber	Warhead Type	Projectile Name	Round	Using Weapon	Projectile Length	Projectile Fuzed Weight	Round Weight	Fuze	Filler Type	Filler Weight	Initial Velocity	Index Effect@1000m	Duration	Firing Range
152mm	Frag-HE CLGP	3OF-38	Santimetr 2K24	2S5/2S19	?	49.50 kg	?	?	RDX	8.50 kg	?	—	05 to 3	2000 to 12000m
152mm	Frag-HE CLGP	3OF-39	Krasnapol 9K25	2S3M/2S5/2S19	1305 mm	19.70 kg	50.80 kg	?	RDX	6.30 kg	?	Laser 200 to 5000	6 to 15 s	3000 to 20000 m
152mm	Frag-HE	3OF-540	?	2S5/2S19	710 mm	43.51 kg	59.50 kg	RGM-2 PD	TNT	6.24 kg	655 m/s	—	—	6710 to 24700 m
152mm	Frag-HE	3OF-45	3VOF58/72/73	2S19	864 mm	43.56 kg	80.80 kg	RGM-2 PD	RDX	7.65 kg	864 m/s	1.5 to 2 x > OF-540	—	6710 to 24700 m
152mm	Frag-HE BB	3OF-61	3VOF91	2S19	864 mm	42.86 kg	80.80 kg	KZ-88 PD	RDX	7.80 kg	828 m/s	1.5 to 2 x > OF-540	—	6710 to 29000 m
155mm	Frag-HE	Germes	?	Germes	?	30.00 kg	60.00 kg	?	RDX	12.00 kg	?	Diving trajectory	54 s	12000 m
203mm	HE	3OF-43	3VOF34	2S7	?	110.00 kg	?	V-491	RDX	17.82 kg	960 m/s	—	—	37287 m
203mm	Frag-HE	?	?	2S7	?	110.00 kg	?	V-491	RDX	17.82 kg	960 m/s	—	—	55000 m

Russian MLRS 300 to 122 mm Rockets

Warhead Designations	Rocket Designations	Mission	Guidance (round/sub-munition)	Range (km) per round	Warhead	Penetration (mm of RHA)	Number per Round	Package	Manufacturer
300 mm SMERCH									
9N152 (SPBE-D)	9M55K-1	Anti-Armor	IR/Photo-Contrast	20-70	EFP	100	5	Sub-munition (MOTIV-3M)	Bazalt
Universal Sub-Munition	9M55K-2	Anti-Armor	IR/MMW	20-70	EFP	60	20	Sub-munition	Splav
9N139	9M55K	Anti-Personnel	Ballistic	20-70	Cluster/Frag	Frag	72	Sub-munition	Splav
9N150	9M55F	Anti-Material/Anti-Personnel	Ballistic	20-70	HE/Frag	Frag	1	Unitary Warhead	Splav
R-90	--	Miniature UAV	Inertial/GPS	20-70	No Warhead/Camera	--	2-3	Camera w/ 30min Flight Time	Eniks
220 mm URAGAN									
9N128F	9M27F	Anti-Material/Anti-Personnel	Ballistic	8-35	EFP	100	1	Unitary Warhead	Bazalt
Universal Sub-Munition	9M27K5	Anti-Armor	IR/MMW	8-35	EFP	60	5	Sub-munition	Splav
9N218K4 w/ PTM-3	9M59	Anti-Armor Mines	Ballistic	8-35	Explosive	Blast	9	Mines	Splav
9N218K3 w/ PFM-1	9M27K3	Anti-Personnel Mines	Ballistic	8-35	Explosive	Frag	312	Mines	Splav
9N218K2 w/ PTM-1S	9M27K2	Anti-Armor Mines	Ballistic	8-35	Explosive	Blast	24	Mines	Splav
9N218K1 w/ N9N210	9M27K1	Anti-Material/AP Mines	Ballistic	8-35	HE/Frag	Frag	30	Bomblets	Splav
122 mm GRAD									
M-21-OF	9M22	Anti-Material/Anti-Personnel	Ballistic	4-20	HE/Frag	Frag	1	Unitary Warhead	Bazalt
Universal Sub-Munition	9M22 Extended Range	Anti-Armor	IR/MMW	4-32	EFP	60	1	Sub-munition	Splav
GRAD Jammer	9M22	Radio Jamming	Ballistic	4-18	Jammer (RF/FM)	--	5	Jammer Round	Splav
GRAD LILIYA-2	9M22	Radio Jamming	Ballistic	4-32	Jammer (UHF/VHF)	--	1	Jammer Round	Splav
GRAD with POM-2S	9M22M Extended Range	Anti-Personnel Mines	Ballistic	4-32	HE/Frag	Frag	5	Mines	Splav
GRAD-PTM-3	9M22M Extended Range	Anti-Armor Mines	Ballistic	4-32	Explosive	Blast	3	Mines	Splav

APPENDIX B

SOVIET/RUSSIAN TANK/ARMORED VEHICLE DEVELOPMENT

Post-War Reorganization

At the end of World War 2, the Soviet Union had two main tank design centers.[1] The SKB-2 (Special Design Bureau-2), headed by General Zhozef Kotin, was the primary design center for heavy tanks such as the KV and IS Stalin series. The bureau had been located at the Leningrad Kirov Plant (LKZ) until October 1941 when it was transferred to the Chelyabinsk Heavy Machinery Plant Number 100, better known as "Tankograd." At the end of the war, Kotin's team was transferred back to Leningrad, but a small engineering team, called the "New Design Bureau" (BNK: *Byuro Novykh Konstruktsiy*) remained at Chelyabinsk to continue its own heavy tank efforts. The BNK was headed by M. F. Balzhi, who had been instrumental in the design of the IS-3 Stalin tank. The main heavy tank design effort in Leningrad became known as the OKBT (Special Tank Design Bureau), but more often was simply called the LKZ design bureau or the Kotin bureau. The division of the bureau into two separate elements was part of a deliberate government policy to induce competition into the weapons acquisition process, nicknamed the "doubling method."[2]

The second major tank design center was headed by Aleksandr Morozov and was responsible for medium tanks such as the T-34, T-44, and T-54. It was originally based at the Kharkov Locomotive Plant (KhPZ Number 183) in eastern Ukraine. Like the SKB-2, in 1941 it was transferred to the Urals because of the German invasion, and relocated at the Urals Railcar Plant (Uralvagon Zavod) in Nizhniy Tagil. Morozov succeeded in transferring the bulk of his design team back to Kharkov in 1951, but as in the case of the "doubling" of SKB-2, a small team of engineers remained behind in Nizhni Tagil to continue evolutionary development of the T-54A medium tank series. Uralvagonzavod was popularly called "Vagonka," and so its design bureau was often nicknamed the Vagonka bureau.

The tank design bureaus were part of the Ministry of Transportation Machinery Production, usually abbreviated in Russian as *Mintransmash*, which controlled the Soviet railroad and heavy transportation equipment industry.[3] Tank design became affiliated with Mintransmash because most tanks were produced at locomotive or railcar production plants. Within the Mintransmash was the Main Tank Design Directorate, nicknamed *Glavtank*, headed in the early 1950s by a former T-34 tank deputy designer, N. A. Kucherenko. Requirements for new tanks were the responsibility of the Main Armored-Tank Directorate (GBTU) based in Moscow. GBTU had its own armored vehicle design effort, the Scientific Research Institute for Armored Technology (NIIBT) located at the Kubinka Proving Ground about fifty kilometer southwest of Moscow. This facility is roughly comparable to Aberdeen Proving Ground in the US or Bovington in the UK, but it also combines the functions of the US Army's Tank & Automotive Command in Warren, Michigan. The NIIBT was responsible for advanced technology development; engineering development was handled by the Mintransmash design bureaus based on GBTU technical-tactical requirements. In the late 1950s, an additional tank design institute was organized in Leningrad, the VNII Transmash (All-Union Scientific Research Institute for the Transportation Machinery Industry), based in large measure on the disbanded elements of Kotin's OKBT tank bureau. The role of "NII" organizations in the Soviet system like NIIBT and VNII Transmash was to conduct basic research and advanced development; engineering development was the responsibility of the design bureaus. The NIIs and design bureaus were supported by several other design bureaus under other ministries specializing in specific subsystem technologies such as Fyodor Petrov's artillery and tank gun design bureau in Sverdlovsk, and Trashutin's diesel engine design bureau in Chelyabinsk. These other bureaus were under other organizations. For example, tank gun and antitank missile development was managed by the GRAU (Main Missile and Artillery Directorate), a counterpart of the GBTU.

The Tank Design Process

Soviet design practices have changed repeatedly over the years but have generally followed a similar process. Programs are generally initiated by the GBTU (Main Armored Directorate), though they are sometimes started by design bureau chief designers. At the earliest stages, a draft design (*eskizniy proekt*) that includes only the basic technical specifications is prepared. This design often forms the basis for further examinations of requirements by the GBTU and General Staff. For a project to proceed to engineering development, a TTT (*Taktiko-tekhnichesskiye Trebovaniya*: Tactical Technical Requirement) is generally required.[4] This is jointly developed between the GBTU, the General Staff and the design bureau. A technical design (*tekhnicheskiy proekt*), which broadly sketches the basic design including its main sub-components, is often part of this process. If the TTT is approved, the working design (*rabochiy proekt*) brings the design to fruition with comprehensive technical drawings and other documentation, suitable for building a prototype. The construction of prototype tanks usually requires higher approval from Mintransmash and the military, though as this study will show, design bureau heads often have used their own authority to test promising ideas. Prototypes are subjected to factory tests and then turned over to the state for trials. The basic trials are conducted at the NIIBT proving ground in Kubinka, but there are additional hot and cold weather test areas elsewhere in the USSR. If successful, the design is accepted by the state for production, an action coordinated by the VPK (Military Industrial Council) in conjunction with representatives the Council of Ministers, the appropriate industrial ministries, and the Soviet Army. This production decision authorizes the start of a small production run of vehicles (from a few dozen to a few hundred), what would be called low rate initial production (LRIP) in the US today. These vehicles were issued to actual Ground Forces units for further trials and to determine any technical problems. Once the operational trials were successfully completed, the tank would be accepted, an event usually occurring about two years after state production acceptance. At this stage the tank would be given its official service designation (for example, the Obiekt 219 becoming the T-80). The army acceptance decision would initiate full scale series production.[5]

Soviet/Russian AFV Izdeliye/Obiekt Numbers

Although Soviet armored vehicles are better known by their service designations that are applied once the vehicle is officially accepted by the state commission for service in the Soviet Army, Soviet (and Russian) vehicles often have another numeric designation applied during the design of the vehicle. This practice began at the Kharkov medium design bureau before World War 2 as a method to identify engineering drawings.

In the post-war years, it became a standardized practice. These numbers are called *izdeliye* (item) or *Obiekt* (object) numbers, so they are sometimes listed as izdeliye 245 or Obiekt 245, etc. In many cases, these designations continue to be used even after a vehicle has entered service. This following list has been compiled from unclassified sources. As is evident from the list, the numbering system follows a pattern, with numeric blocks being assigned to different plants. Although most of these numbers are assigned by the Main Armor Technology Directorate (GBTU), the Main Automotive and Transport Directorate (GAVTU) has issued a separate series of numbers for its projects.

Industrial Index Numbers

The Soviet/Russian defense industry also has a series of industrial index numbers used to identify different military products such as missiles, missile launchers, tank guns, electronics, and other items. These numbers follow the pattern of number-letter-number, such as 2S1. This list also includes three of these categories: the "2A" artillery systems, "2B" mortar systems and "2S" self-propelled systems.

GAVTU Designations

Obiekt/Izdeliye	Development	Description
5	Kharkov Tractor	AT-L light artillery tractor;
5A	Kharkov Tractor	AT-LM
6	Kharkov Tractor	MT-LB
23	Kharkov Tractor	AT-L
26	Kharkov Tractor	2S1 Gvozdika
27	Kharkov Tractor	AT-L Variant
32	Kharkov Tractor	9A34/9A35 (MT-LB for SA-13)
39	Kharkov Tractor	AT-L Variant
40P	GAZ	BRDM
41	GAZ	BRDM-2
49	GAZ	BTR-60
50	GAZ	BTR-70 with BMP-1 turret
51	GAZ	BTR-90 (GAZ-5923)
60	GAZ	BTR-80 (GAZ-5903)
62B	GAZ	BTR-60P Prototype
100	Ishimbai	DT-10P Vityaz All Terrain Vehicle
110	Ishimbai	DT-30P Vityaz All Terrain Vehicle
140	ZiL	BTR-152
141	ZiL	BTR-40
256	Kharkov Tractor	UR-77

GBTU Designations

Obiekt/Izdeliye	Development	Description
003	LKZT-64T	Turbine Testbed
19	Rubtsovsk	BMP Wheel/Track Competitor
78	Murom?	IRM Engineer Reconnaissance Vehicle

Sverdlovsk (Various Pre-War)

Obiekt/Izdeliye	Development	Description
100	Gorlitskiy	GM-100 Transporter
105	Gorlitskiy	SU-100P
108	Gorlitskiy	SU-152G
111	Kharkov	T-46, "Bullet-Proof" Tank
112	Gorlitskiy	BTR-112: APC on GM-100 Chassis
115		T-29: T-28 Replacement
116	Gorlitskiy	SU-152P
118	Sverdlovsk	GMZ
118-2	Sverdlovsk	1980 Vehicle Family, GMZ-3
120	Gorlitskiy	SU-152/2S3 Ancestor
123	Gorlitskiy	9P24 Krug TEL Chassis
124	Gorlitskiy	1S32 Pat Hand Radar Chassis
126SP		T-50 Light Tank
130		T-26 Flamethrower
133		T-26 Flamethrower

Kharkov/Nizhni-Tagil

Obiekt/Izdeliye	Development	Description
134	Morozov	T-34
135	Morozov	T-34-85
136	Morozov	T-44
137	Morozov	T-54
137A	Morozov	T-54A
137P	Morozov	T-54A
137Sh	Morozov	T-54A
137G2	Morozov	T-54B
137M	Morozov	T-54M (Rebuilt T-54s)
138	Morozov	SU-100
139	Morozov	T-54M (Experimental T-54)

Nizhni-Tagil
(KB No. 520 of Plant No. 183 Uralvagonzavod im. D.Z. Dzerzhinskiy)

Obiekt/Izdeliye	Development	Description
140	Uralvagon	T-54B Smoothbore Tank
150	Uralvagon	IT-1 Tank Destroyer
150T	Uralvagon	IT-1 Tank Destroyer with Turbine Engine

Nizhni-Tagil (KB No. 520 of Plant No. 183 Uralvagonzavod im. D.Z. Dzerzhinskiy) (Continued)

Obiekt/Izdeliye	Development	Description
155	Uralvagon	T-55
155M	Uralvagon	T-55M
162	Uralvagon	T-62 Tank
165	Uralvagon	T-62A with D-54 100mm Gun
166	Uralvagon	T-62 with U-5T 115mm Gun
166A	Uralvagon	Disinformation Name for T-64
166M	Uralvagon	T-62 with 5 Wheel Suspension
166MV	Uralvagon	T-62MV
167	Uralvagon	T-67 Improved T-62
167T	Uralvagon	T-67 with Turbine
172	Uralvagon	T-72 Prototype
172M	Uralvagon	T-72 Ural (1971)
172-2M	Uralvagon	T-72 (1974)
172M-1	Uralvagon	T-72 Ural-1
172M-Eh	Uralvagon	T-72 Export (1976)
172M-Eh1	Uralvagon	T-72 Export (1978)
172M-Eh2	Uralvagon	T-72M Export (1978)
172M-1-Eh3	Uralvagon	T-72M Export (1980)
172M-1-Eh4	Uralvagon	T-72M Export (1982)
172M-1-Eh5	Uralvagon	T-72M Export (1983)
172M-1-Eh6	Uralvagon	T-72M1 Export (1983)
172M8	Uralvagon	T-72S Export (1990)
173	Uralvagon	T-64 with V-45K Engine
174	Uralvagon	T-72 Variant
176	Uralvagon	T-72A
177	Uralvagon	Updated Obiekt 140
182	Uralvagon	T-72B Prototype
183	Uralvagon	Experimental Flamethrower
184	Uralvagon	T-72B Olkha
184-1	Uralvagon	T-72B1
184-2	Uralvagon	T-72B2
187	Uralvagon	T-72BM
188	Uralvagon	T-72BU (T-90)
190	Uralvagon	Tank Prototype
195	Uralvagon	Tank Prototype

Leningrad Kirovskiy Plant (LKZ) OKTB Design Bureau

Obiekt/Izdeliye	Development	Description
211	Kotin	T-50 Prototype
212	Kotin	Self-Propelled Vehicle (1939)
216	Popov/Kirov	2S7 Pion
218	Kotin	KV-4
218	Kotin	8U218 R-11 (Scud) Launcher on IS-2 Chassis
219SP1	Popov/Kirov	Similar to Kharkov Gas-Turbine T-64T (GTD-1000T)
219SP2	Popov/Kirov	Early Turbine Testbed for T-64
219	Popov/Kirov	T-80
219A	Popov/Kirov	T-80U1
219AS	Popov/Kirov	T-80U Bereza
219V	Popov/Kirov	T-80 Prototype with 9K119 ATGM
219Ye	Popov/Kirov	T-80B with Shtora-1 and Agava
219M	Popov/Kirov	T-80UM Experimental Tank
219R	Popov/Kirov	T-80B
219RV	Popov/Kirov	T-80BV
219RD	Popov/Kirov	T-80 with A-53-2 Diesel Engine
219RM	Popov/Kirov	T-80 Experimental Variant

Leningrad Kirovskiy Plant (LKZ) OKTB Design Bureau (Continued)

Obiekt/Izdeliye	Development	Description
220	SKB-2	KV-3 Heavy Tank
224	SKB-2	KV-1 Heavy Tank
237	SKB-2	IS-85, IS-1
240	SKB-2	IS-2
240M	SKB-2	IS-2M
241	SKB-2	ISU-152
241M	SKB-2	ISU-152M
242	SKB-2	ISU-122
243	SKB-2	ISU-152-1 with 122mm Gun
244	SKB-2	IS-2 with S-53 85mm Gun
245	SKB-2	IS-2 with 100mm Gun
246	SKB-2	ISU-152-2 with 130mm Gun
247	SKB-2	ISU-130
248	SKB-2	IS-2 with 100mm Gun
249	SKB-2	ISU-122S
250	SKB-2	ISU-122BM
251	SKB-2	ISU-152 Variant with IS-6 Transmission
252	SKB-2	IS-6 with Mech Transmission
260	SKB-2	IS-7
266	SKB-2	T-10 1957 Heavy Tank Variant
268	SKB-2	SU-152; 1956 T-10 Tank Destroyer
271	SKB-2	Oka SM-54 420mm Self-Propelled Weapon
272	SKB-2	T-10M
277	SKB-2	1957 T-10 Heavy Tank (Large Turret)
279	SKB-2	1957 Troyanov nuclear tank
287	SKB-2	T-64-Derived Tank Destroyer
288	SKB-2	T-64-Derived Tank Destroyer

Uraltransmash; Sverdlosk (KB im. Sverdlov)

Obiekt/Izdeliye	Development	Description
300	Tomashov	SAM Chassis
303	Tomashov	2S3 Akatsiya
304	Tomashov	2S3M Akatsiya-M
305	Tomashov	2S4 Tyulpan
307	Tomashov	2S5 Giatsint
307V	Tomashov	2S5 Giatsint
308	Tomashov	Buk Radar Vehicle
316	Tomashov	2S19 Msta-S 152mm SP Artillery
318	Tomashov	GMZ-3 Granilshik Minelayer

Minsk Tractor Plant

Obiekt/Izdeliye	Development	Description
352	Tomashov	2S6 Tunguska
355	Tomashov	9A330 Tor
355M	Tomashov	9A331 Tor-M1 (SA-15 Gauntlet TELAR)

KB No. 60 of the Kharkov Transport Machine Construction Plant No. 75

Obiekt/Izdeliye	Development	Description
401	Morozov	AT-T
405MU	Morozov	BAT-M on T-55
405U	Morozov	BAT-1
405UM	Morozov	BAT-1M
409	Morozov	BTM on T-55
416	Morozov	SU-100 Turret Assault Gun
426U	Morozov	1S12 Long Track Chassis

KB No. 60 of the Kharkov Transport Machine Construction Plant No. 75 (continued)		
Obiekt/Izdeliye	Development	Description
429	Morozov	MT-T Universal Tractor/Transporter
430	Morozov	T-64 Prototype
431	Morozov	Experimental Tank Destroyer
432	Morozov	T-64 Prototype
432A	Morozov	T-64R (T-64 Rebuild)
434	Morozov	T-64
434A	Morozov	T-64R (T-64 Rebuild)
435	Morozov	Modernized Obiekt 430
436	Morozov	T-64 with V-45/UTD-45 Engine
437	Morozov	T-64A
437A	Morozov	T-64B1
437AM	Morozov	T-64B1M
438	Morozov	T-64B1K Command Tank
440	Morozov	T-64 Swimming Tank
445	Morozov	T-64 with V-45 Engine
446	Morozov	T-64AK
447	Morozov	T-64 Prototype with Kobra
447AM	Morozov	T-64BM
447BV	Morozov	T-64BV
447T	Morozov	T-64 Derived Recovery Vehicle
455	Morozov	T-64 Driver Trainer
457	Morozov	5TD Engine
459	Morozov	6TD-2 Engine
476	Morozov	1975 Experimental Standard Tank
477	Morozov	Molot 1984 Experimental Tank

KB No. 60 of the Kharkov Transport Machine Construction Plant No. 75 (continued)		
Obiekt/Izdeliye	Development	Description
478	Morozov	T-80UD Prototype
478B	Morozov	T-80UD
478BE	Morozov	T-80UD for Pakistan
478M	Morozov	1976 Experimental Tank (Shater/Arena (APS))
481	Morozov	OT-54
482	Morozov	OT-55
483	Morozov	T-54 Flamethrower 1959
484	Morozov	T-80UDK Command Tank
494	Morozov	6TDM Experimental Engine

Rubtsovsk Machine Construction Plant (Omsk Plant)

Obiekt/Izdeliye	Development	Description
501	Rubtsovsk	BRM-3K Rys
507	Rubtsovsk	1990 RM-G Recovery Vehicle
520	Omsk	Experimental 1990 ZSU
520M	Rubtsovsk	1990 Experimental MT-L1V Chassis

Obiekt/Izdeliye	Development	Description
530	Omsk	Experimental ZSU
540	Omsk	Expermental Command Post

Mytishchi SKB-40

500	Astrov	ZSU-57-2 Chassis
503	Astrov	Experimental Chassis
520	Astrov	New Model Tracked Chassis

Mytishchi SKB-40 (continued)

Obiekt/Izdeliye	Development	Description
520M	Astrov	1990s Experimental MT-L-1V Chassis
561	Astrov	AT-P
567	Astrov	9S18 Buk (SA-11) Radar Vehicle
568	Astrov	1S91 Straight Flush Chassis
569	Astrov	9P147 Buk Launcher Vehicle
570	Astrov	ASU-76
572	Astrov	ASU-57
573	Astrov	SU-85 ASU-85
574	Astrov	ASU-57P
575	Astrov	ZSU-23-4 (2A6) Chassis
577	Astrov	Buk Transloader Vehicle
578	Astrov	2P25 Kub Chassis
579	Astrov	9S470 Buk Command Vehicle
597-5	Astrov	2S6M2 Tunguska Chassis

Omsk Machine Construction Plant "October Revolution No. 174" (Troyanov OKB; formerly at Voroshilov Plant in Leningrad)

Obiekt/Izdeliye	Development	Description
600	Troyanov	SU-122-54 Assault Gun
601A		Radio Controlled Tank Target
602		MTU-20 Bridgelayer Tank
608		Jet Mine-Sweeper
608P		Armored Recovery Vehicle
609		Armored Recovery Vehicle
630	Omsk	T-80B1K
630A	Omsk	T-80UK
632	Omsk	MTU-72
636	Omsk	Tank Prototype
637	Omsk	IMR-2
639	Omsk	T-55A
644	Omsk	T-80B with V-46-6 Engine

Kurgan Machinery Plant (KMZ) and Blagonravov Design Bureau

Obiekt/Izdeliye	Development	Description
650	KMZ	ATS-59
651	KMZ	AT-S
665	KMZ	Experimental Two Section Coupled Transporter
667	KMZ	ATS-59G
670	Blagonravov	BMP-2 Prototype
675	Blagonravov	BMP-2 IFV
676	Blagonravov	BRM-1K Korshun Scout Vehicle
677	Blagonravov	Experimental BMP
678	Blagonravov	BMP-2K Command
680	Blagonravov	BMP-2 Low Profile Turret Variant 1972
681	Blagonravov	BMP-1G-73mm/2 Man Turret 1977
685	Blagonravov	Airborne Light Tank Prototype 1975
688	Blagonravov	BMP-3 IFV Prototype
688-01	Blagonravov	BMP-3
689	Blagonravov	BMP-3
699	Blagonravov	BMP-3 Multi-Purpose Hull

Appendix B: Soviet/Russian Tank/Armored Vehicle Development

Chelyabinsk Tractor Plant/ChTZ No. 100 and M. F. Balzhiy/Isakov; OKB No. 200 (New Design Bureau (BNK= Biuro novikh konstruktsiy))

Obiekt/Izdeliye	Development	Description
701	Balzhiy	IS-4 (701-1, -2, -5, -6)
703	Balzhiy	IS-3
703M	Balzhiy	IS-3M
704	Balzhiy	ISU-152
710	Balzhiy	Experimental Heavy Tractor
711	Balzhiy	Experimental Heavy Tractor
712	Balzhiy	AT-S Tractor
714	Isakov	BMP-1 Variant
730	Balzhiy	T-10
734	Balzhiy	T-10M
740	Balzhiy	PT-76
740B	Balzhiy	PT-76B
749	Isakov	BMP-2 with 30mm Gun 1974
750	Balzhiy	BTR-50
750K	Balzhiy	BTR-50PU Command Vehicle
750PK	Balzhiy	BTR-50PK
765	Isakov	BMP-1 Korshun
766	Isakov	BMP Variant
767	Isakov	PRP-3 BAL Artillery Reconnaissance Vehicle
768	Isakov	BMP Long Hull, 73mm 2 Man Turret 1974
769	Isakov	BMP Long Hull, 30mm 2 Man Turret
770	Isakov	1957 Heavy Tank
773	Isakov	BMP-1K
774	Isakov	BMP-1KSh
775	Isakov	Low Profile Rubin Missile Tank

Chelyabinsk Tractor Plant/ChTZ No. 100 and M. F. Balzhiy/Isakov; OKB No. 200 (New Design Bureau (BNK= Biuro novikh konstruktsiy)) (continued)

Obiekt/Izdeliye	Development	Description
775T	Isakov	Low Profile Rubin Missile Tank Turbine Engine
778	Isakov	BMP-1KShM Command
779	Isakov	PRP-4 Nard
780	Isakov	Almaz Experimental Tank on 775 Chassis

Leningrad Kirovskiy Plant (Missile Launcher Vehicles)

Obiekt/Izdeliye	Development	Description
830	LKZ	9A82 Launcher: S-300V/SA-12 TEL
831	LKZ	9A83 Launcher: S-300V/SA-12 TEL
832	LKZ	9S15T Radar Vehicle (S-300V)
833	LKZ	9S32-1 Radar Vehicle (S-300V)
834	LKZ	9A84 Loader Vehicle (S-300V)
835	LKZ	9A85 Loader Vehicle (S-300V)
836	LKZ	9S19 Radar Vehicle (S-300V)
837	LKZ	9S457 Command Vehicle

Volgograd (Formerly Stalingrad Tractor Plant/STZ; Gavalov/Shabalin Design Bureau)

Obiekt/Izdeliye	Development	Description
905	Gavalov	BTR-50PN

Volgograd (Formerly Stalingrad Tractor Plant/STZ; Gavalov/Shabalin Design Bureau) (continued)		
Obiekt/Izdeliye	Development	Description
906	Gavalov	PT-85; 1963 Light Amphibious Tank
907	Gavalov	BMP Prototype; PT-76M
911	Gavalov	Gavalov APC 1964
914	Gavalov	1964 APC
915	Gavalov	BMD-1
916	Shabalin	BMD-2
918	Shabalin	BMP Prototype
924	Shabalin	Fialka 122mm SP Howitzer on BMD-1
925	Shabalin	BTR-D
925S	Shabalin	2S9 Nona-S

Volgograd (Formerly Stalingrad Tractor Plant/STZ; Gavalov/Shabalin Design Bureau) (continued)		
Obiekt/Izdeliye	Development	Description
926	Shabalin	BMD-1KSh Soroka
932	Shabalin	BREM-D
934	Shabalin	PT-100 Sudya Light Tank
940	Shabalin	1976 KShM Potok-4 Command Vehicle
950	Shabalin	BMD-3

Bryansk Automobile Plant (BAZ) and Kutaiskiy Plant		
1200		Bryansk Wheeled APC
1800		BTR with BMP-1 Turret

Soviet/Russian Main Missile and Artillery Administration (GRAU) Index Numbers

Gun Tubes
2A3	406mm SM-54 Kondensator Heavy Weapon
2A6	ZSU-23-4 Shilka
2A10	AZP-23 Amur Gun System on ZSU-23-4 Shilka
2A13	ZU-23 23mm Air Defense Autocannon
2A14	ZU-23 23mm Air Defense Autocannon
2A17	M-62-T2 12mm Tank Gun on T-10
2A18	D-30 122mm Howitzer
2A19	T-12 100mm Towed Anti-Tank Gun
2A20	U-5T 115mm Tank Gun
2A21	D-68 115mm Tank Gun (T-64)
2A25	73mm Grom Low Pressure Gun on Obiekt 287
2A26	D-81 125mm Tank Gun (T-64A)
2A28	Grom 73mm Low Pressure Gun
2A29	MT-12 Towed 100mm Anti-Tank Gun
2A31	122mm D-32 Howitzer on 2S1 Gvozdika
2A32	122mm D-32 Howitzer on 2S2 Fialka
2A33	152mm D-22 Howitzer on 2S3 Akatsiya
2A36	152mm Giatsint-B Towed Gun
2A37	152mm Gun on 2S5 Giatsint
2A38	30mm Gun on 2S6 (Derived from AO-17)
2A42	30mm Gun on BMP-2
2A43	152 Giatsint-BK Gun
2A44	203mm Gun on 2S7
2A45	Sprut-B 125mm Towed Anti-Tank Gun
2A46	D-81T 125mm Tank Gun
2A48	100mm Rifled Tank Gun for PT-100
2A51	120mm Weapon 2S9 Nona-S
2A60	120mm Weapon on 2S9 and 2S23 Nona-SVK
2A61	152mm Towed Howitzer
2A64	152mm Gun on 2S19 Msta-S
2A65	152mm Msta-B Towed Gun
2A70	100mm Gun on BMP-3
2A72	30mm Cannon on BMP-3
2A77	30mm Gun on Pantsir
2A80	120mm Weapon on 2S31

Mortar Tubes
2B1	420mm Mortar
2B8	240mm Mortar
2B9	Vasilyek 82mm Auto-Mortar
2B11	120mm Sani Mortar
2B14	82mm Podnos Mortar
2B16	120mm Nona-K Towed 120mm Weapon

Self-Propelled Artillery Systems
2S1	Gvozdika 122mm Self-Propelled Howitzer
2S2	Fialka 122mm Self-Propelled Howitzer on BMD-1 Chassis (Obiekt 924)
2S3	Akatsiya 152mm Self-Propelled Howitzer

Soviet/Russian Main Missile and Artillery Administration (GRAU) Index Numbers (continued)		2S7	Pion 203mm Self-Propelled Gun
		2S9	Nona-S 102mm Self-Propelled Weapon
		2S12	Sani 2B11/GAZ-66 Mortar Portee
		2S17	152 mm Self-Propelled Howitzer
2S4	Tyulpan 240mm Self-Propelled Mortar	2S19	Msta-S 152mm Self-Propelled Gun
2S5	Giatsint 152mm Self-Propelled Gun	2S23	Nona-SVK 120mm Self-Propelled Weapon
2S6	Tunguska 30mm Air Defense Gun/Missile Vehicle	2S30	Iset 152mm Self-Propelled Gun
		2S31	Vena 120mm Self-Propelled Weapon

Russian Tank Design Bureaus and Institutes

RusNIIBT (Scientific Research Institute for Armored Vehicle Technology)
143070 Kubinka, Moskovskiy obl.
Director: Prof. V. Bryzgov

NIIBT is the Russian Army's primary center for basic research and advanced development of armored vehicle technology. It is co-located with the Kubinka proving ground (Poligon) southwest of Moscow, the main Russian armored vehicle test facility. The NIIBT is primarily responsible for the development of advanced technologies for armored vehicles, working in conjunction with design bureaus around Russia.

VNIITransmash
Mobile Vehicle Engineering Institute ul. Zarechnaya 2
198323 St. Petersburg
Tel: 011-7-812-135-98-37, FAX: 135-98-37
General Director: Eduard K. Potemkin, General Designer: N. S. Popov

VNIITransmash is one of two primary tank and armored vehicle design institutes, and is subordinate to the Ministry of Transportation Machinery rather than the army. These two institutes (the other being the Army's NIIBT at Kubinka) are responsible for advanced development and basic research on armored vehicle technology, and roughly correspond to the US Army's Aberdeen Proving Ground and Tank & Automotive Command.

The VNIITransmash has an experimental design plant which produces armored vehicle prototypes. In the 1960s, the facility was headed by General P. K. Voroshilov, and after 1969 by V. S. Starovoytov. The bureau is probably associated with the NII-Military Transport Engineering in Gorelovo in the St. Petersburg suburbs. Potemkin is also the general director of the October Revolution Transport Plant PO, which presumably undertakes prototype construction for the NII.

Popov is the general designer at the Kirov Plant which developed the T-80.

Chelyabinskiy Traktorniy Zavod
Chelyabinsk Tractor Plant
Chelyabinsk, Russian Federation

The Chelyabinsk Tractor Plant was one of the main centers of Soviet heavy tank design until the late 1960s. The design bureau was formed there in the autumn of 1941 when Zh. Kotin's SKB-2 heavy tank design bureau was transferred there from Leningrad. The main portion of the Kotin team returned to the Kirov Plant in Leningrad after the war. However, a remnant of the design bureau remained behind forming the "New Design Bureau" (BNK= *Biuro novikh konstruktsiy*) under the direction of M. F. Balzhiy. After Balzhiy's retirement, the bureau was taken over by Pavel P. Isakov. Isakov's team was initially assigned the development of advanced tank concepts, including the radical Obiekt 775 missile-armed tank and the Obiekt 770 heavy tank. Isakov's team was responsible for the winning design in the BMP competition, beating out alternate designs from Briansk and Rybinsk. In 1968, production of the BMP-1 was intiated at Kurgan and a second design bureau was spun off from Chelyabinsk under A. Blagonravov. In the 1970s, the Chelyabinsk and Kurgan design bureaus competed on BMP upgrades, with Kurgan often winning the key competitions including the BMP-2 and BMP-3. The Chelyabinsk bureau no longer appears to be particularly active in BMP development.

Uralvagon KB
Uralvagonzavod PO imeni F. E. Dzerzhinskiy
622019 Nizhni Tagil
Tel: 011-7-34351-3-17-74
General Director: Vladimir Seryakov
Chief Engineer: Vladimir S. Vernik

The Uralvagonzavod rail car plant (Vagonka) has been a major center for tank design since 1941 when the GKB-T-34 under Aleksandr Morozov was transferred to Nizhni Tagil from the Kharkov Locomotive Plant. The Morozov KB developed the T-34, T-44, and T-54. Morozov returned to Kharkov to head the tank design bureau in December 1951. The design bureau was then temporarily lead by A. V. Kolesnikov, but in January 1953,

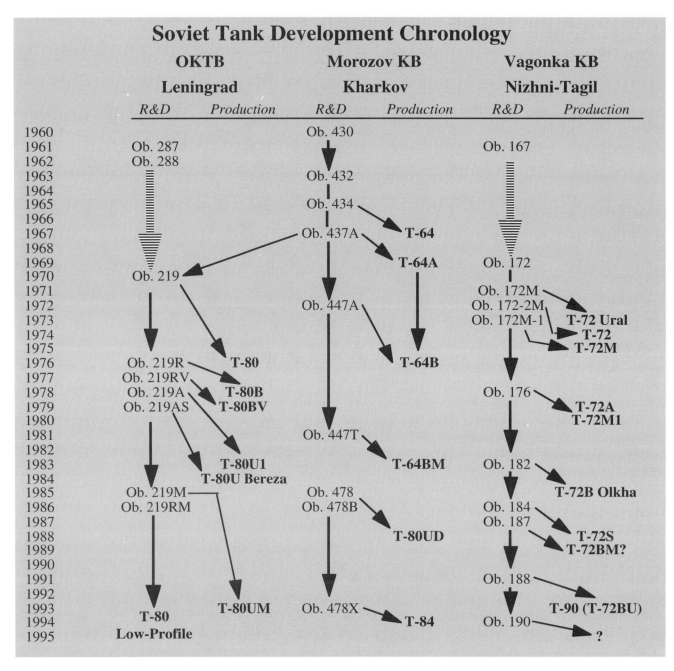

the leadership passed to Leonid Kartsev. Kartsev headed the bureau during the development of the T-62 tank, IT-1 tank destroyer and other designs. Kartsev resigned for health reasons in the summer of 1969, and his place was taken by V. N. Venediktov, who headed the T-72 design effort. The Uralvagon plant has been primarily responsible for T-72 production, and the new T-90 tank.

Kharkovskiy KB po Mashinostroeniyu im. A. Morozova

Malyshev Heavy Machine Building Plant (Kharkov Locomotive Plant)
Kharkov, Ukraine
Chief Designer: Anatoliy Slovikovskiy

The Kharkov Locomotive Plant (KhPZ) was the traditional center for Soviet medium tank development. It was the location of the pre-war Firsov KB which developed the BT cavalry tank and which was succeeded in 1939 by the Koshkin KB after Firsov's arrest during the Purge. In 1941, the design bureau under Aleksandr Morozov was evacuated to Nizhni Tagil where work on the T-34 continued. Morozov returned to Kharkov in 1951, heading the design bureau until his retirement in 1967. The bureau was subsequently headed by Nikolay Shomin. In the 1950s and 1960s, the design bureau developed the T-64 tank. It is currently in Ukraine and has been working on a derivative of the T-80UD, called the T-84.

Kharkovskiy Traktorniy Zavod
Kharkov Tractor Plant
Kharkov, Ukraine

This plant has a small design collective that has been responsible for light armored vehicles manufactured at the plant, including the MT-LB, 2S1 *Gvozdika* and its associated chassis. It is now located in Ukrainian territory and its current status is not known.

Kurgan Design Bureau
Kurgan Machine Building PO (Kurganmash)
Prospekt Mashinostroitelniy 17
640631 Kurgan, Russian Federation
Tel: 011-7-35222; FAX: 011-7-35222
Chief Designer: A. Blagonravov

The Kurgan design bureau was originally formed from one of the heavy tank design collectives located in Chelyabinsk under P.P. Isakov. In 1968, BMP-1 production was started in Kurgan, and as a result, a small engineering design bureau was opened to manage the program. In the 1970s, the Kurgan and Chelyabinsk plants developed competitive versions of the BMP design, with the Kurgan design team eventually emerging as the predoment bureau. The new BMP-3 infantry fighting vehicle was developed by a design team under chief designer A. Blagonravov at the Kurgan Machine Plant in the mid-1970s.

Gavalov OKB
Volgograd Tractor Plant
Vologograd, Russian Federation

The Gavalov design bureau has been primarily involved in the development of light armored vehicles. Development of the BMD-1 and BMD-2 vehicles has been undertaken by a design bureau under the direction of I.V. Gavalov. Gavalov also work on a variety of other light armored vehicle projects including competitors to the BMP, and new light tanks in the 1970s, which were not accepted for production. The new BMD-3 was developed by a team under Arkadiy Shabalin who became the new general designer of the facility after Gavalov's retirement.

GAZ Design Bureau
Arzamasskiy Mashinstroitelniy Zavod "Motor"
ul. 9 Maya 2
Arzamas, Nizhnegorod obl.
Tel: 2-07-80, FAX: 8-83-147-236-56
Director: V. I. Tyurin
Chief designer: Aleksandr Masyagin

The primary design bureau for Soviet and Russian wheeled armored vehicles has been associated with GAZ (Gorkovskiy Avtomobilny Zavod, in Gorki, now Nizhni Novgorod). The design bureau was headed by V.A.Dedkov after World War 2, with V.K.Rubtsov serving as the chief designer on the BTR-40, BTR-152, BRDM and BTR-60. The current chief designer, A. Masyagin, was responsible for the BTR-70 and BTR-80. The primary facility for the military elements of the GAZ series is now in the neighboring Arzamas plant, south of Nizhniy Novgorod. Additional light armored vehicle production has been undertaken at the Izhevsk Mechanical Plant and the Altay Agricultural Machinery Plant in Rubtsovsk.

Spetsmash (OKBT)
Kirovskiy Zavod, Prospekt Stachek 47
198097 St. Petersburg
FAX: 252-04-16
General Designer: Nikolai S. Popov
Director: Piotr G. Semenenko

The Kirov Plant (*Kirovskiy Zavod*) has traditionallly been the home of the main Soviet heavy tank design bureau, dating back to the OKMO of the 1930s. In World War 2, the SKB-2 heavy tank design bureau was transferred to Chelyabinsk. In 1946, portions under the direction of General Zh. Kotin returned to the Kirovskiy plant and the bureau was eventually renamed OKBT (Special Design Bureau of Tank Construction). They were responsible for heavy tank designs such as the IS-4 and T-10, as well as amphibious tanks (PT-76) and tracked missile launcher vehicles for the Luna (FROG) system. Khrushchev's 1960 directive to end tank production led to a cessation of production and a diminution of design work at the plant until the early 1970s. At this time, the plant was again reoriented toward tank work, embarking on the development of the T-80 tank as well as series tank production. The plant eventually manufactured the T-80 tank, finally halting tank production in November 1991. OKBT has been responsible for other types of armored vehicle designs such as the 2S7 Pion heavy self-propelled gun. The Kirov design bureau is now called Spetsmash.

Sterlitamak Machine Building Plant (STEMA)
ul. Gogolya 124
453125 Sterlitamak, Bashkortostan ASSR
Tel: 011-7-34711-4-94-10, FAX: 4-35-77
Director: Viktor A. Sukhanov

STEMA is a new defense plant established in 1983 with production only beginning in 1988. The design bureau is involved in adapting the T-72 and T-80 tank components for a new chassis for the 2S19 Msta-S self-propelled 152 mm gun which the plant produces.

Uraltransmash PO

ul. Frontovikh Brigad 9
Ekaterinburg
Tel: 34-46-61, 34-45-74, FAX: 3432-34-46-42
Chief designer: Yuriy Tomashov
Deputy Chief designer: Mikhail Tretyakov

This production facility in (the former Sverdlovsk Metallist Zavod No. 40) has traditionally been associated with self-propelled artillery production. The bureau was headed by L. Gorlitskiy and developed the SU-122, SU-85 and SU-100 during the war, and other designs such as the abortive SU-100P, SU-152G and SU-152P in the late 1940s. The bureau developed the medium armored chassis used for the Krug (SA-4 Ganef) air defense missile system, which later formed the basis for the 2S3 Akatsiya self-propelled hoitzer, 2S4 self-propelled mortar, GMz armored minelayer and others. The Uraltransmash NPO design bureau is now under the direction of Yuriy Tomashov has developed several current Russian systems including the 2S5 and 2S19. In the case of the 2S19 Msta-S, the chassis was developed in conjunction with the STEMA (Bashkiri) Machine Plant in Sterlitamak. Uraltransmash is currently manufacturing the 2S19 Msta-S self-propelled 152 mm gun, GMZ-3 mine-laying vehicle and 2S5 self-propelled 152 mm gun. The design bureau works in conjunction with the Perm artillery design bureau.

Mytishchi Design Bureau

Mytishchi Machine Building Plant MMZ
ul. Kolontsov 4
141009 Mytishchi, Moskovskiy Obl.
Tel: 011-7-095-582-57-20
Director: Yuri Gulko

The Mytishchi plant is the primary designer and manufacturer of light armored tracked vehicles for air defense applications. It is the contemporary descendent of the World War 2 Astrov OKB-40. Past programs have included the ZSU-57-2, ZSU-23-4, and the chassis for the 2K12 Kub (SA-6 Gainful), Buk-1M (SA-11 Gadfly), 2S6 Tunguska and others. It is also the manfacturer of related light and medium armored chassis including the new MT-SM.

Associated Ordnance Design Bureaus

Titan TsKB

Barrikady Zavod
Prospekt imeni V. I. Lenina,
400071 Volgograd
Tel: 011-7-8442-79-79-00, FAX: 71-39-33

The Titan TsKB at the Barrikady plant is involved primarily in the design of missile launchers for mobile ballistic missiles. It had been involved in the development of the launching system for the 9K72 (SS-1c Scud B), OTR-22 Oka (SS-12 Scaleboard), RSD-10 Pioneer (SS-20 Saber) IRBM and the RS-12M Topol (SS-25). Its chief designer for twenty-nine years has been Valerian M. Sobolev. It has also worked on other specialized combat vehicles on heavy wheeled chassis, for example the new 130 mm Bereg coastal defense artillery system.

Splav NII

Tula
General designer: G. A. Denezhkin

The Splav NII is a spin-off of one of the Tula armaments plants. In 1957, it competed for the development of a new 122 mm artillery rocket system, which later emerged as the BM-21 Grad. The design bureau head at the time was A. N. Ganichev. Other systems developed by Splav include the Prima advanced 122 mm rocket system, the Uragan MLRS and the Smerch MLRS. Production of many of its systems are undertaken by the Lenin Machine Building Plant in Perm.

PO Spetstekhnika

620012 Ekaterinburg, Russian Federation
Tel: (7 3432) 37-17-00
(Associated production plant)
Lenin Machine Building Plant
Uralskaya 21, 614014 Perm
Tel: 011-7-3422-36-41-15
Director: Vladislav Kudryshev

The Artillery Design Bureau of the Sverdlovsk (now Ekaterinburg) Machine Construction Plant has been the Soviet Union's primary artillery and tank gun design facility since the Grabin Central Artillery Design Bureau was closed on Khrushchev's orders in 1957. The bureau designed the D-10 100 mm gun, U-5 115 mm gun, and D-81 125 mm tank gun. The bureau had been headed by Gen. Fedor F. Petrov until 1978. The chief designer is currently Yuriy N. Kalachnikov. The Spetstekhnika design bureau often cooperates in its design work with the Perm Machine Building Plant and the Central Scientific Research Institute for Precision Machine Construction (TsNII TochMash), headed by Avenir G. Novozhilov. The associated machine plant manufactures multiple rocket launchers (including the 9A52 Smerch and 9P140 Uragan), artillery systems (2S23 and 2A65) and the 125 mm D-81T tank gun.

KB Priborstroyeniya (KBP), Shipunov KB

ul. Shcheglovskaya Zaseka
300001 Tula, Russian Federation
Tel: 44-94-74

This bureau was formed in 1962 at Tula to develop

advanced weapons. Arkadiy G. Shipunov was appointed general designer in 1982. The bureau has been involved in two spheres of development: anti-tank missiles and gun systems. In the armored vehicle gun field, the bureau developed the armament system for the BMP-2, BMP-3, and BMD-2. In the tube-fired missile area, a team under Petr Komonov (missile) and Igor V. Aristarkhov (laser guidance) were responsible for laser guided weapons including the Krasnopol 152 mm artillery projectile, as well as tank-gun guided missiles such as Bastion, Svir and Refleks. The bureau also developed the Fagot (AT-4), Konkurs (AT-5), Metis (AT-7), and Kornet (AT-14) missiles. The bureau is also involved in the development of both the naval and army air defense systems using their Treugolnik (SA-19) missiles: the 2S6 Tunguska and naval Kortik/Kashtan system. In the aircraft field, the bureau has developed the GSh-30, GSh-30-1 and GSh-6-30 aircraft cannon. The bureau was also responsible for the development of the Drozd active tank defense system in the 1970s and 1980s.

KB Mashinstyorenie NPO (KBM; Oka OKB; S.P.Nepobidimy OKB)
Kolomna, Russian Federation
Chief Designer: Nikolai Gushchin

The KBM Design Bureau grew from the *Gladkostvolnoi artillerii KB* (mortar design bureau) headed by Boris Shavyrin which was formed on 11 April 1942 at the Zavod Number 4 in Kolomna outside Moscow. Its post-war projects included the 160 mm divisional mortar M-160 Model 1949, the M-240 Model 1950 240 mm heavy mortar, and the 120 mm M-120 Model 1955 regimental mortar. It was also involved in the development of the 82 mm, 107 mm and 120 mm recoilless rifles. Due to Khrushchev's directives, many conventional weapons bureaus were shifted to missiles. Shavyrin's mortar bureau was assigned to anti-tank missiles in competition to Makarov's and Nudelman's OKB-16. This led to the 3M6 Shmel (AT-1 Snapper) and 9M14 Malyutka and in 1969. Sergei Pavlovich Nepobidimy took over the Kolomna bureau in 1965 after Shavyrin's death and himself was replaced by Nikolay Gushchin on his retirement in 1992. The bureau was responsible for the Shturm-S (AT-6 Spiral) and probably more recent types.

The bureau was the subcontractor to OKB-134 in the development of the Strela 2 (SA-7 Grail) manportable air defense missile. The bureau took over further development of this missile, as well as later derivatives such as the Strela 3 (SA-14 Gremlin) and the Igla (SA-16 Gimlet).

It is unclear at what stage the bureau became involved in ballistic missiles, but the bureau was involved with with the development of both the 9M123M Tochka (SS-21 Scarab) and the 9M714/OTR-23 Oka (SS-23 Spider) tactical ballistic missiles. Besides Nepobidimy, another important designer of the center is A.S.Ter-Stepanyan.

Besides its work on anti-tank guided missiles, the bureau's other association with armored vehicles was its development work on the Arena KAZT active tank defense system, first publicly revealed in 1993 and still in development.

NII Stali
ul. Dubinskaya 81a
127411 Moscow, Russian Federation
Teki: 485-35-10

NII Stali (Scientific Research Institute for the Steel Industry) is Russia's primary development center for armor systems. Its work in the past has included advanced metallurgy for steel armor, as well as novel armor technologies. It developed the laminate armor first used on the T-64 tank, the "bra" armor applique used on the T-55 modernization program, and the Kontakt-1 and Kontakt-5 reactive armors. It also develops other forms of protective systems including anti-radar/anti-infrared blankets for armored vehicles and body armor for infantry.

ENDNOTES

Introduction and Overview

1   This book adopted as a convention that when material dealt with the period prior to 1991, it would be referred to as Soviet or the Soviet Union (USSR)

Chapter 1
Evolution of Soviet/Russian Tanks

1   For a more detailed examination of World War 2 Soviet tank development, see: Steven Zaloga and James Grandsen, *Soviet Tanks and Combat Vehicles of World War 2* (Arms & Armour Press: London, 1984).

2   Colonel General Dmitri A. Volkogonov. "What Will the Russian Army Be Like?", *Armiya*, #11-12 1992, pp. 9-16. This was the first time that any Russian official had publicly admitted to the total size of the Soviet tank inventory. There is some confusion over this figure, as other sources suggest that 77,000 is the total number of tanks manufactured post-war by the Soviet Union. Major General I. Ye. Krupchenko, "Time and Tanks (Past, Present and Future)", *Voennaya Mysl*, No. 6, 1993. (Translation in JPRS-UMT-93-009-L p. 30; and *Izvestiya*, 12 June 1992. In 1990, when the Soviet Union reported its total weapons strength to the United Nations, it indicated that its total tank strength as of 1 January 1990 was 63,900 including PT-76 light tanks. It is not clear what this included, for example whether it included tanks used in semi-static defenses in the Far East which the Soviets often exclude from calculations. V. Petrovskiy, *Information on the Total Numbers and Main Categories of Weapons of the Armed Forces of the USSR*, United Nations General Assembly, 15 October 1990, A/C.1/45/4.

3   V.V. Shylkov, "I tanki nashi bystry", *Mezhdunarodoviy Zhizn*, No. 9, 1988, pp. 117-129; V.V. Shylkov, "Bronya krepka: Tankovaya asimmetriya i realnaya bezopasnost", *Mezhdunarodoviy Zhizn*, No. 11, 1988, pp. 39-52. There were a number of responses to Shylkov, including: V.P. Krikunov, "Prostaya arifmetika V.V. Shylkov", *Voenno-Istoricheskiy Zhurnal*, No. 4, 1989, p. 42.

4   N.P. Zolotov and S.I. Isayev, "Boyegotovy byli...", *Voenno-Istoricheskiy Zhurnal*, No. 11: 1993, p. 77.

5   B. Muller-Hillebrand, *Das Heer: Der Zweifrontenkrieg*, Band III, (E.S. Mittler & Sohn, Frankfurt: 1968), unnumbered appendix: Bestand: 1. Panzerkampfwagen.

6   I.P. Shmelev, S.P. Isaykin, "Tanki", *Voenno-Istoricheskiy Zhurnal*, No. 10, 1992, pp. 92-95.

7   For a more thorough discussions of the causes of the defeat in the armored force, see: Steven Zaloga, "Technological Surprise and the Initial Period of War: The Case of the T-34 Tank in 1941," *Journal of Slavic Military Studies*, Vol. 6 No. 4 (December 1993), pp. 634-646.

8   For a more detailed discussion of the Russian post-war tank design reorganization and process, see Appendix B. This appendix also includes a detailed listing of design bureau/institutes involved in armor design and manufacture, including a listing of their locations, function, and product lines.

9   The LB stood for Lavrentiy Beria, the head of the Soviet secret police, and reflected the fact that the gun was designed by one of the "special technical bureaus" organized of engineers imprisoned in the GULAG. It is possible that the team included captured German engineers.

10   L. N. Kartsev, *Moya sudba- Nizhni Tagil* (Kosmos: Nizhni Tagil, 1991), p. 5. This order apparently affected only the three plants assigned to the T-54 program: Nizhni-Tagil, Omsk, and Kharkov, as US Army intelligence concluded from tanks captured in Korea that production of the T-34-85 continued at Gorkiy's Krasnoye Sormovo plant through 1950. See: *The Armored Vehicle Industry of the Sino-Soviet Bloc 1945-1955*, DA-G2 IRP 8969, 17 Sep 1956.

11   OPVT: *Oborudovanie dlya podvodnogo vozhdeniya tankov*, equipment for underwater tank driving.

12   Both Britian and the United States did build heavy tanks in the 1950s, namely the 120mm armed Conqueror and the M103. In both cases these were overreactions to the Soviet T-10 heavy tank, and only a few hundred of each type were built, compared with several thousand Soviet heavy tanks.

13   Leonid Kartsev, Moya sudba- Nizhni Tagil (Nizhil Tagil: Kosmos, 1991). p. 104

14   The LB in the cannon designation stood for Lavrentiy Beria, the head of the Soviet secret police. The name was selected since the gun was designed by a special design bureau located in the GULAG system of imprisoned engineers.

15   Sergei Gryankin, "PT-76," *Tekhnika Molodezhi*, No. 8, 1990, p. 19.

16   According to CFE documents, there were still 602 PT-76 in units west of the Urals in November 1990, 282 on 1 January 1994, 126 on 1 January 1995, 25 on 1 January 1996, and 5 on 1 January 1997.

17   V. Bryzgov, et. al., Bronetankovaya Tekhnika, (Gonchar: Moscow, 1994), p. 106.

18   For a more detailed account, see S. Zaloga, *Target America: The Soviet Union and the Startegic Arms Race 1945-1964* (Novato, CA: Presidio Press; 1993).

19   The Morozov biography provides virtually no detail of the post-war programs. See: V.D. Listrovoi and K. Slobodin, *Konstruktor Morozov* (Moscow: Politizdat; 1983). Some information is contained in: Garry Shizrin, "Methods of Practical Decision-Making in the Soviet Defense Industry: The Kharkov Military Industrial Complex" in *Soviet Defense Decision-Making: An Integrated View*, Volume II (Falls Church, VA: Delphic Associates, 1989).

20   Vitaly Moroz, "T-64: Pervenets vtorogo pokoleniya," *Krasnaya Zvezda*, 2 October 1992.

21   Dmitry Rototaev, "How Reactive Armor Was Created, Military Parade, July-August 1994, pp.90-91.

22   Lt. Col. A. Beskurnikov, "Ne pokhozkiy ni na odin tank v mire," *Voennye Znaniya*, 9, 1993, p. 11.

23   VPK (Voenno-Promyshelenniy Komitet) is the Military Industrial Committee of the Council of Ministers, and the primary state organization for coordinating the requirements of the Ministry of Defense, the industrial ministries, and the Communist Party's governmental agencies. It was headed in the early 1950s by Vyacheslav Malyshev, and in 1957 to 1963 by Dmitri Ustinov.

24   The F.F.Petrov design bureau is located at Artillery Plant Number 9 in Sverdlovsk (renamed in 1991 to its traditional name, Ekaterinburg). In recent years, it was renamed Spetstekhnika. This bureau has designed all major Soviet and Russian tank guns since the 85mm gun on the T-34-85 of 1944.

25   This curious feature was the outcome of a 22 October 1962 demonstration of tank technology to Nikita Khrushchev. Khrushchev, for reasons clear only to himself, blurted out, "A tank has to be able to dig in the earth like a mole." This started the effort to develop a self-entrenchment system.

26   Kartsev in his biography mentions that the separation of the loader would make it difficult to assist him should the tank be damaged in combat. It is also likely that he was concerned that the korzhina loader made access to additional ammunition in the tank nearly inaccessible.

27   The Kharkov bureau was located in Ukraine, while Nizhni Tagil is in the Urals industrial region; the full plant name is Ural Railcar Plant.

28   S. Ptichkin, "Russian Work: an interview with Nikolai Shomin," *Krasnaya Zvezda*, 22 September 1990.

29   This version is described in: B. S. Safanov and V. I. Murakhovski, *Osnovnye Boyevye Tanki* (Moscow: Arsenal Press, 1993), p. 68.

30   Armor thickness figures in the study relating to the T-64, T-72, ands T-80 are taken from the following unclassified

Russian General Staff document, provided during the CFE discussions, and are not necessarily authoritative: Tsentr operativno-strategicheskikh issledovaniy Generalnogo shtaba VS SSSR, *Iskhodnye dannye dlya rascheta koeffitsientov soizmerimocti obrazov vooruzheniya i voennoi tekhniki i voiskovikh formirovaniy SSSR i NATO, Kniga 1: Dannye SSSR* (Moscow: Ministry of Defense, 1991).

31 Dmitri Rototaev, "How Explosive Reactive Armor was Created," *Military Parade,* July-August 1994, p. 91

32 It was not unusual for the Warsaw Pact countries to purchase their tanks from more than one source. In the case of Germany, the first 133 came from the USSR, while 156 came from Poland and 260 from Czechoslovakia.

33 ERAWA stands Explosive Reactive Armor-Adam Wisniewski. Wisniewski holds Polish Patent No. 156,463 for its design.

34 Tomasz Szulc, "PT-91 Twardy," *Nowa Technika Wojskowe*, No. 5, 1993, p. 2.

35 Jan Wojciechoski, "Modernization of the Fire COntrol Systems in T-72 and other T-series Tanks," *Nowa Technika Wojskowa*, Feb-Mar 1994 (Trans: JPRS-UMA-94-032, pp14-22).

36 Giovanni de Briganti, "Czech Consortium Teams with French on Upgrades," *Defense News*, 18-24 October 1993.

37 A short history of the early turbine experiments can be found in: V. Solomay, "Transportnyie gazoturbinnye dvigateli, *Tekhnika i vooruzhenie*, No. 6 1991, pp. 2-3.

38 The Glushenkov design bureau (currently the *Omskoe motostroitelnoe konstruktorskoe byuro*) completed its first gas turbine engine design, the 300 horsepower GTD-1 from 1957 to 1958. Its development of a gas turbine powerful enough for tanks did not take place until 1964. This chronology of the bureau comes from a brochure handed out by the bureau at the 1992 Moscow Air Show.

39 The Russians have been shy of releasing comparative range data for the T-80. In an article in *Tekhnika Molodezhi* (No. 6, 1993) by tank historian Igor Shmelev, the road range of the T-80B on integral fuel is stated to be 335 kilometers. In a promotional brochure by the Kirov Plant, it is stated to be 450 kilometers, but this presumably includes two external fuel drums with 400 additional liters of fuel. These figures suggest that fuel consumption in the T-80's GTD-1000T turbine is about double that for the V-46 diesel, 0.18 to 0.20 kilometers per liter verses about 0.41 kilometer per liter.

40 Igor Shmelev, "T-80 Tank," *Tekhnika Molodezhi*, No. 6, 1993, pp. 2-5. (Trans: JPRS-UMA-93-042, 10 Nov 93, p. 8)

41 Of the 4,876 T-80s reported in the Atlantic-to-the-Urals (ATTU) region in September 1990, only 112 (2.2 percent) were T-80Us. The most common type was the T-80B and its variants which represented 89 percent of the T-80 inventory. The number of T-80s in the ATTU on 1 January 1994 was 3,004, and on 1 January 1995 was 3,144.

42 Mikhail Damchuk, "Czolg T-80," *Nowa Technika Wojskowa*, Special Number 1994, p. 6.

43 NII Stali placard at the September 1994 Nizhni Novgorod arms show. An advertisement by the VNII Transmash on Kontakt-5 suggests that it increases protection against APFSDS by 150 to 200 millimeters of RHA, and 400 to 500 millimeters against HEAT.

44 Dmitry Rototaev, "How Explosive Reactive Armor was Created,"*Military Parade,* August 1994, p. 91.

45 A promotional brochure handed out at Abu Dhabi indicated that T-80U production hadn't begun until 1992; this probably refers to the version with the 1,250 horsepower engine.

46 At the 1994 Nizhni Novgorod arms show, the Tula KBM design bureau displayed the TPN-4 gunner's day/night sight, which bureau representatives indicated was a thermal imaging sight operating in the 8 to 12 micron band. However, this sight appears to be identical to the Buran-PA sight shown in recent Russian articles on the T-80U. See: Vyacheslav Yeliseyev, "Russian Main Battle Tank T-80U," *Military Parade*, September-October 1994, p. 53.

47 Christopher Foss, "Ukrainian T-84 joins tank export push," *Jane's Defence Weekly*, 29 January 1994, p. 10.

48 This design bureau is the former mortar design bureau headed by Boris Shavyrin. In the 1960s, it was headed by S. P.

Nepobidimy and was converted to missile work. It has been responsible for development of tactical ballistic missiles (Tochka/SS-21, Oka/SS-23; antitank missiles Malyutka/AT-3, Kokon/AT-6; and man-portable air defense missiles Strela-2/SA-7; Igla-1M/SA-16).

49   According to a Zenit advertisement, the TShU-1-7 protects a sector 4 degrees in elevation and 20 degrees in azimuth. The light intensity is 20 millicandela, and it draws 1 kilowatt of power.

50   Yuriy Lepskiy, "Armor," *Trud*, 19 September 1992 (Trans.: JPRS-UMA-92-040, p. 47).

51   Christopher Foss, "First T-80U exports could go to Dubai," *Jane's Defence Weekly*, 7 Aug 1993, p. 5.

52   Andrei Volpin, *Russian Arms Sales Policy Toward the Middle East*, Research Memorandum No. 23, The Washington Institute for Near East Policy, October 1993, p. 14.

53   D. Voskoboinkov, et al., "Diplomatic Panorama," *Interfax*, 5 April 1994 (Trans. FBIS-SOV-94-066, 6 Apr 1994, p. 7)

54   Fred Kaplan, "Russian T-80 found to be lacking," *The Boston Globe*, 21 February 1995, p. 4.

55   Colonel General Mayev, "Our malefactors point out the deficiencies of Russian weapons and Technology," *Krasnaya Zvezda*, 20 February 1995.

56   Colonel General Mayev, 20 February 1995.

57   The Scientific and Research Institute for Armor and Technology is located in the Kubinka region, 60 miles southwest of Moscow. This facility is the main testing facility for armor vehicles and technology and is often referred to as the 'Aberdeen of the East' or the 'Bovington of the East'. Kubinka has only recently opened its gates to foreign visitors

58   Christopher Foss, "Revolutionary Russian MBT prototype built," *Jane's Defense Weekly*, 28 January 1995, p. 2.

59   Aleksandr Yegorov, "Russia's Tank Pool," *Krasnaya Zvezda*, 12 September 1992 (Trans.: FBIUS-SOV-92-184, p. 24).

60   Lt. Gen. (Dr.) P. Vyazitskiy, et al., "When Man is Forgotten," *Voenniy Vestnik*, No. 4, 1991, pp. 14-17.

61   Tomasz Szulc, "Wspolczesne czolgi Rosji", *Nowa Technika Wojskowa*, No. 5/1994, pp. 1-6; V. Lykov, "The Problems Still Remain," *Voenniy Vestnik*, No. 5, 1991, pp. 47-48.

62   Examples are: the 45mm gun in 1932 when many countries were still using machine guns; the 76mm gun in 1940 when most countries were using 37mm guns; the 100mm gun in 1945 when most countries were using 76mm to 90mm guns; the 115mm smoothbore with APFSDS in 1962 when most countries were using rifled 105mm guns with HEAT; the autoloaded 125mm smoothbore in 1967; the 73mm low pressure gun on the BMP-1 in 1970 when most infantry vehicles had 12.7 to 20mm guns; the combined 100mm/30mm gun system on the BMP-3 when most countries are using 25mm or 30mm guns on IFVs.

Chapter 2
Anti-Armor Developments

1   For further information on towed anti-tank gun systems, see *Chapter VI. Towed Artillery* of this study.

2   The other exception was Britain which developed a sping-loaded anti-tank launcher, the PIAT. However, the PIAT, like its German and American counterparts, used a shaped-charge warhead.

3   Nikolai Yakubovich, "Pushki Kurchevskogo", *Tekhnika Molodezhi*, 7/1993, p. 28-31.

4   Infantry guns were a common 1930s concept. These weapons were intended to be general purpose weapons capable of providing direct fire support with HE rounds, as well as anti-tank performance. Another good example is the Italian 47mm gun.

5   G. Biryukov and G. Melnikov, *Antitank Warfare*, (Progress: Moscow, 1973), p.45.

6   For further information on towed anti-tank gun systems see the Towed Artillery chapter of this study.

7   On TsNII Tochmash see: A. Yegorov, "TsNII-Tochmash is not sending out an SOS", *Krasnaya Zvezda*, 4 Sep 1993 (Trans: JPRS-UMA-93-037, p. 11).
    On the Shavyrin bureau see: Sergey Ptichkin, "Invincible", *Military Parade*, May-June 1994, p. 98-99; S.P. Nepobidimy, "Takticheskoe raketnoe oruzhie Rossii: Zametki po istorii sozdaniya", *Vooruzhenie, Politika, Konversiya*, 1/1993, pp.19-25; A. Aboronov, "Konstruktoskomu biuro mashinostroeniya-50 let", *Tekhnika i Vooruzhenie*, No. 11/12, 1992, pp.44-45.
    On the Nudelman bureau, see: A. Dolgikh, "Skorostrelniy Nudelman", *Krasnaya Zvezda*, 12 August 1992.

8   The Ptichkin article cited above contains an interview with S.P.Nepobidimy, the principal designer of the Shmel, which hints that the Shmel employed Western technology, probably from the French.

9   An RPO is the Russian designation for a class 3 fuel-air explosive missile system (called thermobric by the Russians) launched from a man-portable tube at bunkers, caves, or buildings.

10 A. Aboronov, "Konstruktorskomu biuro mashinostroeniya- 50 let", *Tekhnika i Vooruzhenie*, No. 11/12, 1992, p. 44.

11 The reference to the naval forces referred to Khrushchev's decision to drastically cut back surface warship production in favor of small missile-armed vessels. Khrushchev was threatening to do the same in the army, eliminating tanks in favor of ATGMs.

12 Available Russian descriptions of this system are not clear on this point. B. A. Kurkov, et. al., *Osnovnye boevye tanki*, (Moscow: Arsenal Press, 1993), pp. 58-59.

13 "Russian Artillery Rounds", *Military Parade*, July-August 1994, pp. 78-79.

14 Transcript of presentation by Lt. Gen. Nikolai A, Zhuravlyov, "The Russian View", Shepard Conference on Armour/Anti-Armour, London, October 1992.

15 The letter "H" is not present in the Russian language and so the letter "G" is often utilized for the letter "H". This is why one will see this system also referred to as "Germes" or "Hermes."

Chapter 3
Armored Infantry Vehicles

1   For example of early APC development, see: Steven Zaloga and James Grandsen, *Soviet Tanks and Combat Vehicles of World War Two*, (Arms & Armour Press: London, 1984).

2   The ZiS factory (Zavod imeni Satlina- Factory named in honor of Stalin), was renamed ZiL (Zavod imeni Likhcheva) after Stalin's death, a name it has retained to this day.

3   The letter V was selected as it is the fourth letter in the Cyrillic alphabet (A,B,V,G,D, etc.)

4   Ye. Prochko, "Mnogonozhna", *Tekhnika Molodezhi*, 12/1983, p. 43.

5   The armored transporter design bureau was originally located at the GAZ plant in Gorkiy, but transferred to neighboring Arzamas when BTR production was shifted to the plant there in the 1970s.

6   "Der Prototyp eines BTR-70/ABC-Aufklarung", *Soldat und Technik*, 10/1991, p. 729.

7   AMZ is the new corporate name for the Arzamas plant where these vehicles are designed and built (*A.O. Arzamasskiy mashinostroitelniy zavod*).

8   Aleksandr Yegorov, "BREM speshit na pomoshch", *Krasnaya Zvezda*, 26 February 1993.

9   *Podvizhniy punkt upravleniya PU-12M6*, sales brochure from the General Technical Department of the Russian Federation Committee for Foreign Economic Relations, 1994.

10 Vladimir Berezko, "Khimlaboratoriya pod broney", *Krasnaya Zvezda*, 22 April 1994.

11 The program may also have been the victim of the anti-Semitic purges going on in the defense industry at the time, as Gorlitskiy was Jewish.

12 Balzhiy's BNK design bureau in Chelyabinsk was a satellite design bureau to Kotin's in Leningrad, and Kotin's bureau is usually credited with the development of both the PT-76 and BTR-50 even though the actual development work was apparently undertaken in Chelyabinsk.

13 The "P" suffix indicated *plavayushchiy*, or amphibious.

14 The Schutzenpanzer Lang HS.30 was developed in the mid-1950s, and series produced for the Bundeswehr in 1959-1962 by Henschel, Hanomog and Leyland. It was armed with a 20mm L/86 autocannon.

15 GRAU (Main Missile and Artillery Directorate) was responsible for the development of tank guns, and anti-tank missiles as well as artillery weapons.

16 V. Vlasov and Yu. Voronov, "Small Caliber for Air and Ground", Military Parade, No. 2, 1994, p. 79.

17 Andrei Maltsev, "Something New from the BMP Stable: The BMP-3", *Military Technology*, 11, 1991, p. 122.

18 N. Bachurin and S. Kuts, "Tankoviy Muzei: Legkie bronirovannnye mashiny", *Tekhnika i Vooruzhenie*, 1-2/1994, p. 4.

19 N. Bachurin and S. Kuts, "Tankoviy Muzei: Legkie bronirovannnye mashiny", *Tekhnika i Vooruzhenie*, 1-2/1994, p. 4.

20 Vadim Vlasov, "Luchshie v mire BMP-eto produktsiya Kurganmaszavod", *Krasnaya Zvezda*, 28 May 1994.

21 For photographic evidence of these changes, see: S. Zaloga, W. Luczak, and B. Beldam, *Armor of the Afghanistan War*, (Concord: Hong Kong, 1992).

22 Valeriy Baberdin, "They have experts in the Urals", *Krasnaya Zvezda*, 30 July 1994, p. 5.

23 V. Vlasov, Yu. Voronov, "Small Caliber for Air and Ground", *Military Parade*, March 1994, p. 80.

24 There is some confusion as to when BMP-2 production began. The Baberdin article cited above suggests that it occured after the outbreak of the Afghanistan war in December 1979. Another article on the Kurgan plant indicates that production started in 1980: "Proizvodstvennoe obedinenie Kurganskiy Mashinostroitelniy Zavod", *Tekhnika i Vooruzhenie*, No. 2, 1993, p. 5.

25 Ye. Prochko, "AT-P Artilleriskiy Traktor", *Tekhnika Molodezhi*, No. 12, 1993, p. 17.

26 Conversations with General V.L. Bryev, Russian Armored Forces Academy, October 1994.

27 General V.L. Bryev, October 1994.

28 Some of the early production BMP-3s may have been limited to the 22 rounds in the autoloader, as several Russian sources give the BMP-3 only 22 rounds of ammunition. The additional 18 round stowage may have been a later addition.

29 S. Fedoseyev, "BMP-3", Voenniye Znaniya, No. 5-6, 1992, p. 6; (Translated: JPRS-UMA-93-004, pp.24-25.)

30 Conversations with General V.L. Bryev, Russian Armored Forces Academy, October 1994.

31 General V.L. Bryev, October 1994.

³² *Iskhodnie dannye dlya rascheta koeffitsientov soimerimosti obraztsov vooruzheniya i voennoy tekhniki i voiskovykh formirovaniy SSSR i NATO, Kniga 1*, (Tsentr operativno-strategicheskikh issledovaniy Generalnogo shtaba VS SSSR: Moscow, 1991).

³³ The Namut sight selected combines the French SAT Athos thermal imager combined with a stabilized sight from the Russian Peleng-Belomo firm. The package is so large that it had to be externally mounted in a sub-turret on the left rear side of the turret. The French thermal imager was selected to ensure parts commonality with the unit used in the LeClerc tank.

³⁴ Andrew Wilson, "Rosvoorouzhenie funds new projects," *International Defense Review*, 11/1994, pp. 66-67.

³⁵ The most critical assessment of this shortcoming was contained in : Gen. Lt. P. Vyazitskiy, "When man is forgotten", *Voenniy Vestnik*, No. 4, 1991, pp. 14-17.

³⁶ A. Yegorov, "Round Table: Military Reform and the Ground Troops: Armor and the People", *Krasnaya Zvezda*, 18 April 1991; (Trans. JPRS-UMA-91-012, p. 29-34). This was a round table discussion by key leaders of the Russian Ground Forces and military industry talking about future trends in AFV development. This sentiment was echoed in another survey which bluntly stated that: "It may be better to reduce the number of tank formations than to weaken them by replacing tanks with lightly armored BMPs and various anti-tank systems." Maj. Gen. I.Ye. Krupchenko, "Time and Tanks-Past, Present and Future", *Voeynnaya Mysl*, June 1993, (Trans: JPRS-UMT-93-009-L, p. 33).

³⁷ These tendencies have been especially noted in the Air Force, which has been the most vociferous in complaining about the tendency of the industrial leaders to decide system requirements. See for example: Aleksandr Velovich, "Changes in the Structure of the (Soviet) Defense Industry", IDA Seminar Series on Changes in the Former Soviet Union, 1992.

³⁸ See the Baberdin article, May 1994 cited above.

³⁹ General V.L. Bryev, October 1994.

⁴⁰ "Santal fit offered for BMP-3s, Piranhas", *Jane's Defence Weekly*, 27 February 1993, p. 10.

Chapter 4
Airborne Armor

¹ The VDV (Vozdushno-Desantnaya Voiska) has traditionally been a semi-independent branch of the Soviet Army, separate from the Ground Forces. It served as a strategic rapid deployment force for the Soviet General Staff, in some respects similar to the role of the US Marine Corps in US military policy.

² Sergey Gryankin, "ASU", *Tekhnika Molodezhi*, 5/1990, p. 17.

³ A. Yegorov, "Combat Readiness: View of a Problem", Krasnaya Zvezda, 11 Nov 1990, p. 1 (Trans.: FBIS-SOV-90-223, p. 72).

Chapter 5
Self-Propelled Artillery

¹ For further detail see, Andreas Geuckler, "The 122mm 2S1 Self-propelled Howitzer," *Jane's Intelligence Review*, April 1992, pp. 178-184.

² Colonel Vitaliy Moroz, "'Arsenal:' 'Akatsiya' That Smells of Gunpowder," *Krasnaya Zvezda*, in Russian, 2 April 1993, p.2 (Translated in JPRS-UMA-93-017), p. 24.

³ Andreas Geuckler, "Soviet 2S3 152mm Self-Propelled Howitzer," *Jane's Intelligence Review*, December 1991, pp. 555-571.

Endnotes

4   For further details see, Christopher Foss, "The 152mm 2S5 Self-propelled Artillery System, " *Jane's Intelligence Review*, December 1992, pp. 542-545.

5   For further detail see, Andreas Geuckler, "The 303mm 2S7 Self-Propelled Gun," *Jane's Intelligence Review*, September 1992, pp. 393 395.

6   Vitaly Moroz, "'Giatsint' and 'Tulpan': Maneuverability and Power," *Military Parade* July/August 1994, pp. 86-89.

7   For further details see, Oleg Anatolyev, "On Target!," *Military Parade*, January/February 1994, pp. 14-17.

8   For more details see, Oleg Anatolyev, "On Target!," *Military Parade*, January/February 1994, 14-17. See also A. Koshchavtsev, "The 2S19 Msta-S," *M-Khobbi*, March 1994, pp. 8-12.

9   The 7 to 8 rounds by 8 howitzers by 43 kilogram average shell weight equals 2,408 to 2,752 kilograms (5,309 to 6,068 pounds) of charge which can be loaded in a semiautomatic fashion.

10 Igor Dubrovin, "Landing Force Won't Get By," *Military Parade*, March/April 1994, pp. 54-58.

11 Vladimir Maryukha, "Ground Dreadnought for Naval Battles," *Krasnaya Zvezda* in Russian, 19 November 1993, p. 2 as translated in JPRS-UMA-93-045, 22 December 1993, pp.21-23.

12 There is no "H" in the Russian alphabet. In English this would be Hermes.

13 Vitaliy Morozov, "Arsenal: There Are No Mortars More Powerful Than 'Tyulpan," *Krasnaya Zvezda*, in Russian, 21 May 1993, p. 2 (Translated in JPRS-AUMA-93-023), p. 19.

14 For further details on 2S9, 2S23, and 2B16 see Sergery Samoylyuk, "The Three Faces of 'Nona'," *Military Parade*, May/June 1994, pp. 34-37.

15 The designation "B" was used to note a towed gun while the designation "S" was utilized for self-propelled guns and howitzers.

16 For further details on 2S9, 2S23, and 2B16 see Sergery Samoylyuk, "The Three Faces of 'Nona'," *Military Parade*, May/June 1994, pp. 34-37.

17 "Missile and Artillery 1L219 'Zoopark-1' Surveillance," *Tekhnika I Vooruzheniye* in Russian, No. 1, January 1993, pp. 14-15 (Translated in JPRS-UMA-93-018), 9 June 1993, p.43.

18 Vitaley Smirnov and Gennady Dolgikh, "The Zoopark-1 Reconnaissance Complex," *Military Parade*, November/December 1994, pp. 24-26.

19 Viktor Litovkin, "Russia's New Weapons," *Izvestiya* in Russian, 8 October 1993, p. 6 translated in JPRS-UMA-93-039, 20 October 1993, p. 5-6.

20 For further details see, Christopher Foss, "Russian Hurricane Multiple Rocket System in Detail, *Jane's Intelligence Review's Pointer*, January 1995, pp. 4-5.

21 Yuri Voronov, "'Vivari' Controls 'SMERCH' Launchers," *Military Parade*, May/June 1994, pp. 22-27.

22 "The Automated System 1K123," *Tekhnika I Vooruzheniye* in Russian, No. 1, January 1993, pp. 33 translated in JPRS-UMA-93-018, 9 June 1993, p.43.

23 It should be kept in mind that NATO terminology confuses the nature of this systems. The designations FROG-3, FROG-4, and FROG-5 implies three radically different systems. Although the missiles were different, they are used as a common launch system.

24 Sergei Morgachev, "Plutonium is a friend, but the truth is more costly", Megapolis Ekspress, 9 January 1992. (Translated FBIS-SOV-92-034, p. 8).

25   Such a warhead has been described in Russian sources. See: Dr. A. Prishchepenko, "Ship Electronic Warfare-the warfare of the future?", Morskiy Sbornik, No. 7, 1993, (Translated in JPRS-UMA-93-042, 10 November 1993).

Chapter 6
Towed Artillery and Anti-Tank Systems

1   For further insight into Soviet/Russian artillery developments, see the first three sections of *Chapter V. Self-Propelled Artillery*.

2   For a complete listing of the ammunition projectiles utilized by this weapon type, see *Appendix A*.

3   For a complete listing of the ammunition projectiles utilized by this weapon type, see *Appendix A*.

4   For a complete listing of the ammunition projectiles utilized by this weapon type, see *Appendix A*.

5   For a complete listing of the ammunition projectiles utilized by this weapon type, see *Appendix A*.

6   For further details see, Christopher Foss, "The 152mm 2S5 Self-propelled Artillery System," *Jane's Intelligence Review*, December 1992, pp. 542-545.

7   For a complete listing of the ammunition projectiles utilized by this weapon type, see *Appendix A*.

8   For a complete listing of the ammunition projectiles utilized by this weapon type, see *Appendix A*.

9   For more details see, Oleg Anatolyev, "On Target!," *Military Parade*, January/February 1994, 14-17. See also A. Koshchavtsev, "The 2S19 Msta-S," *M-Khobbi*, March 1994, pp. 8-12.

10 For further discussion, see *Chapter II. Anti-Armor Development*.

11 G. Biryukov and G. Melnikov, *Antitank Warfare*, (Progress: Moscow, 1973), p. 45.

12 For further information on towed anti-tank gun systems, see *Chapter V. Self-Propelled Artillery*.

13 For further information on through-tube laser guided anti-tank gun rounds, see *Chapter II. Anti-Armor Development*.

14 For a complete listing of the ammunition projectiles utilized by this weapon type, see *Appendix A*.

15 For a complete listing of the ammunition projectiles utilized by this weapon type, see *Appendix A*.

Appendix B
Soviet/Russian Tank/Armored Vehicle Development

1   There was a third tank design bureau, OKB-40, headed by Nikolai A. Astrov at Mytishchi, but it was solely involved in the design of light tanks and derivative light armored vehicles such as the SU-76 light assault gun. After the war, the bureau devoted its attention to light armored vehicles, specializing in air defense vehicle chassis such as the ZSU-23-4 Shilka.

2   G. Shizrin, "Methods of Practical Decision-Making in the Soviet Defense Industry: The Kharkov Military Industrial Complex", in *Soviet Defense Decision-Making: An Integrated View*, Vol. 2, (Falls Church, VA: Delphic Associates 1989), pp. 87-88.

3   From 1941 to 1945, Soviet armored vehicle production was directed by the People's Commissariat for the Tank Industry (NKTP) headed by V.A.Malyshev. On 15 October 1945, as part of the general industrial reorganization following the war, this became the People's Commissariat of Transport Machine Building (NKTMP). In March 1946, the NKTMP became the Ministry of Transport Machine Building (Mintransmash). In 1953-54, the Transport Machine Building Industry was again reorganized,

becoming the Transport and Heavy Machine Building Industry (TiTMP), a name which it retained until the end of the USSR. To avoid confusion, in this study, we refer to the organization as Mintransmash, the nickname which has stuck with the organization throughout its many name changes.

4   This is also sometimes called a TTZ (*Taktiko-tekhnichesskiye zadaniye*: Tactical Technical Assignment)

5   The production decision is usually integrated into the Soviet Army's five-year plan by the General Staff. Once the General Staff has prepared its 5-Year Plan, it is then handled by the Gosplan (Central Planning Commission) Military Directorate. The Military Directorate of the Gosplan must integrate the Ground Force's requirements for armored vehicles within the general 5-Year plan of the Soviet Union. Once this integration takes place, the responsibility for ironing out the responsibilities of the several industrial ministries which provided components to the tank production line, is the work of the Military Industrial Commission (VPK). This standing commission is not a permanent organization, but brings together high state officials responsible for carrying out the industrial production plans reached by Gosplan. The commission has at least three permanent members, the head of the Military Directorate of Gosplan, the head of the General Staff's Department for Organization, Planning and Technology and the Minister of the Defense Industries. Depending on the matters that must be dealt with, additional members come from the Military Departments of the Communist Party Central Committee, the Chief of the Military Department of the Ministry of International Trade (on matters of weapons export), and members of various elements of the Defense Ministry or armed forces.

INDEX

Aborenkov, Vasiliy, 358
Aerazur, 296
Afghanistan, 35, 181, 186, 217, 223, 226, 247, 254, 256, 257, 270, 302, 303, 340, 343, 369, 385
Air Assault Force (VDV), 269, 288, 292, 293, 294, 295, 296, 300, 301, 305, 306, 340, 343, 354, 362
Algeria, 377
Angola, 257
Arena, 137
Artillery Plant Number 9, 9, 21, 393, 396, 400, 402, 407
Arzamas AMZ Plant, 229, 285
Arzamas-16, 87, 375
Astrov, Nikolas, 12, 74, 75, 267, 288, 290, 291

Babadzhanyan, Marshal A., 97
Balzhiy Design Team, 74, 234, 313
Balzhiy, M.F., 239, 240
Barikhin, A.A., 28
Barmin Design Bureau, 314, 358
Barricade Production Association, 331
Barrikady Plant, 377, 381
Bashkiri Machine Plant (STEMA), 325, 350
Bazalt GNPO, 342, 345, 390
Bazalt State Research and Productions Enterprise, 204
Belgium, 134
Bharat Dynamics, 180, 266
Blagonravov, A., 243, 251, 252, 269, 270
Blazer armor, 40, 98, 118
BMD Design Bureau, 269
BMP Design Bureau, 269, 294
Bradley armored personnel carrier, 244, 247, 252, 267, 270, 275, 276
Brazil, 205
Brezhnev, Leonid, 90, 142
Brow armor, 37, 53, 54
Bryansk Automotive Works, 377, 381, 387
Bryev, General V.L., 232
Bulgaria, 51, 138, 180, 191, 193, 195, 343, 383, 385

Canada, 17, 208, 308
Charnko, E.V., 290
Chechnya, 122, 124, 161, 162, 166, 168, 280, 284, 285, 287, 340, 392, 411
Chelyabinsk, 11, 13, 16, 18, 20, 31, 32, 50, 61, 63, 71, 89, 123, 131, 231, 294, 311
China, 29, 30, 31, 43, 78, 139, 161, 167, 191, 266, 267, 361, 380, 393, 394
Christie (Walter) Designs, 12, 18
Chuikov, Marshal V.I., 50, 91, 102
Croatia, 136, 137, 383
Cuba, 257, 258, 377
CVRDE, 266
Cyprus, 280
Czechoslovakia, 21, 31, 32, 39, 40, 51, 132, 133, 135, 136, 220, 237, 264, 265, 310, 319, 332, 364, 369, 383, 385

DANA (152mm SPG), 319

Dedkov Design Bureau, 213, 220, 226
Degtaryev Machinery Plant, 180, 193
Denezhkin, Gennadiy, 361
Didusev, Colonel Boris A., 164
Drozd (KAZT), 41, 43, 56, 137, 168
Dukhov, General Nikolai, 87
DYNA-72 (reactive armor), 135

Egypt, 70, 175, 190, 191, 249, 250, 291, 377, 383, 393, 394
Electron Progress Company (Bulgaria), 205
Elers-Elektron, 127, 195
Erawa, 133

Ferma, 325
Finland, 170, 322, 400
Fitterman, B.M., 210, 211
FMC, 267
France, 7, 17, 20, 21, 61, 63, 65, 66, 70, 89, 110, 160, 163, 171, 208, 238, 248, 270, 279, 307
FSC (Poland), 220
FV432 armored personnel carrier, 217

Galkin, Colonel General Aleksandr, 124, 166
Ganichev, A.N., 314, 358, 377
GAU (Main Artillery Directorate), 308, 313, 314, 391
Gavalov KB, 239, 240, 301, 344
Gavalov, I., 78, 79, 241, 243, 294, 302, 345
GAVTU (Main Auto-Tractor Directorate), 314, 315
GAZ Automotive Plant, 12, 213
GBTU (Main Armor Directorate), 11, 50, 57, 61, 66, 71, 90, 92, 96, 102, 104, 106, 142, 196, 239, 250, 390
German missiles (WW2) 314, 374, 379, 380
Germany, 10, 12, 14, 16, 17, 18, 19, 20, 21, 22, 26, 65, 66, 90, 103, 112, 157, 170, 171, 172, 180, 204, 205, 208, 211, 215, 237, 238, 239, 247, 248, 251, 255, 270, 291, 295, 308, 309
Germany, East, 223, 266, 364
GIPO (State Institute for Optical Applications), 185
Glushenkov Design Bureau, 142
Gobiato, Lenoid, 336
Golents, L.L., 88
Gorbachev, Mikhal, 72, 123, 279
Gorki, 12
Gorlitskiy, Lev I., 233, 311, 317
Grabin Design Bureau, 22, 24, 173, 313, 337, 404
Grachev, Pavel, 124, 161, 162, 166
GRAU (Main Missile and Artillery Directorate), 50, 239, 254, 313, 314, 383, 390
Greece, 258
Gryazev, Vasily, 178, 239, 251
Gvay, Ivan, 358

Hungary, 133, 340, 343, 404

India, 133, 137, 139, 180, 266, 336
Infrared devices, 26
International Harvester, 210
Iran, 133, 191, 257, 383

444

Iraq, 37, 127, 139, 257, 332, 339, 383, 394, 410
Isotov KB Design Bureau, 196, 239, 240, 243, 252, 294, 301
Isotov, Sergey P., 142
Israel, 70, 98, 118, 135, 191, 287, 392
Italy, 170, 205
Izakov, Pavel, 71, 89, 90, 92, 177, 196
Izdekiye 252, 65
Izdeliye 100, 233
Izdeliye 112, 233
Izdeliye 269A, 382
Izdeliye 676, 259
Izdeliye 723, 263
Izdeliye 767, 259
Izdeliye 773, 354
Izdeliye 774, 261

Japan, 8, 11, 205, 291

Kalachinikov, Yuri, 301, 337, 341, 344
Kamaz Plant, 231
Karl Gerat (WW2 German mortar), 337
Kartsev, Leonid, 31, 32, 50, 56, 89, 91, 102, 103, 104, 105, 177, 196
KAZ, 45
KB Priborostroyeniye, 239
KB Tochmash Design Bureau, 96, 197
KB-1 Design Bureau, 56, 173, 177, 196
KB-11 Nuclear Weapons Design Bureau, 375
KB-3 Design Bureau (GSKB-30), 172
Kharkov Locomotive Plant, 11, 21, 87, 88, 91, 93, 94, 103, 104, 110, 115, 123, 144, 145, 151, 152, 164, 181
Kharkov Tractor Plant, 267, 306, 315, 346, 351
Khinikadze, Dr. Alexsandr V., 346
Khrushchex, Nikita, 50, 56, 72, 73, 75, 78, 87, 89, 142, 168, 172, 173, 196, 239, 243, 311, 312, 313, 374, 391, 404
Kirov Plant, 11, 74, 89, 114, 123, 142, 150, 287, 311, 313
Kirov, 12, 63, 70
Kirovskiy Plant, 322
Klimov Design Bureau, 142
KM Mashinostroeniya (Machine Industry Design Bureau), 385, 386, 390
Kolesnikov, A, 22, 371
Kolomna Machine Design Bureau, 157, 180, 182, 185, 186, 189, 239
Komonov Petr, 197
Kompressor Plant, 358
Konovalov, N, 74
Kontakt (reactive armor), 40, 55, 98, 99, 117, 118, 119, 120, 122, 124, 125, 134, 147, 151, 152
Korolev, Sergei, 314
Kostikov, Andrey, 358
Kotin Heavy Tank Design Bureau, 12, 65, 70, 89, 90, 175, 178, 234, 312, 314, 380
Kotin, Zosef, 11, 63, 63, 114, 142, 196, 240
Krasnoye Sormovo, 11
Kravtsev, Colonel A.F., 74, 75, 233, 234, 288, 290
Krayev, Colonel Vladimir, 294
Kritsin, A.I., 93
Kucherenko, N, 22
Kurgan Machine Plant, 79, 243, 251, 254, 270, 272, 284, 285, 286, 307, 346
Kurnosov, V, 178
Kuwait, 136, 257, 276

Lebanon, 339
LeBedev, General I.A., 90, 196
Lend-Lease, 17, 74, 89, 208, 210, 314
Lenin Heavy Machine Industry Plant, 387
Lenin TMS Plant, 385
Libya, 257, 319, 383
Likhachev Motor Vehicle Plant, 365
LNPO (Leningrad Research-Production Association), 142
LWP Workshops (Poland), 238

M113 armored personnel carrier, 217
Makarov, Nikolai. 173
Makayev, Viktor, 380, 382
Malakhit NPO, 299
Malakhit Sterkh tactical mini-UAV, 299
Malyshev Plant, 136, 151, 291
Malyshev, Vyacheslav, 87, 196
Masyagin, Aleksandr, 226
Matyukhin, V., 22
Miass Automotive Plant, 361, 380
Mine Rollers, PT-3, 24
Mine Rollers, PT-54, 23
Mine Rollers, PT-55, 46
Minsk Automotive Plant, 335, 370
Mintransmash (GVTU), 31
MKS-350-9 multi-canopy parachute, 295
Morocco, 160
Morozov Design Bureau, 12, 18, 21, 22, 31, 90, 103, 136
Morozov, Aleksandr, 11, 87, 88, 89, 102, 104, 196
Moscow Engineering Academy, 233
Motovilikha Machine Construction Planr, 341, 344, 345, 370
Mozambique, 257
Mytishchi Machinery Construction Plant, 267, 288

NAMI (Scientific Research Automotive and Auto-engine Institute), 142
Nepobidimy, S.P., 180, 189, 239, 385, 386
NII Armetekh (Automotive and Mechanical and Technology Scientific Research Institute), 203
NII Prometei, 88
NII Splav, 314
NII Stali, 36, 37, 40, 55, 88, 98, 115, 118, 120, 124, 125, 147
NII-13, 337
NII-61, 173
NIIBT (Research Institute of the Main Armor Directorate), 40, 65, 98, 118, 142
NIMI (Mechanical Engineering Research Institute), 126, 202
Nizhni Tagil, 11, 13, 16, 21, 23, 26, 31, 50, 56, 89, 90, 91, 98, 102, 103, 104, 107, 110, 123, 124, 142, 144, 160, 164, 165, 167, 175
NKT (Scientific-Technical Commission), 50
NORINCO, 267, 392, 393, 397
North Korea, 24, 139, 191, 383
North Vietnam, 175, 190
Novozhilov, Avenir G., 341
NPO Zenit, 125, 130, 157
NPP Splav, 387
Nudelman, A.E., 56, 173, 175, 177

Obiekt 19, 239, 240, 243
Obiekt 50, 220
Obiekt 51, 229
Obiekt 60, 220

445

Obiekt 105, 311
Obiekt 108, 311
Obiekt 116, 311
Obiekt 120, 317
Obiekt 1200, 239, 240, 243
Obiekt 123, 317
Obiekt 136, 18, 21
Obiekt 137, 21, 22
Obiekt 140, 28, 31, 32, 50
Obiekt 150, 56, 73, 90, 177, 178
Obiekt 150T, 56, 142
Obiekt 155, 32
Obiekt 165, 50
Obiekt 166, 50
Obiekt 167, 50, 56, 57, 103, 104
Obiekt 1676TD, 57
Obiekt 167T, 142
Obiekt 172, 57, 103, 105, 107
Obiekt 172-2M, 105, 107
Obiekt 172M, 132
Obiekt 172M-1, 108, 110, 133
Obiekt 174, 133
Obiekt 182, 115
Obiekt 184, 115, 116
Obiekt 188, 122, 124
Obiekt 216, 322
Obiekt 218, 380
Obiekt 219, 115, 144
Obiekt 219A, 147
Obiekt 219M, 151
Obiekt 219R
Obiekt 219RV, 147
Obiekt 268, 311
Obiekt 271, 313
Obiekt 277, 70, 71, 142
Obiekt 279, 70, 71
Obiekt 287, 90, 142, 178
Obiekt 288, 142, 144
Obiekt 303, 317
Obiekt 305, 337
Obiekt 307, 320
Obiekt 416, 310
Obiekt 430, 31, 50, 51, 88, 90, 91
Obiekt 432, 91, 92
Obiekt 434, 93, 97
Obiekt 437, 94
Obiekt 447, 96, 97, 164
Obiekt 478B, 151
Obiekt 483, 29
Obiekt 561, 290
Obiekt 573, 310
Obiekt 600, 311
Obiekt 630, 145
Obiekt 675, 252, 254
Obiekt 676, 82
Obiekt 680, 251, 254, 270
Obiekt 681, 252, 254
Obiekt 685, 79, 82, 269, 270
Obiekt 688, 82, 270, 272, 279
Obiekt 701, 61
Obiekt 701-2, 61
Obiekt 701-5, 61
Obiekt 701-6, 61
Obiekt 703, 63, 65

Obiekt 750, 234
Obiekt 750PK, 236
Obiekt 765, 241, 243
Obiekt 768, 254
Obiekt 769, 254
Obiekt 770, 70, 71, 240
Obiekt 775, 90, 96, 196, 240
Obiekt 775T, 142
Obiekt 906, 79
Obiekt 911, 240
Obiekt 914, 240, 243, 294
Obiekt 915, 294
Obiekt 934, 79, 81, 82, 269, 345
Obiekt 940, 79
Obiekt 950, 303
Obukhov, Anatoli, 342
OKB-1, 314, 380
OKB-16 Design Bureau, 56, 175, 177
OKB-40, 267
OKBT (Special Tank Design Bureau), 142, 144
Okunev, I.V., 32, 50, 103
Omsk Plant Number 174, 12, 21
OPVT River Fording Equipment, 25, 26, 29, 96
Orlov, N.I., 211, 213

P-90 aluminum airborne container, 290
Pakistan, 139, 152, 161
PAZ System, 32, 108, 236
PCE (Industrial Optics Center), 134
Perm Machine Building Plant, 337, 371
Persits, Zinoviy M., 383
Petain, Marshal, 279
Petropaulovsk Machinery Plant, 388
Petrov, General F., 16, 21, 24, 28, 87, 90, 94, 291, 311, 314, 315, 391, 393, 396, 400, 407
Petrov, V., 181
Platov, A.I., 88
Podpolianske Strojarne Detva, 264
Pogozhev, I.V., 24
Poland, 21, 28, 31, 32, 33, 39, 45, 51, 132, 133, 134, 136, 170, 189, 220, 237, 265, 319, 383
Poluboyarov, P.P., 196, 197
Popov, Nikolai, 114, 142, 323
Potapov, General Yuri, 97, 197
Priborstroyeniya Design Bureau, 270
Project Kentaur, 295
Project Reaktaur, 296
Project Rhino (India), 139
PRSM-915 rocket parachute system, 296
PSP Bohemia, 136

Radiopribor NPO, 229, 354
Radus-Zenkovich, General A.V., 50
Raspletin, A.A., 56, 173, 177, 196
Reaktivniy NII (Scientific Research Institute for Rockets), 358, 377
Rodionov, V.F., 211, 213
Romania, 31, 46, 136, 189, 191, 219, 220, 224, 317, 364, 383
Rubtsovsk Machine Building Plant, 239, 254, 262, 280, 285, 346, 354, 356
Russo-Japanese War, 336

Saratov Machinery Plant, 181
Scientific and Research Institute for Armor and Technology, 163, 164

Scientific Research Institute for Parachutes, 296
Shabalin, A., 79, 269, 302, 341
Shankarpally Ordnance Factory, 266
Shashmurin, Nikolai, 65, 74
Shavyrin, Boris, 172, 180, 189, 313, 337, 374, 386
Shchit Machinery Plant, 180, 181
Shemerlinsk Specialized Motor Vehicle Plant, 335
Shipunov KBP Design Bureau, 147, 239, 251, 270
Shipunov, Arkadiy, 41, 178, 292, 193
Shomin, Nikolai, 98. 104, 106
Shtora self-defense system, 126, 127, 154, 159, 160, 167, 157, 193, 350
Silin, V., 239
Sivgneubrandenburg, 266
Slovakia, 134, 135, 136, 336
SOFMA, 135
South Africa, 134, 307, 393
South Korea, 280, 336
Soviet Central Auto and Tractor Directorate, 267
Special Mortar Design Bureau, 173
Spetsmash Design Bureau, 123
Splav GNNP (Main Scientific Production Enterprise), 358, 361, 364, 369
Splav State Research and Productions Enterprise, 204, 205
Stalin, Zosef, 67, 207
Stalingrad Tractor Plant, 11
Stalowa Wola, 266
Sverdlovsk, 16, 21, 28
Sweden, 160, 170, 204, 205
Sych, General Aleksandr M., 32
Syria, 133, 247, 249, 250, 257

Taiwan, 191
Tankograd Complex, 11
Tatra Automotive Factory, 220
Thailand, 84
Thomson-CSF (France), 135
Titan Central Design Bureau, 381
Tomashov, Yuriy, 325, 337
Transmash Plant, 123, 270
Transport Machine Plant, 151
Trashutin Main Diesel Design Bureau, 31, 32, 103, 104
Tretyakov, Mikhail, 325
Troyanov, L.S., 71, 74, 75, 311
TsNII Tochmash, 82, 173, 178, 193, 195, 341, 344, 345, 346
TsNIIAG Central Scientific Research for Automation and Hydraulics, 383
Tula KBP, 178, 182, 185, 193, 194, 197, 199, 203
Tula Machine Design Bureau, 41, 191, 255, 305, 336, 356
Tulskiy Armaments Plant, 180, 193
Tveretskiy, Masgen A., 314

Ukraine, 100, 136, 139, 151, 154, 167, 346
United Arab Emirates, 157, 161, 275, 276, 284, 412
United Kingdom, 7, 8, 17, 21, 30, 61, 63, 66, 70, 110, 111, 112, 113, 160, 163, 171, 208, 238, 248, 291, 391
United States, 8, 17, 20, 21, 22, 23, 26, 30, 50, 51, 57, 61, 63, 65, 66, 68, 70, 75, 90, 96, 103, 109, 110, 111, 112, 113, 130, 144, 147, 152, 161, 163, 171, 180, 193, 195, 196, 202, 204, 205, 206, 208, 210, 211, 238, 239, 247, 248, 252, 270, 291, 295, 308, 310, 312, 319, 358, 365, 388, 390, 391
Uralmash Plant, 310
Urals Railcar Plant, 110, 123, 160
Uraltransmash Plant Number 50, 233, 311, 313, 317, 325, 337, 350
Uralvagon KB, 108, 115, 123
Ustinov, D.F., 50

Vagonka Plant, 31, 50, 56, 102, 103, 104, 105, 123, 142
Vasiliev, P., 22
Vayysburd, L.A., 50
Vazov Engineering Plant, 180, 191, 193, 194
Venediktov, V.N., 105
Vishnevskiy, Vladimir, 203
Vladimirov KPV, 211
VNII Signal, 229, 354
VNII Transmash, 118, 127, 142, 144
Volgograd Tractor Plant, 78, 79, 240, 294, 344, 345
Volsk Mechanical Plant, 183
VOP 025, 135
Voronov, Marshal N.N., 313, 391
Voroshilov, Marshal Klimenti, 14
Votkinsk Plant, 385
VPK (Military Industrial Commission), 50
VRZ Number 2, 74, 233

WITU (Military Institute of Armament Technology), 133

Yemen, 257, 383
Yugoslavia, 136, 189, 190, 191, 394, 397, 404

Zaitsev, General of the Army, Mikhail, 254
Zakharov, Mikhail A., 286
Zhukov, Marshal G., 211
Zhukovskiy, N., 74
ZiS Automotive Plant, 211
ZiS-123, 210
ZiS-151, 210
ZTS Detva (Slovakia), 237
ZTS Dubnica, 264
Zuyev, Lieutenant General Leonid, 295
Zverev, Sergei A., 103, 104
Zvyagin, K., 178

Mr. Andrew W. Hull

Mr. Hull is currently a Project Leader at the Institute for Defense Analyses (IDA) and a member of the Board of Advisors of the Patterson School of Diplomacy and International Commerce of the University of Kentucky. He has almost twenty-five years of experience studying the evolution of Soviet/Russian military equipment. Over the last five years, he has attended international arms shows, visited numerous Russian military museums, and talked with many Russian armor and artillery designers. During the course of his career, Mr. Hull has written about Russian military equipment developments in *Armor*, *Field Artillery Journal*, Jane's *Missiles & Rockets* as well as Jane's *Intelligence Review*.

Mr. David R. Markov

Mr. Markov is a research staff member with the Institute for Defense Analyses (IDA). He is the author of numerous articles, studies, and publications on military technology and defense issues. Mr. Markov is also a specialist correspondent with Jane's *Intelligence Review* (*JIR*) and a contributor to *Air Force Magazine's* "Russian Military Almanac" issue. Mr. Markov has visited several of Russia's premier armor museums and collections, and has interviewed countless Russian designers on current trends in armor and artillery design. His experiences resulted in the rare distinction of having examined every armor vehicle type and artillery piece currently in service in the Russian Federation.

Mr. Steven J. Zaloga

Mr. Steven J. Zaloga is the author of over thirty books on Soviet and Russian tank history, and numerous other books on Soviet military history and technology. He serves as a senior analyst covering missile and arms export issues for Teal Group Corporation and is an adjunct staff member of the Institute for Defense Analyses. He is on the editorial board of the *Journal of Slavic Military Studies*, and he is a special correspondent for Jane's *Intelligence Review*. Mr. Zaloga received his BA in history from Union College, his MA in east European history from Columbia University, and he did graduate research and language study at Universitet Jagiellonski.